# Advances in Insect Rearing for Research and Pest Management

# Advances in Insect Rearing for Research and Pest Management

EDITED BY

## Thomas E. Anderson
## and Norman C. Leppla

TECHNICAL EDITORS

Teri Houck
and Tom Knecht

Routledge
Taylor & Francis Group

NEW YORK AND LONDON

First published 1992 by Westview Press, Inc.

Published 2021 by Routledge
605 Third Avenue, New York, NY 10017
2 Park Square, Milton Park, Abingdon, Oxon OX14 4RN

*Routledge is an imprint of the Taylor & Francis Group, an informa business*

Copyright © 1992 Taylor & Francis

Library of Congress Cataloging-in-Publication Data
Advances in insect rearing for research and pest management / edited
   by Thomas E. Anderson and Norman C. Leppla.
      p.   cm. — (Westview studies in insect biology)

   1. Insect rearing.   I. Anderson, Thomas E.   II. Leppla, N. C.
III. Series.
SF518.A38   1992
638—dc20                                                          89-14632
                                                                      CIP

ISBN 13: 978-0-3670-1338-7 (hbk)
ISBN 13: 978-0-3671-6325-9 (pbk)

## DEDICATED TO

# André Van der Vloedt

This book is affectionately dedicated to the late André Van der
Vloedt, who pioneered large-scale in vitro rearing of the tsetse fly and
dedicated his life to controlling this plague of mankind.  Above all,
he was a man of great spirit, passion, and inspiration.

André Van der Vloedt, second from right, demonstrating his tsetse fly trap
near Lake Victoria in Kenya. His colleagues were part of a review team
assembled at the International Center of Insect Physiology and Ecology in
1989 to review the insect rearing program and evaluate its role in
developing new pest suppression technologies.

# Contents

*Preface*                                                                    *xi*

*Acknowledgments*                                                            *xiii*


**PART ONE**
**Historical Perspective**

1    The Insect Rearing Group and the Development of
     Insect Rearing as a Profession, *W. A. Dickerson
     and N. C. Leppla*                                                         3


**PART TWO**
**Insect Rearing Research**

2    Molecular Genetic Mechanisms for Sex-Specific
     Selection, *Alfred M. Handler*                                          11

3    Insect Rearing and the Development of Bioengineered
     Crops, *Terry B. Stone and Steven R. Sims*                              33

4    Development of Artificial Diets for Entomophagous
     Insects by Understanding Their Nutrition and Digestion,
     *I. G. Yazlovetsky*                                                      41

5    Assimilation, Transport, and Distribution of Molecules
     in Insects from Natural and Artificial Diets, *Jeffrey P. Shapiro*      63

6    Using a Systematic Approach to Develop Artificial Diets
     for Predators, *Allen C. Cohen*                                         77

7    Feeding and Dietary Requirements of the Tephritid
     Fruit Flies, *George J. Tsiropoulos*                                    93

8    Flea Rearing in Vivo and in Vitro for Basic and Applied
     Research, *Nancy C. Hinkle, Philip G. Koehler,*
     *and Richard S. Patterson*                                        119


**PART THREE**
**Insect Rearing Support**

9    Insect Rearing Management (IRM): An Operating System
     for Multiple-Species Rearing Laboratories,
     *Pritam Singh and G. K. Clare*                                    135

10   Multiple-Species Insect Rearing in Support of Research,
     *E. G. King, Jr., and G. G. Hartley*                              159

11   Artificial Rearing Technique for Asian Corn Borer, *Ostrinia*
     *furnacalis* (Guenee), and Its Applications in Pest Management
     Research, *Zhou Darong, Ye Zhihua, and Wang Zhenying*             173

12   Comparison of Artificial Diets for Rearing the Sugarcane
     Borer, *J. R. P. Parra and L. H. Mihsfeldt*                       195

13   Rearing Lepidoptera for Plant Resistance Research,
     *F. M. Davis and W. D. Guthrie*                                   211

14   Influence of Artificial Diet on Southern Corn Rootworm
     Life History and Susceptibility to Insecticidal Compounds,
     *Pamela Marrone, Terry B. Stone, and Steven R. Sims*              229

15   Importance of Host Plant or Diet on the Rearing of Insects
     and Mites, *M. J. Berlinger*                                      237

16   Seasonal and Nutritional Influences on the Toxicological
     Response of the First Instar Larvae of *Spodoptera*
     *littoralis* (Boisduval) in a Mass Rearing Culture,
     *V. Flueck, F. Bourgeois, and P. Stoeklin*                        253

17   Evaluating the Role of Genetic Change in Insect Colonies
     Maintained for Pest Management, *Robert L. Mangan*                269

18   Problems with Entomopathogens in Insect Rearing,
     *George G. Soares, Jr.*                                           289

## PART FOUR
## Insect Rearing for Pest Management

19   Straggling in Gypsy Moth Production Strains: A Problem
     Analysis for Developing Research Priorities, *T. M. ODell*          325

20   New Technologies for Rearing *Epidinocarsis lopezi*
     (Hym., Encyrtidae), a Biological Control Agent Against
     the Cassava Mealybug, *Phenacoccus manihoti*
     (Hom., Pseudococcidae), *P. Neuenschwander and T. Haug*              353

21   Microhymenopterous Pupal Parasite Production for
     Controlling Muscoid Flies of Medical and Veterinary
     Importance, *Philip B. Morgan*                                       379

22   Rearing Systems for Screwworm Mass Production,
     *David B. Taylor*                                                    393

23   Mass Rearing Biology of Larval Parasitoids
     (Hymenoptera: Braconidae: Opiinae) of Tephritid Flies
     (Diptera: Tephritidae) in Hawaii, *Tim T. Y. Wong and
     Mohsen M. Ramadan*                                                   405

24   Mass Rearing of *Chrysoperla* Species, *Donald A. Nordlund
     and R. K. Morrison*                                                  427

25   Automated Mass Production System for Fruit Flies Based
     on the Melon Fly, *Dacus cucurbitae* Coquillett (Diptera:
     Tephritidae), *H. Nakamori, H. Kakinohana, and M. Yamagishi*        441

## PART FIVE
## Insect Rearing in the Marketplace

26   The Establishment of Commercial Insectaries,
     *Peter L. Versoi and Lee K. French*                                 457

27   Production and Utilization of Natural Enemies in
     Western European Glasshouse Crops, *W. J. Ravensberg*               465

28   Mass Rearing of Phytoseiid Mites for Testing and
     Commercial Application, *L. A. Gilkeson*                             489

29   Gypsy Moth Parasites: Commercial Production and
     Profitability, *Mark Ticehurst*                                    507

*List of Contributors*                                                  517
*About the Book and Editors*                                           521

# Preface

This volume is a comprehensive international overview of the science and art of insect rearing. The importance of this subject is growing, both in basic research and the applied field of pest management. The concern of the public for alternative, environmentally sound methods of controlling insect pests has produced growing support for management programs employing the release of beneficial species. Moreover, the burgeoning field of biotechnology is now making possible programs that were unimaginable a few short years ago.

Professional entomologists in government, academia, and private industry who deal daily with the rearing of insects for biological testing or release in pest management programs are an obvious audience for this text. There are, however, other equally important audiences. Researchers in other disciplines have found that insects are relatively easy-to-rear, effective models for basic studies on subjects as diverse as artificial intelligence and space medicine. Students reading this text will gain an appreciation of the importance of insect rearing to a broad diversity of research programs. Administrators responsible for program funding and development, and nonbiologists responsible for the design and construction of laboratory facilities, will gain an appreciation of the manpower and resource requirements necessary for successful, efficient insect rearing. Operators of small businesses will benefit from learning of the experiences of others who have successfully taken insect rearing techniques from the laboratory to the marketplace and who sell insects for research and pest control in agriculture and forestry.

This book originated from a symposium of the same name held at the 1988 XVIII International Congress of Entomology in Vancouver, Canada. The symposium provided an excellent international forum for the discussion of many new advances in the field of insect rearing, but certain areas could not be adequately represented. With the support of the Insect Rearing Group, additional contributors were selected for a more comprehensive coverage of insect rearing. The goal was to feature not only academic research reviews but to also balance these basic contributions with descriptions of active government and private-sector programs. A special emphasis has been placed on those entrepreneurs who have made insect rearing a success in the marketplace.

*Advances in Insect Rearing for Research and Pest Management*, a sequel to *Advances and Challenges in Insect Rearing* published in 1984, is meant to complement other excellent books on insect rearing that have a "cookbook" style and have focused on providing step-by-step rearing instructions for specific species. As the reader will see, such how-to books are frequently cited in the contributions included herein and provide an essential basis for any project involving insect rearing. This volume builds on that foundation and provides the reader with an appreciation of the general issues and technical developments that impact on the rearing of arthropod species. In addition, this book illustrates how advances in rearing techniques have influenced breakthroughs in other areas, such as the development of artificial diets for research on the production of pest resistant crop varieties or mass production of insects for management programs.

Anyone conducting research with laboratory-reared insects, producing insects for mass release or sale, managing such programs, or designing rearing facilities will find that *Advances in Insect Rearing* will broaden their perspective. This will permit readers to anticipate problems and improve their rearing systems based on the most current information available.

*Thomas E. Anderson* and *Norman C. Leppla*

# Acknowledgments

We offer our thanks to the many people and organizations without whose help this project would not have been successful. Working with manuscripts from around the world, V. V. Robinson, Phyllis Mayfield, Nichole Hornsby, Mary Miles, and Susan Anderson typed the text and tables for all chapters and keyed in the editorial changes. Teri Houck and Tom Knecht were responsible for technical editing, layout, and preparation of final camera-ready copy. The National Biological Control Institute of APHIS generously provided a grant for final technical editing of the camera-ready copy. The North Carolina Entomological Society contributed time and services in overseeing the proper disbursement of the grant funds. The Agricultural Products Division of BASF Corp. funded production of quality photographic plates for many of the charts and illustrations and supported the many miscellaneous secretarial functions associated with this project.

*T. E. A.* and *N. C. L.*

PART ONE

# Historical Perspective

# 1

# The Insect Rearing Group
# and the Development of
# Insect Rearing as a Profession

## W. A. Dickerson and N. C. Leppla

Insect rearing has been a part of human history for more than 5,000 years, if beekeeping and silk production are taken into account. The ancient Egyptians commonly cultured honeybees for food and embalming materials. Marco Polo, describing his trip to China in 1255 A.D., stated that "no fewer than a thousand carriages and packhorses, loaded with silk, make their way daily into Peking." Except for these two examples, however, insect rearing was essentially a static art from 3000 B.C. until 1900 A.D. Papers on rearing *Drosophila* sp. began to appear in the early 1900s as a consequence of genetic studies. Nutritional and rearing studies of the European corn borer and pink bollworm followed in the 1940s. Beginning in the mid 1950s, insect rearing proliferated, perhaps inspired by publication of the first successful artificial diet for a phytophagous insect by Stanley D. Beck et al. (1949).

The science and profession of insect rearing gained acceptance during the 1950s and grew exponentially during the next two decades. This expansion occurred because insect rearing was required in order to develop and implement new pest control technologies, such as host plant resistance, chemical and microbial pesticides, sterile insect technique, and biological control. The United States Department of Agriculture (USDA) identified a nucleus of scientists dedicated to advancing this field and in 1958 held a national conference entitled "Culture of Insects as a Requisite for Research on Insect Control." This meeting was followed in 1963 by a second USDA national meeting, "Planning and Training Conference for Insect Nutrition and Rearing." This conference included sections on basic rearing research, rearing for experimentation, and mass rearing for control and eradication. This approach was expanded to include international specialists. Many of their programs were described in *Insect Colonization and Mass Production,* edited by Carroll N. Smith in 1964, and *Radiation, Radioisotopes, and Rearing Methods in the Control of Insect Pests,* published by the International Atomic Energy Agency (IAEA) in 1967. A new generation of scientists built on this work during the late 1960s and early 1970s.

By the mid 1970s, virtually every entomological research facility of any appreciable size included an insectary, and most entomological research utilized some laboratory-reared insects. However, concerns began to be expressed regarding the quality of these insects and, as a consequence, some resources were redirected from production to quality control. In 1974, the USDA held a workshop entitled "Genetics of Insect Behavior" to conduct a review, discussion, and planning session on the questions of genetic and behavioral deterioration and amelioration in mass-produced insects. At that meeting, scientists involved in insect rearing recognized the need for improved communication and cooperation to solve common problems, so they established the international insect rearing newsletter, *FRASS* (Fig. 1). It was intended to facilitate interaction among rearing personnel and promote efficiency in their respective research and production programs. They established a directory of personnel that evolved into the Insect Rearing Group and addressed the following issues:

- Acquiring inexpensive, high-quality supplies
- Fostering cooperative research and publication
- Locating sources of useful insect species
- Defining requirements for rearing facilities
- Standardizing diets, techniques, and facilities
- Establishing insect quality control
- Developing systems analysis for insect rearing
- Identifying engineering expertise
- Incorporating insect genetics
- Providing effective sanitation and quarantine
- Exchanging pertinent literature
- Supporting insect rearing as a science.

The Insect Rearing Group immediately circulated a list of inexpensive, high-quality rearing supplies, and compiled the booklet *Arthropod Species in Culture,* by Willard A. Dickerson et al. in 1980, updated by Dennis R. Edwards et al. in 1987, in response to a request from the National Academy of Sciences. Membership grew to more than 500. Leadership began to move from government to the private sector, and the group became an important influence on the Entomological Society of America (ESA), International Congress of

**Fig. 1.** *FRASS* **(Insect Rearing Group Newsletter) letterheads representing issues from 1975 to 1991.**

Entomology (ICE), International Organization for Biological Control (IOBC), USDA, IAEA, and a host of other entomological institutions. Informal influence was exerted on education and training, employment, funding, health and safety, facilities, equipment, materials, and other kinds of support for insect rearing.

Probably the most important accomplishment was collaboration to document the field in publications, such as the *Journal of Economic Entomology, Entomophaga, Facilities for Insect Research and Production,* edited by Norman C. Leppla and Thomas R. Ashley, *Advances in Insect Rearing,* edited by Edgar

G. King and Norman C. Leppla, and the *Handbook of Insect Rearing,* edited by Pritam Singh and Raymond F. Moore. The Insect Rearing Group has conducted symposia at ESA national and regional meetings since 1975, and the first international insect rearing conference, sponsored by the USDA, was held in 1980. Thus, within its first five years, the group helped move insect rearing from almost purely technical support to an established science.

Curiously, the theoretical basis for quality control in insect rearing evolved in parallel rather than being integrated with rearing techniques and applications. It was addressed formally for the first time at the 1971 symposium of IOBC on implications of mass-rearing operations. This was followed in 1972 by two provocative articles in *Entomophaga:* "Behavioral Aspects of Mass-Rearing of Insects," by Ernst F. Boller and "Genetic Aspects of Insect Production," by Manfred Mackauer. Intense interest in this topic began with two symposia at the Fifteenth International Congress of Entomology, "Characterization and Evalation of Insect Colonies" and "Natural Enemies." *Quality Control: An Idea Book for Fruit Fly Workers,* edited by Ernst F. Boller and Derrell L. Chambers, was published in 1977.

Many very useful articles and manuals were produced during the ensuing years, but none have unified quality control and brought it into routine practice in insect production. Quality control is actually a system that incorporates production control (materials, equipment, and operations), process control (unfinished insect products), product control (finished insect products), and field evaluation with feedback. A comprehensive quality control system includes insect rearing and monitors everything from colonization effects through colony management to use in experimentation or field control.

The success of insect rearing and quality control research plus the widespread implementation of rearing programs has transformed entomology into a highly scientific and technical field. It has provided for expansion of entomology in ways that would not have been possible without a dependable supply of laboratory-reared insects. Much of this success, directly or indirectly, is due to contributions fostered by the Insect Rearing Group. This book is the latest example of their nurturing influence on the science of insect rearing.

## References

Beck, S. D., J. H. Lilly & J. F. Stauffer. 1949. Nutrition of the european corn borer, *Pyrausta nubilalis* (Hbn.). I. Development of a satisfactory purified diet for larval growth. Ann. Entomol. Soc. Amer. 42: 483-496.

Boller, E. F. 1972. Behavioral aspects of mass-rearing of insects. Entomophaga 17: 9-25.

Boller, E. F. & D. L. Chambers, editors. 1977. Quality control: An idea book for fruit fly workers.

Dickerson, W. A., J. D. Hoffman, E. G. King, N. C. Leppa & T. M. ODell. 1980. Anthroped species in culture in the United States and other countries. Entomological Society of America, College Park, Maryland.

Edwards, D. R., N. C. Leppla & W. A. Dickerson. 1987. Arthropod species in culture. Entomological Society of America.

International Atomic Energy Agency. 1967. Radiation, radioisotopes, and rearing methods in the control of insect pests.

King, E. G. & N. C. Leppla, editors. 1984. Advances and challenges in insect rearing. U.S. Dept. Agric., Agric. Res. Service.

Machauer, M. 1971. Genetic aspects of insect production. Entomophaga 17: 27-48.

Singh, P. & R. R. Moore, editors. 1985. Handbook of insect rearing, vols. 1 and 2. Elsevier, Amsterdam.

Smith, C. N. 1964. Insect colonization and mass production. Academic Press, New York.

PART TWO

# Insect Rearing Research

# 2

# Molecular Genetic Mechanisms for Sex-Specific Selection

## *Alfred M. Handler*

### Introduction

One of the more effective means of controlling insect pest populations without the use of noxious agents has been the sterile insect technique (SIT) originally proposed by Knipling (1955). This method has proven effective in eradicating several insect pests, notably the screwworm fly from Mexico and the Mediterranean fruit fly from Mexico and Guatemala (Snow 1988). At present, programs are being evaluated to target other insects, including various tephritid fruitflies, mosquitoes, the boll weevil, and lepidopterans. In general, SIT takes advantage of the fact that for insect females that mate once or twice in a breeding season or lifetime, mating with a sterile male results in inviable progeny. Upon large-scale sterile-male release, indigenous pest populations can either be eradicated within several generations or finally eradicated when initially suppressed by insecticides.

Theoretically, SIT programs would be most efficient and effective if a breeding population of insects could give rise solely to sterile males capable of reproductive competition with native males in the field. In practice, because of the lack of efficient sexing systems, most current SIT programs require the rearing, sterilization, and release of females as well as males. This problem results in a doubling of costs for rearing and sterilization, which can be considerable. It also decreases effectiveness because the released females cause crop damage by foraging and egg-laying, and they compete with the targeted native females in mating with the sterile males (Robinson et al. 1986). In addition, female sterilization requires higher irradiation doses than males in insects such as the boll weevil and screwworm (LaChance 1979), causing somatic chromosomal damage, decreasing viability, and producing more general

11

systemic effects.   For these reasons, the possibility of using genetic manipulations to select against females early in their development and the possibility of sterilizing only males have been the subject of considerable interest and study in recent years (Robinson 1983).

Several classical genetic methodologies can be used for genetic sexing, including the manipulation of sex-ratio distorter and sex-determination genes, meiotic drive, hybrid effects, maternal effect lethals, sex-chromosome translocations, and sex-specificities in development and behavior.  When either altered or subjected to selective pressure, these mechanisms can act to bias the survival of the gametes for or zygotes of a particular sex.  The specifics of these mechanisms and of the environmental influences on sex-specific selection have been reviewed elsewhere (Robinson 1983) and, for the most part, will not be discussed here.

Of these methods, one of the most widely studied has been the use of genes with selectable phenotypes being linked by means of chromosomal translocation to the male-determining Y chromosome.  Although genetic-sexing strains have proven effective in mosquitoes, large-scale rearing of similar strains of other insects, tephritid fruit flies in particular, have yet to be proven successful. Difficulties include both the identification of appropriate genes with easily selectable phenotypes and the genetic breakdown of strains under large-scale rearing conditions (Foster et al. 1980, Rössler 1985, Hooper et al. 1987).  The basis of these problems lies not only in the difficulty of genetically manipulating these species, but also in the inherent genetic instability and/or inviability of strains that have undergone mutagenesis and chromosomal rearrangements. With recent molecular biological analyses in *Drosophila melanogaster,* it has been realized that a possible resolution of these difficulties would take advantage of recombinant DNA technology (Louis et al. 1987, Robinson et al. 1988, Handler & O'Brochta 1991).  Chimeric gene fusions could be created in vitro to link a transcriptional unit encoding a selectable gene product to a regulatory region that confers sex-specific and/or conditional control.  The integration of such a gene fusion into a host genome would result in sex-specific expression of the selectable gene but, in general, creates almost no discernible genetic damage and would be no more susceptible to genetic instability than other resident genes.  The purpose of this article is to review the current state of research in sex-specific selection, both under way and prospective, that takes advantage of molecular genetic techniques.  A full appreciation of the potential effectiveness of these methods and the rationale upon which they are based will come into focus by also reviewing, as a prelude, some classical genetic sexing schemes.  Although molecular genetic studies to create and test genes for genetic sexing are under way, it should be noted that the implementation of these techniques will depend upon the ability to integrate these gene fusions, by germ line transformation, into appropriate host genomes.  Thus far, efficient gene transfer is possible in only one insect, *Drosophila melanogaster* (Rubin & Spradling 1982, Spradling & Rubin 1982), and current efforts to develop gene transfer methods for nondrosophilids are reviewed elsewhere (Handler & O'Brochta 1991).

## Classical Genetic Sexing Methods

A straightforward approach toward achieving genetic sexing is to produce a strain where expression of a selectable phenotype is sex limited, which may be achieved in several ways. The particular method most exploited to date involves linking a dominant-acting selectable gene to the male-limited Y chromosome (in males that are heterogametic). Thus, the dominant gene phenotype is expressed only in males and is distinguishable from the nontranslocated recessive alleles present in both sexes. For example, a wild type (normal) brown pupal color allele may be Y-linked in a strain where the resident autosomal genes are recessive white color mutations, resulting in wild type brown males and white females (Fig. 1). However, for some genes mutant alleles have phenotypes dominant to wild type because of a gain of function, such as neomorphs, antimorphs, or hypermorphs. This would include, for example, mutations resulting in some types of chemical resistance. In this case the dominant mutant allele would be Y-linked, with the autosomal recessive being wild type.

The simplest method to create Y-linkage to a dominant gene for genetic sexing involves irradiating males that are heterozygous for the particular gene (i.e., having both dominant and recessive alleles) and then mating the irradiated males to females that are homozygous recessive for the gene (Fig. 1). The resultant male offspring are then backcrossed to homozygous recessive females as single- pair matings for one generation. A putative Y-linkage translocation is inferred for those single-pair mating lines in which all male offspring show dominant expression of the gene, with all females being recessive. In the absence of the appropriate Y-autosomal translocation, both dominant and recessive expression would be observed in both sexes.

Creating the appropriate translocation is also possible by having the original irradiated males homozygous for the gene, eliminating the possibility of having the recessive allele translocated, but selection would be more difficult, possibly requiring successive single-pair matings to select lines without the homologous untranslocated dominant allele. Without specific Y-linked markers, final determination of the desired translocation would require cytogenetic analysis.

Translocation stability may be a function of the amount and/or location of the rearrangement, but a more consistent problem with translocation lines is that they are invariably aneuploid, being deficient or duplicated for chromosomal material including and surrounding the translocated gene. A sex-ratio distortion toward males is expected as well because of an increased frequency of lethal homozygous deficiencies in females. As long as the translocation remains stable, males will not be homozygous deficient, although in successive generations increasing numbers of males will be duplicated for the region, which is also deleterious to viability. For males to be euploid, the translocation would have to be reciprocal (from exchange between only two chromosomes) and present in a balanced condition. This can be achieved in no more than 50% of male progeny in any generation, with lower frequencies occurring in succeeding generations without parental selection. The most genetically consistent, or true, breeding translocation strain would require homozygous recessive alleles on homologous nontranslocated chromosomes in both males and females, with the

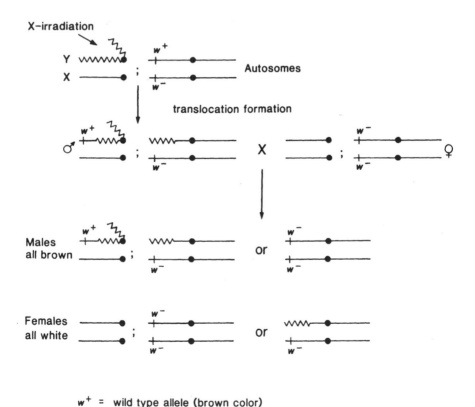

Fig. 1. **A method to create a genetic sexing strain by translocating an autosomal wild type gene for pupal color onto the male-limited Y chromosome.**

dominant gene being Y-linked. This would result in males consistently being duplicated for the translocated region, with a partial Y deficiency. Females would be euploid.

Thus, while translocation strains can be selected for optimal stability and viability, it may be a formidable task. It is unlikely that they will ever have the fitness of wild type insects, which could negatively impact the effectiveness of SIT. Nevertheless, Y-translocation strains may be able to generate enough numbers of males to offset their decreased fitness and decrease costs to a level making their use effective. The development of specific translocation strains for genetic sexing are reviewed below.

*Pupal Color*

The initial use of Y-autosome translocations to select males specifically was suggested by Whitten (1969) in the context of creating strains in the sheep blowfly, *Lucilia cuprina,* having sex-specific pupal coloration. The rationale was that coloration differences would allow high-speed sorters to automatically separate the sexes. Following a similar strategy, Rössler (1979) created the first Y-autosome translocation for genetic-sexing in the Mediterranean fruitfly, *Ceratitis capitata,* linking the Y to the wild type gene for brown pupal color. This allowed the distinction of male brown pupae from mutant female black pupae. More recently, sexing strains distinguishing male brown pupae from female mutant white pupae have been created. The initial strain apparently was subject to high levels of recombination under mass-rearing conditions, leading to a rapid breakdown of the sex-specific coloration (Hooper et al. 1987). Another pupal coloration strain was recently constructed and its genetic integrity was analyzed after many generations under both small-scale laboratory rearing and mass-rearing conditions (Busch-Petersen et al. 1988). Although decreased male viability was observed, only low-level instability of the male-specific brown coloration was detected under mass-rearing, this being attributed largely to strain contamination. The major drawback of sexing based upon pupal color is that the costs required to rear the females are not eliminated, although the females can be used for continued breeding.

*Temperature-Sensitive Lethals*

Selectable phenotypes may be expressed during the larval, pupal, or adult stages affecting viability, an external body structure, or the insects' biochemistry or physiology. For the purposes of genetic sexing, the most efficient type of selection would occur early in development, directly affecting female viability. Such selection would eliminate female rearing costs and the need for automatic sorters or discriminating chemicals, which may be costly and toxic. Viability selection can be accomplished by having a Y-linked wild type allele of a temperature-sensitive lethal (tsl) mutation. In an otherwise tsl mutant background, embryonic or first larval instar males would be able to survive, while females would die at the nonpermissive temperature.

Temperature sensitivity is due to mutations in a general class of genes whose protein product is heat labile, becoming altered in function in response to increased temperature. For lethal genes, normal function is necessary for survival. Although originally discovered in bacteriophage (Edgar & Lielausis 1964), temperature-sensitive lethal mutations are relatively common in insects, representing approximately 10% of the lethal genes in *Drosophila* (Suzuki 1970), and *tsls* have been recovered at a 1% frequency of screened chromosomes in an ethyl methanesufonate (EMS; induces point mutations) screen of *Musca domestica* (McDonald & Overland 1972). The potential utility and abundance of *tsl* genes encouraged Busch-Peterson (1988) to undertake a screen for them in *Ceratitis capitata*. After EMS mutagenesis, putative *tsls* affecting either the embryonic or first larval instar stage were recovered at an approximate frequency of 1%. Unfortunately, almost all of these lines have lost their temperature sensitivity for undetermined reasons. Although there may be inherent instability for these types of genes, stable lines do exist in *Drosophila*, and their relative ease of isolation should encourage continued study.

## Chemical-Selection Genes

**Alcohol Dehydrogenase.** Another general approach for using Y-linked translocations to achieve genetic sexing is based on chemical selection, whereby the translocated gene confers resistance to the selective agent. A potentially useful selection system is based on alcohol dehydrogenase (ADH) activity, which in *Drosophila* allows both positive and negative selection. Wild type insects having normal ADH activity are resistant to ethanol treatment but die when challenged by the secondary alcohols pentenol or allyl-alcohol because of their metabolism to toxic ketones (Sofer & Hatkoff 1972, O'Donnell et al. 1975). Conversely, insects homozygous for an alcohol dehydrogenase (adh⁻) mutation die when challenged with 4% ethanol, although they survive pentenol treatment because of their inability to metabolize it. Pentenol selection of mutagenized insects has provided an effective means for selecting *adh⁻* mutants in *D. melanogaster*.

Robinson and van Heermert (1980) tested the use of ADH selection in *D. melanogaster* by translocating the wild-type *adh⁺* gene to a Y-chromosome and crossing it into an *adh⁻* mutant strain. This resulted in an efficient selection against female larvae fed ethanol in their diet. The application of this method to *C. capitata*, though, first required the isolation of ADH mutants from this species. Although *adh* null mutants were finally isolated with some difficulty, their use was problematic since none were viable as homozygotes (Riva & Robinson 1986), possibly because of preexisting or induced recessive lethal mutations on the mutant chromosome. These ADH mutations were therefore not useful for the suggested selection scheme (Riva & Robinson 1983), which required females being homozygous *adh*-null. Some success was found, though, in a strain in which males had an *adh*-null allele translocated to the Y and an autosomal *adh*$^+$ allele making it heterozygous viable, though with reduced ADH activity (Robinson & Riva 1984). When fed with allyl-alcohol the males with reduced ADH activity survived, while females, being *adh*$^+$, were killed. Although genetic sexing was achieved with this strain, it has not found practical use because of problems with viability, sex-ratio bias for females, *adh*⁻ loss by recombination, and probably high embryo mortality resulting from the strain's double translocation (Wood et al. 1985), which renders a majority of the zygotes aneuploid lethals.

It has since been discovered that much of the difficulty in isolating *adh*-null mutants by pentenol screening was probably a result of the medfly having two distinct genes encoding ADH (Malacrida et al. 1988), as opposed to *D. melanogaster*, which has a single locus. Thus, isolating a complete ADH null mutant in a single screen would require a double mutation (which would be very rare) in order to survive pentenol selection. If it is discovered that the two genes have different temporal specificities, it may be possible to isolate mutations for each gene independently by screening at different developmental times. Nevertheless, for ADH and other enzyme systems that consist of duplicate or overlapping gene activities (or for any multifactorial selectable characteristic), the development and maintenance of genetic-sexing strains would remain difficult.

**Xanthine Dehydrogenase.** One of the best-characterized loci in *D. melanogaster* is the rosy gene complex (Chovnick et al. 1980). This gene encodes the enzyme xanthine dehydrogenase, which is involved in purine metabolism, and as a result rosy mutants are lethally sensitive to high levels of dietary purines. In the course of a mutant screen of *C. capitata*, Saul (1982) isolated a brownish-eyed mutant, designated as *rosy-like*, that showed sensitivity to purines. Although it was never determined if this mutation affected the xanthine dehydrogenase structural locus, the purine sensitivity of the strain nevertheless suggested its use for genetic sexing (Saul 1982). Saul (1984) subsequently isolated four lines having male-limited expression of the *rosy*$^+$ allele (when crossed into a *rosy*⁻ strain), presumed to be Y-autosome translocations but not cytogenetically identified. Upon purine selection only 10-15% *rosy*⁻ females survived, though general viability was decreased and more so in response to purines, compared with wild type flies. It was determined that these lines would not be ideal for genetic sexing, although the general scheme remains promising.  It should be noted that although nontoxic, purines are relatively expensive, limiting practical use of this technique for large-scale selection.

**Insecticide Resistance.**  When gene mutations are isolated for known enzymes, it is possible to make a directed study of chemical sensitivity or resistance, as has been done with *adh* and *rosy* mutants.  Since relatively few gene-enzyme systems have been characterized in nondrosophilids, such directed studies are  limited.  An alternative approach toward chemical selection is to simply screen natural populations or mutagenized strains with a lethal chemical, selecting for insects that are either resistant or sensitive at a threshold concentration.  In this way gene mutations allowing chemical selection may be isolated without regard, at least initially, to the particular gene product affected.

A primary candidate for this type of chemical selection has been insecticides, initially because of the availability of resistance mutants arising in the field.  Indeed, the development and field application of Y-autosome translocation genetic-sexing strains has met its greatest success with insecticide resistance selection in mosquitoes.  The impetus for this effort has been due in part to female mosquitoes being transmitters of disease, prohibiting their release. Effective genetic sexing strains using a Y-linked dieldrin resistance gene have been developed for *Anopheles gambiae* (Curtis et al. 1976) and *A. culicifacies* (Baker et al. 1981).   A propoxur-resistance strain was developed for *A. albimanus* (Seawright et al. 1978), and a malathion-resistance strain for *A. quadrimaculatus* (Kim et al. 1987).  Indeed, the *A. albimanus* strain was the first genetic sexing strain to undergo large-scale rearing and field testing where it met with some success (Dame et al. 1981).  Unlike *C. capitata* (Cladera 1981, Rössler 1982) or *D. melanogaster* (Hiriazumi 1971) in which male recombination does not occur or occurs at very low frequencies, male recombination does occur in these mosquito species, which would result in the eventual loss of the resistance gene.  This has been ameliorated by either selecting for a tight linkage between the resistance gene and the translocation breakpoint (where recombination is suppressed) or by subsequently selecting for an inversion in the translocation chromosome, as was done with *A. albimanus* (Seawright et al. 1978).  Thus far, the development of insecticide resistance sexing strains in *C. capitata* have not progressed as well as in mosquitoes. Although dieldrin resistance has been localized to a single gene (Busch-Petersen & Wood 1986), efforts to isolate a stable Y-translocation have been unsuccessful (Wood et al. 1987).

Despite the varied success with insecticide resistance, clearly the widespread use of such a selection technique would pose difficulties in terms of the possible release of insecticide-resistant insects and in the handling and disposal of large quantities of toxic chemicals.  For these reasons, and the need to broaden the possibilities for chemical selection in *C. capitata,* Rössler (1988) has begun a study to isolate resistance genes for chemicals toxic to insects but normally having low toxicity in mammalian systems. An analysis of potassium sorbate, Avermectin, and Cyromazine has indicated resistance regulated by a single gene only for Cyromazine. Further studies on this resistance strain for genetic sexing are anticipated.

## Molecular Genetic Sexing Methods

The genetic sexing methods discussed thus far are based upon classical genetic mechanisms, often resulting from innovative approaches in an effort to manipulate and genetically stabilize insect species not optimally amenable to such procedures. Nevertheless, in the 35 years since Knipling (1955) first proposed the use of SIT, and the 20 years since the first genetic-sexing strain was developed (Whitten 1969), a genetic-sexing strain for mass rearing and release on a continued basis has not been forthcoming. The use of Y-autosome translocations has met some success in allowing genetic sexing under laboratory conditions, but beyond a few cases with mosquitoes, the integrity of these strains has not been maintained under mass rearing.

The development of genetic-sexing strains based upon recombinant DNA technology, at least theoretically, should be able to ameliorate many of the limitations discussed. At this point, though, it is important to note that there are no specific recombinant DNA molecules available that would assuredly result in effective genetic sexing, and the technology to transfer such molecules into non-drosophilid host genomes (i.e., germline transformation) is still being developed. Thus, this discussion will generally be prospective and to a large extent based on our knowledge of gene isolation, function, and manipulation in *D. melanogaster.* The power and efficiency of recombinant DNA techniques encourages the belief that once gene transfer is possible in nondrosophilid insects, the implementation of genetic-sexing techniques similar to those proposed here and also more novel techniques will be rapidly forthcoming.

### *Gene Structure*

Before the use of recombinant DNA techniques to manipulate genetic material for genetic sexing or other means of insect control can be appreciated, it is useful to review some of the basic molecular properties of eucaryotic genes (Fig. 2). These genes are generally divided into two functional domains: the transcriptional unit, which encodes the gene product, and the regulatory region, which controls the temporal, spatial, and quantitative specificities of transcription.

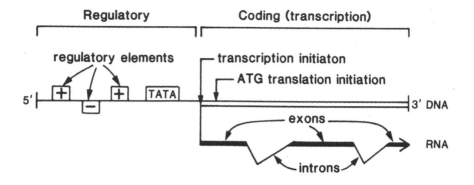

**Fig. 2.  The major DNA regulatory and coding components necessary for eucaryotic gene expression.**

The transcriptional unit, beginning with a transcriptional start site, proceeds downstream in a 5' to 3' orientation as an untranslated leader sequence of variable length before reaching the ATG translational start site.  The translated portion of the transcriptional unit consists of nucleotide sequences in an open reading frame, or exons.  The open reading frame is often, but not always, interspersed by intervening sequences, or introns, which are spliced out of the RNA transcripts before translation, allowing the restoration of the reading frame among the exons.  The open reading frame ends where nonsense codons are encoded, creating a translational stop signal.  Downstream to the translational stop are sequences encoding the mRNA poly adenylate tail and the transcription termination site.

The regulatory, or promoter, region is generally found in nontranscribed flanking sequences 5' to the transcriptional unit.  Usually adjacent to the transcriptional start is the TATA box, which is required for transcription, providing a polymerase binding site.  Upstream (or 5') from the TATA box are nucleotide sequences of varying length and position known as regulatory elements, which can act to influence transcription positively or negatively, often as a result of binding to regulatory proteins produced elsewhere in the genome.  Beyond defining temporal and spatial activity, regulatory elements can specify sex-specific activity; feedback regulation; and responses to hormones, metals, or environmental influences such as temperature or stress, among others types of regulation. These regulatory elements may have the same or similar sequences when present in different genes, although their physical location may not be well defined.  They may exist in an interspersed relation to one another in various orientations hundreds of base pairs apart and sometimes on opposite strands of DNA. Although generally found in the 5' flanking sequences, regulatory elements may also be found (or function when transposed) in 3' regions, and sometimes in introns.

Another means of regulating gene expression, but not by influencing transcriptional activity, is by alternative intron splicing. For reasons not well elucidated, genes may have alternative splice sites resulting in different translated products in terms of size and function from a single gene. In some cases a translational stop signal within an intron may result in a truncated product if revealed by alternative splicing. Indeed, such is the case in *D. melanogaster*, where alternative splicing affects the sex-specific function of three sex-determination genes (Baker 1989).

A major discovery in molecular genetic analysis was the finding that regulatory and transcriptional sequences from different genes derived from different organisms can be spliced together to form functional chimeric gene fusions when introduced into a host genome (Kelly & Darlington 1985). Furthermore, gene expression usually occurs according to the specificities of the regulatory region. Since regulatory element function may depend upon interactions with regulatory proteins or cofactors, whose availability may be further regulated species-specifically, the fidelity of regulatory region function is best preserved in the species, or one closely related, from which it was isolated. On the other hand, production and activity of the gene product can occur in widely diverse heterologous systems. Indeed, bacterial coding regions are often used as markers, or reporter gene products, for analyzing promoter activity in eucaryotic systems.

The ability to create chimeric genes is thus the basis for using recombinant DNA methods for genetic sexing. Sex-specific expression of a selectable gene may be achieved by linking its coding region to a promoter from a sex-specific gene instead of creating chromosomal rearrangements in order to link the gene to a sex-limited chromosome. Furthermore, additional control of gene expression may be achieved by linking conditional regulatory elements (responding to exogenous factors) to the promoter. In this way a breeding culture may be maintained under permissive conditions, with sex-specific selection occurring in the progeny under nonpermissive conditions.

It is important to note that sex-specific gene promoters may also have regulatory elements for tissue and temporal specificity, which may result in the gene product being produced in a tissue inappropriate for selection or at a less than optimal time, such as adulthood. These influences may be ameliorated by isolation of a sex-specific gene acting in a wide range of tissues early in development. Alternatively, if the required regulatory elements for sex, tissue, and temporal specificity, possibly from different genes, can be defined and isolated, they may be linked together to create a promoter tailored for a specific type of selection.

## Sex-Specific Selection

**Male-specific Expression.** One use of chimeric genes to achieve male-specific selection utilizes the same basic rationale as Y-linked selectable genes—that is, linking the regulatory region from a male-specific gene to the coding or transcriptional region of the selectable gene. Gene integration could occur in most any euchromatic location in the genome with the transformed strain displaying male-specific expression of the selectable gene product. Thus, in terms of the classical schemes discussed, a strain mutant for either alcohol dehydrogenase, xanthine dehydrogenase, temperature-sensitive lethality, or pupal coloration could be transformed with the appropriate wild type transcriptional unit linked to a male-specific promoter to achieve the appropriate selection. Since it would be better not to rely on mutant strains, a preferable strategy would be to introduce a selectable gene with a phenotype dominant to wild type. Such genes would include insecticide-resistance genes that have been cloned in the mosquito, *Culex quinquefasciatus* (Mouches 1988), and *D. melanogaster* as well as other genes conferring chemical or drug resistance. Notable among these is the bacterial gene for neomycin phosphotransferase, which, after transformation, has been shown to confer resistance to a neomycin analog, geneticin in *D. melanogaster* (Steller & Pirrotta 1985).

A major limitation of the male-specific promoter strategy at present is the lack of suitable regulatory regions. Most male-specific genes isolated to date in *D. melanogaster* are active in male-specific tissues, and apparently their sex-specific activity is a function of tissue specificity and not direct interactions with sex-determining mechanisms (DiBenedetto et al. 1987, Wolfner 1988). Thus, these genes may not have a discrete male-specific regulatory element that can be linked with other elements for tissue and temporal regulation to achieve appropriate expression. Unless selection can occur by the male-specific gene product acting solely in such tissues, which are usually part of the reproductive system, the use of male-specific promoters like those available at present from *D. melanogaster* may not be useful. An exception to this restriction might be use of genes encoding enzymes being nonspecific in terms of tissue activity and substrate availability. For example, ADH is normally active in the fat body, although in transformant strains where *adh* expression is limited to the ovaries or salivary glands (due to linkage with tissue specific promoters), ethanol resistance is observed.

In lieu of appropriate male-specific promoters, male-limited expression could also be achieved by selecting for gene integration in the Y chromosome or by translocating an autosomal integration onto the Y. Although the goal of molecular genetic techniques is to avoid use of chromosomal rearrangements, until male-specific promoters are available this strategy at least allows the use of isolated genes known to be selectable with defined activity, rather than depending on the laborious selection of spontaneous or induced mutations.

In *Drosophila,* gene transfer into the highly heterochromatic Y has not been reported, although greater amounts of euchromatin in other species might allow Y integration (McInnis et al. 1988). If translocating autosomal integrations onto the Y is necessary, this could result in many of the problems already observed in translocation strains. To alleviate this condition, it may be possible to create a highly viable strain having a small Y-autosome translocation tested for stability. The autosomal euchromatin could then act as a target site for gene integration by germline transformation.

**Female-specific Expression.** Sex-specific selection may also be achieved by female-specific expression of a negatively selectable gene product. Classical means of achieving this have not been pursued in those insects without a female-specific chromosome. Similar to chimeric gene fusions discussed for male-specific expression, the sex-specific regulatory element from female-specific genes may be isolated and linked to a selectable transcriptional unit. Analogous to males, many female-specific genes function in female reproductive tissue, implying that their promoters do not have female-specific regulatory elements. An exception to this, at least in *D. melanogaster*, is the promoter region for the yolk protein (YP) genes. In most insects YP is made by the fat body, although in various dipterans the ovary also contributes a significant proportion of the YP. In *D. melanogaster* YP gene expression in the fat body is primarily regulated by the sex-determination gene hierarchy throughout development. Thus, a change in sexual state, caused by the manipulation of a sex-determination gene, can turn YP synthesis on or off in a fully differentiated adult insect regardless of the chromosomal sex (Belote et al. 1985, McKeown et al. 1987). This indicates that for the YP genes at least, sex-specific regulation is direct, or at least does not depend upon tissue differentiation, and that a sex-specific regulatory element exists which responds to sex-determination gene activity. YP promoter regions have already been linked to the *adh* coding region (YP-Adh), resulting in ADH activity consistent with YP regulation, in particular, being under female-specific and hormonal control (Shirras & Bownes 1987).

When present in an *adh⁻* background, the YP-Adh fusion gene could conceivably allow genetic sexing based upon female-specific sensitivity to pentenol or allyl-alcohol (Fig. 3A). Although the YP female-specific regulatory element has not been isolated, *adh* expression is nevertheless normal because it shares the same tissue and developmental specificities in the adult fat body as YP. Thus, until the regulatory element is defined and isolated its use will be limited to selectable genes normally expressed in the adult fat body. At present,

Sex–specific ADH Selection

(A) Female-specific promoters

YP 5' | Adh coding

– Female-specific ADH in adults
– Females lethally sensitive to pentenol

(B) Anti–sense RNA

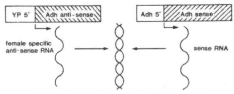

YP 5' | Adh anti-sense        Adh 5' | Adh sense

female specific
anti-sense RNA                                    sense RNA

– RNA heteroduplex formation inhibits translation
– Female-specific lack of ADH
– Females lethally sensitive to ethanol

(C) Sex–specific intron splicing

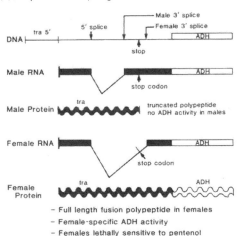

Male 3' splice
5' splice        Female 3' splice
tra 5'                                              ADH
DNA
stop

Male RNA                                    ADH
stop codon

tra                 truncated polypeptide
Male Protein                     no ADH activity in males

Female RNA                                    ADH
stop codon

tra                 ADH
Female
Protein

– Full length fusion polypeptide in females
– Female-specific ADH activity
– Females lethally sensitive to pentenol

**Fig. 3. Three methods by which molecular genetic manipulation could be used to achieve genetic sexing using the expression of an alcohol dehydrogenase gene for positive or negative selection.**

this is not practical for genetic sexing in terms of relying on adult selection and difficulty in developing ADH mutant strains (see above).

A more useful strategy would rely on dominant female-specific ectopic expression of a product resulting in lethality—for example, expression of the *Bacillus thuringiensis* toxin, inappropriate expression of chitinase, or ADH expression in insects that normally do not have ADH, allowing pentenol selection. Tissue-specific activity would be less important to the lethal function of such products, although maintenance of females for breeding would depend upon additional conditional regulation, such as heat shock control. It is important that such strains would not depend upon preexisting mutations.

Another approach not relying on mutations, although speculative at present, would be the use of female-specific expression of antisense RNA. Antisense RNA results from the transcription of the opposite (complementary) noncoding DNA strand of a particular gene. Antisense genes are created simply by inverting part or all of a transcriptional unit with respect to the promoter. When present together in the cell, antisense RNA can block, to varying degrees, the translation of the normal sense RNA, thereby prohibiting gene expression and in effect creating a mutation for that gene. The mechanism by which this blocking occurs is unknown, though possibly RNA heteroduplex formation blocking translation plays a role. Antisense RNA has proven most effective in cell lines in eucaryotic systems (Green et al. 1986), although embryonic phenocopies in *Drosophila* have been produced for the Krüppel embryonic mutant (Rosenberg et al. 1985). For genetic sexing in insects having normal *adh* expression, female-specific expression of antisense *adh* (using the YP promoter) could result in female lethality in response to ethanol (Fig. 3B). This strategy could be quite efficient, more simply, by using antisense lethal genes under conditional (i.e., heat shock) control as well as female-specific control.

Female-limited gene expression can also be achieved without a female-specific promoter. As mentioned previously, intron splicing occurs sex specifically for three sex-determination genes in *D. melanogaster*. For these genes, one of their introns contains a single 5' splice acceptor site, but two 3' splice acceptor sites with a translational stop signal encoded within the two sites. If the first 3' acceptor site (5' to the second site) is utilized, as occurs specifically in males, then the stop signal is revealed in the transcript, resulting in a truncated nonfunctional gene product (Fig. 3C). If the second 3' acceptor splice site is utilized, as occurs in females, the stop signal is excised with the intron from the transcript, resulting in a full-length functional product. If a selectable gene is linked in frame to one of these sex-determination genes in the exon immediately downstream (3') to the intron, then its expression should occur only in females where the open reading frame is maintained. The resulting gene product would be a fusion protein containing part of the sex-determination gene product and the full-length selectable gene. Various enzymes produced as fusion proteins are able to maintain their function. A major advantage over a female-specific promoter, such as the YP promoter, is that the processes resulting in sex-specific intron splicing (due to other sex-determination genes) are active in almost all tissues throughout development, allowing selection early in development. Female-limited expression of lethal-product genes may also be achieved this way, though they would require linkage of the sex-specifically spliced intron to a conditional promoter. This should not be problematic since other introns whose splicing is tissue specific maintain this specificity when placed in the context of another gene (Laski & Rubin 1989). A limitation of this strategy is that, while promoter regions have been shown to maintain normal function between widely divergent species (e.g., *Bombyx* chorion gene promoter function in *D. melanogaster*; Mitsialis & Kafatos 1986), sex-determination genes have been isolated only in *D. melanogaster,* and at present it is not possible to determine their regulation in other species. However, once gene transformation in nondrosophilid insects is possible, which will be required for any of these recombinant DNA-based schemes, heterologous sex-determination gene activity will be easily tested, and identification of species-specific sex-determination genes will be facilitated.

**Sex-determination Genes.** In addition to taking advantage of differential intron-splicing mechanisms, it is possible to manipulate the sex-determination genes themselves to achieve sex-specific development (Fig. 4). In *D. melanogaster* the sex-determination genes *transformer* (tra) and *transformer*-2 (tra-2) are required in chromosomal females for female development (Baker& Belote 1983). In the absence of the normal function of either of these genes, chromosomal females develop as morphological males that are sterile but capable of mating. In chromosomal males these genes have no apparent function (i.e., mutants remain morphologically male), except that *tra-2* is required for fertility. Thus, it is conceivable that if *tra-2* activity (or an analogous gene in another insect) could be conditionally eliminated throughout development, then a fertile strain could give rise to an entire population of sterile males. This has already been achieved in *D. melanogaster* by means of a temperature-sensitive mutation of *tra-2* (tra-2[ts]; Belote & Baker 1982), although in this mutant strain females raised at the permissive temperature are only marginally fertile, and transformed females (into males at 29° C) have lowered viability. Theoretically, elimination of *tra-2* activity could also be achieved without inducing mutations by having an antisense *tra-2* gene linked to a heat-shock promoter integrated into the genome.

At present, this type of scheme for genetic sexing is highly speculative for nondrosophilid insects, but it nevertheless illustrates one of the greatest potential strengths for the use of recombinant DNA techniques. Its enormous efficiency compared to other methods discussed is demonstrated by the fact that all the

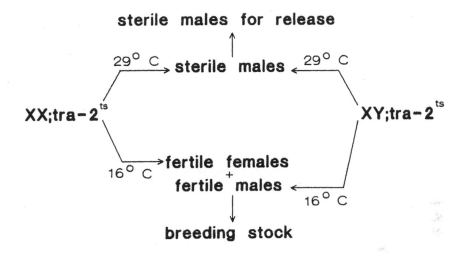

**Fig. 4.** Diagram of how conditional manipulation of the temperature-sensitive *transformer-2ᵗˢ* allele from *Drosophila melanogaster* can be used to create a breeding stock of flies yielding only sterile males.

resultant zygotes are allowed to develop and can be used for release, requiring maintenance of a 50% smaller breeding population. Chemical or mechanical male selection and sterilization mechanisms would be unnecessary, eliminating the considerable time, labor, and equipment costs associated with these procedures. Since the basic methodology does function in *Drosophila*, it is quite conceivable that similar sex-determining mechanisms exist in other insects (Nöthiger & Steinmann-Zwicky 1985) and may be similarly manipulated to achieve male-limited progeny and sterilization. The expenditure of resources and effort required to identify and isolate sex-determining genes and the ability to perform gene transformation efficiently (which in turn would facilitate achieving the former goal) in economically important insects would be quite worthwhile in view of the eventual efficiency and savings realized.

### References

Baker, B. S. 1989. Sex in flies: the splice of life. Nature 340: 521-524.
Baker, B. S. & J. M. Belote 1983. Sex determination and dosage compensation in *Drosophila melanogaster*. Annu. Rev. Genet. 17: 345-397.

Baker, R. H., R. K. Salai & K. Raana. 1981. Genetic sexing for a mosquito sterile-male release. J. Hered. 72: 216-218.

Belote, J. M. & B. S. Baker. 1982. Sex determination in *Drosophila melanogaster*: analysis of transformer-2, a sex-determining locus. Proc. Natl. Acad. Sci. USA 79: 1568-1572.

Belote, J. M., A. M. Handler, M. F. Wolfner, K. J. Livak & B. S. Baker. 1985. Sex-specific regulation of yolk protein gene expression in *Drosophila*. Cell 40: 339-348.

Busch-Petersen, E. 1988. Development of induced sex-separation mechanisms in *Ceratitis capitata* (Wied.): EMS tolerance and suppression of female recombination, pp. 209-216. *In* A. P. Economopoulos [ed.], Fruit Flies (Proceedings of the Second International Symposium, Colymbari, Crete, Greece 1986). Elsevier Science Publishers, Amsterdam.

Busch-Petersen, E. & R. J. Wood. 1986. The isolation and inheritance of dieldrin resistance in the Mediterranean fruit fly *Ceratitis capitata* (Wiedmann) (Diptera: Tephritidae). Bull Ent. Res. 76: 567.

Busch-Petersen, E., J. Ripfel, A. Pyrek & A. Kafu. 1988. Isolation and mass rearing of a pupal genetic sexing strain of the Mediterranean fruit fly, *Ceratitis capitata* (Wied.) (IAEA-SM-301/35), pp. 210-220. *In* Modern Insect Control: Nuclear Techniques and Biotechnology (Proceedings International Symposium) Vienna, Austria.

Chovnick, A., M. McCarron, H. Clark, A. J. Hilliker & C. A. Rushlow. 1980. Structural and functional organization of a gene in *Drosophila melanogaster*, pp. 3-23. *In* O. Siddiqi, P. Babu, L. M. Hall & S. C. Hall [eds.], Development and neurobiology of *Drosophila*. New York, Plenum.

Cladera, J. L. 1981. Absence of recombination in the male of *Ceratitis capitata*. Experientia 37: 342-343.

Curtis, C. F., J. Akiyama & G. Davidson. 1976. A genetic sexing system in *Anopheles gambiae* species A. Mosq. News 36: 492-498.

Dame, D. A., R. E. Lowe & D. L. Williamson. 1981. Assessment of released sterile *Anopheles albimanus* and *Glossina morsitans* morsitans, pp. 231-242. *In* J. B. Kitzmiller & T. Kanda [eds.], Cytogenetics and Genetics of Vectors. Elsevier Biomedical, Amsterdam.

Dibenedetto, A. J., D. M. Lakich, W. D. Kruger, J. M. Belote, B. S. Baker & M. F. Wolfner. 1987. Sequences expressed sex-specifically in *Drosophila melanogaster* adults. Develop. Biol. 119: 242-251.

Edgar, R. S. & I. Lielausis. 1964. Temperature-sensitive mutants of bacteriophage T4D: Their isolation and genetic characterization. Genetics 49: 649-662.

Foster, G. G., R. H. Maddern & A. T. Mills. 1980. Genetic instability in mass-rearing colonies of a sex-linked translocation strain of *lucillia cuprina*

(Wiedemann) (Diptera: Calliphoridae) during a field trial of genetic control. Theor. Appl. Genet. 58: 164-175.

Green, P. J., O. Pines & M. Inouye. 1986. The role of antisense RNA in gene regulation. Annu. Rev. Biochem. 55: 569-597.

Handler, A. M. & D. A. O'Brochta. 1991. Prospects for gene transformation in insects. Annu. Rev. Entomol. 36: 159-183.

Hiriazumi, Y. 1971. Spontaneous recombination in *Drosophila melanogaster* males. Proc. Natl. Acad. Sci. USA 68: 268-278.

Hooper, G. H. S., A. S. Robinson & R. P. Marchand. 1987. Behaviour of a genetic sexing strain of Mediterranean fruit fly, *Ceratitis capitata*, during large scale rearing, pp. 349-362. *In* A. P. Economopoulos [ed.] Fruit flies (Proceedings of the 2nd International symposium, Colymbari, Crete, Greece, 1986). Elsevier Publishers, Amsterdam.

Kelly, J. H. & G. J. Darlington. 1985. Hybrid genes: molecular approaches to tissue-specific gene regulation. Annu. Rev. Genet. 19: 273-296.

Kim, S. S., J. A. Seawright & P. E. Kaiser. 1987. A genetic sexing strain of Anopheles quadrimaculatus species A. J. Am. Mosq. Cont. Assoc. 3: 50.

Knipling, E. F. 1955. Possibilities of insect control of eradication through the use of sexually sterile males. J. Econ. Entomol. 48: 459-462.

Lachance, L. E. 1979. Genetic strategies affecting the success and economy of the sterile insect release method, pp. 8-19. *In* M. A. Hoy & J. J. McKelvey [eds.], Genetics in relation to insect management. The Rockefeller Foundation, New York.

Laski, F. A. & G. M. Rubin. 1989. Analysis of the cis-acting requirements for germ-line-specific splicing of the P-element ORF2-ORF3 intron. Genes Devel. 3: 720-728.

Louis, C., C. Savakis & F. Kafatos. 1987. Possibilities for genetic engineering in insects of economic interest, pp. 45-57. *In* A. P. Economopoulos [ed.], Fruit Flies (Proc. 2nd Int. Symp. Colymbari, Crete, 1986). Elsevier, Amsterdam.

Malacrida, A. R., G. Gasperi, L. Baruffi & G. F. Biscaldi. 1988. Updating the genetics of *Ceratitis capitata* (Wied.), pp. 221-227. *In* Proceedings, Modern Insect Control: Nuclear Techniques and Biotechnology. International Symposium, Vienna, Austria 1987.

McDonald, I. C. & D. E. Overland. 1972. Temperature-sensitive mutations in the house fly: the characterization of heat-sensitive recessive lethal factors on autosome III. J. Econ. Entomol. 65: 1364-1368.

McInnis, D. O., S. Y. T. Tam, C. R. Grace, L. J. Heilmann, J. B. Courtwright & A. K. Kumaran. 1988. The Mediterranean fruit fly: Progress in developing a genetic sexing strain using genetic engineering methodology (IAEA-Sm-301/28), pp. 251-256. *In* Proceedings, Modern

Insect Control: Nuclear Techniques and Biotechnology. International Symposium, Vienna, Austria 1987.

McKeown, M., J. M. Belote & B. S. Baker. 1987. A molecular analysis of transformer, a gene in *Drosophila melanogaster* that controls female sexual differentiation. Cell 48: 489-499.

Mitsialis, S. A. & F. C. Kafatos. 1986. Regulatory elements controlling chorion gene expression are conserved between flies and moths. Nature 317: 453-456.

Mouches, C. 1988. Génétique chez les insectes: Clonage et transgénose de génes de résistance aux insecticides (IAEA-SM-301-23), pp. 257-261.. *In* Proceedings, Modern Insect Control: Nuclear Techniques and Biotechnology. International Symposium, Vienna, Austria 1987.

Noethinger, R. & M. Steinmann-Zwicky. 1985. A single principle for sex-determination insects. Cold Spring Harbor, Symp. Quant. Biol. 50: 615-621.

O'Donnell, J., L. Gerace, F. Leister & W. Sofer. 1975. Chemical selection of mutants that affect alcohol dehydrogenase in Drosophila. II. Use of 1-pentyne-3-ol. Genetics 79: 73-83.

Riva, M. E. & A. S. Robinson. 1983. Studies on the Adh locus in *Ceratitis capitata*, pp. 163-170. *In* R. Cavalloro [ed.] Fruit flies of economic importance (Proceedings International Symposium, Athens, Greece 1982.) Balkema, Rotterdam.

Riva, M. E. & A. S. Robinson. 1986. Induction of alcohol dehydrogenase null mutants in the Mediterranean fruit fly *Ceratitis capitata*. Biochem. Genet. 24: 765-774.

Robinson, A. S. 1983. Sex ratio manipulation in relation to insect pest control. Annu. Rev. Genet. 17: 191-214.

Robinson, A. S. 1984. Unexpected segregation ratios from male-linked translocations in the Mediterranean fruit fly, *Ceratitis capitata* (Diptera: Tephritidae). Genetica 62: 209-213.

Robinson, A. S. & M. E. Riva. 1984. A simple method for the isolation of allelic series using male-linked translocations. Theor. Appl. Genet. 67: 305-306.

Robinson, A. S. & C. Van Heemert. 1980. Genetic sexing in *Drosophila melanogaster* using the alcohol dehydrogenase locus and a Y-linked translocation. Theor. Appl. Genet. 59: 23-24.

Robinson, A. S., M. E. Riva & M. Zapater. 1986. Genetic sexing in the Mediterranean fruit fly, *Ceratitis capitata,* using the Adh locus. Theor. Appl. Genet. 72: 455-457.

Robinson, A. S., C. Savakis & C. Louis. 1988. Status of molecular genetic studies in the medfly, *Ceratitis capitata*, in relation to genetic sexing (IAEA-SM-301/38), pp. 241-250. *In* Proceedings, Modern Insect Control:

Nuclear Techniques and Biotechnology. International Symposium, Vienna, Austria 1987.

Rosenberg, U. B., A. Preiss, T. E. Seifert, H. Jackle & D. C. Knipple. 1985. Production of phenocopies by Krüppel antisense RNA injection into *Drosophila* embryos. Nature 313: 703-706.

Roessler, Y. 1979. Automated sexing of *Ceratitis capitata*: The development of strains with inherited sex-limited pupal colour dimorphism. Entomophaga 24: 411-416.

Roessler, Y. 1982. Genetic recombination in males of the Mediterranean fruit fly, and its relation to automated sexing methods. Ann. Entomol. Soc. Am. 75: 28-31.

Roessler, Y. 1985. Effect of genetic recombination in males of the Mediterranean fruit fly (Diptera: Tephritidae) on the integrity of "genetic sexing" strains produced for sterile-insect releases. Ann. Entomol. Soc. Am. 78: 265-270.

Roessler, Y. 1988. Selection for resistance in the Mediterranean fruit fly for genetic sexing in sterile insect technique programmes (IAEA-SM-301/36), pp. 229-240. *In* Proceedings, Modern Insect Control: Nuclear Techniques and Biotechnology. International Symposium, Vienna, Austria 1987.

Rubin, G. M. & A. C. Spradling. 1982. Genetic transformation of *Drosophila* with transportable element vectors. Science 218: 348-353.

Saul, S. H. 1982. Rosy-like mutant of the Mediterranean fruit fly, *Ceratitis capitata* (Diptera: Tephritidae), and its potential for use in a genetic sexing program. Ann. Entomol. Soc. Am. 75(4): 480-483.

Saul, S. H. 1984. Genetic sexing in the Mediterranean fruit fly, *Ceratitis capitata* (Wied) (Diptera: Tephritidae): conditional lethal translocations that preferentially eliminate females. Ann. Entomol. Soc. Am. 77: 280-283.

Seawright, J. A., P. E. Kaiser, D. A. Dame & C. S. Lofgren. 1978. Genetic method for the preferential elimination of females of *Anopheles albimanus*. Science 220: 1303-1304.

Shirras, A. & M. Bownes. 1987. Separate DNA sequences are required for normal female and ecdysone-induced male expression of *Drosophila melanogaster* yolk protein 1. Molec. Gen. Genetics 210: 153-155.

Snow, J. W. 1988. Radiation, insects and eradication in North America: An overview from screwworm to bollworm (IAEA-SM-301/29), pp. 3-14. *In* Proceedings, Modern Insect Control: Nuclear Techniques and Biotechnology. International Symposium, Vienna, Austria 1987.

Sofer, W. & M. Hatkoff. 1972. Chemical selection of alcohol dehydrogenase negative mutants in *Drosophila*. Genetics 72: 545.

Spradling, A. C. & G. M. Rubin. 1982. Transposition of cloned P-elements into *Drosophila* germ line chromosomes. Science 218: 341.

Steller, H. & V. Pirotta. 1985. A transposable P vector that confers selectable G418 resistance to *Drosophila* larvae. EMBO J. 4: 167-171.

Suzuki, D. T. 1970. Temperature-sensitive mutations in *Drosophila melanogaster.* Science 170: 695-706.

Whitten, M. J. 1969. Automated sexing of pupae and its usefulness in control by sterile insects. J. Econ. Entomol. 62: 272-273.

Wolfner, M. F. 1988. Sex-specific gene expression in somatic tissues of *Drosophila melanogaster.* Trends in Genetics 4: 333-337.

Wood, R. J., S. S. Saaid, R. M. Shahjahan & D. I. Southern. 1987. Excess male production in lines of the Mediterranean fruit fly *Ceratitis capitata* (Wied.) isolated after X irradiation followed by outcrossing. *In* A. P. Economopoulos [ed.], Fruitflies (Proceedings International Symposium Crete, 1986). Elsevier, Amsterdam.

Wood, R. J., D. I. Southern, S. S. Saaid & G. S. Proudlove. 1985. Progress towards developing a genetic sexing mechanism for the Mediterranean fruit fly *Ceratitis capitata* (Wied.). (Report on Joint FAO/IAEA Research Coordination Meeting on the Development of Sexing Mechanisms, 1985 July 15-19, Vienna.)

# 3

## Insect Rearing
## and the Development
## of Bioengineered Crops

*Terry B. Stone and Steven R. Sims*

### Introduction

The naturally occurring spore-forming microorganism *Bacillus thuringiensis* (Bt) produces delta-endotoxin proteins. When properly formulated and applied, endotoxin-spore combinations are potent insecticides against many Lepidoptera, Coleoptera, and Diptera pests. Bt products such as Dipel, Thuricide, and more recently Condor and Javelin, have been commercially available for more than 20 years, although only about 1-2% of world pesticide use involves this group of microbials. These levels of use may now be on the threshold of a major increase. The rapid development of molecular biology techniques has permitted incorporation of genes coding for expression of Bt delta-endotoxin proteins into both plants and other microbe species. This has opened new commercial vistas for the pest control industry. Acceptance of this new technology will be aided by increasing widespread concern over continued use of conventional toxic pesticides, plus the relentless development of pesticide resistance.

The essential role of insect rearing in transgenic plant commercialization is often overlooked amidst the glamour of the technology. For example, Western Blot and ELISA assay procedures are routinely used to determine levels of endotoxin expression in transgenic plants and to prioritize constructs for further evaluation (Fuchs et al. 1990). However, results from these assays do not always correlate with or accurately predict levels of insecticidal activity. Insects of uniform high quality must therefore be available for initial screening of plant constructs. In addition, insects are required for secondary whole-plant assays and artificial infestation of plants under field conditions where natural pest population levels are low or absent.

Studies on the resistance of target insects to recombinant plants/microbes require considerable insect rearing support. Recent documentation of the potential for insect resistance to delta-endotoxin expressed in a recombinant *Pseudomonas fluorescens* (Stone et al. 1989) required substantial numbers of *Heliothis virescens* (Lepidoptera:Noctuidae). Similar insect quantities will be required for anticipated selection of insect lines resistant to the effects of plant expressed endotoxins. Ultimately, resistant insect lines will serve several valuable functions. These include comparative studies on endotoxin mechanism(s) of action in the insect midgut (Hofmann et al. 1988) and genetic analysis of the resistance trait. Details of resistance genetics will play an important role in future resistance management strategies (Roush & McKenzie 1987, Gould 1988).

Surveys of target pest populations from geographic regions of proposed transgenic plant use can provide valuable information on pre-existing genetic variability for resistance and may also influence resistance management and product use decisions. These surveys involve considerable effort in sample collection, colony establishment, and bioassay. The vital role of insect culture in such surveys and other aspects of transgenic species commercialization cannot be overemphasized. In this paper we describe the establishment and maintenance of insect colonies for the insect-resistant recombinant plant discovery effort at Monsanto.

## Establishment of Insect Colonies

During 1988-1989, we established more than 25 colonies of *Heliothis virescens* and *H. zea* from field collections. These collections were evaluated for their susceptibility to *B. thuringiensis* subsp. *kurstaki*. Procedures for routine maintenance of existing lab-adapted *Heliothis* colonies are well known (Burton 1969, King & Hartley 1985, Pantana 1985). We found less information to guide us in the initiation of a new colony from wild individuals. Numerous factors are involved in the collection-establishment process but the following seem particularly important: adequate sampling from several sites within a particular population, proper shipment of the appropriate life-history stages, quarantine procedures, adaption of the strain to laboratory conditions, and prolonged maintenance.

Before establishing insect cultures in the laboratory, however, it is useful to determine their intended research purpose(s). Hoy (1976) identified several purposes for insect colonies and their required attributes. For the development of bioengineered or synthetic pesticides, it is necessary to maintain colonies that provide a uniform and consistent response when bioassayed. Therefore, colonies in culture for many generations are preferred over those newly established from the field. An excellent reference for available lab-adapted insect and mite colonies is the publication *Arthropods in Culture* (Edwards et al. 1987). Conversely, to better reflect the wild type of response when monitoring field populations for pesticide resistance, freshly collected insects or insects in culture for only 1-2 generations are preferred.

In the development of sampling methods for initiating field collected colonies, it is useful to consider the ideas of Levins (1969) and Southwood (1978). They proposed that organisms adopt strategies of adaption and distribution according to the way they experience their environment. For sedentary populations "patchiness" results, making it essential that several patches be sampled to form a colony more representative of the entire population. Highly mobile insects, such as the *Heliothis* spp., tend to experience their environment as an average of smaller sub-patches that is the same for all members of their population. This results in genetic adaptation that may be more similar across subpopulations. Consequently, sampling from one or a few populations will result in a representative sample of the population (Bartlett 1985). In establishing *Heliothis* we were interested in determining if a geographic trend in Bt susceptibility was evident in populations from the states representing the cotton growing region i.e., AL, AR, AZ, CA, FL, GA, LA, MO, MS, NC, SC, and TX. Therefore, for our sampling to be representative of the native populations from this region, it was not necessary to collect from several sites within each state but rather from several geographically distinct areas within the region.

Once insects are collected or a colony identified, proper preparation, shipment, and receipt is critical to their establishment as a viable new colony. It is important that a stage amenable to transport be shipped. For *Heliothis*, almost any development stage, other than adults, can be shipped successfully using an overnight mail service. From an established colony, newly oviposited eggs or pupae are preferred. Raulston et al. (1976) found 1 pint cardboard containers with moist vermiculite to be most suitable for shipping pupae. These containers allow for the dissipation of metabolic heat and air transfer which, when not provided, can result in significant pupal mortality. For field collected insects, eggs or pupae are also preferred. If not available, however, 3rd to 5th instar larvae can be shipped, without food, individually in one ounce cups in an ice chest. The 24-28 period of starvation during shipment will kill those insects infected with a pathogen, or of poor vigor, leaving the hardiest individuals readily disposed to feed on an artificial diet once in the laboratory. We have had less success when plant material or artificial diet is provided as an interim food source. Often larval mortality results as the food material becomes fouled with feces or microbial contamination.

Quarantine procedures are essential before integrating a newly established colony with existing colonies in the laboratory. The environment should be isolated and as sterile as possible to minimize transmission of pathogens or parasites. Eggs and pupae should be surface sterilized with chlorine bleach or formaldehyde. For *Heliothis*, eggs on an oviposition cloth can be easily sterilized with 1.0% clorox (.052% sodium hypochlorite) using a plant sprayer set to release a fine mist (A. Johnson, personal communication). Eggs treated in this manner can be stored, without rinsing in water, with no loss in viability. Pupae can be surface sterilized by immersion in 5.0% clorox (.262% sodium hypochlorite) for 5 min.

Larvae should be kept separate and closely observed for the appearance of diseased individuals. Goodwin (1984) provides an excellent review for the recognition and diagnosis of diseases occurring in laboratory colonies. Diagnostic services may also be available through academic institutions or private consultants. Those insects exhibiting symptoms of microbial or viral infection, or whose vigor is questionable, should be discarded to prevent the remaining individuals from becoming infected. Because some pathogens build up slowly in a population, a newly introduced colony should remain isolated for several generations prior to integrating it into existing cultures.

Rearing parameters for newly collected insects are similar to those for established colonies. However, unselected field insects may not adapt readily to standard practices. Temperature, relative humidity, photoperiod, and diet may have to be manipulated before the desired rate of larval development and level of oviposition are achieved. In general, *Heliothis* larvae may be reared optimally between 26 and 29°C with 60-70% relative humidity (Burton 1969, King & Hartley 1985, Pantana 1985). Since variability in development time may be greater for field collected insects, it may be necessary to retard the development of some individuals to coincide with that of others in the cohort. Neonate larvae can be stored on oviposition cloths up to 2 days at 10°C. Older instar larvae on diet and pupae can be stored at the same temperature for up to 7 days and eggs up to 5 days without a significant reduction in hatch. Adults should not be stored.

King and Hartley (1985) found 25°C with 70-80% relative humidity and a 14:10 light:dark photoperiod to be optimum for pupae and adult rearing. We fed adults 8% sucrose in an 18.9-liter cardboard drum lined with paper toweling for oviposition. Approximately 100 moths in a 1:1 male:female sex ratio were added per container. Under these conditions, laboratory adapted females will deposit 500 - 1,000 eggs over 7 to 9 days (King & Hartley 1985). First generation field collected insects, however, typically mate later, produce fewer eggs, and have a longer period of oviposition than females of laboratory adapted colonies (Roush 1986, Roulston 1975). After several generations in culture, .

both researchers subsequently found that the mating and oviposition behaviors of recently field collected insects were similar to laboratory strains.

For *Heliothis* reluctant to oviposit under conditions such as those given above, we have found the addition of an indirect light source during the dark hours beneficial. Apparently the light simulates the natural twilight the moths are accustomed to during mating and oviposition. Increasing the female:male ratio will also provide greater numbers of progeny. This is acceptable for evaluating wild insects within 1-2 generations from the field. However, if maintaining the colony for long term use, this practice is not recommended. Bartlett (1985) found that although the number of progeny is increased, the effective population size is reduced to the level of the number of males. Rather, it is desirable to initiate the colony with a greater number of insects from diverse areas.

## Enhancing Genetic Heterozygosity

A bane of maintaining insect colonies for prolonged periods of time is the gradual decrease in mating, oviposition, larval survivorship, and pupal eclosion that can result from an inbreeding depression. Roush (1986) defined inbreeding depression as the deleterious effects on fitness that result from increased homozygosity as caused by consanguineous matings. For laboratory adapted *Heliothis* colonies, depending on the number of individuals mating and producing fertile ova, these symptoms may not be a problem for 10 or more generations. However, offspring of wild moths one generation from field can show the same characteristics of inbreeding depression as those exhibited by an established lab colony (Roush 1986). This natural tendency for increased homozygosity necessitates the periodic introduction of native or other distinct genetic material to enhance vigor.

A common practice for increasing genetic heterozygosity involves outcrossing laboratory females to wild males. Wild males are preferred for matings, rather than females, as they are less likely to transmit disease to progeny (Hamm et al. 1971). Young et al. (1975) demonstrated increased competitiveness among laboratory *H. zea* when females were crossed with wild males from St. Croiz, Virgin Islands. In field trials, the resulting offspring were more attractive and mated more frequently than their laboratory counterparts. Young et al. (1976) crossed black light-collected *H. zea* males with disease free laboratory females to improve the vigor of a laboratory colony suffering from inbreeding depression and a severe infestation of *Nosema heliothidis*. Several lines were established on the basis of mating, oviposition, egg hatch, and normal pupal eclosion. Finally, a series of crosses were systematically made, as depicted in Table 1, that resulted in maintenance of the

Table 1. Sample crosses for minimizing further inbreeding in a corn earworm colony

| Parent line[1] | 1st generation, 1st cross | 2nd generation, 2nd cross | 3rd generation, 3rd cross |
|---|---|---|---|
| A | A x B and<br>B X A = 1. | 1 x 3 and 3 x 1 or<br>(AB) x (CD) = $A_1$. | $A_1$ x $B_1$ = 1. |
| B | B x C and<br>C x B = 2. | 2 x 4 and 4 x 2 or<br>(BC) x (DE) = $B_3$. | B x C = 2. |
| C | C x D and<br>D x C = 3. | 3 x 5 and 5 x 3 or<br>(CD) x (EA) = C. | C x D = 3. |
| D | D x E and<br>E x D = 4. | 4 x 1 and 1 x 4 or<br>(DE) x (AB) = D. | D x E = 4. |
| E | E x A and<br>A x E = 5. | 5 x 2 and 2 x 5 or<br>(EA) x (BC) = E. | E x A = 5. |

[1]Line designations were used for convenience in making crosses and do not imply that parent lines were maintained.
Source: (from Young et al. 1976).

colony for over 70 generations without the development of characteristics typical of inbreeding depression. Similar crossing schemes have also been used to form heterogenous strains for insecticide resistance selection (Payne et al. 1985) and parasite release programs (Ashley et al. 1974).

## Conclusion

The techniques of insect rearing are similar whether it be to provide insects for investigating the mode of action of an insecticide, the selection of insecticide resistance, or evaluating susceptibility to recombinant plant tissue. Colonies must be established that suit the purpose for which they are intended, are free of disease, and of good vigor. In so doing, the rearing manager is able to provide a high-quality insect that meets the needs of the investigator.

## References

Ashley, T. R., D. Gonzalez & T. F. Leigh. 1974. Selection and hybridization of *Trichogramma*. Environ. Entomol. 3: 43-48.

Bartlett, A. C. 1985. Guidelines for genetic diversity in laboratory colony establishment and maintenance. *In* P. Singh & R. F. Moore [eds.], Handbook of insect rearing, vol. 1. Elsevier, Amsterdam.

Burton, R. L. 1969. Mass rearing the corn earworm in the laboratory. USDA-ARS 33-134.

Edwards, D. R., N. C. Leppla & W. A. Dickerson. 1987. Arthropod Species in Culture. Entomol. Soc. Amer., College Park, Maryland.

Fuchs, R. L., S. C. MacIntosh, D. A. Dean, J. T. Greenplate, F. J. Perlak, J. C. Pershing, P. G. Marrone & D. A. Fischhoff. 1990. Quantification of *Bacillus thuringiensis* insect control protein as expressed in transgenic plants, pp. 105-113. *In* L. A. Hickle & W. L. Fitch [eds.], Analytical chemistry of *Bacillus thuringiensis*. ACS Symp. Ser. 432.

Goodwin, R. H. 1984. Recognition and diagnosis of diseases in insectaries and the effects of disease agents in insect biology, pp. 96-129. *In* E. G. King & N. C. Leppla [eds.], Advances and challenges in insect rearing. USDA Handbook.

Gould, F. 1988. Evolutionary biology and genetically engineered crops. Bioscience. 38: 26-33.

Hamm, J. J., R. L. Burton, J. R. Young & R. T. Daniel. 1971. Elimination of *Nosema heliothidis* from a laboratory colony of corn earworm. Ann. Entomol. Soc. Am. 64: 624-627.

Hofmann, C., H. Vanderbruggen, H. Hofte, J. van Rie, S. Jansens & H. van Mellaert. 1988. Specificity of *Bacillus thuringiensis* delta-endotoxins correlated with the presence of high-affinity binding sites in the brush border membrane of target insect midguts. Proc. Natl. Acad. Sci., USA. 85: 7844-7848.

Hoy, M. A. 1976. Genetic improvement of insects: fact or fantasy. Environ. Entomol. 5: 833-839.

King, E. G. & G. G. Hartley. 1985. *Hiliothis virescens. In* P. Singh & R. F. Moore [eds.], Handbook of insect rearing, vol. 2. Elsevier. Amsterdam.

Levins, R. 1969. Some demographic and genetic consequences of environmental hetergeneity for biological control. Bull. Entomol. Soc. Am. 15: 237-240.

Pantana, R. H. 1985. *Heliothis zea* and *Heliothis virescens. In* P. Singh & R. F. Moore [eds.], Handbook of insect rearing, vol 2. Elsevier, Amsterdam.

Payne, T., B. Disney & T. M. Brown. 1985. Field surveillance and laboratory selection for pyrethrioid resistance in the tobacco budworm in South Carolina. J. Agric. Entomol. 2: 85-92.

Raulston, J. R. 1975. Tobacco budworm: observations on the laboratory adaptation of a wild strain. Ann. Entomol. Soc. Am. 69: 139-142.

Raulston, J. R., J. W. Snow & H. M. Graham. 1976. Large-scale shipping techniques for tobacco budworm pupae. USDA-ARS Prod. Res. Rep. No. 166.

Roush, R. T. 1986. Inbreeding depression and laboratory adaptation in *Heliothis virescens*s (Lepidoptera: Noctuidae). Ann. Entomol. Soc. Am. 79: 583-587.

Roush, R. T. & J. A. McKenzie. 1987. Ecological genetics of insecticide and acaracide resistance. Ann. Rev. Entomol. 32: 361-380.

Southwood, T. R. E. 1978. Ecological methods. Wiley, New York.

Stone, T. B., S. R. Sims & P. G. Marrone. 1989. Selection of tobacco budworm for resistance to a genetically engineered *Pseudomonas fluorescens* containing the delta-endotoxin of *Bacillus thuringiensis* subsp. *kurstaki*. J. Invert Pathol. 53: 228-234.

Young, J. R., J. W. Snow, J. J. Hamm, W. D. Perkins & D. G. Haile. 1975. Increasing the competitiveness of laboratory-reared corn earworm by incorporation of indigenous moths from the area of release. Ann. Entomol. Soc. Am. 68: 40-42.

Young, J. R., J. J. Hamm, R. L. Jones, W. D. Perkins & R. L. Burton. 1976. Development and maintenance of an improved laboratory colony of corn earworms. USDA-ARS (rep.) S-110.

# 4

---

# Development of Artificial Diets
# for Entomophagous Insects
# by Understanding Their
# Nutrition and Digestion

## *I. G. Yazlovetsky*

### Introduction

The development of effective and economically profitable technologies for mass rearing of entomophagous insects is at present a key goal for successful biocontrol programs. Conditions for the nutrition of insect predators and parasitoids are of special value in the success of insect mass rearing programs. Inadequate nutrition usually results in great changes in the metabolism, behavior, and other characteristics of insect vital activity. These changes inevitably depreciate subsequent insect release. An ability to provide optimal nutrition will greatly effect both expenditures for entomophage production as well as colony quality and sometimes determines economic expediency and indeed the very possibility for mass culture (e.g., in the case of rearing species with narrow food specialization) (Chambers 1977, Yazlovetsky 1986). The main developmental prospects for entomophage mass production for inundative releases are, therefore, associated with the development of cheap and adequate artificial diets (Mellini 1975, Waage et al. 1985, Yazlovetsky 1986).

Successful development of such technologies based on artificial media will, however, require a thorough knowledge of entomophage physiology and metabolism, and an understanding of the peculiarities of their interactions with insect hosts and prey (Thompson 1986). In spite of some promising results obtained in the development of artificial diets for entomophages, the use of artificial diets in mass propagation programs is currently limited to only a few species of predators and parasitoids (Gao et al. 1982, Hagen 1987, Ridgway et

41

al. 1970, Slansky & Rodriges 1987, Vinson & Barbosa 1987, Waage et al. 1985, Yazlovetsky & Nepomnyashaya 1981).

One can say with certainty that the current theoretical base sharply limits the number of insect predators and parasitoids for which it is possible to develop artificial diets. This limit is largely the result of narrowly specialized parasitoids strongly adapted to their hosts. Most successes in this field are currently based on the development of artificial media for wide range oligo or polyphagous predators and parasitoids (Gomez et al. 1982, Mellini 1975, Singh 1977, Thompson 1986, Waage et al. 1985). However, even in such cases one has to overcome considerable difficulties to successfully develop artificial food substrates capable of competing successfully with natural ones, regarding nutritive value and cost. Polyphagous predators and parasitoids may often lack many of the specialist's behavioral barriers that impeded the replacement of predation and parasitism targets by alternative artificial ones. Polyphagous species, however, simultaneously possess specific peculiarities of nutrition, digestion, respiration, and other aspects of vital activity that must be considered in the process of artificial media development. The absence of defecation in larval developmental stages, for example, is a peculiarity manifested by many entomophagous insects, including all endoparasitoids (Ermitcheva et al. 1988, Mellini 1975, Thompson 1986). This peculiarity places very strict limitations on artificial media. First of all, they must maximize, regarding the form available and the concentration, all nutrients required for growth development and reproduction with minimal quantities of non-digestive admixtures. This limitation necessitates careful definition of their requirements in the main dietary components (i.e., proteins, amino acids, fats, and carbohydrates) and also in additional dietary components (i.e., mineral salts, vitamins, growth factors, and water). This requires a direct search for cheap media nutrients and the study of their chemical composition, regarding the ability of the species to utilize these dietary components.

In addition, while preparing artificial media for the majority of entomophages, one has to avoid the use of inert, non-digestive fillers (agar, cellulose, etc.) widely utilized to obtain the consistency required in nutritional diets for many other insects (Singh 1977). Hence, one must develop formulation variants of liquid artificial media lacking fillers for insect predators and in particular for parasitoids. Diet encapsulation into different polymeric materials looks most promising in this respect. The size, form, and characteristics of such capsule envelopes are in each case determined by the specific peculiarities of certain predatory or parasitoid species.

The third group of strict limitations hampering the development of artificial media is peculiar only to insect endoparasitoids. The host body is considered as a medium, ovipositional substrate, and a habitat. Therefore, it is necessary to provide an artificial diet with all nutrients required in a form acceptable for digestion and to create conditions supporting oviposition and both embryonic and larval development. Thus, the concentrations of different inorganic and low-molecular organic substances (salts, amino acids, polypeptides, organic acids, mono, and oligosaccharides) as well as associated pH and osmotic pressure values in artificial media are the most important factors ensuring success of oviposition and subsequent embryonic development. If larval parasitoids lack tracheae and obtain oxygen by cuticular diffusion (which is characteristic for many insect endoparasitoids) the content of oxygen present in a liquid nutritional diet is an important factor influencing larval survival in a food substrate. When larval endoparasitoids have tracheae and respire either through air or association with the host's tracheal system, artificial media must accommodate this. However, no one has succeeded in doing this until recently (Mellini 1975, Thompson 1986).

To overcome the technical difficulties associated with mass rearing predators and parasitoids, it is necessary to accumulate knowledge on their physiology, nutritional biochemistry, and ecology. Development and successful utilization of new artificial diets in biological pest control programs will be determined by the intensity of studies in this field.

The following investigation trends are of special interest for the development of artificial media for entomophagous insects (Yazlovetsky 1986):

1. Studies on the biochemical composition of hosts and prey of insect parasitoids and predators to evaluate requirements for the latter in different nutrient regimes.

2. Investigation of mechanisms for the digestion of different natural medium components by entomophages; i.e., studies of pattern peculiarities of predator and parasitoid alimentary canals, composition and substrate specificity analysis, and other characteristics of digestive enzymes.

3. Studies of changes in parasitoid and predator composition depending on that of food substrates.

Modern instrumental methods of insect biochemistry and physiology promote rather quick solutions to the above problems, producing objective and exact information. Even in the initial stages of investigations on entomophagous insects, analytical instrument assays are a good supplement (and sometimes a complete alternative) to the traditional laborious empirical methods for defining the nutritional requirements of insect predators and parasitoids based on axenic breeding of entomophages on artificial media of chemically defined composition (Ermitcheva et al. 1987, 1988; Hagen 1987; Kaplan et al. 1986; Kejser & Yazlovetsky 1988; Moontyan & Yazlovetsky 1988; Thompson 1986).

# Development of Artificial
## Diets for *Chrysoperla carnea* Steph.

Earlier, we reported first attempts to apply some elements of the above listed non-traditional approaches to the development of artificial substrates for *Chrysoperla carnea* Steph. larvae (Yazlovetsky et al. 1979a, 1979b). Development of the liquid nutritional medium was preceded by an evaluation of predator dietary requirements on the basis of total biochemical parameters and a study of the qualitative and quantitative amino acid composition of the two natural prey species, *Meqoura viciae* Buckt. and *Aphis fabae* Scop. We also considered the alternative natural food, *Sitotroga cerealella* Oliv. eggs, used in laboratory propagation of *Chrysoperla carnea* (Tables 1 and 2) and some protein components of artificial media including water soluble fractions of commercial protein concentrates. (Tables and figures begin on page 49).

Mathematical methods were applied to optimize nutritional diet compositions. This investigation produced standard methods to search for optimal ratios of different known components in artificial media. It promoted the directed use of single low-molecular substances in addition to their compound mixtures and, more importantly, natural biopolymers (proteins, polysaccharides, nucleic acids, etc.) in artificial diet recipes. Use of this method optimized the amino acid composition and also other parameters of the nutritional media. This, together with the analysis of literary data on the presence of trypsin-like proteolytic enzymes in larval guts, resulted in the development of rather simple and inexpensive liquid media for *C. carnea* larvae. One of these media (Table 3) in an encapsulated formulation provided normal development of larval predators from hatching to the first moult, allowing the release of 2nd instar larvae for use against aphids on different greenhouse crops. Costs to provide larvae with food using microencapsulated media were 25-30 times lower than similar expenditures for *C. carnea* breeding on *Sitotroga cerealella* eggs (Yazlovetsky & Nepomnyashaya 1981).

Further improvement of nutritional media for the mass culture of this predator is associated with the necessity to identify larval requirements in other classes of nutrients (fats, carbohydrates, sterols, nucleic acids, mineral salts, and vitamins). A successful diet must utilize inexpensive sources of these substances. We managed to develop a number of recipes for inexpensive liquid nutritional diets ensuring the complete cycle of *Chrysoperla carnea* development.

Accumulated information on the chemical composition of several aphid species and similar methods for the selection of components were successfully applied to develop optimal compositions of artificial media that assured normal larval development of another aphidophagous predator *Chrysopa septempunctata* Wesm. (Table 4).

Estimation of the rate of hydrolysis of high-molecular weight nutrients (proteins, polysaccharides, glycoproteins, and nucleic acids) by corresponding enzymic systems in larval entomophages yielded data on the degree of absorption of these materials. This indicates the possibility of developing specific methods for estimating the acceptability of different artificial media components even before the initiation of whole insect investigations. We initiated systematic investigations on digestive enzymes composition, properties, and localization for some larval predators and parasitoids showing promise for biological pest control. Data have already been obtained on the distribution of different enzymes in the alimentary canals of several larval species. Details on digestive patterns have been revealed and described. This has contributed to the understanding of nutrition and digestion mechanisms of the entomophagous insects under study.

Activity values of endopeptidases and aminopeptidases found in predator foreguts are low. The totality of the data obtained allows one to disclaim the presence of any considerable digestion of the protein beyond the guts of *C. carnea* larvae (Ermitcheva et al. 1987). Investigations of *C. carnea* larvae have shown that pH values of the foregut contents were 6.1; those of the midgut were 6.7. The activity optimum of midget endopeptidases is within a pH of 4 - 10 maximum activity values of 7.0 and 8.2 (Fig. 1). The highest activity was observed in the alkaline zone. Two peaks in endopeptidase activity indicate the presence of at least two proteolytic enzymes in the gut. The main proteolytic activity volume is focused in midguts of *C. carnea* larvae. Foreguts were found to display only 4.8% of endopeptidase activity and midguts 1.5% of leucinaminopeptidase (Table 5).

Polyacrylamide gel electrophoresis has been used to study intraspecific changeability of the main digestive proteases of *C. carnea* larvae. It is possible to use these enzymes for population genetics in addition to predator food specialization studies (Fig. 2). The change of a food substrate *(Sitotroga cerealella* eggs, different aphid species, artificial media) does not lead to that in the endopeptidase composition of *C. carnea* larvae.

Investigation on alimentary canal carbohydrases of larval predators have demonstrated that invertase and amylase are concentrated mainly in the midgut. Small quantities of these enzymes (0.1% of total invertase and 0.2% of that of amylase) have been identified in salivary glands. Twelve to eighteen percent of carbohydrases have been found in foreguts. The main activity volume of invertase and amylase was found to concentrate in larval midguts (82 and 84% respectively) (Table 6). These data also confirm the functional role of *C. carnea* larvae midguts as the main zone for food digestion (Moontyan & Yazlovetsky 1988).

Studies on distribution patterns for lipolytic enzymes yielded conclusions similar to studies of endopeptidases, aminopeptidases, and carbohydrases. The foregut and midgut of *C. carnea* larvae have been found to display, respectively, 3 and 97% of lipolytic activity of this predator's digestive enzymes.

When *C. carnea* larvae fed on colored diets enveloped in transparent microcapsules, food penetrated into the mandibular canals soon after they pierced the microcapsules. Liquid uptake was continuous, rapid, and without regurgitation. Apparently, only homogeneous liquids and nonviscous nutritional mixtures provide optimal conditions for feeding of this larvae predator. Attempts to create powder or paste-like nutritional media (Ferran et al. 1981) as well as those containing solid admixtures (Bigler et al. 1976) do not look promising.

High invertase activity found in the alimentary canals of *C. carnea* larvae facilitates the use of saccharose as an inexpensive source of carbohydrates in artificial media (Table 7). This sugar can be used instead of expensive fructose as a carbohydrate component of artificial diets. Replacement of fructose by saccharose does not result in the reported sharp deterioration of *C. carnea* larval development (Vandersant 1973, 1974). The change of a food substrate presumably affects the activity of the insect digestive system, especially that of digestive enzymes. We examined the effect of different types of food for *C. carnea* larvae on the activity levels of enzymes responsible for digestion of the main medium components (proteins, carbohydrates, and fats). Long term maintenance of the predator on *Sitotroga cerealella* eggs and *Schizaphis graminum* aphids as well as transition of larvae from one type of food to another were found to not significantly influence activity levels of lipases and proteases in the alimentary canal. *C. carnea* bred on aphids had higher activity of digestive carbohydrase (two to three times) than those cultured on *Sitotroga cerealella* eggs (Table 8). We used the activity values of the main digestive enzymes of *C. carnea* larvae to evaluate the degree of nutrition specialization to other food substrates and to determine the suitability of artificial media and the individual components.

## Development of Artificial
## Diets for *Trichogramma* sp.

*Trichogramma pintoi* Voeq. and *T. evanescens* Westw. have been chosen for the investigation of the nutrition and digestion of insect parasitoids in order to develop artificial diets for their mass culture. Our choice was based on the great practical importance of *Trichogramma* as a biocontrol agent and on specific characteristics of these small wasps. However, the small sizes of *Trichogramma* larvae impede studies on the mechanisms of their nutrition and digestion and on the estimation of their nutritional requirements. Therefore, *Trichogramma* artificial media have been developed empirically and contain insect hemolymph as an obligatory component. This limits the development of inexpensive food substrates (Gomez et al. 1982, Thompson 1986, Yazlovetsky 1986). The development of artificial media for *Trichogramma* requires a careful determination of larval dietary requirements and a thorough investigation of the nutrition of parasitoid preimaginal stages (Ermitcheva et al. 1987, Hawlitzky & Boulay 1982).

The evaluation of *Trichogramma* larval requirements for amino acids, polypeptides, proteins, and fats was preceded by studies on the qualitative and quantitative composition of these compounds in eggs of some lepidopterans (Kaplan & Yazlovetsky 1980, Sumenkova et al. 1980) (Tables 9 and 10). The relative deficiency of some amino acids, oleic acid, and a considerable excess of the linolenic acid in eggs of *Hyphantria cunea* Drury as compared to other lepidopterans hampers *Trichogramma* development. The content of the major nutritional components in eggs of favorable (*Mamestra brassicae* L., *Mamestra oleraceae* L.) and unfavorable (*Hyphantria cunea*). *Trichogramma* hosts allows designation of permissible fluctuations of these components in artificial media, as is necessary for optimizing diet composition (Yazlovetsky et al. 1979a).

*Trichogramma* females apparently parasitize host eggs by injecting them with a secretion whose mode of action has not been completely studied (Voegele 1974). Electrophoretic studies indicate that this secretion contains different hydrolytic enzymes that stimulate the production of low-molecular weight fragments of protein, polysaccharides, nucleic acids, and fats. As a result, polymerized molecules of the host embryo become more accessible for parasitoid embryo nutrition (Ermitcheva et al. 1988).

To investigate in detail the mechanism of *Trichogramma* larval nutrition, we studied changes in the activity of their proteolytic enzymes during larval development. Proteolytic activity was not detected in the 1st and 2nd instar larval homogenates. Third instar larvae had activity variation in time, with maximal values being recorded after 62 h of larval development (Fig. 3). Maximal protease activity was observed at pH 8.0. Hence, the uptake of amino acids, peptides, and proteins is the most important process for 1st and 2nd instar larvae while digestion and assimilation are emphasized in 3rd. The 3rd instar reveals the acceptability of nutritional substrates consumed by larval parasitoids and the specificity of their digestive enzymes. *Trichogramma* development in unfavorable *(Hyphpantria cunea)* host eggs has shown that only single parasitoids completed their development and that adults had unproportionally hooved abdomens, rendering them incapable of finding host eggs (Hoffman et al. 1975). Considerable mortality of the 3rd instar larvae and pupae, as well as adult deformities, seemed to be associated with nutritional deficiencies in Hyphantria cunea eggs and artificial media. When proteins isolated from eggs of different lepidopterans were compared according to the rate of their hydrolysis with *Trichogramma* larval proteases, the protein from *Barathra brassicae* eggs was 20 times more accessible for hydrolysis than that from *Hyphantria cunea*, and three times more accessible than the protein from *Sitotroga cerealella* eggs. Less significant differences have been revealed from the hydrolysis rate of lepidopteran egg polysaccharides using *Trichogramma* larval amylases. Our results agree with many data on different *Trichogramma* preferences to lepidopteran eggs (Lewis at al. 1976) and could become the initial stage in biochemical studies on parasitoid nutritional specialization. Similar techniques can be used for estimations of acceptability of nutritional media components (Ermitcheva et al. 1988).

Currently the mechanism of *Trichogramma* larval nutrition is imagined to be as follows. Female parasitoids inject a special secretion with their eggs that results in host embryo mortality and tissue lysis, providing favorable conditions for the parasitoid embryo. *Trichogramma* eggs start consuming low-molecular weight nutrients by osmotic pressure. First instar larvae start feeding with their mandibles, the process being most active in the 2nd instar. During the 3rd instar practically all of the host egg contents have moved to the larval parasitoid alimentary canals and intensive food uptake is replaced by digestion and assimilation. The lack of essential nutrients, and the substrate specificity of *Trichogramma* digestive enzymes result in parasitoid mortality or developmental abnormalities.


## Conclusions

Similar investigations are being carried out with other insect predators and parasitoids showing promise for biological pest control. We are studying the composition and properties of proteolytic, amylolytic, and lypolytic enzymes of the hematophagous braconid ectoparasitoid, *Bracon hebetor* Say, and defining characteristics and digestive mechanisms. We are also determining the chemical composition of its natural host (*Heliothis armigera* Hubner) and laboratory host (*Galleria mellonella* L.). Predatory bugs, entomophages of *Leptinotarsa decemlineata* Say, are another subject of investigation. Finally, we are studying the composition and properties of digestive enzymes of larval and adult *Podisus maculiventris* Say and *Perillus bioculatus* Fabr., the mechanism of their nutrition and digestion, and the composition of their natural food, Colorado potato beetle eggs and larvae.

**Table 1. Percentage wet weight of some natural food substrates for larvae of *Chrysoperla carnea***

| Substances | Aphids[a] | | Eggs[b] |
|---|---|---|---|
| | A | B | |
| Total amino acids | 12.40 ± 1.2 | 12.60 ± 1.1 | 15.60 ± 1.4 |
| Free amino acids | 0.42 ± 0.17 | 0.66 ± 0.15 | 0.52 ± 0.25 |
| Total carbohydrates | 4.93 ± 0.34 | 12.19 ± 0.94 | 6.94 ± 0.89 |
| Free carbohydrates | 0.25 ± 0.04 | 7.45 ± 0.43 | 0.68 ± 0.11 |
| Lipids | 2.30 ± 0.50 | 6.00 ± 1.10 | 8.50 ± 1.70 |
| Nucleic acids | 0.08 ± 0.01 | 0.28 ± 0.01 | 1.12 ± 0.11 |
| Mineral substances | 1.40 ± 0.4 | 2.00 ± 0.3 | 1.60 ± 0.4 |
| Water | 78.55 ± 1.48 | 64.80 ± 2.76 | 66.03 ± 0.48 |
| Total nitrogen | 1.93 ± 0.39 | 2.20 ± 0.26 | 2.55 ± 0.15 |
| pH | 6.8 - 6.9 | 6.8 - 6.9 | 6.6 - 6.7 |

Means in columns are $\bar{x} \pm t_{0.05}$*SE (n=4), t - Student criterion
[a]A, *Megoura viciae;* B, *Aphis fabae.*
[b]*Sitotroga cerealella.*

**Table 2. Percentage wet weight of amino acids in some natural food substrates for larvae *Chrysoperla carnea***

| Amino acids[b] | Aphids[a] | | Eggs[c] |
|---|---|---|---|
| | A | B | |
| Lysine | 0.88 | 1.05 | 1.00 |
| | 0.03 | 0.01 | 0.02 |
| Histidine | 0.39 | 0.02 | 0.02 |
| | 0.02 | 0.02 | 0.02 |

*continues*

50

**Table 2, continued**

| Amino acids[b] | Aphids[a] | | Eggs[c] |
|---|---|---|---|
| | A | B | |
| Arginine | 2.00 | 0.72 | 3.01 |
| | 0.02 | 0.02 | 0.04 |
| Aspartic acid | 0.76 | 1.60 | 1.42 |
| | 0.01 | 0.05 | 0.04 |
| Threonine | 1.32 | 0.63 | 0.78 |
| | 0.05 | 0.18 | 0.02 |
| Serine | 0.60 | 0.74 | 0.74 |
| | 0.08 | 0.05 | 0.04 |
| Glutamic acid | 1.54 | 0.38 | 2.04 |
| | 0.08 | 0.13 | 0.15 |
| Proline | 0.55 | 0.40 | 0.71 |
| | 0.04 | 0.01 | 0.05 |
| Glycine | 0.50 | 0.64 | 0.80 |
| | 0.01 | 0.02 | 0.02 |
| Alanine | 0.71 | 0.80 | 0.61 |
| | 0.01 | 0.06 | 0.04 |
| Valine | 0.56 | 0.32 | 0.80 |
| | 0.02 | 0.03 | 0.02 |
| Methionine | 0.14 | 0.13 | 0.32 |
| | traces | traces | traces |
| Isoleucine | 0.52 | 0.62 | 0.71 |
| | 0.01 | 0.01 | 0.01 |
| Leucine | 0.80 | 1.22 | 1.06 |
| | 0.01 | 0.02 | 0.02 |
| Tyrosine | 0.76 | 0.85 | 0.76 |
| | 0.02 | 0.03 | 0.01 |
| Phenylalanine | 0.47 | 0.59 | 0.62 |
| | 0.01 | 0.02 | 0.02 |
| Tryptophane | 0.12 | 0.16 | 0.16 |
| | traces | traces | traces |
| Total | 12.62 | 12.92 | 16.13 |
| | 0.42 | 0.66 | 0.52 |

Note: Values are means of four to five determinations.
[a] A, *Megoura viciae;* B, *Aphis fabae.*
[b] Upper line is total amino acid content, lower line is free amino acid content.
[c] *Sitotroga cerealella.*

**Table 3.** Composition of liquid homogeneous nutritional diet for *Chrysoperla carnea* larvae

| Component | Amount (g/30 g of medium) |
|---|---|
| Enzymatic casein hydrolyzate | 0.71 - 0.93 |
| Water wheat germ extract | 1.42 - 1.74 |
| Water extract of brewer's yeast | 0.47 - 0.67 |
| Sucrose | 1.5 - 3.5 |
| Lipid mixture[a] | 0.3 - 0.5 |
| $NaH_2PO_4 \cdot H_2O$ | 0.020 |
| $K_2HPO_4$ | 0.040 |
| Choline chloride | 0.0125 |
| Cobalamin | 0.000001 |
| Ascorbic acid | 0.025 |
| Tween 80 | 0.03 |
| Water | to 30 g |

[a]Mixture of soya oil, sunflower lecithin, and cholesterol in a ratio of 5:5:1.

**Table 4.** Composition of liquid homogeneous diet for *Chrysopa septempuncta* larvae

| Component | Amount (g/120 g medium) |
|---|---|
| Enzymatic casein hydrolyzate | 3.72 |
| Water wheat germ extract | 6.96 |
| Water extract of brewer's yeast | 2.68 |
| Sucrose | 6.00 |
| Lipid mixture[a] | 0.12 |
| Trimyristin[a] | 0.05 |
| $NaH_2PO_4*H_2O$ | 0.08 |
| $K_2HPO_4$ | 0.16 |
| Choline chloride | 0.05 |
| Cobalamin | 0.000004 |
| $MgSO_4*7H_2O$ | 0.05 |
| $FeSO_4*7H_2O$ | 0.05 |
| Nicotinic acid | 0.002 |
| Ca Patothenate | 0.002 |
| Folic acid | 0.0005 |
| Thiamine hydrochloride | 0.0005 |
| Inositol | 0.02 |
| Biotin | 0.00004 |
| Riboflavin | 0.0001 |
| Pyridoxine hydrochloride | 0.0005 |
| Tween 80 | 0.12 |
| Water | to 120 g |

[a]Mixture of soya or sunflower oil, sunflower lecithin, and - sitosterol in the ratio of 5:5:1.
[b]Addition of trimyristin improves indices of larvae development on artificial diet.

**Table 5.** Proteolytic activity in different intestinal regions of *Chrysoperla carnea* larvae (nmoles NH₂/min per organ)

| Enzyme | Foregut | Midgut | | Entire |
| | | Wall | Content | |
| --- | --- | --- | --- | --- |
| Endopeptidase gelatin substrate | 2.50 ± 0.4 | traces | 58.3 ± 6.5 | 51.2 ± 5.2 |
| Leucinamino-peptidase L-leucine-p-nitro-anilide substrate | 0.85 ± 0.01 | 30.0 ± 5.2 | 24.2 ± 4.9 | 57.2 ± 9 |

Note: Means in columns are ± $t_{0.05}$*SE (n=4), t - Student criterion

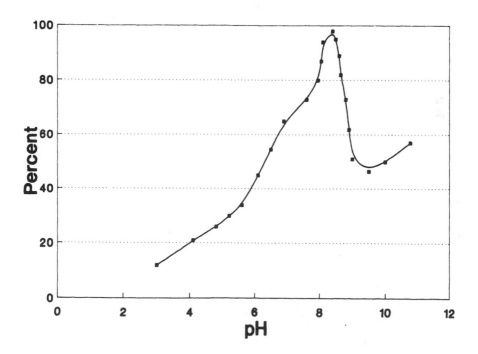

**Fig. 1.** Activity of midgut endopeptidases in the gut of *Chrysoperla carnea* larvae.

**Table 6. Carbohydrase activity in different regions of alimentary tracts of *Chrysoperla carnea* larvae (nmoles of reducing sugars/min per organ)**

| Enzyme | Tract region | Wall | Content | Entire | Salivary glands |
|--------|--------------|------|---------|--------|-----------------|
| Invertase | Foregut | 4.86 ± 0.3 | 187.98 ± 5.2 | 193.8 ± 25 | 1.06 ± 0.03 |
| | Midgut | 19.20 ± 0.5 | 943.70 ± 0.1 | 962.0 ± 5.3 | |
| Amylase | Foregut | 0.26 ± 0.05 | 15.40 ± 1.2 | 18.0 ± 2.5 | 0.21 ± 0.04 |
| | Midgut | 9.52 ± 0.4 | 104.20 ± 5.4 | 113.6 ± 5.5 | |

Note: Means in columns are x ± $t_{0.05}$*SE (n=4), t- Student criterion

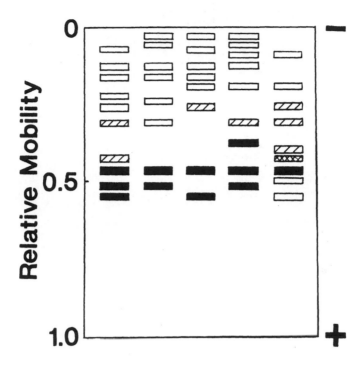

**Fig. 2.** Polyacrylamide gel of *C. carnea* larval digestive proteases.

**Table 7.** Activity of carbohydrases of homogenates in midguts of *Chrysoperla carnea* larvae as related to different substrates

| Enzyme | Substrate | Relative activity[1] |
|---|---|---|
| Disaccharidase | Sucrose | 10.0 |
| | Maltose | 1.18 |
| | Trehalose | 0.86 |
| Amylase | Starch | 1.0 |
| | Amylopectin | 1.3 |
| | Glycogen | 0.78 |

[1]Amylase activity on starch has been used as the standard.

**Table 8. Activity of the main digestive enzymes of *Chrysoperla carnea* larvae feeding on different food**

| Enzyme | Substrate | Food | | | |
|---|---|---|---|---|---|
| | | A | B | C | D |
| Proteases | Protein 1[a] | 22.6 ± 3.4 | 18.0 ± 3.7 | 22.5 ± 2.5 | 23.01 ± 1.5 |
| | Protein 2[b] | 23.0 ± 3.8 | 20.4 ± 4.0 | 26.9 ± 2.9 | 25.80 ± 2.0 |
| | Gelatine | 47.7 ± 2.5 | 24.7 ± 0.6 | 42.8 ± 3.9 | 37.00 ± 3.0 |
| | Azocasein | 19.9 ± 2.4 | 18.2 ± 3.6 | 24.0 ± 6.1 | |
| Carbo-hydrases | Starch | 63.0 ± 2.1 | 127.4 ± 27.0 | 109.1 ± 2.0 | 49.00 ± 3.0 |
| | Trehalose | 10.7 ± 1.1 | 18.1 ± 5.0 | 24.3 ± 2.0 | 6.50 ± 1.0 |
| | Maltose | 58.5 ± 4.1 | 113.0 ± 2.4 | 62.6 ± 1.0 | 28.30 ± 2.0 |
| Lipases | Triolein | 15.9 ± 1.0 | 16.6 ± 1.8 | 20.6 ± 4.4 | 17.20 ± 1.5 |

Note: Enzyme activity is expressed for proteases in nmoles $NH_2$min/individual (azocasein in $E_{340}$/min), for carbohydrases in nmoles glucose/min/individual, and for lipases in nmoles of fatty acids/min/individual. Means in columns are x ± t0.05*SE (n=4), t- Student criterion.

[a]Protein from *Sitotroga* eggs.
[b]Protein from *Schizaphis graminum* aphids.

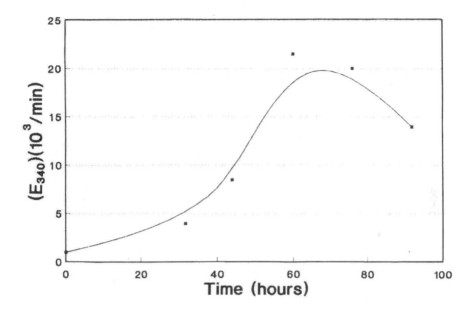

**Fig. 3. Proteolytic enzyme activity in 3rd instar larvae.**

**Table 9. Amino acids content of some lepidopteran eggs, hosts of** *Trichogramma*

| Amino acid[b] | Amount in Lepidopteran eggs[a] (% of wet weight) | | |
| --- | --- | --- | --- |
| | A | B | C |
| Lysine | 0.74 | 0.95 | 1.05 |
| | 0.04 | 0.03 | 0.05 |
| Histidine | 0.49 | 0.49 | 0.55 |
| | 0.08 | 0.07 | 0.04 |
| Arginine | 2.22 | 0.60 | 0.92 |
| | 0.10 | 0.12 | 0.09 |

*continues*

58

Table 9, continued

| Amino acid[b] | Amount in Lepidopteran eggs[a] (% of wet weight) | | |
|---|---|---|---|
| | A | B | C |
| Aspartic acid | 0.79 | 1.08 | 1.26 |
| | 0.07 | 0.05 | 0.04 |
| Threonine | 0.60 | 0.48 | 0.67 |
| | 0.13 | 0.08 | 0.08 |
| Serine | 0.54 | 0.03 | 0.05 |
| | 0.12 | 0.03 | 0.05 |
| Glutamic acid | 1.50 | 1.98 | 1.94 |
| | 0.19 | 0.16 | 0.11 |
| Proline | 0.51 | 0.67 | 0.69 |
| | traces | trace | traces |
| Glycine | 0.59 | 0.65 | 0.76 |
| | 0.03 | 0.02 | 0.02 |
| Alanine | 0.59 | 0.79 | 0.61 |
| | 0.03 | 0.01 | 0.05 |
| Valine | 0.59 | 0.62 | 0.92 |
| | 0.04 | 0.02 | 0.03 |
| Methionine | 0.24 | 0.27 | 0.20 |
| | traces | traces | 0.02 |
| Isoleucine | 0.56 | 0.40 | 0.67 |
| | 0.01 | 0.01 | 0.02 |
| Leucine | 0.78 | 1.04 | 0.70 |
| | 0.01 | 0.01 | 0.03 |
| Tyrosine | 0.56 | 0.79 | 0.80 |
| | traces | 0.02 | 0.05 |
| Phenylalanine | 0.36 | 0.49 | 0.69 |
| | 0.01 | 0.01 | 0.02 |
| Tryptophane | 0.16 | 0.22 | 0.134 |
| | traces | traces | traces |
| Total | 11.38 | 12.14 | 13.37 |
| | 0.85 | 0.68 | 0.77 |

Note: Values are means of four to five determinations.

[a]A, *Barathra brassicae*, B, *Mamestra oleracia*, C, *Hyphantria cunea*.

[b]Upper line is total amino acid content, lower line is free amino acid content.

Table 10. Fatty acid composition of some lepidopteran eggs, hosts of *Trichogramma* (% of total fatty acids)

| Fatty acid | Amount in Lepidopteran eggs[a] (% of total fatty acids) | | | | |
|---|---|---|---|---|---|
| | A | B | C | D | E |
| Myristic acid (C)14.0 | 1.60 ± 0.8 | 0.3 ± 0.1 | 0.4 ± 0.2 | 0.6 ± 0.2 | traces |
| Palmitic acid (C)16.0 | 41.10 ± 2.6 | 29.2 ± 1.2 | 43.0 ± 0.8 | 36.9 ± 1.9 | 31.7 ± 2.2 |
| Palmito-oleic acid (C)16.1 | 2.60 ± 0.1 | 1.7 ± 0.2 | 2.8 ± 0.9 | 4.1 ± 0.7 | 3.1 ± 0.6 |
| Stearic acid (C)18.0 | 3.60 ± 0.2 | 1.3 ± 0.1 | 1.2 ± 0.1 | 3.5 ± 0.2 | 4.5 ± 0.3 |
| Oleic acid (C)18.1 | 43.08 ± 2.3 | 48.6 ± 2.8 | 44.1 ± 2.1 | 39.6 ± 2.4 | 21.5 ± 1.3 |
| Linoleic 18.2 | 4.00 ± 0.6 | 15.2 ± 1.0 | 6.2 ± 1.3 | 9.3 ± 0.9 | 10.8 ± 1.4 |
| Linolenic acid (C)18.3 | 3.30 ± 0.5 | 3.6 ± 0.1 | 1.5 ± 0.1 | 5.0 ± 0.2 | 26.3 ± 1.3 |
| Unidentified acid | - | - | - | 0.8 ± 0.1 | 2.0 ± 0.1 |

Means in columns are $x \pm t_{0.05}*SE$ (n=4), t - Student criterion,
[a] A, *Sitotroga cerealella*; B, *Ephestia kuhniella*; C, *Barathra brassicae*; D, *Mamestra oleracea*; E, *Hyphantria cunea*.

# References

Bigler, F., A. Ferran & J. Lyon. 1976. L'elevage larvaire de deux predateurs aphidiphages (*chrysopa carnea* Steph., *Chrysopa perla* L.) a l'aide de differents milieux artificiels. Ann. Zool. Ecol. Anim. 8: 551-558.

Chambers, D. L. 1977. Quality control in mass rearing. Ann. Rev. Entomol. 22: 289-308.

Ermitcheva, F. M., V. V. Sumenkowa & I. G. Yazlovetsky. 1987. Localization of proteases in guts of *Chrysopa carnea* larvae. Izvestia Acad. Sci. Moldavian SSR, Ser. Biol. Chem. Sci. 4: 49-52 (in Russian).

Ermitcheva, F. M., A. B. Vereschagina, L. I. Ageeva & I. G. Yazlovetsky. 1988. On the mechanism of *Trichogramma* larvae nutrition, pp. 63-73. *In* N. M. Golyshin & SH. M. Greenberg [eds.], *Trichogramma* v zaschite rastenii. Agropromizdat, Moscow (in Russian).

Ferran, A., J. Lyon & M. Larroque. 1981. Essai d'elevage de differents predateurs aphidiphages *(Coccinellidae, Chrysopidae)* a l'aide de poudre lyophilisee de couvain de reines d' abeiles. Agronomie. 1 (7): 579-586.

Gao Yi-Guang, Dai Kai-Jia, & Shong Lien-Shing. 1982. Studies on the artificial host eggs for *Trichogramma*, p. 181. *In* Less Trichogrammes. Les Colloques de 1' INRA, INRA Publ., Paris (9).

Gomez, A. N., L. C. Estrada & R. B. Galan. 1982. Dietas artificiales en insecta (Excepto *Lepidoptera Coleoptera*). Un compendio de referencias. INIA Publ., Madrid.

Hagen, K. S. 1987. Nutritional ecology of terrestrial insect predators, pp. 533-577. *In* F. Slansky & J. G. Rodrigues [eds.], Nutritional ecology of insects, mites, spiders and related invertebrate. Wiley, New York.

Hawlitzky, N. & C. Boulay. 1982. Regimes alimentaires et developpement chez *Trichogramma maidis* dans l'oeuf *Anagasta kuehniella*, pp. 101-106. *In* Les richogrammes, Les Colloques de l'INRA, INRA Publ., Paris (9).

Hoffman, J. D., C. M. Ignoffo & W. A. Dickerson. 1975. In vitro rearing of the endoparasitic wasp, *Trichogramma pretiosum*. Ann. Entomol. Soc. Amer. 68 (2): 335-336.

Kaplan, P. B., A. M. Nepomnyashaya & I. G. Yazlovetsky. 1986. Fatty acids composition of the predatory larvae of the aphidophagous insects and its dependence on the diet. Entomologicheskoje obozrenie. 65 (2): 262-268.

Kaplan, P. B. & I. G. Yazlovetsky. 1980. Fatty acids of eggs of some Lepidoptera, the hosts of *Trichogramma*, pp. 29-32. *In* N. A. Filippov [ed.], *Trichogramma*, part 2, Kishinex, (in Russian, summary in English).

Kejser, L. S. & I. G. Yazlovetsky. 1988. Comparative studies on sterols of *Chrysopa carnea* larvae and different types of their food. Journal of Evolutional Biochemistry and Physiology, Leningrad. 24 (2): 157-164.

Lewis, W. J., D. A. Nordlund & H. R. Gross. 1976. Production and performance of *Trichogramma* reared on eggs of *Heliothis zea* and other hosts. Environ. Entomol. 5 (3): 449-452.

Mellini, E. 1975. Possibilita de allevamento de Insetti entomophagi parassiti su diete artificiali. Bull. 1st. Entomol. Univ. Bologna. 32: 257-290.

Moontyan, E. M. & I. G. Yazlovetsky. 1988. Digestive carbohydrases of *Chrysopa carnea* larvae. Izvestia Acad. Sci. Moldavian SSR, Ser. Biol. Chem. Sci., (3): 61-64 (in Russian).

Ridgway, R. L., R. K. Morrison & M. Badgley. 1970. Mass rearing of a green lacewing, *Chrysopa carnea Steph.* J. Econom. Entomol. 63: 834-836.

Singh, P. 1977. Artificial diets for insects, mites, and spiders. Plenum, New York.

Slansky, F., Jr. & J. G. Rodrigues. 1987. Nutritional ecology of insects, mites, spiders and related invertebrates: An Overview, pp. 1-69, *In* F. Slansky & J. G. Rodrigues [eds.], Nutritional ecology of insects, mites, spiders, and related invertebrate. Wiley, New York.

Sumenkova, V. V., L. I. Ageeva & I. G. Yazlovetsky. 1980. Studies on feeding requirements of Trichogramma: Amino acids and proteins of some *Lepidoptera* eggs, pp. 20-28. *In* N. A. Filippov [ed.], Trichogramma, Kishinev, part 2, (in Russian, summary in English).

Thompson, S. N. 1986. Nutrition and in vitro culture of insect parasitoids. Ann. Rev. Entomol. 31: 197-219.

Vandersant, E. S. 1973. Improvement in the rearing diet for *chrysopa carnea* and the amino acid requirements for growth. J. Econ. Entomol. 66 (2): 336-338.

Vandersant, E. S. 1974. Development, significance and application of artificial diets for insects. Ann. Rev. Entomol. 19: 139-160.

Vinson, S. B. & P. Barbosa. 1987. Interrelationships of nutritional ecology of parasitoids, pp. 673-695. *In* F. Slansky & J. G. Rodriges [eds.], Nutritional ecology of insects, mites, spiders and related invertebrate. Wiley, New York.

Voegele, J. 1974. Less Trichogrammes. Modalites de la prise de possession et de l'hote le parasite embryonnaire *Trichogramma brasilinsis*. Ann. Soc. Entomol. France. 10 (3): 757-761.

Waage, J. K., K. P. Carl, N. J. Mills & D. J. Greathead. 1985. Rearing entomophagous insects, pp. 45-66. *In* P. Singh & R. F. Moore [eds.], Handbook of insects rearing, vol. 1. Elsevier, Amsterdam.

Yazlovetsky, I. G. 1986. Current tendencies in the development of artificial nutritional media for mass rearing of insects. Information Bulletin EPS IOBC (15): 21-42 (in Russian, summary in English).

Yazlovetsky, I. G. & A. M. Nepomnyashaya. 1981. Essai d'elevage massif de *Chrysopa carnea* sur mileux artificiels microen capsules et de son application dans la lutte contre les pucerons en serrs, pp. 51-58. *In* Lutte biologique et integree contre pucerons. INRA Publ., Paris.

Yazlovetsky, I. G., E. M. Mencher, A. M. Nepomnyashaya & V. V. Sumenkova. 1979. New approach to elaboration of artificial nutritive diets for mass rearing of entomophagous insects. Izvestia Acad. Sci. Moldavian SSR, Ser. Biol. chem. Ski. (1): 55-63 (in Russian).

Yazlovetsky, I. G., V. V. Sumenkova & A. M. Neponmyashchaya. 1979. Biochemical investigation of feeding requirements of aphidophagous insects. General characterization and amino acid composition of some aphid species and the Angoumois grain moth eggs, pp. 19-28. *In* I. S. Popushoe [ed.], Biochemistry and physiology of insects. Stiintsa Publ., Kishinev.

# 5

## Assimilation, Transport, and Distribution of Molecules in Insects from Natural and Artificial Diets

*Jeffrey P. Shapiro*

### Insect Rearing and Progress in Insect Control

Rearing systems have provided research and development laboratories with the bases for numerous bioassays, usually in the form of raw materials—the insects—to be used in vivo or in vitro. Modified rearing systems themselves are sometimes used for in vivo bioassays. In any case, insect rearing has been the foundation for the discovery and development of most conventional insecticides and biological control agents. Now, biotechnology is offering new tools for the discovery and implementation of insect control methods. A strong argument can be made that biosynthetic compounds and engineered plants, microorganisms, and other agents will gradually supplement and supplant our diminishing arsenal of synthetic organic insecticides (Table 1) (Meeusen & Warren 1989). Although all the examples in Table 1 are proteinaceous, and therefore macromolecules, simple organic compounds may also eventually be engineered into a system by introducing genes coding for enzymes that either synthesize simple compounds de novo or modify existing compounds. Rearing systems will play a vital role in these developments and will reflect changing ideas about the physical, chemical, and biological interfaces between insect and control agent.

As novel control agents are developed, emphasis in research will change from organic compounds, synthesized and formulated by chemists, to macromolecular or organic natural products, synthesized by organisms and usually formulated as components of those same organisms. Modes of penetration, translocation, and effects at target sites will differ from those of conventional insecticides. Digestion, penetration, absorption, and transport

63

**Table 1.** Examples of insecticidal or insect-bioregulatory molecules with potential for use as insect control agents, or components of such agents, through genetic engineering

| Molecule | Source | Reference |
|---|---|---|
| δ-Endotoxin | *Bacillus thuringiensis* | Obukowicz et al. 1986a,b Ahmad et al. 1989 |
| Cowpea Trypsin Inhibitor | *Vigna unguiculata* | Hilder et al. 1987 |
| Proteinase Inhibitors I/II | *Solanaceae* | Johnson et al. 1989 |
| Juvenile Hormone Esterase | *Heliothis virescens* | Hammock et al. 1990 |

through the hemolymph from the digestive tract to target organs will become especially important in making control agents effective. Conventional insecticides generally enter an insect through the cuticle and must be able to readily penetrate the cuticle to be effective. (The chemical and physiological determinants of penetration are reviewed by Gilby 1984 and Welling & Paterson 1985.) Topical application has been very useful for testing the toxicities of cuticle-penetrating substances. With the advent of biologically engineered control agents, the insect alimentary canal will become the prime site for penetration of novel agents, limiting the value of topical assays. Instead, future assays for biologically derived agents will be based on dietary systems.

The insect diets used in rearing and assay are central to discovering, developing, and enhancing the efficacy of novel agents, and to understanding their natural roles. However, the complications of dietary interactions among allelochemicals and nutrients in artificial and natural diets sometimes dramatically affect nutrient utilization (Reese 1979). Conversely, nutrients can affect the efficacy of allelochemicals. Understanding the interactions between nutrients and allelochemicals or engineered agents in the context of digestion, assimilation, translocation, and mode of action can save effort and increase rates of product discovery and development.

This article will briefly summarize the complex assimilative and translocative processes in insects and note the most important characteristics of dietary components to consider in the rearing and use of insects for assay of control agents. Putative roles of hemolymph proteins in the processes of absorption and transport will also be discussed. As background to these physiological processes, reviews from several fields of study will be cited throughout this work.

In the most prominent examples of engineered control agents, genes for *Bacillus thuringiensis (B.t.)*, δ-endotoxins are being successfully incorporated into a variety of microbial and plant species (Meeusen & Warren 1989). After ingestion, the δ-endotoxin proteins, in the form of isolated protein or genetically incorporated into host organisms, act at the insect midgut. The endotoxins enter and affect the insect only through the digestive system. They are activated by physical and biochemical factors in the midgut and act directly on midgut cells. Cellular penetration may not even be necessary for toxicity, since binding of the endotoxins to high affinity sites occurs at the brush border membrane (Hilder et al. 1987). Despite distinct chemical contrasts between the macromolecular endotoxins and small compounds, the δ-endotoxins can serve to contrast conventional and engineered insect control agents.

## Formulations: An Analogy

Toxicologically, the transformed microbial or plant host of the δ-endotoxin gene serves two roles, each of which can be compared to an analogous role in the manufacture of a conventional synthetic insecticide (Fig. 1):

1. The host is the actual producer of the δ-endotoxin, biosynthesizing it from the genetic template. The synthetic chemist fulfills this role in the manufacture of conventional insecticides.
2. The transformed host serves as the formulation for delivery of the δ-endotoxin to the insect. Formulation of a conventional insecticide includes carriers, wetting agents, synergists, etc., in addition to a small percentage of active agent, and can turn an isolated compound of moderate activity into a potent mixture through effects on the external and internal physiology of the target insect. The toxicity and toxico-kinetics of a formulated insecticide may therefore differ considerably from those of the isolated insecticide.

Since the δ-endotoxin in an engineered system is ingested as an integral component of the host organism, that formulation complicates comparisons between the isolated toxin administered in a feeding assay and the host endotoxin

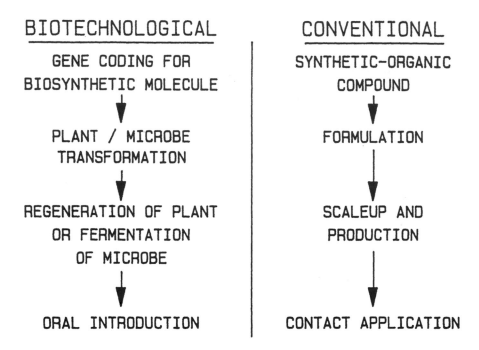

**Fig. 1. Analogy between steps in derivation of biotechnological and synthetic insect control agents.**

system administered in the greenhouse or field. In the engineered system, calculations of basic parameters such as absolute dose and rate of exposure to the toxin are complicated by levels of expression of the toxin and rate of feeding of an insect on the host plant. Even harder to calculate or predict, however, are the effects of a host plant's biochemistry on efficacy of an engineered effector.

To compare results from laboratory assays with applications in the field, an enhanced understanding of insect digestive physiology and biochemistry will be required. Among the processes to be understood, our knowledge is limited regarding stability of an agent in the gut, penetration through (or effect upon) the midgut epithelium, and transport to active sites via the hemolymph. However, we should be able to derive operative concepts from a parallel understanding of synthetic chemical control agents and their formulations.

### Artificial Versus Natural Diets

Artificial diets are simplified, optimized versions of natural diets. Simplification of the natural milieu leaves out numerous compounds and their polymers that would otherwise be absorbed to both the detriment and benefit of the insect. Optimization yields a diet that aims to achieve maximal growth rate and final size of the insect; optimization thus refers to conditions that "benefit" the insect within a restricted, subjective, utilitarian definition of the term.

While optimized diets yield maximal growth and production of an insect species, they do little for comparing various biochemical control agents. Many studies on effects of natural products are conducted with artificial rearing systems. Isolated natural products or mixtures of them are combined with artificial diet or layered over the diet, forcing ingestion by the insect under study. Other studies compare insects fed an artificial diet with those fed a natural diet. In either case, the comparison is limited by the complexities of and differences among the dietary matrices. When examining the effects of a substance or mixture on an insect, many factors 1) influence digestion, absorption, transport, and toxicities, and 2) complicate comparisons between control and experimental groups and between laboratory and greenhouse or field experiments. The following factors can distinguish activities in artificial diets from those in natural or genetically engineered diets.

**Active Biotic Factors.** Microbes or active biochemical factors such as enzymes that are usually present in natural systems will rarely be present in artificial systems. Such factors may alter the apparent chemical composition of an engineered agent once it is ingested. One example of this involves midgut symbionts. In aphids and certain other insects, destruction of symbionts by antibiotics can result in depletion of necessary products such as sterols that are synthesized solely by the symbionts, resulting in inhibition of growth, development, and survival (Mittler 1971a, 1971b). Since antibiotics and antiseptics are commonly used in artificial diets, their effects on putative control agents must be considered.

**Complexation of Active Components.** Within the matrix of either natural or artificial diet, compounds or macromolecules may complex with other compounds or macromolecules. Some may become less available for digestion and absorption, while others may be more readily absorbed through cooperative effects. The tannins and $\alpha$-tomatine are examples of the former. Tannins bind and precipitate proteins, and their ingestion may result in significant decrease in protein digestibility (Feeny 1968, Reese 1979, Duffey 1980). Binding of activated $\delta$-endotoxin by high concentrations of tannins in some plant species could prevent binding of the toxin to midgut cells, decreasing its activity; binding of proteolytic enzymes by tannins could inhibit proteolytic activation of the $\delta$-endotoxin.

$\alpha$-Tomatine, an alkaloid in tomato plants, apparently exerts toxicity against insect herbivores and their parasites by binding ß-sterols. $\alpha$-Tomatine might be useful as an allelochemical component engineered or bred into crops if it were not for its potential impact on useful parasites, parasitoids, and predators (Campbell & Duffey 1979). Its toxicity can be alleviated by high dietary content of ß-sterols (Campbell & Duffey 1981), possibly complicating interpretation of studies in which sterol content varies between natural and artificial diets.

**Changes in Midgut Environment.** The biochemical and physical environments of the digestive system may be changed appreciably by the presence, absence, or quantity of a compound or mixture of compounds. With ingestion, induced or introduced changes in the gut environment may result in changes in pH, reducing potential, conductivity of specific ions, etc., or in changes in activities of many enzymes such as proteases or mixed function oxidases (MFOs) in the midgut. MFOs oxidize toxic compounds, reducing their activity and making them more excretable. Some secondary plant products significantly induce MFO activity in the insect midgut and can result in decreased toxicity of plant-derived allelochemicals, such as nicotine (Brattsten & Wilkinson 1977).

**Internal Sensitivity to Active Agents.** Coabsorption of dietary compounds with an active agent may alter the internal response or responsiveness of an insect at sites targeted by the active agent. This may occur directly, through synergism or antagonism toward the agent at the target site, or indirectly, through induction or repression of specific cellular systems such as membrane receptors. Synthetic organic insecticide formulations often include synergists and carriers.

The most common synergists, inhibitors of mixed function oxidases, increase the susceptibility of an insect to an active compound by inhibiting the enzymes that metabolize and thus inactivate that compound (Brattsten 1979). Inadvertently, some solvents used in insecticide formulation may do the opposite, inducing enzyme synthesis and effectively increasing resistance of the target insect (Brattsten et al. 1977). Applebaum (1985) gives examples of many naturally occurring substances that inhibit the digestive process, especially regarding proteolytic activity.

## The Roles of Absorptive and Transport Systems

The need to understand the processes of digestion, absorption, and transport of compounds (as summarized in Fig. 2) in the insect should now be manifest. Although processes contributing to the removal of compounds from the insect (e.g., enzymatic metabolism) have been well studied, processes involved in addition of compounds to the system (e.g., absorption, transport, and binding)

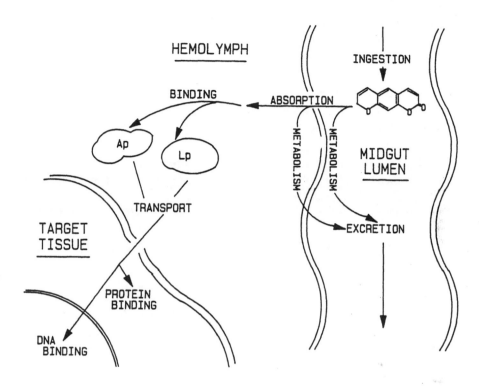

**Fig. 2. Physiological processes involved in adding to or removing from a compound's availability to target sites.**

are not as well understood. The following generalizations about these processes can aid in understanding, predicting, and comparing effects of compounds in rearing, assay, and field systems.

## Digestion

Following ingestion, the process of digestion exerts the initial influence on absorption of a compound. Digestion involves a large number of enzyme-catalyzed biochemical reactions: hydrolysis of proteins to peptides and peptides to amino acids; breakdown of some complex carbohydrates (generally excluding cellulose, except in termites) to mono- or di-saccharides, and of polynucleotides (DNA, RNA) to purines and pyrimidines; and derivatization of low molecular weight compounds. (See Applebaum 1985 for descriptions of enzyme classes and the digestive process.) These processes usually result in products of increased polarity (Brattsten 1979). Digestion therefore results in decreasing concentrations of macromolecules and increasing concentrations (and thus rates of absorption) of low molecular weight compounds.

Though digestion can result in either activation or inactivation of compounds, it is usually thought of as an inactivating process. Mixed function oxidases, glutathione transferase, and other enzymes act to increase polarity and excretability of numerous compounds at the midgut epithelium; proteolytic enzymes hydrolyze and inactivate protein toxins in the midgut lumen. However, digestion can also activate toxins, as with the protease- and pH-activated *B.t.* δ-endotoxin.

Digestive tissues and processes can be direct targets of agonists and antagonists. Proteolytic inhibitors are widespread in plants and may contribute to resistance against insect pests in many species (Ryan 1979, Gordon 1968). They usually act by deactivating the catalytic site or sites of a proteolytic enzyme and are among the agents being actively explored for use in biologically engineered plant defenses (Meeusen & Warren 1989, Haunerland & Bowers 1986). In an engineered system, the natural presence of such inhibitors in host plants might decrease efficacy of an incorporated protein such as *B.t.* δ-endotoxin by preventing necessary proteolytic activation. Rearing and assay systems, lacking the inhibitors, would not detect such a problem, which would appear in final stages of testing. Alternatively, protease inhibitors may synergize some proteinaceous agents by decreasing their digestion in the midgut.

## Absorption

Two physiochemical factors have an immediate impact upon absorption rate: the size of a compound or molecule, and the polarity or lipophilicity of lower molecular weight compounds. The gut epithelium, and to a lesser extent the peritrophic membrane (if present in the species or stage of insect in question), act as molecular sieves. Above a certain molecular weight, penetration from lumen through the epithelium into the hemocoel is impossible, prohibiting passage of macromolecules such as complex carbohydrates, proteins, large nucleic acids, and other polymers.

Absorption of lower molecular weight compounds occurs through passive diffusion of apolar (lipophilic) compounds and some ions, through facilitated diffusion aided by shuttle proteins in membranes, or through active transport of compounds by energy-driven processes. The lipid-protein matrix of epithelial cell membranes inhibits penetration by polar compounds, except through facilitated diffusion or active transport by membrane proteins. Once a polar compound crosses the epithelial membrane, it is readily dissolved in the aqueous phase and taken into circulation. On the other hand, lipophilic compounds tend to diffuse readily into and across cell membranes, but their uptake into circulation is limited by poor solubility in the aqueous phase of the hemolymph. Despite this fact, most effective insect toxins, both synthetic and natural, are moderately to highly apolar. These concepts sparked spirited debates as to whether toxic doses of insecticide are absorbed by vertical diffusion through cuticle into hemolymph or by horizontal diffusion into the spiracles and along tracheae to target organs. Though the debate was not conclusively settled, popular opinion supports vertical diffusion and transport to target organs through hemolymph (see Gilby 1984 or Welling & Paterson 1985 for complete discussions).

## Transport

The apparent paradox of the high toxicities of insoluble apolar compounds may be resolved by realizing the biophysical nature of hemolymph: although an aqueous medium is not conducive to transport of apolar compounds, hemolymph is not purely aqueous. It contains high concentrations of proteins, and proteins are amphophilic ("loving both sides") macromolecules, i.e., they are compatible with both lipid and aqueous phases. As alluded to by Campbell and Duffey (1981), transport of lipophilic compounds by mammalian proteins is well known, but analogous modes of transport in insects are more obscure. Hemolymph proteins that bound insecticides in vivo and in vitro were first observed in the mid-1970s (Welling & Paterson 1985; Shapiro et al. 1988a), but specific proteins were not identified until 1984-1986.

The best known insect transport proteins are the lipophorins (Shapiro et al. 1988a). Lipophorins are lipoproteins, or spherical fluid particles of lipid and protein found in virtually all species of insects examined to date. Circulating freely in hemolymph, they absorb lipids from midgut and release them into fat body for storage and absorb lipid from fat body for delivery to sites of utilization. Firm evidence from several species shows lipophorins to function as lipid shuttles (see Chino et al. 1981 and Tsuchida & Wells 1988). Their protein moieties recirculate, while lipid components are transported unidirectionally. Insecticides are also bound by insect lipoproteins (Shapiro et al. 1988a), identified recently as lipophorins (Kawooya et al. 1985).

Another class of hemolymph protein, the arylphorins, has also recently been shown to bind insecticides when mixed in vitro with hemolymph from *Heliothis zea* (Haunerland & Bowers 1986). Arylphorins are known as amino acid storage proteins, thought to act as a sink for amino acids utilized in cuticle synthesis during metamorphosis. Binding of a range of insecticides by an arylphorin from *Heliothis zea* was a novel discovery. Perhaps of more consequence was that the proportion of a compound bound to arylphorin versus lipophorin was dependent upon the partition coefficient of the compound, representing its lipophilicity. Another example of xenobiotic binding to hemolymph proteins has recently been discovered and described in vivo and in vitro, although the binding protein has yet to be identified with any known class of insect protein.    In the root weevil larva *Diaprepes abbreviatus*, a model fluorescent compound (7-amino-3-phenyl coumarin, or coumarin-10) was absorbed from a semidefined diet (Shapiro et al. 1988b), and from a force-fed mixture in oil (Shapiro 1989), into hemolymph.   A large protein of 480,000 molecular weight, present at concentrations approaching 100 mg/ml, bound 95% of the coumarin-10 found in hemolymph; lipophorin bound 5%.  A dissociation constant of $1.5 \times 10^{-6}$ M was determined in vitro.

These few examples of xenobiotic binding by proteins and lipoproteins in hemolymph supply a link between absorption and target site interaction (Fig. 2). The kinetics of toxicity are critical: A toxic compound, whether an insecticide (Welling & Paterson 1985) or phytochemical (Duffey et al. 1978) must interact with a target site at a concentration high enough to produce the desired effect. An increased rate of transfer from the midgut serosa into hemolymph and subsequent transport to target sites may increase availability of the compound in the critical concentration at the target site.

Binding of apolar compounds may involve proteins in hemolymph other than the arylphorins and lipophorins.   Even regarding lipophorins, binding and transport of compounds other than the native lipids are poorly described, and the kinetics of uptake and transport are virtually unknown.  The roles of hemolymph proteins relative to intoxication versus detoxification are also unknown. Knowledge of their roles in the uptake and transport processes can aid in the discovery and increased efficacy of control agents.

## Conclusion and Summary

Although assimilative processes are complex, awareness of the biochemistry in a natural or engineered system modeled by an assay system can highlight critical relationships among components, and problems may be alleviated during system design.  For example, if one is testing candidates for a proteinaceous toxin to engineer into a plant variety that contains some known proteolytic enzymes, similar enzymes can be included in a diet assay system.  Comparing conditions for assimilation of natural versus artificial diets can therefore reveal pitfalls between the processes of discovery and implementation of bioregulatory agents, or between observations in a laboratory versus a field environment. Both quantitative (e.g., significant differences in $LD_{50}$) and qualitative (e.g., toxicity versus nontoxicity) differences in activity of an agent may be confusing to an investigator.

To aid in the design of dietary systems, additional knowledge of assimilative processes is necessary.  The basic enzyme classes and their roles in digestion are well defined, and selected enzyme activities in an insect can be readily studied. However, absorption is more difficult to study. General principles are given for absorption of specific biochemical classes (Turunen 1985), but the range of compounds to be studied is much too diverse and methods too painstaking for thorough study in any one insect. Although general principles for transport of polar compounds are clear, those for transport and final disposition of lipophilic compounds are still obscure and deserve further attention.

# References

Ahmad, W., C. Nicholls & D. J. Ellar. 1989. Cloning and expression of an entomocidal protein gene from *Bacillus thuringiensis* galleriae toxic to both Lepidoptera and Diptera. FEMS Lett. 59: 197-202.

Applebaum, S. W. 1985. Biochemistry of digestion, pp. 279-311. *In* G. A. Kerkut & L. I. Gilbert [eds.], Comprehensive insect physiology biochemistry and pharmacology, vol. 4. Pergamon Press, New York.

Brattsten, L. B. 1979. Biochemical defense mechanisms in herbivores against plant allelochemicals, pp. 199-270. *In* G. A. Rosenthal & D. H. Janzen [eds.], Herbivores: Their interaction with secondary plant metabolites. Academic Press, New York.

Brattsten, L. B. & C. F. Wilkinson. 1977. Insecticide solvents: Interference with insecticidal action. Science 196: 1211-1213.

Brattsten, L. B., C. F. Wilkinson & T. Eisner. 1977. Herbivore-plant interactions: Mixed function oxidases and secondary plant substances. Science 196: 1349-1352.

Campbell, B. C. & S. S. Duffey. 1979. Tomatine and parasitic wasps: Potential incompatibility of plant antibiosis with biological control. Science 205: 700-702.

Campbell, B. C. & S. S. Duffey. 1981. The alleviation of alpha tomatine induced toxicity to the parasitoid *Hyposoter exiguae* by phytosterols in the diet of the host *Heliothis zea*. J. Chem. Ecol. 7: 927-946.

Chino, H., H. Katase, R. G. H. Downer & K. Takahashi. 1981. Diacylglyerol-carrying lipoprotein of hemolymph of the American cockroach: purification, characterization, and function. J. Lipid Res. 22: 7-15.

Duffey, S., M. S. Blum, M. B. Isman & G. G. E. Scudder. 1978. Cardiac glycosides: a physical system for their sequestration by the milkweed bug. J. Insect Physiol. 24: 639-645.

Duffey, S. S. 1980. Sequestration of plant natural products by insects. Annu. Rev. Entomol. 25: 447-477.

Feeny, P. P. 1968. Effect of oak leaf tannins on larval growth of the winter moth *Operophtera brumata*. J. Insect Physiol. 14: 805-817.

Gilby, A. R. 1984. Cuticle and insecticides, pp. 694-702. *In* J. Bereiter-Hahn, A. G. Matoltsy & K. S. Richards [eds.], Biology of the integument. Springer-Verlag, Berlin.

Gordon, H. T. 1968. Quantitative aspects of insect nutrition. Amer. Zool. 8: 131-138.

Hammock, B. D., B. C. Bonning, R. D. Possee, T. N. Hanzlik & S. Maeda. 1990. Expression and effects of the juvenile hormone esterase in a baculovirus vector. Nature 344: 458-461.

Haunerland, N. H. & W. S. Bowers. 1986. A larval specific lipoprotein: Purification and characterization of a blue chromoprotein from *Heliothis zea*. Biochem. Biophys. Res. Communication 134: 580-586.

Hilder, V. A., A. M. R. Gatehouse, S. E. Sheerman, R. F., Barker & D. Boulter. 1987. A novel mechanism of insect resistance engineered into tobacco. Nature 330: 160-163.

Johnson, R., J. Narvaez, G. An & C. Ryan. 1989. Expression of proteinase inhibitors I and II in transgenic tobacco plants: Effects on natural defense against *Manduca sexta* larvae. Proc. Natl. Acad. Sci. USA 86: 98719875.

Kawooya, J., P. S. Keim, J. H. Law, C. T. Riley, R. O. Ryan & J. P. Shapiro. 1985. Why are green caterpillars green? A.C.S. Symp. Series 276: 511-521.

Meeusen, R. L. & G. Warren. 1989. Insect control with genetically engineered crops. Annu. Rev. Entomol. 34: 373-381.

Mittler, T. E. 1971a. Dietary amino acid requirements of the aphid *Myzus persicae* affected by antibiotic uptake. J. Nutr. 101: 1023-1028.

Mittler, T. E. 1971b. Some effects on the aphid *Myzus persicae* of ingesting antibiotics incorporated into artificial diets. J. Insect Physiol. 17: 1333-1347.

Obukowicz, M. G., F. J. Perlak, M. Kusano, E. J. Kretzmer & L. S. Watrud. 1986a. Integration of the deltaendotoxin gene of *Bacillus thurengiensis* into the chromosome of root-colonizing strains of pseudomonads using Tn5. Gene 45: 327-331.

Obukowicz, M. G., F. J. Perlak, K. Kusano Kretzmer, E. J. Mayer, S. L. Bolten & L. S. Watrud. 1986b. Tn5-mediated integration of the delta-endo-toxin gene from *Bacillus thurengiensis* into the chromosome of root-colonizing pseudomonads. J. Bacteriol. 168: 982-989.

Reese, J. C. 1979. Interactions of allelochemicals with nutrients in herbivore food, pp. 309-330. *In* G. A. Rosenthal & D. H. Janzen [eds.], Herbivores: Their interaction with scondary plant metabolites. Academic Press, New York.

Ryan, C. A. 1979. Proteinase inhibitors, pp. 599-618. *In* G. A. Rosenthal & D. H. Jansen [eds.], Herbivores: Their interaction with secondary plant metabolites 599-618.

Shapiro, J. P. 1988. Isolation and fluorescence studies on a lipophorin from the weevil *Diaprepes abbreviatus*. Arch. Insect Biochem. Physiol. 7: 119-131.

Shapiro, J. P. 1989. Xenobiotic absorption and binding by proteins in hemolymph of the weevil *Diaprepes abbreviatus*. Arch. Insect Biochem. Physiol. 11: 65-78.

Shapiro, J. P., J. H. Law & M. A. Wells. 1988. Lipid transport in insects. Annu. Rev. Entomol. 33: 297318.

Shapiro, J. P., R. T. Mayer & W. J. Schroeder. 1988. Absorption and transport of natural and synthetic toxins mediated by hemolymph proteins. *In* F. Sehnal, A. Zabza, & D. L. Denlinger [eds.], Endocrinological frontiers in physiological insect ecology. Wroclaw Technical Univ. Press, Wroclaw.

Tsuchida, K. & M. A. Wells. 1988. Digestion, absorption, transport and storage of fat during the last larval stadium of *Manduca sexta*. Changes in the role of lipophorin in the delivery of dietary lipid to the fat body. Insect Biochem. 18: 263-268.

Turunen, S. 1985. Absorption, pp. 241-277. *In* Kerkut & L. I. Gilbert [eds.], Comprehensive insect physiology biochemistry and pharmacology, vol. 4. Pergamon Press, New York.

Welling, W. & G. D. Paterson 1985. Toxicodynamics of insecticides, pp. 603-645. *In* Kerkut & L. I. Gilbert [eds.], Comprehensive insect physiology biochemistry and pharmacology, vol. 12. Pergamon Press, New York.

# 6

## Using a Systematic Approach to Develop Artificial Diets for Predators

*Allen C. Cohen*

### Introduction and Historical Perspective

The greatest barrier to mass production of predatory insects as a major force in pest control is the lack of suitable artificial diets. Singh (1977) noted that 754 species of arthropods had been reared on artificial diets by 1977; of these, 27 were arachnids. The remaining are insect species spanning 10 orders consisting of 19 families of Coleoptera, 24 of Diptera, 11 Hemiptera, 8 Hymenoptera, and 27 Lepidoptera. Singh and Moore (1985) reported similar numbers. However, Waage et al. (1985) pointed out that no suitable artificial diets have been developed for predators. These authors described diets that incorporated insect materials, but practical and economical considerations rule against production programs that involve rearing at least two trophic levels.

An oligidic diet free of insect components (A. C. Cohen 1985) was developed and has sustained over 50 continuous generations of predators *(Geocoris punctipes* Say Lygaeidae, Heteroptera) for the past 6 years. Over the past several years, we have reared cultures of 500-2,000 adults per week using this diet. Efforts to produce larger numbers are hindered by the time-consuming nature of diet presentation and other nonautomated culture techniques. Since this diet has had a more sustained record of production than any other artificial diet for predators, I offer here a perspective on development of the diet.

The status of in vitro culturing of parasitic insects (Parasitoids) has been reviewed by House (1977), Greany (1986), Thompson (1986), and Nettles (1989). Of all the families and species studied, diets for only three families and about 10 species of predators are described, and nearly all such diets contain insects as components. A few reports on *Chrysoperla (Chrysopa)* claim success in completion of development with artificial diets devoid of insect parts. Only

Vanderzant (1973) reported continuous (18) generations on diets without insect ingredients.

Similarly, diets that support continuous generations of coccinelids contain insect materials (honey bee drone brood or pulverized aphids), but Attallah and Killebrew (1967) described a casein and wheat germ diet that supported eight generations of *Coleomegilla maculata* De Geer.

It is striking that so few successful diets have been developed for predators and that those reported even two decades ago have not found their way into use for the mass production of predators. In light of current needs, it seems that mass production of predators would be most helpful and welcome in pest control. This disappointing record raises questions about the basis of this failure. The comments of Waage et al. (1985) express the difficulty encountered by students of predator nutrition: "Predators would be expected to be more easily reared on artificial diets than parasitoids, because their development is less dependent on the physiology of their hosts, and because they are generally less host specific. In fact, the contrary is the case." My present treatment of this subject will explain many pitfalls and some ways to circumvent them in future work on predators.

My work over the past decade has been dedicated to the trophic biology of predaceous insects. My impression is that the progress in dietetics for predators has been overstated and that the difficulties have been underestimated. These difficulties as they pertain to phytophagous and other nonpredatory insects are detailed by other authors such as House (1974 & 1977) and Singh and Moore (1985). I will discuss sources of difficulty in predator dietetics and the systematic tactics that I have found useful in dealing with these problems.

First, many predators will eat only living, moving prey. Thus, many potentially useful predators such as praying mantis and dragon flies cannot be induced to feed on artificial diets because the diets do not provide appropriate visual stimuli. Next, predators that use chemical cues for prey search and acceptance must find those phagostimulants associated with artificial diets to ensure appropriate patterns of feeding behavior. Finally, even after proper physical and chemical stimuli have been presented, the nutrients in the diet must be qualitatively and quantitatively appropriate to support normal growth, fecundity, and behavior (e.g., pheromones), etc.

Rationale for diet development described in the literature range from presentation of previously described diets, formulation of diets based on detailed chemical analyses of the natural foods (Akey & Beck 1971, 1972), or whole carcass analysis of the insects being studied (Rock 1967). The first method is essentially a trial and error approach where foods of assumed nutritional quality are offered with the hope that they will be acceptable. The second method is highly mechanistic, based on the assumption that if we analyze all components of the insect's natural food and if we present all components in the diet, it should support the insect in question.

Some nutritionists (e.g., Rock & Hodgson 1971) use radioisotopes to determine the metabolic capabilities of an insect. For example, a labeled carbon source such as $C_{14}$ acetate is provided in a diet deficient in several amino acids but adequate in a nitrogen source. After a period of time, the insect is analyzed to determine whether or not it produced the missing amino acids. If the insect succeeded in producing them, it is assumed that those amino acids are not required. This approach has been used extensively for determining lipid synthesis abilities (Stanley-Samuelson et al. 1988). This is a useful alternative to the classical deletion-addition approach described by Waldbauer (1968), wherein diets lacking in certain nutrients are offered while growth and/or development are measured to determine whether these nutrients are required.

Another recent approach is the dietary self-selection technique or the "cafeteria" method (Waldbauer & Bhattacharya 1973, Waldbauer et al. 1984, R. W. Cohen et al. 1987, 1988). In this method there is an assumption of internal "nutritional wisdom" whereby the insect can sense deficiencies of specific nutrients in its body and consumes foods rich in those nutrients (Greenstone 1979). In the work of Waldbauer et al. (1984), simple combinations of nutrients were offered such as casein, sugars, or starches. For such a system to work, there must be a mechanism for sensing a deficiency (e.g., the internal malaise concept discussed by Waldbauer et al. 1984) and a recognition of the nutrient in a food source that the organism is selecting. This seems reasonable where the nutrient is some simple substance such as a sugar or free amino acid, but more complex models of feedback are required to explain consumption of bound substances such as amino acids within proteins (Waldbauer et al. 1984). Moore (1986) used similar methods to determine feeding preferences and utilization efficiencies to optimize diets for *Heliothis zea* Boddie (corn earworms). This approach coupled with video image analysis techniques shows promise for future work on insect dietetics.

It is recognized that the presence of a nutrient in the diet does not necessarily mean that the nutrient is available to the insect's metabolic pool. Turunen (1983) pointed out, for example, that certain lepidopteran larvae that consume oil-rich seeds are adept at absorption of oils. However, leaf-eating lepidopteran larvae are much less efficient at absorbing oils added to their artificial diets and instead require polar lipids that are readily absorbed. This could be a very important issue in lipid nutrition of predatory insects whose foods are rich in polyunsaturated fatty acids occurring as moieties of polar lipids.

# A Case Study:
## Developing an Artificial Diet for *Geocoris punctipes*

It took 5 years to develop an adequate artificial diet for the predaceous heteropteran *G. punctipes*. Much of this time was spent in presenting nutrients in a liquid form, in response to misconceptions in the literature that hemipterans require liquid food. A thorough understanding of heteropteran feeding mechanisms would have prevented these misdirected efforts. This supports the retrospectively self-evident maxim that a thorough knowledge of an insect's biology is a requisite to the practical management of that insect.

Scrutiny of the feeding behavior of these predators revealed that they eat far more than prey's hemolymph. Using rasping and cutting stylets (Cobben 1978, Sweet 1979, A. C. Cohen 1989b) they macerate the solid material in their prey and inject digestive enzymes that liquify prey solids (A. C. Cohen 1989b). Careful observation showed that the prey is a virtual "cafeteria" of different materials, including lipid-rich fat body and protein/glycogen-rich muscles. Indeed, the predators were not confining their ingestion to hemolymph but were utilizing at least 60 to 80% of the prey's body contents (A. C. Cohen 1989a) and over 90% of the eggs' contents (A. C. Cohen 1984). Cellular membranes known to be rich in phospholipids (Lehninger 1982) are available after mechanical and partial chemical breakdown. The diameter of the predator's food canal (about 1 u in *G. punctipes* and 5-20 u in a reduviid, *Sinea confusa*, Candell [A. C. Cohen 1989b]) has bearing on how finely the particulates or lipid masses must be rendered before they are accessible to ingestion.

My approach to formulating diets for *G. punctipes* changed dramatically once I learned that materials could be presented in forms other than as homogeneous liquids and that predaceous heteropterans did not require a membrane to feed through. Indeed, a dried beef liver-hamburger-sugar diet supported two generations of *G. punctipes*. However, a wet diet mixture proved more appropriate (A. C. Cohen 1985). We present the diet in stretched Parafilm packets, allowing the predators access and simulating their feeding through prey cuticle, upon which they construct their salivary sheath collars (Fig. 1). This presentation also keeps the diet moist and somewhat protected from microbial contamination.

## Diet Ingredients

Initially, I used the hemolymph of *Spodoptera exigua* Hubner (beet armyworm, or BAW) as a model for artificial diet because we initially supported *G. punctipes* on heat-killed BAW larvae. Analysis of hemolymph and whole BAW (A. C. Cohen & Patana 1982) and other *G. punctips* prey (A. C. Cohen et al. 1985) showed high concentrations of K and Mg and low levels of Na.

This led to the formulation of diets with higher phagostimulatory potential than traditional "vertebrate salt mixtures." Using these K, Na, and Mg ratios, I devised a diet that supported complete development of one generation of *G. punctipes* (A. C. Cohen 1981), the first such diet to support development of a predaceous heteropteran. I tested a vertebrate salt mixture in this basic diet against the phyotophagous insect salt mixture and found the latter to be superior in both feeding preference and survival (A. C. Cohen, unpublished data). However, inability to produce continuous generations on this diet prompted further changes and new directions. The nutrient-poor hemolymph was an unfortunate choice to serve as a model for profiles of artificial diets.

**Fig. 1. A flange of stylet secretion glued to the surface of a Parafilm covered diet packet demonstrating "normal" feeding response of a predator *(G. punctipes)* feeding on artificial media.**

We cultured *G. punctipes* on a variety of insect prey and plant materials (e.g., coddled beet armyworms and green beans as described by Champlain & Sholt 1966 or on aphids and sunflower seeds as described by Tamaki & Weeks 1972). We tested different insect materials (live and coddled lepidopterous larvae, aphids, 1st-3rd instar lygus bugs) and concluded that insect eggs and free water were optimal (A. C. Cohen & Debolt 1983). With this optimal "natural diet," I had an excellent model upon which to base the chemical composition of my diet. A. C. Cohen and Patana (1985) analyzed the chemical content of *Heliothis virescens* (F.) eggs and used these chemical profiles as a basis for selecting foods that conformed to these profiles. Using the nutrient composition of commonly available foods (Adams 1975), I found that 45% beef liver, 45% beef hamburger, and 10% sucrose solution closely fit the profile of nutrients in *H. virescens* eggs (Fig. 2).

I formulated this diet reasoning that: 1) materials of animal origin would be suitable for a predatory insect, 2) beef mineral profiles (including high cellular potassium levels) were similar to those in phytophagous insects, 3) the nutrient profile resembled that of *Heliothis* eggs, and 4) the heterogeneous character of the diet would allow pre-oral selection observed in predator feeding on natural diet. This diet formulation proved to be successful, promoting 6 years and over 50 continuous generations of *G. punctipes* and also supporting certain other predator species for several generations (A. C. Cohen, unpublished data).

It is important to note that a "bottleneck" effect (an initial reduction in population size, followed by an increase) was observed in every effort to offer a new diet to any of the predators studied. This initial reduction was observed when predators were recently field-captured and kept on insect prey or artificial diet, as well as when laboratory-adapted insects were offered an alternative food or prey species.

Although essential amino acids are key factors in metazoan nutritional homeostasis (House 1974), tests of carcass composition of *G. punctipes* indicated that this predator seems to possess considerable dietary independence with respect to amino acid metabolism. Profiles of amino acid composition from whole carcass analyses were nearly identical in *G. punctipes* from several field sites and in laboratory-reared insects fed artificial diet, *Heliothis* eggs, or BAW larvae (A. C. Cohen, unpublished data). It is unlikely that these generalist predators from distinct geographical regions and from laboratory cultures ate diets with nearly identical amino acid compositions. It is more reasonable to infer from these data that *G. punctipes'* amino acid pattern is endogenously set and adhered to by metabolic mechanisms.

Preliminary tests with radioisotopes and classical dietary deletion techniques indicate that there are no noticeable deviations in *G. punctipes* from the "rat essential" pattern of amino acid requirements found in many species of insects. It appears that these predators simply use metabolic and excretory selection to accomplish the observed nutrient balance and conformation to the strict profile described above.

This type of independence is not evident in the pattern of fatty acid profiles (Fig. 3). The pattern is influenced by environmental factors such as differences in the dietary fatty acids (A. C. Cohen 1989c). Similar dependence of carcass fatty acid content on diet profiles was described in another heteropteran, *Lygus hesperus* (Knight) (A. C. Cohen 1989c). Tests using radioisotopes (including

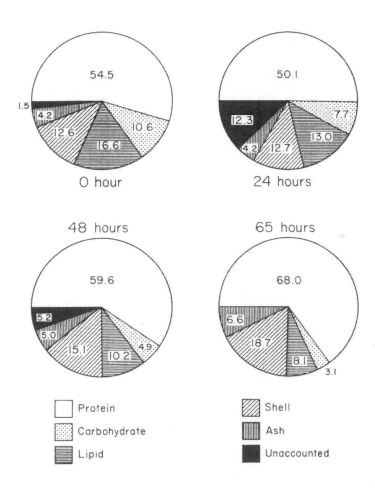

**Fig. 2. A profile of the chemical composition of *Heliothis virescens* eggs demonstrating the model used (0 hour) for formulation of *G. punctipes* diets.**

precursors such as $^{14}$C acetate and $^{14}$glucose) are currently being conducted in my laboratory to determine the ability of heteropteran predators to synthesize de novo polyunsaturated fatty acids. Other tests of absorption of various lipid types show that polar lipids are more readily absorbed than neutral lipids (A. C. Cohen, unpublished data). However, with the appropriate emulsifiers present in the diet, absorption of triacylglycerols and sterols can be enhanced to nearly that of polar lipids. This became evident in experiments with radiolabeled lipids ($^{14}$C cholesterol, $^{14}$C triacylglycerol, and $^{14}$C phosphatidylcholine) where the following respective absorption efficiencies were noted: 95.3 $\pm$ 0.32% S.D., 93.3 $\pm$ 1.7% and 95.4 $\pm$ 0.7%) (A. C. Cohen, unpublished data). I attribute these uniformly high absorbencies to the presence of bile salts in the beef liver diet used to deliver the radiolabeled lipids.

### Quality Control

There is growing awareness (King & Leppla 1984) that laboratory-produced insects may lack qualities found in field strains of the same species and that artificially produced insects may not serve the functions for which they were

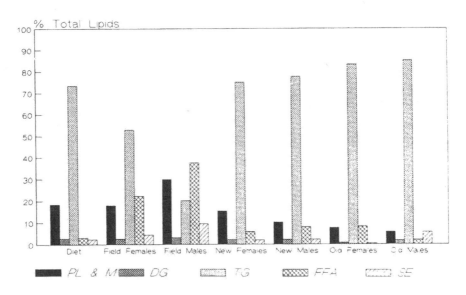

PL & M—phospholipids & monoacylglycerides; DG—diacylglycerides; TG—triacylglycerides; FFA—free fatty acids; SE—steryl esters

**Fig. 3. Lipid profiles of whole carcasses of laboratory-reared versus field-captured *Lygus hesperus*, demonstrating the dependence of this hemipteran insect's lipid content upon dietary composition.**

intended (such as sterile male releases, predation, and parasitism). Thus, those involved in rearing (especially in vitro production) must attend to questions of quality control beyond the standard measures of productivity, growth rate, size, and disease control in cultures.

This laboratory is extensively committed to comparative studies of behavioral and metabolic quality of our predators. Fig. 4 illustrates a summary of metabolic tests that we perform comparing materials transfer in predators consuming natural diets *(Heliothis* eggs) with those consuming artificial diet (from data presented in A. C. Cohen 1984 and A. C. Cohen & Urias 1988).

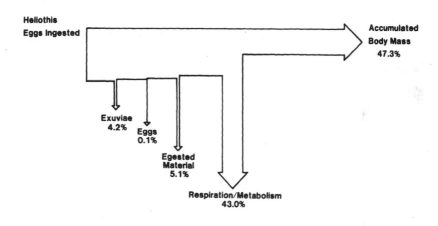

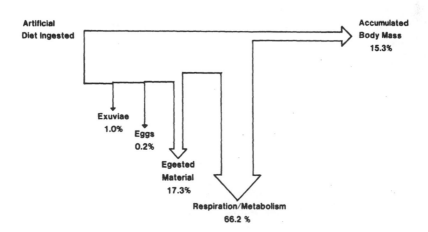

**Fig. 4. Comparison of metabolic fate and efficiencies of conversion of two diets, *Heliothis* eggs (top) and artificial diet (bottom), as nutrients for the predator *Geocoris punctipes*.**

Evidently food conversion is less efficient with artificial diets than it is with insect eggs. It is further evident from Fig. 4 that losses in respiration and egestion (excretion and defecation) are responsible for the lower efficiency in predators fed artificial diets. Such comparisons provide feedback for dietary adjustments to increase the efficiency of total predator biomass. I assume that diets which foster greater metabolic efficiency will also produce behaviorally higher quality predators than will a less nutritive diet. Also, an increase in biomass produced per unit of diet weight would enhance the economic efficacy of the rearing program.

In general, the issue of quality control of insects reared on artificial diets is in its infancy and that of quality control of entomophages produced on artificial diets is in its early embryonic stages. ODell (1984) pointed out that most quality control efforts are directed toward production rather than the health or success potential of the insects produced. Obviously, quality control in entomophages involves validation of the predators' and parasites' vigor in the field.

This laboratory is involved with establishing tests of this capability. Hagler and A. C. Cohen (unpublished data) have performed behavioral, immunological, and electrophoretic tests of *G. punctipes* cultured in vitro for nearly 6 years while comparing the performance of these predators with field-collected *G. punctipes*. We found that antigens in their food could be detected 48 h after prey consumption. A method of marking our predators with rubidium-spiked artificial diet was developed (A. C. Cohen & Jackson 1989). This research background will permit comparisons to be made in the field between augmentatively released predators and their wild counterparts.

As a result of studies illustrated in Fig. 5, we have embarked on an extensive investigation to discover the weak points in the nutritional "chain." This laboratory's current focus on lipid and protein digestion and absorption has revealed important basic information about the trophic physiology of predaceous hemipterans (A. C. Cohen 1989a, 1989b). Demonstration of weak salivary triacylglycerol lipases and strong phospholipases will be a useful guide for making dietary adjustments that suit these differences in enzyme systems. For example, inclusion of polar lipids (such as lecithin) in place of neutral lipids should be a useful adjustment in diets for predaceous hemipterans because they are more readily absorbed (A. C. Cohen, unpublished data).

There are some important points to be raised about programs dedicated to the study of arthropods' trophic biology. Multilateral approaches are necessary for understanding the basic trophic biology of the insects that we wish to manipulate. An understanding of the nutritional requirements of an insect cannot be fully achieved without appreciable knowledge of its behavior; neither of these areas can be well understood without fundamental knowledge of digestive and absorptive physiology. Accordingly, the physiological studies depend upon morphological information as well as sensory biology, genetics, ecology, and evolution. Programs that recruit cooperators trained in interdisciplinary approaches will be most successful in contributing to long-term management of both the pests and beneficial arthropods that we wish to control.

Second, with many newer approaches to controlling arthropods (such as host plant resistance and splicing bacterial toxin-producing genes into plants) the first line of attack is at the trophic level rather than the cuticular interface between target species and their environment. For example, Shapiro (in this volume)

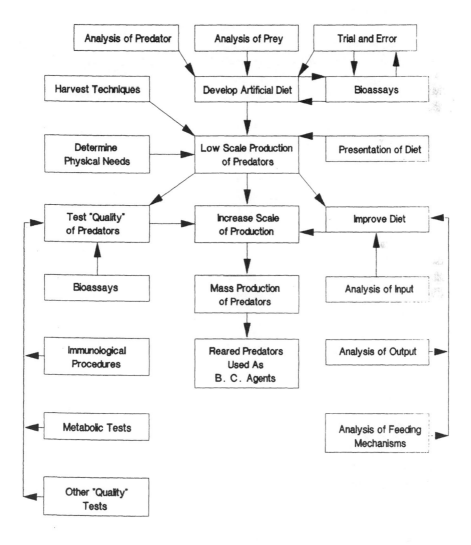

**Fig. 5. A flow diagram that describes the development of an in vitro rearing program for the predator *Geocoris punctipes*.**

points out that efforts are currently being made to produce plants that are resistant to pests through genetic engineering. He also further asserts that for such efforts to be effective, we need a thorough understanding of the mechanisms of digestion, absorption, and transport of the materials that we aspire to use as control agents. Furthermore, if these efforts at pest management are truly performed in an integrated fashion, we must also have an understanding of these mechanisms as they pertain to arthropods in higher trophic levels. For example, Campbell and Duffey (1979) have demonstrated that some plant secondary compounds such as tomatine may be useful in host plant resistance programs but may interfere with biological control efforts by adversely affecting entomophages. The same authors have shown (1981) that certain plant nutrients can alleviate the adverse effects of tomatine. These examples illustrate that a thorough knowledge of trophic biology is not only needed in rearing programs but is equally necessary in other aspects of pest management.

The flow chart in Fig. 5 summarizes the thoughts offered here and the basic mechanism that I followed. All components that lead to the "Increase Scale of Production" were useful in developing my program. In terms of time frameworks where these methods are to be applied to a new subject, many of the activities can be accomplished simultaneously (e.g., "Analysis of Predator" and "Analysis of Prey"). Many of these components represent ongoing processes, such as developing improved tests for quality of predators, metabolic tests, and diet improvement based on new nutritional discoveries. Finally, funding agencies and management must be apprised that, while this multilateral approach is productive and useful in solving many of our pest problems, it is also expensive. However, there should be substantial spin-off from one program to another.

## Acknowledgment

I thank Drs. P. Greany, G. Hoffman, J. Hagler, and W. S. Nettles for helpful reviews of this manuscript and Dr. Ward-Medley for manuscript preparation.

## References

Adams, C. F. 1975. Nutritive value of american foods. Agriculture Handbook No. 456. USDA, ARS, Washington, D.C.

Akey, D. H. & S. D. Beck. 1971. Continuous rearing of the pea aphid, *Acyrthosiphon pisum*, on a holidic diet. Ann. Entomol. Soc. Am. 64: 353-356.

Akey, D. H. & S. S. Beck. 1972. Nutrition of the pea aphid, *Acyrthosiphon pisum*: requirements for trace metals, sulfur and cholesterol. J. Insect Physiol. 18: 1901-1914.

Attallah, Y. H. & R. Killebrew. 1976. Ecological and nutritional studies on *Coleomegilla maculata* (Coleoptera: Coccinellidae). IV. Amino acid requirements of the adult determined by use of $C^{14}$-labelled acetate. Ann. Entomol. Soc. Am. 60: 186-188.

Campbell, B. C. & S. S. Duffey. 1979. Tomatine and parasitic wasps: Potential incompatibility of plant antibiosis with biological control. Science 205: 700-702.

Champlain, R. A. & L. L. Sholt. 1966. Rearing *Geocoris punctipes*, a lygus bug predator, in the laboratory. J. Econ. Entomol. 59: 1301.

Cobben, R. H. 1978. Evolutionary trends in Heteroptera. Part II. Mouthpart structures and feeding strategies. Agric. Res. Reports, Wageningen 78-5: 1-407.

Cohen, A. C. 1981. An artificial diet for *Geocoris punctipes* (Say). Southwest Entomol. 6: 109-113.

Cohen, A. C. 1984. Food consumption, food utilization, and metabolic rates of *Geocoris punctipes* (Het:Lygaeidae) fed *Heliothis virescens* (Lep: Noctuidae) eggs. Entomophaga 29 (4): 361-367.

Cohen, A. C. 1985. Simple method for rearing the insect predator *Geocoris punctipes* on a meat diet. J. Econ. Entomol. 78: 1173-1175.

Cohen, A. C. 1989a. Ingestion efficiency and protein consumption by a heteropteran predator. Ann. Entomol. Soc. Am. 82(4): 495-499.

Cohen, A. C. 1989b. Feeding adaptations of predaceous heteropterans. Ann. Entomol. Soc. Am. (in press).

Cohen, A. C. 1989c. Fatty acid distribution as related to adult age sex and diet in the phytophagous heteropteran, *Lygus hesperus*. J. Entomol. Sci. (in press).

Cohen, A. C. & J. W. Debolt. 1983. Rearing *Geocoris punctipes* (Say). Southwest Entomol. 8: 61-4.

Cohen, A. C. & C. G. Jackson. 1989. Using rubidium to mark a predator, *Geocoris punctipes* (Hemiptera: Lygaeidae). J. Entomol. Sci. 24(1): 57-61.

Cohen, A. C. & R. Patana. 1982. Ontogenetic and stress-related changes in hemolymph chemistry of beet armyworms. Comp Biochem. Physiol. 71A: 193-198.

Cohen, A. C. & R. Patana. 1983. Efficiency of food utilization by *Heliothis zea* fed artificial diets or green beans. Can. Entomol. 116: 171-176.

Cohen, A. C. & R. Patana. 1985. Chemical composition of tobacco budworm eggs during development. Comp. Biochem. Physiol. 81B: 165-169.

Cohen, A. C. & N. M. Urias. 1986. Meat-based artificial diets for *Geocoris punctipes* (Say). Southwest. Entomol. 11(3): 171-176.

Cohen, A. C. & N. M. Urias. 1988. Food utilization and egestion rates of the predator *Geocoris punctipes* (Hemiptera: Heteroptera) fed artificial diets with rutin. J. Entomol. Sci. 23(2): 174-179.

Cohen, A. C., J. W. Debolt & H. A. Schreiber. 1985. Profiles of trace and major elements in whole carcasses of *Lygus hesperus* adults. Southwest Entomol. 10: 239-243.

Cohen, R. W., S. Friedman & G. P. Waldbauer. 1988. Physiological control of nutrient self-selection in *Heliothis zea* larvae; the role of serotonin. J. Insect Physiol. 34(10): 935-940.

Cohen, R. W., G. P. Waldbauer, S. Friedman & N. M. Schiff. 1987. Nutrient self-selection by *Heliothis zea*: A time-lapse film study. Entomol. Exp. Appl. 44: 65-73.

Cohen, R. W., S. L. Heydon, G. P. Waldbauer & S. Friedman. 1987. Nutrient self-selection by the omnivorous cockroach *Supella longipalpa*. J. Insect Physiol. 33(2): 77-82.

Greany, P. D. 1986. *In vitro* culture of hymenopterous larval endoparasitoids. J. Insect Physiol. 32: 403-408.

Greenstone, M. H. 1979. Spider feeding behavior optimises dietary essential amino acid composition. Nature 282(5738): 501-503.

House, H. L. 1974. Nutrition, pp 1-62. *In* M. Rockstein [ed.], The physiology of Insecta, vol. 5. Academic Press, New York.

House, H. L. 1977. Nutrition and natural enemies. *In* R. L. Ridgeway & S. B. Vinson [eds.], Biological control by augmentation of natural enemies. Plenum, New York.

King, E. G. & N. C. Leppla (editors). 1984. Advances and challenges in insect rearing. USDA-ARS. New Orleans, Louisiana.

Lehinger, A. L. 1982. Principles of biochemistry. Worth, New York.

Moore, R. F. 1986. Feeding preferences and utilization studies as tools in developing an optimum diet for *Heliothis zea* (Lepidoptera: Noctuidae). J. Econ. Entomol. 79: 1707-1710.

Nettles, W. C. 1989. In vitro rearing of parasitoids: role of host factors in nutrition. Arch. Insect Biochem. Physiol. (in press).

O'Dell, T. M. 1984. Production use and quality testing in insect rearing, p. 155. *In* E. G. King & N. C. Leppla [eds.], Advances and challenges in insect rearing. USDA-ARS. New Orleans, Louisiana.

Rock, G. C. 1971. Utilization of D-Isomers of the dietary indispensable amino acids by A*argrotaenia velutinana* larvae. J. Insect Physiol. 17: 2157-2168.

Rock, G. C. & E. Hodgson. 1971. Dietary amino requirements for *Heliothis zea* determined by dietary delection and radiometric techniques. J. Insect Physiol. 17: 1087-1097.

Rock, G. C. & K. W. King. 1967. Amino acid composition in hydrolysates of the red-banded leaf roller, *Argyrotaenai velutinana* (Lepidoptera: Tortricidae). Ann. Entomo. Soc. Am. 59: 273-277.

Singh, P. 1977. Artificial diets for insects, mites, and spiders. Plenum, New York.

Singh, P. & R. Moore. 1985. Handbook of insect rearing, vol. 1 and 2. Elsevier, Amsterdam.

Srivastava, P. N., Y. Gao, J. Levesque & J. L. Auclair. 1985. Differences in amino acid requirements between two biotypes of the pea aphid, *Acyrthosiphon pisum*. Canadian J. Zool. 63(3): 603-606.

Stanley-Samuelson, D. W., R. A. Jurenka, C. Cripps, G. J. Blomquist & M. deRenobales. 1988. Fatty acids in insects: composition, metabolism, and biological significance." Arch. Insect Biochem. Physiol. 9: 1-33.

Sweet, M. H. 1979. On the original feeding habits of Hemiptera. Ann. Entomol. Soc. Am. 72: 575-579.

Tamaki, G. & R. E. Weeks. 1972. Biology and ecology of two predators *Geocoris punctipes* (Stal) and *G. bullatus* (Say). USDA Tech. Bull. 1446: 46.

Thompson, S. N. 1986. Nutrition and in vitro culture of insect parasitoids. Ann. Rev. Entomol. 31:197-219.

Turunen, S. 1983. Absorption and utilization of essential fatty acids in lepidopterous larvae: metabolic implications, pp. 57-71. *In* T. E. Mittler & R. H. Dodd [eds.], Metabolic aspects of lipid nutrition in insects. Westview Press, Boulder, Colorado.

Vanderzant, E. S. 1973. Improvements in the rearing diet for *Chrysopa carnea* and the amino acids requirements for growth. J. Econ. Entomol. 66: 336-338.

Waage, J. K., K. P. Carl, N. J. Mills & D. J. Greathead. 1985. Rearing entomophagous insects, pp.45-66. *In* P. Singh & R. Moore [eds.], Handbook of insect rearing, vol. 1. Elsevier, Amsterdam.

Waldbauer, G. P. 1968. The consumption and utilization of food by insects, pp. 229-288. *In* Advances in insect physiology, vol. 5. Academic Press, New York.

Waldbauer, G. P. & A. K. Bhattacharya. 1973. Self-selection of an optimum diet from a mixture of wheat fractions by the larvae of *Tribolium confusum*. J. Insect Physiol. 19: 407-418.

Waldbauer, G. P., R. W. Cohen & S. Friedman. 1984. Self-selection of optimal mix from defined diets by larvae of the corn earworm, *Heliothis zea* (Boddie). Physiol. Zool. 57: 590-597.

# 7

# Feeding and Dietary Requirements of the Tephritid Fruit Flies

*George J. Tsiropoulos*

## Introduction

Insects, being heterotrophs like all animals, require exogenous nutrients for both tissue construction and satisfaction of their energy requirements. However, research on insect nutrition did not begin until the onset of this century. During the past 40 years, over 30 reviews of insect nutrition have been published; the most recent and general of those are House (1972, 1974), and Dadd (1970, 1973, 1977, 1985).

The first nutritional studies were a byproduct of research in genetics. Geneticists working with a "false" fruit fly, the well-known *Drosophila*, have made great contributions in insect nutrition. *Drosophila* was the first multicellular invertebrate to be cultured nonaxenically (Delcourt & Guyenot 1910); it was also the first to be grown on a chemically defined diet (Schultz et al. 1946). The very first studies by Delcourt and Guyenot were concerned with reduction of variability in genetic experiments and standardization of the "nutritional environment." At the same time, Back and Pemberton (1917) working in Hawaii with a true fruit fly, *Dacus cucurbitae* Coquillett, realized that adult nutrition influenced sexual activity and egg maturation. When the flies were fed fresh cucumber slices, they mated and deposited eggs in a much shorter period than when they were fed papaya and water. However, Back and Pemberton did not speculate on what nutrient factors might be affecting fecundity. Up to the second world war, several scientists, mainly in the New World, produced a great amount of information regarding the nutrition of fruit fly species of economic importance, like the apple maggot, *Rhagoletis pomonella* (Walsh), and the walnut husk fly, *Rhagoletis completa* Cresson. Fluke and Allen (1931) first attempted to rear the apple maggot on artificial diet and discovered that by adding yeast to a honey-water mixture, longevity was prolonged and copulation and oviposition occurred. The first meridic and holidic adult fruit fly diets were developed after 1945 by Hagen and Finney (1950), who first used yeast hydrolysates instead of yeast, plus a carbohydrate to feed adults of the Oriental fruit fly, *Ceratitis capitala* (Wiedemann). This discovery enabled research on fruit fly nutrition and biology all over the world. For Tephritids, there are works by Steiner and Mitchell (1966), Bateman (1981), and Tsitsipis (1987). Tephritids, a comparatively young family of Diptera, arose in the middle Tertiary, about 50 million years ago. Today, about 5,000 species have been described. As larvae, they feed and live in fruits, leaves, stalks, flowers, or seeds, while some form galls in stems, crowns, and rhizomes (Novak & Foote 1980). As adults, all fruit flies are polyphagous, obtaining the necessary metabolites from a variety of natural food sources (Bateman 1972).

Research on tephritid feeding and dietary requirements is very active since a number of fruit fly species are of great economic importance. This includes the following topics: fruit fly mass rearing for application of the Sterile Insect Technique (SIT) (Singh 1977); fruit fly survey and chemical control, fruit fly detection, monitoring, and control are based on the use of food attractants (Chambers 1977); nutritional ecology and feeding behavior for development of more efficient integrated pest management (IPM) (Prokopy 1977, Prokopy & Roitberg 1984).

To facilitate the discussion, we shall follow the steps shown in Fig. 1. Feeding includes attraction of the fly to food and all necessary mechanisms and behavior for appropriate food to reach its stomach. This paper describes fruit fly dietary requirements and presents some thoughts about using nutritional information for more effective fruit fly control.

## Feeding Requirements

An adult tephritid has a life filled with "decisions." Actually, the fly makes a decision when after "evaluating" all environmental inputs, it reacts in some manner (Dawkins & Dawkins 1973, McFarland 1977). The reactions can include changes in the amount, rate and/or timing of food foraging behavior, food acceptance, metabolism enzyme synthesis, and other feeding and nutritional parameters. Thus, tarsal contact with the food source initiates "phagostimulation." This is followed by the third step, "ingestion," which brings food into the fly's stomach.

All fruit fly feeding responses and decisions involve some underlying nutritional components by which the flies achieve and maintain an optimum "set of states" for maximal fitness in their specific environment.

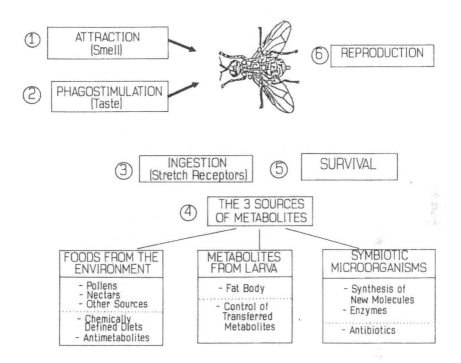

**Fig. 1. Factors affecting fruit fly feeding and nutrition.**

Feeding is a dynamic and active process with many feedback interactions, affecting and being affected by variables, such as movement, phagostimulation, ingestion, survival, and reproduction (Fig. 1).

Adult fruit flies, when deprived of water or food, initiate flights with a high frequency, to discover new locations rich in essential metabolites (Ripley et al. 1940, Monro 1966, Economopoulos et al. 1982). Tephritids respond to both the visual and chemical stimuli of water and food. The associated behavioral responses to locate a food source are governed by "attraction," the term used in its broad meaning, including both oriented (taxes) and non-oriented movements. Additionally, females of *D. cucurbitae* move on a daily basis from melon fields, which they visit for oviposition, to nearby wild vegetation, where they find food and shelter (Nishida & Bess 1957). A similar behavior has been reported for *D. dorsalis* (Bess & Haramoto 1961), *D. zonatus* (Syed et al. 1970), and *R. completa* (Barnes 1959). Contrary to that, *C. capitata* does not show daily foraging rhythm (Sigwalt et al. 1968). *R. pomonella* has five factors that shape its foraging behavior (Prokopy 1982): resource patch structure, population of origin, physiological state, presence of competitors and natural enemies, and previous experience.

Prokopy and Roitberg (1984) ask, "How does a forager go about sampling resources to arrive at an average value for a locale?" and "How does it make trade-offs in attempting to satisfy all its resource requirements?" To answer questions like these, it is necessary to identify the types of resources utilized by adult tephritids in nature, and to estimate the quantity, quality, and distribution of each resource in their normal habitat. Moreover, the feeding activity pattern of the fly, and the specific stimuli to which it responds, must be known. Tree size and shape, as well as the parallactically shifting pattern of individual leaves, have been found to be the principal environmental stimuli attracting *R. pomonella* adults on trees (Moericke et al. 1975, Prokopy 1977). Color or silhouette of foliage are involved in *C. capitata* and *D. oleae* attraction to trees (Sanders 1968, Prokopy & Haniotakis 1975). In terms of light energy spectra, tephritid visual receptors are most sensitive in the band of 500-600 nm (Agee & Parks 1975). Green tree leaves reflect, within the insect-visible spectrum, maximally within this same band, while outside the insect-visible spectrum, tree leaves reflect a peak energy in the infrared band, 750-1350 nm (Callahan 1975). On the other hand, transmitted or reflected light energy below 500 nm may be repulsive to landing tephritids (Moericke et al. 1975, Prokopy & Haniotatkis 1975). Prokopy (1977) concluded that "adult tephritid visual location of vegetation may involve the dual processes of attraction to foliar reflected light and repulsion by nonfoliar reflected or transmitted light, such as skylight." Discovery of the powerful attractancy of various daylight fluorescent yellow colors, reflecting light energy between 500-600 nm, produced a supernormal foliage type stimulus, used in various shapes, sizes and hues to trap several species of fruit flies (Prokopy & Boller 1971, Prokopy 1972, Prokopy & Economopoulos 1976, Prokopy et al. 1975, Remund 1971, Russ et al. 1973). After landing on or near a food source, a fruit fly precisely locates the food by characteristic patterns of movement (Prokopy 1976).

Chemical stimuli in the environment can be received either from a distance or by direct contact via the fly's special tarsal receptors. Distant chemical stimuli may involve odors emitted from adult fruit fly food sources, derived from their natural breakdown process. The main gas is ammonia, but secondary gaseous products, mainly from amino acid breakdown, are also involved (Jarvis 1941, McPhail 1939, Hodson 1943, Gow 1954, Barnes & Osborn 1958, Cirio & Vita, 1980, Bateman & Morton 1981, Tsiropoulos & Zervas 1985). The role of microbes in fruit fly attraction to various food sources has been studied by Gow (1954) and Morton and Bateman (1981). The breakdown, besides microbial, can be chemical and/or photochemical (Bateman & Morton 1981).

Odors emitted from blossoms or foliage of nonhost plants have also been found attractive to several fruit fly species (Kawano et al. 1968, Metcalf et al. 1975, 1979, 1981, Shah & Patel 1976, Fletcher et al. 1975).

Guerin et al. (1982), have studied the responses of *C. capitata, D. oleae,* and *R. cerasi* to some 30 volatiles by electrophysiological and field tests. The volatiles, ranging from C5 to C12 with alcohol, aldehyde, and ester terminal functional groups, were screened using the electroantennogram (EAG). A strong degree of conformity was observed in the EAG responses to compounds in the series of saturated aliphatic aldehydes, with heptanal, octanal, and nonanal evoking the highest responses in all three species. Female medflies were attracted to traps baited with heptanal, while *D. oleae* flies were attracted to traps baited with hexanol, octanal, nonanol, and nonanal. Males of several *Dacus* species are attracted to plant species emitting o-methyl-eugenol (Kawano et al. 1968, Fletcher et al. 1975, Metcalf et al. 1975, Drew & Hooper 1981). Since these plants do not provide adult food, as Fletcher et al. (1975) suggest, this may serve for just male aggregation.

The successive steps of a fruit fly's feeding behavior are shown in Fig. 2. A potential nutritional substrate, to reach a fly's stomach, depends on and is the result of an interplay between a) an external sensory excitation, coming from its peripheral tarsal and/or labeller chemoreceptors and b) an internal, post ingestive inhibition coming from the gut and body wall stretch mechanoreceptors, as well as from the thoracic ganglion locomotor center (Detheir 1976, Bernays & Simpson 1982). Thus, during feeding, the central nervous system (CNS) of a fly is receiving, at the same time, a vast and complex amount of positive and negative information, which ultimately is integrated to produce the appropriate final motor output. To explain the function of the CNS, by which the fly recognizes, accepts, and ingests a food, the concept of "across-fibre patterning" has been developed (Dethier 1976, Boeckh 1980, Van Drongelen et al. 1978, Gelperin 1971, Bernays & Simpson 1982).

The first step of the external, peripheral sensory input, refers to contact of the tarsal chemoreceptors (Angioy et al. 1978, Gothilf et al. 1971). If the stimulation is positive, proboscis extension takes place. If it is negative, the fly takes off searching for another food source. The next step involves contact of the proboscis chemoreceptors with the substrate. Positive stimulation causes the labeller lobes to spread, allowing contact of their special chemoreceptor papillae with the substrate. Here again, if the input is negative, the fly flies away. Positive stimulation of the chemoreceptor papillae leads to the fourth step, i.e., the pharyngeal pump starts sucking, while food quality is continuously monitored by the special chemoreceptor sensilla (Rice 1973, Tsiropoulos, unpublished data).

For continuous fly feeding, i.e, long first meal duration and high sucking rate, the positive peripheral input must be continuous, since sensory adaptation

98

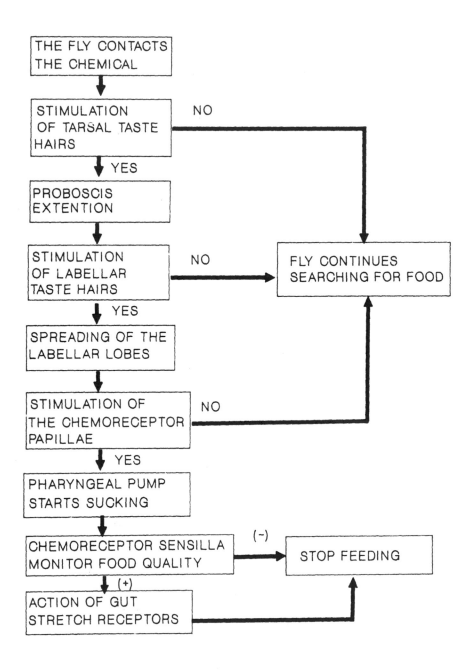

**Fig. 2. Fruit fly feeding behavior sequence.**

might occur rapidly and make the fly stop feeding. This is accomplished by special molecules present in the food substrate, the phagostimulants, which secure a continuous positive input. Several types of molecules act as fruit fly phagostimulants. Various carbohydrates have been found phagostimulatory in *C. capitata* (Gothilf et al. 1971, Fuchs 1974, Galun et al. 1981) and in *A. suspensa* (Sharp & Chambers 1984, Galun et al. 1985). Amino acids have also been found phago-stimulatory in *C. capitata* (Galun et al. 1978b, 1981, 1985, Tsiropoulos & Zervas 1985), in *A. suspensa* (Sharp & Chambers 1983), in *D. oleae* (Tsiropoulos 1984), and in *D. tryoni* (Morton & Bateman 1981). Finally, salts have been found phagostimulatory in *C. capitata* (Gothilf et al. 1971).

Feeding termination, step 5, is directed by the net input of the negative post ingestive feedback over the positive peripheral sensory input, which for Diptera is generally accepted as coming from three sources: crop and/or foregut stretch receptors, body-wall stretch receptors, and the inhibitory activity of the locomotor center in the thoracic ganglion (Dethier 1976). The feeding process might also be influenced negatively by molecules known as "feeding deterrents" affecting both the initiation of feeding and/or the amount of food eaten. To this category belong various alkaloids, terpenoids, glucosides, phenolics, and amines. However, no research has been done on feeding deterrents for fruit flies.

## Natural Diets

The metabolites necessary for adult fruit fly survival and reproduction come from three sources (Fig. 1): foods taken directly from the environment, microflora associated with the fly, and storage in the larval fat body. In nature, adult fruit flies exploit a great variety of food sources, like honeydews, nectars, plant exudations, fruit juices, pollens, fruit and leaf surface bacteria, fungal spores, bird droppings, and others (Christenson & Foote 1960, Steiner & Mitchell 1966, Bateman 1972). Silvestri (1914) reported that *D. oleae* as well as other Tephritids were seen feeding on honeydews. Since then several reports of fruit fly honeydew feeding appeared (Middlekauff 1941, Hagen 1958, Matsumoto & Nishida 1962, Neilson & Wood 1966, Boush et al. 1969, Tsiropoulos 1977a). Most of the honeydews, however, have been found deficient in nutritional constituents, thus being "incomplete." For this reason, Boush et al. (1969) failed to develop a chemically defined diet for adult *R. pomonella* based on amino acid analysis of apple aphid honeydew. Hagen and Tassan (1972) discussed the role of symbiotic bacteria in providing the missing metabolites in honeydews, while Tsiropoulos (1977a) determined that honeydews trap pollen grains that germinate and release their proteinaceous contents into the honeydew by diffusion. He also showed that *D. oleae* flies can survive and reproduce on honeydews of the black scale, *Saissetia oleae* (Bern), the coccoid,

*Filippia oleae* (Costa), and the olive psylla, *Euphyllura olivina* (Costa). Pollens from five wind-pollinated plant species (olivetree, *Olea europaea* L.; grapevine, *Vitis vinifera* L.; pistacio, *Pistacia lentiscus* L.; pine, Pinus sp.; and holm oak, *Quercus coccifera* L.) diluted in a sucrose solution were able to support survival and reproduction of adult *D. oleae* (Tsiropoulos 1977a). In Hawaii, Nishida (1963) showed that two *Dacus* species feed on the secretions of extra floral glands of several plants. Leaf and fruit surface bacteria have received special attention as a natural food source for adults of *D. tryoni* (Drew et al. 1983), while specific leaf surface bacteria in nature have been reported as regulating the abundance of the tropical fruit flies, *D. tryoni* and *D. neohumeralis* (Courtice & Drew 1984).

Microbial leaf isolates, cultured from honeydew on which adult *R. pomonella* were feeding, produced volatiles that elicited olfactory responses in the flies (McCollom & Rutkowski 1986).

## Meridic Diets

Most fruit flies must obtain from extrinsic sources, as adults, nearly all the nutrients required for optimal survival, fecundity, and fertility (Hagen 1952). All species require, besides water, carbohydrate as an energy source. Although adult needs vary among species, this is mainly the result of differences in the quality and quantity of metabolites transferred from the immature stages. Fruit fly larvae have not been reared on holidic diets, so the degree of metabolite transfer remains unknown. Starved *D. oleae*, survived 1.3 days, indicating limited metabolite transfer from the larval stage (Tsiropoulos 1980a). In the same study, fly survival in days on each diet ingredient, when offered singly with water, for males and females respectively, was as follows: sucrose 21.3 and 27.5, amino acids 1.4 and 1.7, minerals 1.6 and 1.4, vitamins 2.8 and 3.1, RNA 1.7 and 1.5, cholesterol 1.4 and 1.5, and water only 1.6 and 1.7. Sucrose without extrinsic water allowed 10-day fly survival (Tsitsipis 1975).

For sexual maturation and reproduction, most fruit fly species require a source of nitrogen and a variety of additional nutrients. The need for nitrogen was first shown by Fluke and Allen (1931), who found that the inclusion of an unhydrolyzed yeast in a honey-water diet of *R. pomonella* increased both survival and fecundity. In 1933, they omitted the water and thus prevented fermentation problems. However, Dean (1938) found that, to live and produce eggs, *Rhagoletis pomonella* flies must get carbohydrates, nitrogen, and water; his diet was artificial but not chemically defined. Carbohydrates plus water increased longevity, but very few, if any, eggs were deposited. Protein, added to the carbohydrate-water diet, greatly increased fecundity. Of the various proteins he tested, proteose-peptone plus yeast were the best in stimulating high egg production. Testing various salt mixtures, he found that a mixture containing more or less the inorganic salts found in cow's milk gave the highest fecundity when added to urea. His conclusion on inorganic salts was that they do not affect oviposition markedly and that copper, zinc, and iron chlorides reduce egg production.

Boyce (1934), working with the husk fly, reported that the highest fecundity of 30 eggs per female was achieved with flies feeding on technical grade dry sucrose. He found that dry yeast in the adult diet decreased both longevity and fecundity. He suggested that technical grade sucrose may contain certain impurities that are essential for egg production, since other diets containing chemically pure sucrose gave only 3.7 eggs per female. He also found that glycocoll in the diet decreased longevity but increased fecundity and that both urea and ammonia increased fecundity but did not affect longevity. Comparing carbohydrates, he found that fructose and honey reduced longevity slightly in comparison with sucrose, although fecundity was increased (glucose reduced longevity in comparison with sucrose without affecting fecundity) and that flies cannot survive on dextrine alone. He also experimented with two salt mixtures (A and B) containing the minerals used in plant nutrition research. Pure sucrose plus mineral mixture A gave an average of 9.3 eggs per female as compared with 4.5 eggs obtained from flies fed on a mixture made up with the same grade of sucrose and mineral mixture B. He concluded that increased longevity and fecundity were the result of zinc and copper, since mineral mixture B was identical to A but did not include Zn and Cu.

As already mentioned, until 1945 only crude (aligidic) diets had been developed. The first meridic diets were for *D. dorsalis*, *D. cucurbitae*, and *C. capitata* (Hagen & Finney 1950); the first chemically defined diet (holidic) was developed for the same species (Hagen 1952, 1958; Hagen & Tassan 1972). Discovery of the importance of yeast hydrolyzates, which besides amino acids and peptides provide minerals and vitamins, allowed the development of diets for several fruit fly species. More specifically meridic diets have been formulated for the following species:

- *D. oleae* (Moore 1962, Hagen et al. 1963, Cavalloro 1967, Economopoulos & Tzanakakis 1967, Tsitsipis 1975, Tsitsipis & Kontos 1983);
- *D. dorsalis* (Mitchell et al. 1965, Tagushi 1966);
- *D. tryoni* (Monro & Osborn 1967);
- *C. capitata* (Mitchell et al. 1965, Nadel 1970, Tanaka et al. 1970);
- *C. rosa* (Barnes 1976);
- *A. ludens* (Rhode & Spishakoff 1965);
- *A. suspensa* (Burditt et al. 1974);
- *R. cerasi* (Haisch 1968, Prokopy & Boller 1970);
- *R. pomonella* (Boush et al. 1969); and
- *R. completa* (Tsiropoulos 1974).

All of these diets contain a carbohydrate source that provides necessary energy and a protein source that provides necessary amino acids, vitamins, minerals, and sterols. Some diets contained egg yolk. As energy sources, sucrose and fructose were used, while protein sources were obtained from various types of Brewer's yeast, yeast hydrolysates, casein hydrolysates, or combinations of these ingredients.

The state of the diet and the alimentary canal of the flies greatly affects feeding and assimilation. Tzanakakis et al. (1967) reported that *D. oleae* females from a humid geographical area, with a liquid diet, laid more eggs and had a shorter preoviposition period than those fed a solid diet. Conversely, females from a dry area laid more eggs on solid than on liquid diet. To overcome the problem of ratios between diet components, Mitchell et al. (1965) offered water, sucrose, and yeast hydrolysate separately. The flies performed much better, since they were able to selectively ingest the proper amount of each ingredient. Irradiation causes damage to the alimentary canal epithelium. In *C. capitata* irradiation reduced olfactory responsiveness to total food intake by flies of both sexes, aggregation on and intake of protein hydrolysate by females, and sugar consumption by males. Also, *A. suspensa* irradiation produced a reduction in olfactory response of females to yeast hydrolysate, while in both sexes aggregation on and consumption of yeast hydrolysate were reduced (Galun et al. 1985).

## Chemically Defined Diets

Hagen's discovery that "the further the protein is hydrolysed toward free amino acids, the shorter the preoviposition period," enabled the development of chemically defined diets for tephritids. He formulated the first chemically defined diets (Hagen 1952, 1958) for *C. capitata*, *D. dorsalis*, and *D. cucurbitae*. Chemically defined diets have also been developed for *R. pomonella* (Boush et al. 1969), *D. oleae* (Tsiropoulos 1977b) and *R. completa* (Tsiropoulos 1978). All of these diets contained an amino acid mixture, a carbohydrate, a vitamin mixture, a salt mixture and, all but *R. pomonella's*, a sterol and RNA. The amino acid formulations for all species were based on analyses of yeast hydrolysates, except for *R. pomonella*, which was based on apple aphid, *Aphis pomi*, honeydew supplemented with tryptophane and cysteine.

Hagan (1952, 1958) and Hagen and Tassan (1972), showed that the two *Dacus* species, *dorsalis* and *cucurbitae*, carry very little, if any, nutrients from the larval stage. These species require exogenous metabolites to produce about 20 nonviable eggs per female. Amino acids were utilized only when minerals were present in the diets, which was also true for *D. oleae* and *R. completa* (Tsiropoulos 1977b, 1978). Vitamins were also a requirement for high egg fertility, since nearly all eggs deposited in their absence were malformed. Vitamin E and nitrogen were required by males of *D. dorsalis* for effective mating.

For *R. pomonella* (Neilson & McAllan 1965, Boush et al. 1969), the reproductive capacity of flies fed on defined diets was double that on undefined ones. Lack of vitamins caused a sharp decrease in fecundity, while lack of minerals caused a smaller but definite decrease. It was also found that fatty acids and sterols were not a requirement for the apple maggot.

In *D. oleae* and *R. completa* (Tsiropoulos 1977b, 1978, 1980a), a completely defined diet with the 10 amino acids essential for the rat was as good as one with a mixture of 19 amino acids, as far as survival, fecundity, and fertility were concerned.   For satisfactory survival, the most important ingredient was sucrose, which, with metabolites transferred from the larval stage, allowed production of 2 eggs per female in *D. oleae* but not in *R. completa*. Amino acids supported a substantial increase in egg production, in both species, only when minerals were present in the diets.   Fertility was increased in both species, by the addition of vitamins, but it was not by the addition of only vitamin E, cholesterol, and RNA in *R. completa*.

Single amino acid deletion studies showed that the 10 essential ones plus valine were indispensable for *D. oleae* fecundity, while male survival was also affected by alanine, hydroxyproline, and tryptophane omission. Also, individual amino acids, given as a sole nitrogen source, were determined to *D. oleae* adult survival and reproduction (Tsiropoulos 1983a).

The ratio between nitrogen (N) and carbohydrate (C) is an important parameter for the success of a chemically defined diet. The best N/C ratio for *D. oleae* reproduction was found to be 1.6/40, while higher nitrogen content in the diet reduced egg deposition. Increase of the nitrogen content above 1.6%, up to 6.4%, shortened adult life span of *D. oleae* by 50% (Tsiropoulos 1981).

The 15 vitamins tested individually or in groups, in *D. oleae* (Tsiropoulos 1980b), showed that the survival of both sexes was affected by pyridoxine, riboflavin, or all vitamin omission, while female survival, in addition, was affected by omission of folic acid.   Reproduction of *D. oleae* females was significantly reduced when vitamin E, biotin, choline chloride, inositol, nicotinic acid, and riboflavin were individually omitted or when all vitamins, B-complex vitamins, or folic acid plus RNA were deleted.   Moreover, egg hatch was reduced with single omissions of calcium pantothenate, folic acid, pyridoxine, and thiamine (Tsiropoulos 1980b).

Vitamin overdosing of *D. oleae* influenced its biological performance. Survival was significantly reduced by higher than normal dosages of biotin and pyridoxine, with females being more sensitive to biotin. Fecundity was reduced by surplus biotin, pyridoxine, inositol, and vitamin C, while fecundity was reduced by high dosages of biotin and calcium pantothenate (Tsiropoulos 1982).

Antimetabolites blocking the utilization of specific metabolites have been tested with *D. oleae* (Tsiropoulos 1985) and *R. completa* (Tsiropoulos & Hagen 1987). In *D. oleae,* survival was significantly affected by the antivitamins, the antagonized vitamin in parenthesis: benzimidazole (B-12), apicolinic acid (nicotinic acid), deoxypyridoxine (pyridoxine), neopyrithiamine (thiamine), and a-tocopherol quinone (Vitamin E). Egg production was reduced by the same antivitamins, plus, dl-desthiobiotin (biotin), aminopterin (folic acid), isonicotinic acid (nicotinic acid), and dehydroascorbic acid (Vitamin C). *R. completa* male survival was negatively affected by canavanine and indole, the antimetabolites of arginine and tryptophane, respectively. Fecundity was significantly reduced by a-picolinic acid (nicotinic acid), oxythiamine (thiamine), desoxypyridoxine (Vitamin B6), aminoacetic acid (alanine), L-canavanine (arginine), dl-methionine (glutamic acid), indole (tryptophane), and dehydrocholesterol (cholesterol).

## The Importance of Microflora
## Associated with Fruit Flies

It seems that every insect, like higher animals and man, is associated with a number of microorganisms that play various roles in its nutrition. However, determination of the normal microflora of an insect and its role in the insect's economy is a very difficult task. Many parameters influence the system (insect-microbes-host plants), such as insect life stages, geographical and ecological locations, seasonal variability, microbial mutations, varieties of plant hosts, and microbial culturing techniques.

One of the main difficulties in making accurate nutritional assessments in most insects arises from the presence of various sorts of microorganisms, intracellular and extracellular. Microorganisms may not only play a part in the digestion of particular types of food but may also serve as a source of raw materials (amino acids and vitamins, especially) by virtue of their ability to synthesize a much greater variety of organic molecules than their hosts. Eventually, these molecules become available to the host and disintegrate within the host's alimentary canal. Therefore, it may be possible for a fruit fly to sustain normal life, and even reproduction, on a diet which in reality is deficient in certain nutrients.

The first observations on symbiosis of insects with microorganisms were made much earlier than research on insect nutrition (Buchner 1965). Robert Hooke (1665), using a primitive "microscope," described in his *Micrographia* the symbiotic organ of the human louse as its liver. His contemporary, Jay Swammerdam (1669), also saw this lemon-yellow organ, which glistens through the abdominal integument of *Pediculus* and sketched it as a stomach gland firmly fused with gut and surrounded by tracheae. Petri (1904-1907), one of the pioneers in symbiosis, described the first symbiotic association in Diptera, that of the olive fly, *Dacus oleae*, with the olive knot bacterium, *Pseudomonas savastanoi*. Stammer (1929) examined 37 species of fruit flies, representatives of all subfamilies, and found symbiotic bacteria in all of them. He described the types of symbiotic localization and transmission devices and classified the examined species according to the degree of perfection of these devices. Members of the Dacinae hold the first rank while the Trypetinae hold the lowest. Hellmuth (1956) examined 43 species of fruit flies and found symbiotic bacteria in all of them. She recognized various degrees of coevolution of host and symbiote, the most advanced being in the genus *Dacus* of the Dacinae subfamily.

Hagen et al. (1963) and Hagen (1966), working with the olive fly in Greece, found that adult flies fed artificial diets consisting of fructose, enzymatic protein hydrolysate of brewers; yeast, choline chloride, and water produced normal progeny; but, when he added streptomycin to the diets, the flies' progeny failed to develop. The eggs hatched in olive fruits but the larvae died during the 1st instar. Streptomycin had an adverse effect on the symbiote rather than directly on the insect. Further, the bacteria may play a role in protein hydrolysis in the olive flesh, making proteins available to the young larvae. Moreover, analysis of the olive fruit tissues, by one and two dimensional paper chromatography, revealed the absence of the two essential amino acids, methionine and threonine. Thus, the symbiotic bacteria besides affecting protein breakdown, may synthesize amino acids.

In a comparative study of normal and aposymbiotic *D. oleae* flies, (Tsiropoulos 1989), it was possible to determine the biosynthetic activity of microflora by the use of antibiotics and radiolabeled food (Glucose-U-14C). Four amino acids (alanine, hydroxyproline, proline, and tyrosine), which according to previous studies (Tsiropoulos 1984) had been characterized dispensable, proved to be products of the microflora's biosynthetic ability. Elimination of the walnut husk fly's microflora (Tsiropoulos 1981), adversely affected the fly's survival, when antibiotics were added to three incomplete diets. Also, antibiotics added to complete diets had a negative effect on reproduction. Fytizas and Tzanakakis (1966) confirmed Hagen's findings and showed that the degree of larval growth inhibition depends on the condition of the olive fruit, since an increase in quantity and quality of amino acids occurs with fruit maturation.

Yamvrias et al. (1970), contrary to findings of the other researchers, reported that they did not identify *Pseudomonas savastanoi* among the 37 bacterial species isolated from cephalic crypts of olive flies. They also stated that the chances of finding *P. savastanoi* as a symbiotic bacterium of *D. oleae* are very remote.

Tsiropoulos (1983b), studying wild and laboratory mass-produced flies, found that the microflora composition of the later was more constant. Considerable variation of the microflora composition occurred in wild flies throughout the year. Moreover, a total of 16 different bacteria were associated with the wild flies, compared to only 8 in the laboratory flies. In *R. pomonella,* Dean (1933, 1935) noted anatomical modifications in the adult apple maggot similar to those described by Petri for the olive fly, but he did not try to explain their functions, because he probably was not aware of Petri's work done in Europe. More than thirty years later, Baerwald and Boush (1968) demonstrated the symbiotic relationship of the apple maggot and the apple rot bacterium using the fluorescence antibody technique for the first time in a study of an insect bacterium symbiotic relation. They found that the bacterium is present in all the insect's stages except the egg.

In a more recent study (Rossiter et al. 1982) symbiotic bacteria of *R. pomonella* are suspected to play a role in host race formation and speciation.

Boush and Matsumura (1967) presented another concept related to symbiosis, insecticide degradation by the bacterial symbiote. The apple rot bacterium proved particularly active in degrading organophosphates, such as dichlorvos and DFP. Using insecticides labelled with radioactive $^{14}C$ or $^{3}H$ and autoradiography, they characterized a powerful degradation system in the bacterium; the main pathway of degradation appears to be hydrolytic, through strong esterases. The ecophysiological meaning of such a degradation property of the bacterium may be that such enzymes are needed to sustain a balanced host-microbe relationship by metabolizing certain naturally occurring allelochemicals and thereby providing nutrients to their host. These esterases also could possibly eliminate certain unwanted components of the ingested food through hydrolysis.

Miyazaki et al. (1968) reported that symbiotic bacteria of the apple maggot are capable of synthesizing 18 amino acids, three of which (cystine, methionine, and hydroxyproline) were not found in analyzed apple tissues. They suggested that these three amino acids might be necessary for the fly, but no chemically defined diet was available to test this hypothesis.

## Feeding Behavior, Dietary Requirements, and Fruit Fly Population Management

Knowledge of the requirements of a species for specific metabolites, i.e., amino acids or vitamins, can be used to adversely affect its metabolism by using antimetabolites. Tsiropoulos (1985) and Tsiropoulos and Hagen (1987), based on nutritional information regarding *D. oleae* and *R. completa*, were able to control survival and reproduction of the species by creating detrimental nutritional deficiencies.

Modern fruit fly management technology is increasingly dependent on food attractants, singly or in combination with sex attractants (Jones 1986). Fruit fly food attractants are used to detect invasions in new areas, monitor population fluctuations and time other methods of control, or effect control by bait sprays (Steiner et al. 1961, Chambers 1977). Food attractants of one kind or another have been used for the control of almost every pest species, and the history of fruit fly attractants is quite long. Berlese, an Italian entomologist, published a paper in 1905 on the control of *C. capitata* and other trypetids by food attractants. At about the same time, the poison-bait concept was used for control of the medfly in Hawaii by Maush (1910) and in S. Africa by Malley (1912). A few years later, Howlet (1912) reported on the effect of citronella oil on *D. dorsalis* and *D. zonatus*. The main categories of substances use were various molasses, wheat bran, various ammonium salts, McPhail's mixture of sucrose plus brewers yeast, and a great variety of protein hydrolysates. To date, many fruit fly attractants have been used (Beroza & Green 1963, Chambers 1977), and all of them duplicate the natural foods of Tephritids (e.g., honeydews and nectars).

However, food attractants produce variable results because of variation in the amount, distribution, and types of the natural foods with which they compete. Also, since the amount of movement of a fruit fly population is influenced by the natural food available, the degree of fruit fly control depends on the proportion of the population's habitat treated with the bait.

Any substance, to be an effective bait, should function on a dual basis. First, as an *olfactory attractant*, it must guide the target population to the insecticide treated area or bring the flies inside or onto the traps. Since attraction involved odor perception of the various gases (ammonia mainly) produced by attractant degradation, the role of the decomposing bacteria is of prime importance. Gow (1954), using antibiotics to control bacterial activity, showed that the attractants in the proteinaceous baits were produced by microbiological action. Only a few years ago, Bateman and Morton (1981) showed that a progressive increase in the proportion of flies attracted to traps with a fruit fly attractant occurs when no microorganism inhibitor or preservative (Boric acid-sodium azide solution) was added. The ammonia evolution rate in such traps increased dramatically, from 0.14 on day 1, to 63.24 lh[1]/100 ml on day 8. Morton and Bateman (1981) and Tsiropoulos and Zervas (1985) found that tephritid capture in McPhail traps was increased substantially by the synergistic effect of ammonia attractancy to phagostimulation provided by amino acids. Second, it must function as an arrestant *and* phagostimulant causing flies to ingest the bait. Thus, both smell and taste are involved, and both determine the effectiveness of the bait. If a bait is missing the specific phagostimulant(s), it acts as an attractant only and may not cause the flies to contact the insecticide. On the other hand, if the phagostimulant(s) are present, the insecticide will be ingested along with the bait. The insecticide will act as both a contact and stomach poison.

Contact taste receptors involved in adult feeding behavior have been studied for *R. cingulata* (Frings & Frings 1955), *C. capitata* (Gothilf et al. 1971), and *D. oleae* (Angioy et al. 1978). Carbohydrates, being present in most fruit fly baits, constitute effective phagostimulants. Sucrose has been found highly phagostimulatory in *A. suspensa* (Sharp & Chambers 1984), *C. capitata* (Galun et al. 1981, 1985), and *D. oleae* (Tsiropoulos 1980c). Amino acids, present in all kinds of hydrolysates used as fruit fly baits, have also been found phagostimulatory. Thus, *A. suspensa* adults arrested and aggregated more on arginine, glutamine, and phenylalanine than on other amino acids (Sharp & Chambers 1983). *C. capitata* was phagostimulated by phenylalanine, glutamic acid, arginine, glycine, serine, and methionine and *D. oleae* by arginine and tyrosine. Males were also strongly responsive to aspartic acid, cystine, glycine, lysine, and threonine and females to isoleucine, leucine, methionine, tryptophane, and valine (Tsiropoulos 1984). Certain mineral salts have been found phagostimulatory to *C. capitata* (Gothilf et al. 1971) and *D. oleae* (Angioy et al. 1978). Contrary to that, certain molecules of carbohydrates amino acids and salts exert a phagoinhibitory effect. To be effective and specific for a certain fruit fly species, a bait should include all the known phagostimulatory molecules and at the same time lack all known phagoinhibitory ones.

Priority should be given to improve efficiency, i.e, attraction and phagostimulation and selectivity, since a great array of insect species, including many beneficials and biological control agents, are attracted and killed in bait sprays or mass-trapping applications.

Rachel Carson, in the last chapter of her book *The Other Road* said, "The use of food attractants for insect pest control is one of the most promising, imaginative, and creative approaches to the problem of sharing our earth with other creatures." Since then, our knowledge on feeding and dietary requirements, as well as on the development and use of food attractants for several key pests, has increased tremendously. However, more research on attractants is needed to drastically reduce the huge amount of insecticides used for fruit fly control.

# References

Agee, H. R. & M. L. Parks. 1975. Use of the electroretinogram to measure the quality of vision of the fruit fly. Environ. Letters 10: 171-6.

Angioy, A. M., A. Liscia & P. Pertra. 1978. The electrophysiological response of labellar and tarsal hairs of *Dacus oleae* to salt and sugar stimulation. Boll. Soc. Ital. Biol. Sper. 54: 2115-21.

Back, E. A. & C. E. Pemberton. 1917. The melon fly in Hawaii. U.S. Dept. Agr. Bull. No 491.

Baerwald, R. J. & G. M. Boush. 1968. Demonstration of the bacterial symbiote *Pseudomonas melophthora* in the apple maggot, *Rhagoletis pomonella*, by fluorescent-antibody technique. J. Invert. Path. 11: 251-9.

Barnes, B. N. 1976. Mass rearing the Natal fruit fly by *Pterandrus rosa* (Ksh.) (Diptera:Trypetidae). J. Entomol. Soc. of S. Africa. 39: 121-4.

Barnes, M. M. 1959. Radiotracer labeling of a natural tephritid population and flight range of the walnut husk fly. Ann. Entomol. Soc. Amer. 52: 90-2.

Barnes, M. M. & H. T. Osborn. 1958. Attractants for the walnut husk fly. J. Entomol. 51: 686-98.

Bateman, M. A. 1972. The ecology of fruit flies. Ann. Rev. Entomol. 17: 493-518.

Bateman, M. A. & T. C. Morton. 1981. The importance of ammonia in proteinaceous attractants for fruit flies (Family Tephritidae). Aust. J. Agric. Res. 32: 883-903.

Berleze, A. 1905. Probabile method de lotta efficace contro la *Ceratitis capitata, Rhagoletis cersai* ed altri Tripetidi. Redia 3: 386-8.

Bernays, E. A. & S. J. Simpson. 1982. Control of food intake. Adv. Insect Physiol. 16: 59-118.

Beroza, M. & N. Green. 1963. Materials tested as insect attractants. USDA Agr. Handbook No. 239.

Bess, H. A. & F. H. Haramoto. 1961. Contributions to the biology and ecology of the oriental fruit fly, *Dacus dorsalis*. Hawaii. Agr. Exp. Sta. Tech. Bull. 44.

Boeckh, J. 1980. Ways of nervous coding of chemosensory quality at the input level, pp. 113-122. *In* H. van der Starre [ed.], Alfaction and taste VII. IRL Press, London.

Boush, G. M. & F. Matsumura. 1967. Insecticidal degradation by *Pseudomonas melophthora*, the bacterial symbiote of the apple maggot. J. Econ. Entomol. 60: 918-20.

Boush, G. M., R. J. Baerwald & S. Miyazaki. 1969. Development of a chemically defined diet for adults of the apple maggot based on amino acid analysis of honeydew. Ann. Entomol. Soc. Amer. 62: 19-21.

Boyce, A. M. 1934. Bionomics of the walnut husk fly, *Rhagoletis completa*. Hilgardia 8: 363-79.

Buchner, P. 1965. Endosymbiosis of animals with microorganisms. Interscience.

Burditt, A. K., D. F. Lopez, L. F. Steiner, D. L. Windeguth, R. Baranowski & M. Anwar. 1974. Application of sterilization techniques to *Anastepha suspensa* in Florida, pp. 93-101. *In* Sterility principles for insect control. Proc. Panel organized by FAO/IAEA.

Callahan, P. S. 1975. Insect antennae with special reference to the mechanism of scent detection and the evolution of the sensilla. Intern. J. Insect Morphol. and Embryol. 4: 381-430.

Cavalloro, R. 1967. Orientamenti sull allevanmento permenente di *Dacus oleae* (Diptera:Trypetidae) in laboratorio. Redia 50: 337-344.

Chambers, D. L. 1977. Attractants for the fruit fly survey and control, pp. 327-344. *In* Chemical control of insect behavior: Theory and application. Wiley, New York.

Christenson, L. D. & R. H. Foote. 1960. Biology of fruit flies. Ann. Rev. Entomol. 5: 171-92.

Cirio, U. & G. Vita. 1980. Fruit fly control by chemical attractants and repellents. Boll. Lab. Entomol. Agri. "Fillippo Sylbestri," Portici 37: 127-39.

Courtice, A. C. & R. A. I. Drew. 1984. Bacterial regulation of abundance in tropical fruit flies. Aust. Zool. 21: 251-68.

Dadd, R. H. 1970. Arthropod nutrition. Chem. Zool. 5: 35-95.

Dadd, R. H. 1973. Insect nutrition: Current developments and metabolic implications. Ann. Rev. Entomol. 18: 381-420.

Dadd, R. H. 1977. Qualitatiave requirements and utilization of nutrients: Insects, pp.305-346. *In* Handbook series in nutrition and food, section D, vol. 1. CRC Press, Cleveland.

Dadd, R. H. 1985. Nutrition: Organismus, pp. 313-390. *In* Comprehensive insect physiology, biochemistry and pharmacology, vol. 4. Pergamon, Oxford.

Dawkins, R. & M. Dawkins. 1973. Decisions and the uncertainity of behaviour. Behaviour 45: 83-103.

Dean, R. W. 1933. Morphology of the digestive tract of the apple maggot fly, *Rhagoletis pomonella*. N.Y. St. Agr. Exp. Sta. Bull. 215.

Dean, R. W. 1935. Anatomy and postpupal development of the female reproductive system in the apple maggot, *Rhagoletis pomonella*. N.Y. St. Agr. Exp. Sta. Bull. 229.

Dean, R. W. 1938. Experiments raising apple maggot adults. J. Econ. Entomol. 31: 241-244.

Delcourt, A. & E. Guyenot. 1910. De la posibilite d' etudier certains Dipteres en milieu defini (Drosophila). C. R. Acad. Sc. Paris T. 151: 255-257.

Dethier, V. G. 1976. The hungry fly. Harvard Univ. Press, Boston.

Drew, R. A. I. & G. H. S. Hooper. 1981. The responses of fruit fly species (Diptera:Tephritidae) in Australia to various attractants. J. Aust. Ent. Soc. 20: 201-205.

Drew, R. A. I., A. C. Courtice & D. S. Teakle. 1983. Bacteria as a natural source of food for adult fruit flies (Diptera: Tephritidae). Oecologia 60: 279-284.

Drongelen, W. van, A. Holley & K. B. Doving. 1978. Convergence in the olfactory system: Quantitative aspects of odor sensitivity. J. Theor. Biol. 71: 39-48.

Economopoulos, A. P. & M. E. Tzanakakis. 1967. Egg yolk and olive juice supplements to the yeast hydrolysate-sucrose diet for adults of *Dacus oleae*. Life Sci. 6: 2409-2416.

Economopoulos, A. P., G. Haniotakis, S. Michelakis, G. Tsiropoulos, G. Zervas, I. Tsitsipis, A. Manoukas & A. Kiritsakis. 1982. Population studies with the olive fruit fly in western Crete. Z. ang. Entomol. 93: 463-476.

Fletcher, B. S., A. Bateman, N. K. Hart & J. A. Lamberton. 1975. Identification of a fruit fly attractant in an Australian plant, *Zieria smithii*, as omethyl-eugenol. J. Econ. Entomol. 68: 815-816.

Fluke, C. L. & T. C. Allen. 1931. The role of yeast in life history studies of the apple maggot, *Rhagoletis pomonella*. J. Econ. Entomol. 24: 77-80.

Frings, H. & M. Frings. 1955. The location of the contact chemoreceptors of the cherry fruit fly, *Rhagoletis cingulata* and the flesh-fly, *Sarcophaga bullata*. Am. Midl. Nat. 53: 431-435.

Fuchs, H. 1974. Intensity discrimination between sugar solutions by the Mediterranean fruit fly. M.S. thesis, Bar-Ilan University, Ramat-Gan, Israel.

Fytizas, E. & M. E. Tzanakakis. 1966. Some effects of streptomycin, when added to the adult food on the adults of *Dacus oleae* and their progeny. Ann. Entomol. Soc. Amer. 59: 269-273.

Galun, R., S. Gothilf & S. Blondheim. 1978a. The phago-stimulatory effects of protein hydrolysates and their role in the control of Medflies. FAO meeting in Sardenia, May 1978.

Galun, R., Y. Nitzan, S. Blondheim & S. Gothilf. 1978b. Responses of the Mediterranean fruit fly, *Ceratitis capitata*, to amino acids. pp. 55-63. XVI Int. Congress of Entomology, Kyoto and Naha, Japan.

Galun, R., S. Gothilf, S. Blondheim & A. Lachman. 1981. Protein and sugar hunger in the Mediterranean fruit fly, *Ceratitis capitata*, pp. 245-251. *In* Determination of Behavior by Chemical Stimuli, Proc. 5th European Chemorecption Res. Organizations Symposium.

Galun, R., S. Gothilf, S. Blondheim, J. L. Mazor & A. Lachman. 1985. Comparison of aggregation and feeding responses by normal and irradiated fruit flies, *Ceratitis capitata* and *Anastrepha suspensa* (Diptera:Tephritidae). Environ. Entomol. 14: 726-732.

Gelperin. A. 1971. Regulation of feeding. Ann. Rev. Entomol. 16: 365-378.

Gothilf, S., R. Galun & M. Bar-Zeev. 1971. Taste reception in the Mediterranean fruit fly: Electrophysiological and behavioral studies. J. Insect Physiol. 17: 1371-1384.

Gow, P. A. 1954. Proteinaceous bait for the oriental fruit fly. J. Econ. Entomol. 47: 153-160.

Guerrin, P. M., V. Remund, E. F. Boller, B. Katsoyanos & G. Delrio. 1982. Fruit fly electroantenogram and behavior responses to some generally occuring fruit volatiles, pp. 248-251. CEC/IOBC Symposium, Athens, Nov. 1982.

Guyenot, E. 1917. Researches experimentales sur la vie aseptique et le development d'un organism en fonction deu milieu. Bull. Biol. Fr. Belg. 51: 1-330.

Hagen, K. S. 1952. Influence of adult nutrition upon fecundity, fertility and longevity of three tephritid species. Ph.D. dissertation, Univ. Calif., Berkeley.

Hagen, K. S. 1958. Honeydew as an adult fruit fly diet affecting reproduction. Proc. In. Congr. Ent. 10: 25-30.

Hagen, K. S. 1966. Dependence of the olive fly, *Dacus oleae*, larvae on symbiosis with *Pseudomonas savastanoi* for the utilization of olive. Nature 209: 423-424.

Hagen, K. S., & G. L. Finney. 1950. A food supplement for effectively increasing the fecundity of certain tephritid species. J. Econ. Entomol. 43: 735.

Hagen, K. S. & R. L. Tassan. 1972. Exploring nutritional roles of extra-cellular symbiotes on the reproduction of honeydew feeding adult Chrysopids and Tephritids, pp. 323-351. *In* J. G. Rodriquez [ed.], Insect and mite nutrition.

Hagen, K. S., L. Snatas & A. Tsecouras. 1963. A technique of culturing the olive fly, *Dacus oleae*, on synthetic media under xenic conditions. *In* Radiation and Radioisotpes Applied to Insects. Int. Atomic Energy Agency.

Haisch, A. 1968. Preliminary results in rearing the cherry fruit fly, *Rhagoletis cerasi* on a semi-synthetic medium. In Radiation, Radioisotopes and Rearing Methods in the Control of Insect Pests. Panel organized by FAO/IAEA Div. of Atomic Energy in Food and Agriculture.

Hellmuth, J. 1956. Untersuchungen zur Bakteriensymbiose der Trypetiden. Z. Morphol. Okol. Tiere 4.

Hodson, A. C. 1943. Lures attractive to the apple maggot. J. Econ. Entomol. 36: 545-548.

Hodson, A. C. 1948. Further studies of lures attractive to the apple maggot. J. Econ. Entomol. 41: 61-66.

Hooke, R. 1665. Micrographia. London.

House, H. L. 1972. Insect nutrition, pp. 513-573 *In* N. T. W. Fiennes [ed.],

Biology of Nutrition, Intern. Encyclopedia of Food and Nutrition, vol. 18. Pergamon Press, Oxford.

House, H. L. 1974. Nutrition, pp. 1-62. *In* M. Rockstein [ed.], The physiology of insects, vol. 5. Academic Press, New York.

Howlet, F. M. 1912. The effect of oil of citronella on two species of *Dacus*. Trans. Ent. Soc. London 60: 412-418.

Jarvis, H. 1931. Experiments with a new fruit fly lure. Queensland Agr. J. 36: 485-491.

Jones, D. T. 1986. The use of behaviour modifying chemicals in the integrated pest management of selected fruit fly species, pp.451-458. II. Intern. Symp. Fruit Flies/Crete, Sept. 1986.

Kawano, Y., W. C. Mitchell & H. Matsumoto. 1968. Identification of male oriental fruit fly attractant in the golden shower blossom. J. Econ. Entomol. 61: 986-988.

Malley, R. Cited in Christenson, L. D. 1958. Recent progress in the development of procedures for eradicating or controlling tropical fruit flies. Proc. 10th Int. Congr. Entomol. 3: 11-16.

Marsh, H. D. 1910. Report of the assistant entomologist, pp 152-159. Board of Commissioners Agric. and Forestry, Hawaii.

Matsumoto, B. & T. Nishida. 1962. Food preference and ovarian development of the melon fly, *Dacus cucurbitae*, as influenced by diet. Proc. Hawaii, Entomol. Soc. 28: 137-144.

McCollom G. B. & A. A. Rutkowski. 1986. Microbial leaf isolates associated with *Rhagoletis pomonella*, pp. 251-253. II Intern. Symp. Fruit Flies/Crete, Sept. 1986.

McFarland, D. J. 1977. Decision making in animals. Nature 269: 15-21.

McPhail, M. 1939. Protein lures for fruitflies. J. Econ. Entomol. 32: 758-761.

Metcalf, R. L., W. C. Mitchell, T. R. Fukuto & E. R. Metcalf. 1975. Attraction of the Oriental fruit fly, *Dacus dorsalis,* to methyl eugenol and related olfactory stimulants. Proc. Nat. Acad. Sci. (USA). 72: 2501-2505.

Metcalf, R. L., E. R. Metcalf, W. C. Mitchell & L. W. Y. Lee. 1979. Evolution of olfactory receptor in the Oriental fruit fly, *Dacus dorsalis*. Proc. Nat. Acad. Sci. (USA). 76: 1561-1565.

Metcalf, R. L., E. R. Metcalf & W. C. Mitchell. 1981. Molecular parameters and olfaction in the Oriental fruit fly, *Dacus dorsalis*. Proc. Nat. Acad. Sci. (USA). 78: 4007-4010.

Middlekauff, W. W. 1941. Some biological observation of the adults of the apple maggot and the cherry fruit flies. J. Econ. Entomol. 34: 621-624.

Mitchell, S., N. Tanaka & L. Steiner. 1965. Methods of mass culturing melon flies and oriental and Mediterranean fruit flies, pp. 33-104. U.S. Dept. Agr. Res. Serv.

Miyazaki, S., G. M. Boush & R. J. Baerwald. 1968. Amino acid synthesis by *Pseudomonas melophthora* bacterial symbiote of *Rhagoletis pomonella*. J. Insect Physiol. 14: 513-518.

Moericke, V., R. J. Prokopy, S. Berlocher & G. L. Bush. 1975. Visual stimuli eliciting attraction of *Rhagoletis pomonella* flies to trees. Ent. Exp. Appl. 18: 497-507.

Monro, J. 1966. Population flushing with sexually sterile insects. Science 151: 1563-1568.

Monro, J. & A. W. Osborn. 1967. The use of sterile males to control populations of Queensland fruit fly, *Dacus tryoni* I. Methods of mass rearing, transporting, irradiating and relaeasing sterile flies. Aust. J. Zool. 15: 461-473.

Moore, I. 1962. Further investigation on the artificial breeding of the olive fly, *Dacus oleae*, under aseptic conditions. Entomophaga 7: 53-57.

Morton, T. C. & M. A. Bateman. 1981. Chemical studies of proteinaceous attractants for fruit flies, including the identification of volatile constituents. Austr. J. Agr. Res. 32: 905-916.

Nadel, D. J. 1970. Current mass-rearing techniques for the Mediterranean fruit fly. *In* Sterile Male Technique for Control of Fruit Flies. Proc. Panel Organized by FAO/IAEA.

Neilson, W. T. A. & J. W. McAllan. 1965. Artificial diets for the apple maggot, III. Improved, defined diets. J. Econ. Entomol. 58: 542-543.

Neilson, W. T. A. & A. Wood. 1966. Natural source of food of the apple maggot. J. Econ. Entomol 59: 997-998.

Neuenschwander, P. 1982. Beneficial insects caught by yellow traps in mass trapping of olive fly, *Dacus oleae*. Ent. Exp. Appl. 32: 286-296.

Nishida, T. 1963. Zoogeographical and ecological studies of *Dacus cucurbitae* in India. Hawaii Agr. Exp. Sta. Tech. Bull. 54.

Nishida, T. & H. A. Bess. 1957. Studies on the ecology and control of the melon fly, *Dacus cucurbitae*. Hawaii Agr. Exp. Sta. Tech. Bull 34.

Novak, J. A. & B. A. Foote. 1980. Biology and immature stages of fruit flies: The genus Eurosta (Diptera:Tephritidae). J. Kansas Entomol. Soc. 53: 205-219.

Petri, L. 1904-7. Sopra la particolare localizzazione diuna colonia batterica nel tubo digerente della larva della mosca olearia. Atti reale Accad. Linnei (5), CL. Sci. fis mat. e nat. 13(1904); also ibid 14 (1905); 15 (1906); 16 (1907).

Prokopy, R. J. 1972. Response of apple maggot flies to rectangles of different colors and shades. Environ. Entomol. 1: 720-726.

Prokopy, R. J. 1976. Feeding, mating and oviposition activities of *Rhagoletis fausta* flies in nature. Ann. Entomol. Soc. Amer. 69: 899-904.

Prokopy, R. J. 1977. Stimuli influencing trophic relations in Tephritidae. Collog. Int. CNRS 265: 305-336.

Prokopy, R. J. 1982. Tephritid relationships with plants, pp. 230-239. CEC/IOBC Symposium, Athens, Nov. 1982.

Prokopy, R. J. & E. F. Boller. 1970. Artificial egging system for the European cherry fruit fly. J. Econ. Entomol. 63: 1413-1417.

Prokopy, R. J. & E. F. Boller. 1971. Response of European cherry fruit flies to colored rectangles. J. Econ. Entomol. 64: 1444-1447.

Prokopy, R. J. & G. E. Haniotakis. 1975. Responses of wild and lab-cultured *Dacus oleae* flies to host plant color. Ann. Entomol. Soc. Amer. 68: 73.

Prokopy, R. J. & A. P. Economopoulos. 1976. Attraction of laboratory cultures and wild *Dacus oleae* flies to sticky-coated McPhail traps of different colors and odors. Environ. Entomol. 4: 187-192.

Prokopy, R. J. & B. D. Roitberg. 1984. Foraging behaviors of true fruit flies. Am. Scientist 72: 41-49.

Prokopy, R. J., A. P. Economopoulos & M. W. McFadden. 1975. Attraction of wild and lab-cultured *Dacus oleae* flies to small rectangles of different hues, shades, and tints. Ent. Exp. 18: 141-152.

Remund, U. 1971. Anwendungsmoglichkeiten einer wirksamen visuellen Wegwerffalle fur die Kirschenfliege *(Rhagoletis cerasi)*. Schweiz Z. bsst. Weinban 107: 196-205.

Rhode, R. H. & L. M. Spishakoff. 1965. Technicas usadas en la cultivo de *Anastrepha ludens* II. Memorias del dia Parasitologica. Departo de Parasitologia. Esculoa Nac. Agr. Chapingo (1964): 23-28.

Rice, M. J. 1973. Cibarial sense organs of the blowfly, *Calliphora erythrocephala*. Int. Insect Morph. Embryol. 2: 109-116.

Ripley, L. B., G. A. Hepburn & E. E. Andersen. 1940. Fruit fly migration in the kat river valley. S. Afr. Dept. Agr. Forestry Plant Ind. Ser. 49.

Rossiter, M. C., D. J. Howard & G. L. Bush. 1982. Symbiotic bacteria of *Rhagoletis pomonella*, pp. 77-84. CED/IOBC symposium, Athens, Nov. 1982.

Russ, K., E. Boller, V. Vallo, A. Haisch & S. Sezer. 1973. Development and application of visual traps for monitoring and control of populations of *Rhagoletis cerasi*. Entomophaga 18: 103-116.

Sanders, W. 1968. Die Eiablagehandlung der Mitelmeerfruchfliege *Ceratitis capitata*. Ihre Abhangigkeit von Farbe und Gliederung des Umfeldes. Z. Tierpsychol. 25: 588-607.

Schultz, J., P. S. Lawrence & D. Newmeyer. 1946. A chemically defined medium for the growth of *Drosophila melanogaster*. Anat. Record 96: 540.

Shah, A. H. & R. C. Ratel. 1976. Role of Tulsi plant *(Ocimum santum)* in control of mango fruit fly, *Dacus correctus*. Current Science 45: 313-314.

Sharp, J. L. & D. L. Chambers. 1983. Aggregation response of *Anastrepha suspensa* to proteins and amino acids. Environ. Entomol. 12: 923-928.

Sharp, J. L. & D. L. Chambers. 1984. Consumption of carbohydrates, proteins and amino acids by *Anastrepha suspensa* in the laboratory. Environ. Entomol. 1: 768-773.

Sigwalt, B., F. Soria, A. Yana & C. Baldy. 1968. Deplacement diurne apparrent d' une population de *Ceratitis capitata* sur une parcelle d' agrumes. Ann. Epiphyt. 19: 169-171.

Silvestry, F. 1914. Expedition to Africa in search of the natural enemies of fruit flies. Board of Agric. and Forestry, Hawaii, Bull No. 3.

Singh, P. 1977. Artificial diets for insects, mites, and spiders. Plenum, New York.

Stammer, H. J. 1929. Die Bakteriensymbiose der Trypetiden (Diptera) Z. Morphol. Okol. Tiere 15: 481-523.

Steiner, L. F. & S. Mitchell. 1966. Tephritid fruit flies, pp. 555-583. *In* C. N. Smith [ed.], Insect colonization and mass production. Academic Press, New York.

Steiner, L. F., G. G. Rohwer, E. L. Ayers & L. D. Christensen. 1961. The role of attractants in the recent Mediterranean fruit fly eradication program in Florida. J. Econ. Entomol. 54: 3-5.

Stenersen, J. H. V. 1965. DDT-metabolism in resistant and susceptible stable-flies and in bacteria. Nature. 207: 660-661.

Swammerdam, J. 1669. Algemeene Verhandeling von bloedloose Diertjens. Utrecht. 1669.

Syed, R. A., M. A. Ghami & M. Murtaza. 1970. Studies on the trypetids and their natural enemies in West Pakistan. Pakistan Comm. Inst. Biol. Control, Tech. Bull. 13: 17-30.

Tagushi, T. 1966. Influence of different diets upon the longevity of the adult oriental fruit fly, *Dacus dorsalis*. Res. Bull. Plant Protection Agency, Japan, 20: 16-19.

Tanaka, N., R. Okamot & D. L. Chambers. 1970. Methods of mass rearing the Mediterranean fruit fly currently used by the USDA, pp. 19-23. *In* Sterile Male Technique for control of Fruit Flies. Proc. Panel Organized by FAO/IAEA.

Tsiropoulos, G. J. 1974. Ecophysiology of the adult reproduction of the walnut husk fly, *Rhagoletis completa*. Ph.D. dissertation, Univ. of California, Berkeley.

Tsiropoulos, G. J. 1977a. Reproduction and survival of the adult *Dacus oleae* feeding on pollens and honeydews. Environ. Entomol. 6: 390.

Tsiropoulos, G. J. 1977b. Survival and reproduction of *Dacus oleae* fed on chemically defined diets. Z. ang. Entomol. 84: 192-197.

Tsiropoulos, G. J. 1978. Holidic diets and nutritional requirements for survival and reproduction of the walnut husk fly. J. Insect Physiiol. 24: 239-242.

Tsiropoulos, G. J. 1980a. Major nutritional requirements of adult *Dacus oleae* Ann. Entomol. Soc. Amer. 73: 251-253.

Tsiropoulos, G. J. 1980b. The importance of vitamins in adult *Dacus oleae* nutrition. Ann. Entomol. Soc. Amer. 73: 705-707.

Tsiropoulos, G. J. 1980c. Carbohydrate utilization by normal and y-sterilized *Dacus oleae*. J. Insect Physiol. 26: 633-637.

Tsiropoulos, G. J. 1981. Effects of varying the dietary nitrogen to carbohydrate ratio upon the biological performance of adult *Dacus oleae*. Arch. Intern. Physiol. Bioch. 89: 101-105.

Tsiropoulos, G. J. 1982. Effects of vitamin overdosing on adult *Dacus oleae*, pp. 85-90. *In* Proc. Int. Symp. on Fruit Flies of Economic Importance CEC/IOBC, Athens, Nov. 1982.

Tsiropoulos, G. J. 1983a. The importance of dietary amino acids on the reproduction and longevity of adult *Dacus oleae*. Arch. Intern. Physiol. Bioch. 91: 159-164.

Tsiropoulos, G. J. 1983b. Microflora associated with wild and laboratroy reared adult olive fruit flies, *Dacus oleae*. Z. ang. Entomol. 96: 337-340.

Tsiropoulos, G. J. 1984. Effect of specific phagostimulants on adult *Dacus oleae* feeding behavior, pp. 95-98. *In* Proc. OILB ad hoc meeting on Fruit Flies of Economic Importance, Hamburg, W. Germany, Aug. 1984.

Tsiropoulos, G. J. 1985. Dietary administration of antivitamins affected the survival and reproduction of adult *Dacus oleae*. Z. ang. Entomol. 100: 35-39.

Tsiropoulos, G. J. 1989. Biosynthetic activity of the microflora associated with the olive fruit fly, *Dacus oleae*. Intern. Colloq. Microbiology in Poikilotherms, Paris, July 1989.

Tsiropoulos, G. J. & G. Zervas. 1985. Trapping *Ceratitis capitata* in McPhail traps baited with amino acids and ammonium salts, pp. 384-391. *In* Proc. Experts meeting in Acireale, Italy, March 1985, Integr. Pest Control in Citrus Groves.

Tsiropoulos, G. J. & K. S. Hagen. 1987. Effect of nutritional deficiencies produced by antimetabolites on the reproduction of *Rhagoletis completa*. Z. ang. Entomol. 103: 351-354.

Tsitsipis, J. A. 1975. Mass rearing of the olive fruit fly, *Dacus oleae* at "Demokritos," pp. 93-100. *In* Controlling Fruit Flies by the Sterile insect technique. Proc. Panel Organized by FAO/IAEA. STI/PUB/392.

Tsitsipis, J. A. 1987. Requirements. *In* Robinson & Hooper [eds.], Fruit flies: Their biology, natural enemies and control. WCR vol. 3A, Elsevier, Amsterdam.

Tsitsipis, J. A. & A. Kontos. 1983. Improved solid adult diet for the olive fruit fly, *Dacus oleae*. Entomol. Hellenica 1: 24-29.

Tzanakakis, M. E., J. A. Tsitsipis & L. F. Steiner. 1967. Egg production of the olive fruit fly fed solids or liquids containing protein hydrolyzate. J. Econ. Entomol. 60: 352-354.

Yamvrias, C., C. G. Panagopoulos & P. G. Psalidas. 1970. Preliminary study of the internal bacterial flora of the olive fruit fly. Annls. Inst. Phytopath. Beneki N.S. 9: 201-206.

# 8

## Flea Rearing in Vivo and in Vitro for Basic and Applied Research

*Nancy C. Hinkle, Philip G. Koehler,*
*and Richard S. Patterson*

### Introduction

The need for long-term cat flea colony maintenance resulted in a system whereby cat fleas can be sustained on host cats in the laboratory. This system uses six cats for a weekly production of over 10,000 adult fleas.

The artificial rearing technique uses a blood-warming system and cages to confine individual flea populations, permitting maintenance of separate strains (e.g., insecticide-resistant). Having caged fleas also permits determination of fecal and egg production, feeding, and mortality.

### Rearing Fleas on Host Cats

#### *Hosts for the Cat Flea Colony*

Adult cat fleas can be maintained on various mammals, but cats serve the purpose particularly well in that they are litter-trained and thus urine and feces do not contaminate the flea eggs.

#### *Egg Production and Larval Rearing*

Cats are confined in cages overnight so that flea eggs and feces fall from their coats into the pans beneath (Fig. 1). This material is collected on Mondays, Wednesdays, and Fridays. To separate the eggs and feces from the

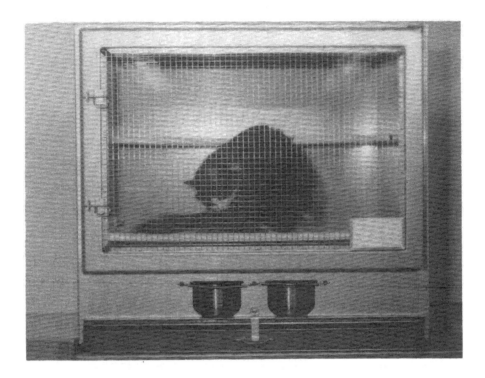

**Fig. 1. Cat cage with pan beneath to collect flea eggs.**

other debris, they are sieved through two screens (number 20 and 35). Cat hair, litter, food, and other large fragments collect in the number 20 sieve and are discarded, while smaller pieces of litter are retained by the number 35 sieve. The eggs may be counted at this point.

The eggs and feces that pass through the number 35 sieve are poured into a large plastic tray (56 x 44 x 8 cm) containing about 1.5 liter of white hobby sand. This sand is fine and of uniform size so that it can be easily sieved. The mixture of flea eggs and adult feces is sprinkled over the top of the sand and supplemented with about 40 ml of dried bovine blood (Fig. 2).

Various other flea larval diets have been described (Bruce 1948, Hudson & Prince 1958, Gilbert 1964), but results of developmental studies show that dried blood alone performs as well as any other diet, including those that incorporate bonemeal, dog chow, brewer's yeast, etc. (Moser 1989).

The dried blood is prepared by pouring citrated bovine blood (Bailey et al. 1975), obtained from the slaughterhouse, into a large metal tray to a depth of

**Fig. 2. Larval rearing tray with powdered blood on white sand.**

about 1 cm. This tray is placed in an oven at 50°C overnight to dry the blood. It is essential that the blood be dried slowly to prevent hardening and the breakdown of nutrients. Once the blood is dry, it is ground in a hammermill (Straub model 4E grinding mill, Straub Co., Philadelphia, PA 19020) and sieved through a number 80 sieve to produce uniformly sized particles. It is stored in the laboratory in a sealed container.

To maintain the trays of larvae at a higher humidity than the rest of the room, separate rearing chambers (80 x 79 x 202 cm) were constructed of plastic sheeting.   Silverman (1981) found optimal conditions for development of *Ctenocephalides felis* to be 27-32°C and 75-92% relative humidity.   Trays of saturated sodium chloride solution in these chambers maintain the humidity at 78%, while the temperature is the same as that of the larger room, 30°C.   This room is maintained on a 12:12 photoperiod throughout the year.   Care is taken in cleaning the rearing room as flea larvae are sensitive to vapors from commonly used cleaning products.   We vacuum, swab the floor with a damp mop, and clean the counter tops with a moistened paper towel.

## Pupal Collection

Larvae begin pupating 3 or 4 days after the eggs are added to the trays; most pupation occurs around the 7th day and ends within 2 weeks.   The contents of each tray are sieved through a number 14 sieve to collect the cocoons.   The remaining larvae are returned to the tray of sand, which is placed back in the rearing chamber.   Larval trays are sieved on Mondays, Wednesdays and Fridays, and each tray is discarded after 2 1/2 weeks.

After collection, the cocoons are retained in the same room, with 78% relative humidity.   Cocoons are placed in emergence containers (Fig. 3) (Cole et al. 1972), which are constructed of glass utility jars (25 x 23 cm diameter) and plastic cookie jars (20 x 20 cm diameter).   The glass utility jar serves as the bottom of the device into which the adult fleas emerge from the cocoon.   The bottom of the plastic cookie jar is cut out and three plastic supports are glued to the inside to hold a stage made like a basket with a screen bottom.   The cocoons are poured onto the screen, the stage is placed in the device, and the lid is placed on top so that emerging fleas cannot hop out but instead fall into the jar beneath.

Wade & Georgi (1988) found that cats were able to remove about 50% of their flea burden in 1 week, so once a week about 100 fleas (approximately 10 days old) from the emergence cage are placed on each animal to ensure infestation levels for providing ample flea egg production (Hudson & Prince 1958).   Female fleas emerge from the cocoons first, so at least 1-week-old fleas are used to ensure a good sex ratio for infesting the cats (Silverman 1981).   Fleas are aspirated from the holding jar and placed directly on the animals.   If egg production declines, additional fleas may be added.

We estimate that our system maintains an average of about 125 fleas per animal.   From each animal we obtain an average of 569 eggs per day.   These produce about 465 cocoons per cat per day (82% successfully develop from egg to pupa), yielding about 15,000 newly emerged adults each week for reinfestation of the animals and testing.

**Fig. 3. Adult emergence container.**

## Introducing a New Animal into the Colony

When a new cat is used for raising fleas, it should be held in isolation until checked for feline leukemia, worms, ear mites, and other ectoparasites. The animal may be de-fleaed using a chemical registered for on-animal use that has a short residual efficacy, such as pyrethrins. Animals should be vaccinated annually for leukemia, panleukopenia, and other diseases.

## Maintenance of Host

The appropriate animal care committee should be consulted regarding suitable housing, cages, and maintenance for the host cats (APHIS 1985). References on the use of cats as laboratory animals include *The Laboratory Cat* (Hurni & Rossbach 1976) and publications of the U.S. Public Health Service (1985). We use large (74 x 66 x 77 cm) stainless steel cages (Unifab Corporation, Kalamazoo, Mich.) with a grated floor (Fig. 1). The cage has two polyethylene resting boards upon which the cat may perch, while the litter pan is placed on the floor. Secure holders for food and water bowls are constructed in the door of the cage to prevent their being tipped over. Beneath the floor grate is a stainless steel pan, equal in length and width to the cage itself (84 x 66 x 6 cm). Flea eggs collect in the pan and are separated from the other debris by sieving.

Environmental conditions in the cat room are important for optimal survival of the ectoparasites. We try to maintain above 60% relative humidity at a temperature of 26°C. As our cats (and thus our fleas) are subjected to ambient lighting, it is important to remember the effect of photoperiod on circadian rhythms. Also, crashes in populations may occur due to changing seasons.

Cats display various degrees of grooming competence. Some are able to rapidly remove a large proportion of the fleas that are put on them, while others are not so successful in their scratching and biting (Silverman 1981). After observation, some cats may be removed from the colony and replaced with others that tolerate higher populations of fleas and thus produce more flea eggs. If a cat displays symptoms of flea allergy dermatitis, it should be replaced. Use of steroids is not encouraged.

As with any parasite, the host has a profound effect on the flea, so maintenance of the host is an important aspect of successful rearing of ectoparasites (Nelson 1984). The host supplies the habitat of the ectoparasite, regulating such things as temperature, humidity, photoperiod, diet, and even hormonal condition in some situations (Prasda 1987). The physical condition of the host affects the vitality of the ectoparasite, and an undernourished or otherwise stressed host can adversely affect the population of ectoparasites.

As with any laboratory colony, disease management is an important factor. When the room holding the emergence containers was maintained at a higher humidity, there was a problem with fungus developing on the cocoons. Because of cat food crumbs contaminating the larval rearing trays, there have been some incidents of grain mites infesting the larval rearing chambers. But with the intimate association of the ectoparasites and their feline hosts, the main disease organism of concern is *Dipylidium caninum*, the double-pored tapeworm.

Periodically, an infestation will sweep the host animals, and proglottids will appear beneath the cat cages. Anthelminthics have always managed the infestation admirably.

Good hygiene is essential in the animal rearing room to ensure the health of the cat colony and thus a thriving flea colony. Infestation of the cat food with beetles can be handled by heating the food or by fumigation. We have done some in-house fumigation using carbon dioxide gas and found that, with sufficient exposure, the technique will eliminate all stages of the beetle. Fresh water and food are provided daily. Litter is changed three times a week. The floor is vacuumed and washed once a week and swept twice a week. Cages are taken out and sanitized every other week.

Our laboratory strain of cat fleas was captured in Gainesville in 1985 from a natural infestation in dogs which had been treated with a variety of insecticides. Without subsequent challenge in the lab, this colony has maintained high levels of insecticide resistance for over 4 years (Koehler et al., in press). As with any laboratory colony, there are some caveats inherent in making comparisons between insects in field populations and those that have been cultured for some time. The lab strain has been held at a much more consistent temperature and relative humidity. For 4 of the past 6 years, they were maintained on a 12:12 photoperiod throughout the year. The animals are now held in a building with natural lighting so that photoperiod changes seasonally. Because of the way in which we collect the cocoons, a portion of the population that pupates more rapidly has been selected. That is, any larva that does not pupate within 16 days of oviposition is discarded. This has very likely narrowed the genetic variation relative to larval development time. There are probably other, more subtle, factors that have been selected for as well.

## A System for
## in Vitro Rearing of Cat Fleas

Several procedures have been developed for rearing fleas in the lab (Cerwonka & Castilla 1958, Kartman 1954, Bar-Zeev & Sternberg 1962, Wade & Georgi 1988). Some of these rely on a living host, with various means of allowing the arthropod to feed (Hastriter et al. 1980); however, this entails maintenance of the host and can be very labor intensive. Consequently, Wade and Georgi (1988) developed an artificial feeding system for adult cat fleas that utilizes bovine blood warmed by a water jacket. We incorporated the advantages of the Cornell (Wade-Georgi) system while decreasing the cost, labor, and time involved in maintenance. It was also desirable to use sturdier materials than the glass Rutledge insect blood feeder to reduce breakage.

## Flea Cages

The device we are currently using has cages similar to those described by Wade and Georgi (1988). The bottom of a plastic pill bottle (5 cm diameter) is cut off (leaving about 3.2 cm) and the open end thus created is covered with nylon screen (300 mesh). A hole (3 cm diameter) is cut in the lid of the bottle and covered with screen (500 mesh). The bottle is then inverted for use so that the finer mesh (300) is on top and the fleas must insert their mouthparts through it to obtain their bloodmeal. The larger mesh on the bottom of the cage allows the eggs and feces to fall through and collect below (Fig. 4). To make it easier

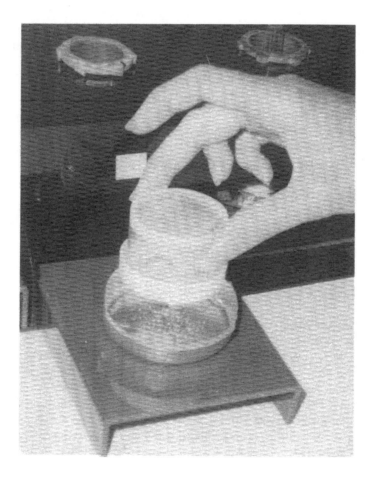

Fig. 4. Flea cage with eggs and feces in dish below.

for the fleas to remain near the feeding surface, a fine layer of clean cat hair is placed against the 300 mesh and supported by a circle of hardware cloth (4 cm diameter) on a hollow pedestal (a 3.0-cm-high plastic cylinder; Fig. 5). The hair provides the fleas a surface to grip while posing minimal obstruction to the eggs and feces.

### Blood Tubes

The tubes containing the blood are constructed of 15 dram plastic vials with a 0.5-cm hole drilled in the bottom. A short section (1 cm, 0.25-cm inner diameter) of plexiglas tubing is cemented to the hole, permitting injection of the blood and providing a handle. The open end of the vial is covered with Parafilm and about 10 ml of bovine blood is added through the plexiglas tubing (Fig. 6). Tubes are changed daily. The old blood is discarded, the tubes are washed and sterilized, new parafilm is stretched over the open end, and fresh blood is added.

**Fig. 5. Parts of flea cage: hardware cloth stage supported by hollow column.**

   Blood temperature is maintained by a copper collar that evenly distributes
the heat produced by the surrounding heat band cable.  The blood-containing
vial fits snugly into the copper collar, constructed from a washer and a copper
coupling joint.  Eight of these slots are contained in a plexiglas module (91 x 13
x 12.5 cm) incorporating the heat band cable and the copper collars, beneath
which are placed the flea cages and aluminum dishes for collecting the flea eggs
and feces.

**Fig. 6.  Inserting blood tube into warming collar.**

## Temperature Control

The copper collars that distribute the heat to the blood tubes are surrounded by an electrical heat band cable (TPI model TPT-6, W.W. Grainger, Inc. Chicago, IL 60648) that contains resistance wires housed in separately channeled, flexible vinyl insulation (rated 6 watts per linear foot). Performance of the system may be monitored by a clear remote pilot lamp that glows when the thermostat actuates the heat band.

Blood temperature is maintained at approximately 41°C by an Omega CN5000 digital temperature controller (Omega Engineering, Inc., Stamford, CT 06907), using a thermistor inserted into one of the wells. The Omega temperature controller has a 10-amp mechanical relay with proportional control, LED display of temperature and output status, and a push-to-engage set point knob. It is adaptable to either J, K, T, or E thermocouple types. The unit is calibrated initially by adjusting the set point until the appropriate temperature range is maintained. An up-scale break protection circuit prevents temperature overshoot.

## Larval Rearing

Eggs and feces are placed in trays augmented with powdered bovine blood and left undisturbed for 5 days. Then they are transferred to trays of white sand to provide a substrate for pupation. Trays are sieved and cocoons collected every other day for 2 weeks, after which the contents of the tray are discarded. This process, continued over several months, has selected for fleas that pupate within 2 weeks, so that there are very few larvae left in the trays when they are discarded.

## Restocking Cages

Cocoons are placed in an emergence container similar to that described by Cole et al. (1972). A screen stage is placed in the jar to hold the cocoons; when the adult fleas emerge they fall through the screen and collect in the utility jar below. Fleas are aspirated from the emergence vessel and poured into the plastic (3.2 x 5 cm diameter) cage.

## Production on the Artificial System

Five generations were run on the artificial system with two or three cages of each generation operational at any given time. The first filial generation was not monitored but the second produced 12,063 pupae, the third 6,226, the fourth

7,200, and the fifth 2,700. Average length of life on the caged fleas averaged 2 days (range 1-86 days) and small groups of caged fleas averaged two eggs per day per female over 1 week (range 0-9.6). On the animal, confined in microcells (Osbrink & Rust 1984), fleas had a mean longevity of 11.2 days for females and 7.2 days for males. The average egg production was 13.5 per female per day. Hudson & Prince (1958) estimated that the average on-animal production was 28 eggs per female per day, and Smit (1973) concurred with an estimation of 25 eggs per female per day.

While production of the in vitro system does not approach that of on-animal rearing, it does provide an alternative to the use of laboratory animals. With the continuing social and political pressures to reduce use of vertebrates in research, such systems may become a viable alternative to on-animal rearing.

## References

APHIS. 1985. Subchapter A—Animal Welfare. Animal and plant health inspection service, USDA.

Bailey, D. L., T. L. Whitfield & G. C. LaBrecque. 1975. Laboratory biology and techniques for mass producing the stable fly, *Stomoxys calcitrans* (L.) (Diptera: Muscidae). J. Med. Entomol. 12(2): 189-193.

Bar-Zeev, M. & S. Sternberg. 1962. Factors affecting the feeding of fleas *(Xenopsylla cheopis* Rothsch.) through a membrane. Ent. Exp. & Appl. 5: 60-68.

Bruce, W. N. 1948. Studies on the biological requirements of the cat flea. Ann. Entomol. Soc. Am. 41: 346-352.

Cerwonka, R. H. & R. A. Castillo. 1958. An appartus for artificial feeding of Siphonaptera. J. Parasitol. 44: 565-566.

Cole, M. M., D. L. VanNatta, W. Ellerbe & F. Washington. 1972. Rearing the oriental rat flea. J. Econ. Entomol. 65(5): 1495-1496.

Gilbert, I. H. 1964. Laboratory rearing of cockroaches, bed-bugs, human lice and fleas. Bull. W. H. O. 31: 561-563.

Hastriter, M. W., D. M. Robinson & D. C. Cavanaugh. 1980. An improved apparatus for safely feeding fleas (Siphonapter) in plague studies. J. Med. Entomol. 17(4): 387-388.

Hudson, B. W. & F. M. Prince. 1958. A method for large-scale rearing of the cat flea, *Ctenocephalides felis felis* (Bouche). Bull. W. H. O. 19(6): 1126-1129.

Hurni, H. & W. Rossbach. 1987. The laboratory cat, pp. 476-492. *In* UFAW (Universities Federation for Animal Welfare) [ed.], The UFAW handbook on the care and management of laboratory animals, 6th ed. Churchill Livingstone: New York.

Kartman, L. 1954. Studies on *Pasturella pestis* in fleas. I. An apparatus for the experimental feeding of fleas. Exp. Parasitology 3: 525-537.

Koehler, P. G., L. A. Lemke, R. S. Patterson & L. M. El-Gazzar. 1989. Insecticide susceptibility of the cat flea (Siphonaptera: Pulicidae). Fla. Entomol. (in press).

Moser, B. A. 1989. Evaluation of several larval diets on cat flea developmental times and adult emergence. M.S. thesis, University of Florida, Gainesville.

Nelson, W. A. 1984. Effects of nutrition of animals on their ectoparasites. J. Med. Entomol. 21(6): 621-635.

Osbrink, W. L. A. & M. K. Rust. 1984. Fecundity and longevity of the adult cat flea, *Ctenocephalides felis felis* (Siphonaptera: Pulicidae). J. Med. Entomol. 21(6): 727-731.

Prasda, R. S. 1987. Host dependency among haematophagous insects: a case study on flea-host association. Proc. Indian Acad. Sci. (Anim. Sci.) 96(4): 349-360.

Silverman, J. 1981. Environmental parameters affecting development and specialized adaptations for survival of the cat flea, *Ctenocephalides felis* (Bouche). Ph.D. dissertation, University California, Riverside.

Smit, F. G. A. M. 1973. Siphonaptera (fleas), pp. 325-371. *In* K. G. V. Smith [ed.], Insects and other arthropods of medical importance. British Museum (Natural History).

U.S. Public Health Service, Department of Health and Human Services, National Institutes of Health. 1985. Guide to the Care and Use of Laboratory Animals. Animal Resources Program, Division of Research Resources, National Institutes of Health, Bethesda, Maryland.

Wade, S. E. & J. R. Georgi. 1988. Survival and reproduction of artificially fed cat fleas, *Ctenocephalides felis* Bouche (Siphonaptera: Pulicidae). J. Med. Entomol. 25(3): 186-190.

# PART THREE

## Insect Rearing Support

# 9

# Insect Rearing Management (IRM): An Operating System for Multiple-Species Rearing Laboratories

*Pritam Singh and G. K. Clare*

### Status and Image of Insect Rearing

Insect Rearing Management (IRM) is a complex field that encompasses science, art, and business management. It requires forecasting, planning, organization, personnel and financial management, and application of flow charts and production schedules. The IRM plan discussed here is flexible, consultative, comprehensive, and reliable. An IRM plan for each species can be drawn that ensures economic production of quality insects to meet program goals. The degree of success of an IRM program can be measured by whether or not all of the objectives of the user group have been met. If they have, the program may be considered successful.

The protocol described here is common to most multiple-species rearing laboratories, no matter how small or large. It leads to effective control of orders, production, supply, maintenance, and costs. The example used in this paper is from a multiple-species rearing facility, but the principles can be effectively applied to any scientific laboratory (e.g., Toxicology, Microbiology, Pharmacology, and Post-harvest) to increase their efficiency and productivity.

Observations on the operation of rearing practices in several multiple-species rearing laboratories in industry, international research organizations, and federal and state departments have indicated that:

1. Insect rearing practices can be wasteful. Once the rearing program is set up it is a continuous operation, irrespective of whether the insects produced are regularly used or not. There are few or no regulatory mechanisms to avoid excessive colony maintenance and supply.

2. There is a lack of biological data on colonization, insect life cycle, yield, and fertility. Records on numbers of generations, rearing 'bottlenecks', or diseases are rarely maintained. Figures for cost per insect are usually not available. Research in insect rearing is either discouraged, or no time is available owing to the pressure of production.

3. In most laboratories, quality control procedures are lacking and little consideration is given to the end use of the insects.

4. The image of insect rearing is that it does not require much science. Many believe that it is easy work; anybody can put some insects in a jar and keep them in the corner of a rundown building where they will grow. As a result, rearing programs and personnel are underappreciated by administrators. Scientists and technicians have difficulty in justifying the importance of their work and getting promotions.

5. Courses in insect rearing management systems are not available at any university in the United States, Canada, or anywhere else in the world. Most insectary managers are either self-trained or have on-the-job experience. The supply of insects is therefore dependent on guesswork. There is no certainty that the numbers and stages required will be available on specific dates. Thus, user groups are unable to plan work in advance.

To improve the status and image of insect rearing, production management must be based on scientific facts coupled with business management practices.

## Background to IRM

It is common practice to rear from 5 to 10 species at the same time in a multiple-species rearing laboratory. Each species has three or four stages, and varied numbers may be required on specific dates. Sometimes, different stages of the same species or of several different species have to be synchronized to meet a program requirement. It is common to have 20 to 40 orders that must be managed simultaneously. How is the accounting maintained for the supply and production system? What kind of protocols are used so that the supply of each order is assured?

We have conceived, developed, and practiced a system that is based on scientific facts. It provides information that enables us to make decisions so that any uncertainty and guesswork can be completely eliminated from the rearing system. It ensures reliability and provides quality insects of known age, stage, and numbers on specified dates to meet the user's objectives and program goals.

This system is called *Insect Rearing Management* (IRM), a term that describes the step-by-step operational management system in a multiple-species rearing laboratory.

The IRM concept was first proposed by Singh and Ashby (1985). It was defined as the efficient utilization of resources for the production of insects of standardized quality to meet program goals. In other words, IRM is the organization of the rearing laboratory, research and development, planning and manipulation of insect colonies, liaison with user groups, and administration of staff and finances in order to meet the precise requirements of the user groups. The terms used in describing IRM have specific meanings that have been clarified and defined by Ashby and Singh (1987) in the *Glossary of Insect Rearing Terms*.

## Phases of IRM

Phases of IRM associated with any production program leading to the eventual supply of insects were identified by Singh and Ashby (1985). These are summarized here to provide a theoretical background to the understanding of the management system.

1. **Objectives.** These must be clearly defined from the outset. Objectives can be for either a new rearing facility or for an insect order in an already established colony in a multiple-species rearing facility.

2. **Laboratory design.** This should take into account the types and numbers of insect species to be reared, their behavior and food requirements, the space and equipment needed, environmental conditions, and hygiene standards. The design should be flexible to allow for future program changes.

3. **Colony establishment and maintenance.** Knowledge of each insect's behavior and natural habitat is useful. The best food and rearing method should be selected. The initial rearing is done in quarantine, and the parental stock should be genetically diverse and disease free. Once established, a colony maintenance system is set up with standard operating procedures (SOP).

4. **Research and development techniques.** The lifecycle must be thoroughly researched under normal laboratory conditions using the best diet, rearing containers, and methods. This provides developmental data that are essential for colony maintenance and production. Studies of the rate of development at various temperatures allows the calculation of day-degrees, which can be a useful tool in the manipulation and storage

of various insect stages. Time-motion studies on standard operating procedures should be performed to calculate labor and costs.

5. **Resources.** These include finance, personnel, equipment, materials, facilities, and insect colonies. The major areas of resource consumption are building, equipment, and colony maintenance; research and development; and staff salaries.

6. **Quality control and biological performance.** Three types of quality control standards have been recognized in relation to the insects' end use.

   - Performance of laboratory-reared insects in a field release situation; for example, sterile insect release programs.
   - Particular traits or sets of traits whereby laboratory reared insects are compared with field-collected insects.
   - Comparison of the laboratory colony with itself over time.

   Before implementing a quality control procedure in a rearing system, three conditions must be met:

   - the methods must be reproducible in the laboratory,
   - standard operating procedures must maintain uniform quality, and
   - filial populations must be capable of reproducing life cycles under specific conditions.

7. **Production.** This occurs through production scheduling and forward planning related to the utilization of resources. Prerequisites for this phase are standard operating procedures, knowledge of life cycle stages, time-motion analyses, and adequate materials and labor. A rearing facility with several environmental control units is required. The guidelines for a production process include:

   - continuous examination of each step to improve the production system,
   - maintenance of good hygiene and cleanliness standards,
   - monitoring of microbial contamination,
   - keeping biological records,
   - checking quality control procedures, and
   - regular assessment of resource consumption.

## Application of IRM

This paper focuses on insect production, the final phase of the insect rearing management process, and explains how an insect order is processed using the IRM protocol. The necessary prerequisites are set forth, and the steps to be followed in processing the order are described.

The protocol is initiated upon receipt of an insect order form from a user group. This form must be submitted to the insectary manager well in advance of the desired delivery date.

The appendix presents an example of an order form for the codling moth, *Cydia pomonella,* shows how the prerequisites would be determined, and illustrates how the order would be processed. The steps are denoted by roman numerals in the text, in Fig. 1, and in the appendix for ease of comparison.

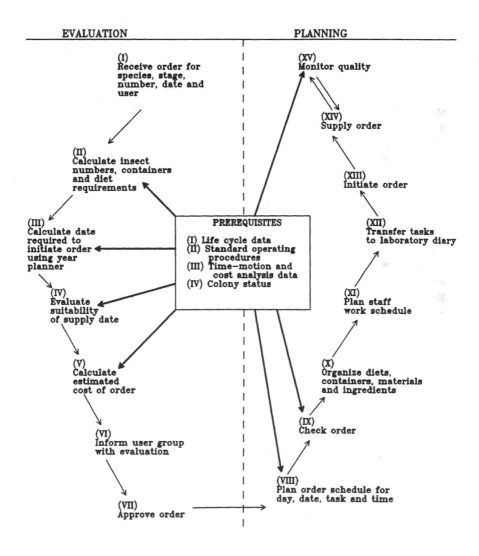

Fig. 1. Steps in processing an insect order.

*Prerequisites Necessary to Process an Insect Order*

**I. Life Cycle Data.** This includes development times, mortality, and yields for all stages at the required temperatures. It also includes data on fecundity and fertility, usable eggs, and eggs per female per day under standard operating conditions. Other specific data may be required, such as the average number of eggs laid per female per day on different varieties of fruits. These data form the basis for calculation of insect numbers required for colony maintenance, colony buildup, insect orders, and supply.

**II. Standard Operating Procedures.** This refers to standard rearing tasks, the amount of diet, number of eggs required per larval rearing container, number of adults per oviposition cage, and other routine laboratory procedures.

**III. Time-motion and Cost Analysis Data.** This requires a comprehensive study of the standard operating procedures. It involves an average time and cost for completion of rearing, laboratory, and administration procedures. An average time and cost required to produce units of 100, 1000, and 10,000 of various stages of each insect are useful indicators for estimating costs. For *Cydia pomonella*, it costs $212 to produce units of 1,000 insects (as shown in appendix).

**IV. Colony Status Data.** This refers to the size and stage of development of a colony or subcolony at any given time and may also include any insects held in storage. It also includes the quality of each generation of the colony, which is checked by comparison against a predetermined standard. This standard is obtained by monitoring previous generations over time under standard operating conditions.

*Steps in Processing an Insect Order*

A typical insect order is processed in two stages, *Evaluation* and *Planning*, as shown in Fig. 1. Step I is reviewing the order. Steps II-VII are followed for a preliminary evaluation of the order to establish whether the user request can be met on specific dates. It is possible for an experienced production manager to cut short or eliminate most of the evaluation steps. The manager can quickly work out colony status, resources, and costs. However, it is recommended that all steps be followed in the initial stages of a production program or if the manager is inexperienced. Steps VIII-XIII are involved in planning and execution, and Step XIV is the eventual supply of the order. Step XV deals with quality control aspects to assure that the insects meet the users' specifications.

● *First Stage—Evaluation*

I. **Review the Order.** The insect order is received from the user on an insect order form (see appendix). This form gives the details of the program; the species, stage, and number required; and dates of supply. It also gives any special instructions, such as quality, rearing conditions or responsibility of pickup or delivery. Large orders should be placed as soon as possible, preferably 3 months in advance to ensure the supply.

II. **Determine Insect Stages, Numbers, Containers, and Diet Requirements.** The main purpose of this step is to calculate the *actual numbers of insects* that need to be produced to supply the order. During the process the different stages, rearing containers, and amount of diet needed are also estimated. Life cycle data combined with standard operating procedures are required for calculations. This is done by working back from the actual number requested, until a stage of development and number of insects is reached that can be obtained from the colony.

III. **Calculate Date to Initiate the Order.** This is *the day* on which the first task to start the order is to be completed. It is calculated by first placing the date of supply requested by the user on a yearly planner (see appendix, Table 1A). The next step is to work back through all the different developmental stages until synchronization with the maintenance colony is achieved. Sometimes manipulation of the colony may be necessary. Then, the date and first task to start the order is obtained and written on the yearly planner.

IV. **Verify Suitability of Original Supply Date.** This is the original date on which insects are to be supplied to the user. If the date requested cannot be met by the production group, an alternative supply date must be negotiated.

V. **Estimate Cost.** This is calculated by using cost analysis data and multiplying the actual number of insects required to process the order by the cost factor.

VI. **Confirm the Order.** The user group is informed of the supply dates and the estimated cost. The order is confirmed in writing.

VII. **Approve the Order.** The order is approved and filed by the insect production manager, and the planning process initiated.

● *Second Stage—Planning*

   **VIII. Plan the Insect Order Schedule.** The schedule is a plan of day-to-day insect handling and tasks necessary to prepare the insect order for supply. This includes a detailed analysis of all tasks, insect stages and numbers, diet, material requirements, labor and costs involved. An example of a schedule is presented in the appendix and includes considerations such as dates, tasks, time requirements, and costs involved.

   **IX.   Check the Order.** Before the order is actually initiated, the schedule is checked by another member of the production group for any errors or omissions.

   **X.    Organize the Diets, Containers, Materials, and Ingredients.** A stock of materials, rearing containers, and diet ingredients is taken and a diet preparation schedule prepared. The items can be pre-ordered if in short supply.

   **XI.   Plan the Staff Work Schedule.** This step involves checking the time required for the order and ensuring that enough staff members will be available on specific days to complete all the tasks. This is done at least 1 week in advance, but 3 to 4 weeks is preferable.

   **XII.  Transfer of Tasks to the Laboratory Diary.** At this stage all the tasks on the order schedule are transferred to the laboratory diary.

   **XIII. Initiate the Order.** The order is initiated by copying the tasks as they appear in the laboratory diary to the weekly task sheet. This is prepared a week in advance, usually on Friday afternoon.

   **XIV.  Supply the Order.** The order is supplied on the specified dates and times to the user group with the quality assessment form.

   **XV.   Monitor Insect Quality.** The predetermined specifications of the order are independently checked by the production and user groups. Any differences between their results are indicative of changes in quality due to handling or environmental conditions. Periodic performance tests are done on the intended use of the insects.

## Conclusion

We believe that the IRM concept has put insect rearing on a sound and scientific footing. It has ensured a continuous and reliable supply of quality insects to meet program challenges. It has earned the respect of administrators. It is efficient, economical, and inspires team confidence. We recommend that all multiple-species rearing laboratories practice IRM.

In summary, a good insect rearing management system must provide the following:

- A system that everyone in the insectary fully understands and faithfully follows.
- Avoidance of people carrying unnecessary information in their heads and using this to make guesses.
- Procedures that are flexible and easy to adapt, improve, or use in part to suit other requirements.
- A record of past performance as a guide to more accurate estimates for future work.
- Apparent profitability of the insectary work.
- Early warning when things go wrong.
- An indication of waste, shortages of equipment and ingredients, and conditions of staff and working space.
- An account of the materials from the time they are ordered until they are used.
- Assurance of maximum productivity with a minimum of work.
- An operating system that functions just as well when the insectary manager is away.

## Acknowledgments

Staff members who worked in the Rearing Section over the past several years have contributed ideas to develop the IRM concept. We especially acknowledge Mr. M. Ashby, Ms. Judith Somerville, Ms. Shelly Dalzell, Mr. J.P.R. Ocieng'-Odero, and Mr. D. J. Rogers.

## References

Ashby, M. D. & P. Singh. 1987. A glossary of insect rearing terms. DSIR Bulletin No 239. DSIR Science Information Publishing Centre, Wellington, New Zealand.

Ashby, M. D., P. Singh & G. K. Clare. 1985. Rearing *Cydia pomonella*, pp. 237-248. *In* P. Singh & R. F. Moore [eds.], Handbook of insect rearing, vol. 2. Elsevier, Amsterdam.

Singh, P. & M. D. Ashby. 1985. Insect rearing management, pp. 185-215. *In* P. Singh & R. F. Moore [eds.], Handbook of insect rearing, vol 1. Elsevier, Amsterdam.

# Appendix:
# Processing an Insect Order

### Example of Order for *Cydia pomonella*

The following is an example of a typical order submitted on the standard insect order form by the user group. This order is used in this appendix to illustrate how orders are processed using the IRM protocol.

INSECT ORDER FORM          Order number: <u>CM/21/87</u>
    Name of user:      <u>Dentener/Waddell</u>
    Section:       <u>Horticulture/Post-harvest</u>
    Program:      <u>Cherries/Japan</u>
    Date order placed:<u>9/9/87</u>

    Species:      <u>*Cydia pomonella*</u>      Strain: <u>Lab</u>

| Suborder Number | Stage Required | Date Required | Number Required |
|---|---|---|---|
| 1 | Mated adults | 12/14/87 | Adults to lay 16,500 eggs on cherries in 24 h. |
| 2 | Mated adults | 12/15/87 | Reuse adults from suborder 1 and supply extra to lay 16,500 eggs on cherries in 24 h. |

**Special Instructions:**
(rearing conditions, diet required, diapause, etc.)

Adults supplied should be individually collected. Oviposition cages must be supplied in multiples of five, placing equal numbers of adult pairs per cage, with a maximum of 50 pairs/cage.

Order Approved by: <u>Pritam Singh</u>          Date <u>9/10/87</u>

## Prerequisites for Processing the Order

### *I. Life Cycle Data*

#### A. Larval rearing containers at 25°C.
- First adult emergence occurs on day 26.
- Mean adult emergence occurs on days 29-31.

#### B. Group oviposition cage data (50 pairs/cage) at 25°C.
- Mean preoviposition period = 1 day.
- Eggs collected from day 3 to 7.
- Mean fecundity/female/day from day 3 to 7 = 20 eggs/female.
- Mean number of eggs/egg sheet/day from day 3 to 7 = 1,000.
- After egg sterilization 50% useable eggs/egg sheet.
- Total number of egg sheets produced/cage = 5.
- Mean fecundity/cage from day 3 to 7 = 5,000 eggs or 2,500 useable eggs.
- Egg development: (7 days at 25°C; 10 days at 20°C).
- Mean oviposition on cherries = 15 eggs/female/24 h.
- Before supplying, adults held for 2 days for mating at 25°C in oviposition cages without egg sheets.

### *II. Standard Operating Procedures*

- Amount of artificial diet per larval rearing tray = 4.3 kg.
- Number of adults per group oviposition cage = 50 pairs.
- Number of egg sheets required to inoculate 1 larval rearing tray of diet = 4 egg sheets containing 500 useable eggs/sheet
- Mean number of adults produced per larval rearing tray = 500.

### *III. Time-motion and Cost Analysis Data*

|  | Time (min) | Cost ($) |
|---|---|---|
| • Preparation, 1 oviposition cage | 6 | 0.80 |
| • Collection, 50 pairs of adults |  |  |
|     Vacuum and transfer to cage | 10 | 1.33 |
|     Hand with a test tube | 50 | 6.66 |

|                                                                 | Time (min) | Cost ($) |
|-----------------------------------------------------------------|:----------:|:--------:|
| • Transfer individually collected adults to oviposition cage, 50 prs/cage at 6 moths/m | 17 | 2.27 |
| • Egg collection, 1 egg sheet/oviposition cage                  | 5          | 0.67     |
| • Egg sterilization/egg sheet                                   | 6          | 0.80     |
| • Diet preparation/larval tray                                  | 70         | 9.33     |
| • Inoculation, 1 larval tray, egg sheets                        | 20         | 2.66     |
| • Cleaning and sterilization, 1 larval tray                     | 15         | 2.00     |
| • Cleaning, 1 oviposition cage                                  | 5          | 0.67     |
| • Cost of artificial diet/larval tray at $2.04 per kg with 4.3 kg per tray |  | $8.77 |
| • Average wages/hour                                            |            | $8.00    |
| • Unit cost to produce 1,000 insects                            |            | $212.00  |

## IV. Colony Status Data

Two subcolonies, A and B were in progress about 3 weeks apart when checked on 10 September (see Table 1A).

### A. Subcolony generation 160 A inoculated on 5 August.

- Size of colony A = 3 larval rearing trays plus 600 individually reared larvae in test tubes.
- Stage of colony A = mated adults.
- Next generation A, 161 A due to be inoculated on 21 September.

### B. Subcolony generation 160 B inoculated on 28 August.

- Size of colony B = same as A.
- Stage of colony B = Larvae 13 days old.
- Next generation B, 161 B due for inoculation on 12 October.

**Table 1A.  Calculation of dates required to initiate the order**

**YEAR PLANNER  1987**

| AUGUST | SEPTEMBER | OCTOBER | NOVEMBER | DECEMBER | DAY |
|---|---|---|---|---|---|
|  | 1 |  |  | 1 | Tue |
|  | 2 |  |  | 2 | Wed |
|  | 3 | 1 |  | 3 | Thu |
|  | 4 | 2 |  | 4 | Fri |
| 1 | 5 | 3 |  | 5 First adult expected 162 A | Sat |
|  |  |  |  |  |  |
| 2 | 6 | 4 | 1 Collect eggs | 6 | Sun |
| 3 | 7 | 5 | 2 Collect eggs | 7 | Mon |
| 4 | 8 | 6 | 3 | 8 Collect adults | Tue |
| 5 Subcolony 160 A inoculated | 9 Order received | 7 | 4 | 9 Collect adults | Wed |
| 6 | 10 Access status of 2 subcolonies | 8 | 5 | 10 Collect adults (also on day 11) | Thu |
| 7 | 11 | 9 | 6 Sterilize eggs (days 6-9) | 11 Transfer adults to cages at 15°C | Fri |
| 8 | 12 | 10 | 7 | 12 Move adults to 25°C | Sat |

| AUGUST | SEPTEMBER | OCTOBER | NOVEMBER | DECEMBER | DAY |
|---|---|---|---|---|---|
| 9 | 13 | 11 | 8 Inoculate 162 A (6 trays) | 13 | Sun |
| 10 | 14 | 12 Inoculate Subcolony 161 B | 9 Inoculate 162 A | 14 Supply suborder 1 | Mon |
| 11 | 15 | 13 | 10 Inoculate 162 A | 15 Supply suborder 2 | Tue |
| 12 | 16 | 14 | 11 | 16 | Wed |
| 13 | 17 | 15 | 12 | 17 | Thu |
| 14 | 18 | 16 | 13 | 18 | Fri |
| 15 | 19 | 17 First adult expected 161 A (days 17-20) | 14 | 19 | Sat |
| 16 | 20 | 18 | 15 | 20 | Sun |
| 17 | 21 Inoculate Subcolony 161 A | 19 | 16 | 21 | Mon |
| 18 | 22 Inoculate 3 trays (21-22) | 20 | 17 | 22 | Tue |
| 19 | 23 | 21 | 18 | 23 | Wed |
| 20 | 24 | 22 | 19 | 24 | Thu |
| 21 | 25 | 23 | 20 | 25 | Fri |

| AUGUST | SEPTEMBER | OCTOBER | NOVEMBER | DECEMBER | DAY |
|---|---|---|---|---|---|
| 22 | 26 | 24 | 21 | 26 | Sat |
| 23 | 27 | 25 | 22 | 27 | Sun |
| 24 | 28 | 26 Collect and transfer adults | 23 | 28 | Mon |
| 25 | 29 | 27 | 24 | 29 | Tue |
| 26 | 30 | 28 | 25 | 30 | Wed |
| 27 | | 29 Collect eggs | 26 | 31 | Thu |
| 28 Subcolony 160 B Inoculated | | 30 Collect eggs | 27 | | Fri |
| 29 | | 31 Collect eggs | 28 | | Sat |
| 30 | | | 29 | | Sun |
| 31 | | | 30 | | |

## Evaluation of the Order

The following steps are carried out to evaluate the order and determine whether the desired quantities of insects can be supplied on the desired date and, if so, what the estimated cost will be.

### *I.   Review the Order*

This step involves reviewing the insect order received (shown in Fig. 1A).

### *II.  Determine Insect Stage, Number, Containers, Materials, and Diet Requirements for the Order*

**Suborder 1.** States that adults are to lay 16,500 eggs on cherries in 24 h.

#### A. Adults and oviposition cages
- Based on life cycle data and standard operating procedures (see section 2.1, 2.2.).
- Allow an average of 15 eggs/female/24 h on cherries.
- To obtain a minimum of 16,500 eggs, 1100 females (2,200 adults) will be required.
- Allowing 1100 pairs at 50 pairs/cage, will require 2 oviposition cages.
- To supply oviposition cages in multiples of 5, (as per order) will require 25 cages.
- Actual insects to be supplied (1,100 pairs of males and females) evenly divided into 25 cages, gives 44 pairs/cage.
- For monitoring quality control, 2 additional oviposition cages of 44 pairs/cage will be required, bringing total pairs required to 1188 (2376 adults).
- A total of 27 oviposition cages will be required.
- To obtain exactly 1,188 males and 1,188 females an extra 10% may need to be collected. This will compensate for any sex ratio imbalance and adult emergence pattern.
- Therefore, about 2600 adults in order to produce the 1188 pairs of males and females.

#### B. Larval rearing trays of diet.
- Allowing average production of 500 adults per tray will require 6 trays to produce 2,600 adults.

## C. Number of egg sheets and oviposition cages to start the order.

- Allowing 4 egg sheets per larval rearing tray, will require 24 egg sheets to inoculate 6 trays of diet.
- Therefore, to produce 24 egg sheets, 5 oviposition cages (5 egg sheets/cage) will be required.
- Adults needed to start the order will be held at 15°C for a short period before use.
- To compensate for any decrease in fecundity due to holding, 1 extra oviposition cage is allowed; total 6 cages (1 oviposition cage/tray).

## D. Adults required from maintenance colony to start the order.

- Six cages at 50 pairs/cage, will require 300 pairs of males and females (600 adults).

**Suborder 2.** States that adults will be used from suborder 1 and extra adults will be needed to lay 16,500 eggs on cherries in 24 h.

## A. Adults and oviposition cages.

- During the handling procedure of transferring adults from suborder 1 to new cherries, some adults are lost and exposure to low temperature may reduce fertility.
- This is compensated for by supplying extra pairs, based on 10% of the original supply for the suborder 1.
- Therefore, 110 extra pairs will be supplied.
- Allowing 10% for imbalance in sex ratio a total of 240 adults may need to be collected to supply the 110 pairs.
- As per special instructions oviposition cages to be supplied in multiples of 5 with adults divided evenly.
- 110 pairs are divided into 5 cages at 22 pairs per cage.

## B. Larval rearing trays of diet.

- 240 adults required is less than half the expected mean from one rearing tray.
- Adults will therefore be obtained from surplus adults from suborder 1.
- No extra trays will be required.

### III. Calculate Dates Required to Initiate Order (Table 1A)

- These dates are calculated by using a year planner, life cycle development data, the supply dates requested, and the current colony status.
- Enter the dates of supply requested on the year planner (14-15 December). Work back from these dates through the developmental stages (as calculated in section 2.1) until the date required to inoculate the 6 larval rearing trays (8-10 November) is reached.
- To obtain adults to start the order, continue to work back to the 26 October.
- Now place on the year planner the status of the 2 subcolonies 160 A and 160 B (dates of previous inoculation, 160 A on 5 August and 160 B on 28 August; dates due for next inoculation, 161 A on 21 September and 161 B on 12 October).
- Status of the 2 subcolonies on the year planner shows that the adults for this order have to come from subcolony 161 A.
- Subcolony 161 A is due for inoculation on 21 September.
- Subcolony A needs to be manipulated to bring it into synchronization with the order.
- This is done by placing the larval rearing trays of 161 A at 15°C on emergence of the first adult, starting 17 October and ending 20 October.
- This is the date that the first task to initiate the order must be completed.

### IV. Verify Suitability of Supply Date

- Comparing the status of the maintenance colony with the dates to initiate the order and supply dates, it is confirmed that the supply dates requested are suitable to process the order.

### V. Estimate Cost of Order

To produce 1,000 insects it costs $212 (21.2 cents per insect).

- Suborder 1: To produce 2600 adults (2600 x .212) = $551.20.
- Suborder number 2: To produce 240 adults (240 x .212) = $50.88.
- Estimated cost of order = $602.

## VI. Confirm the Order

Notify the user group in writing that the estimated supply date can be met and that the cost is estimated to be $602.

## VII. Approve the Order

Approve and file the order and initiate the planning process as illustrated in the next section.

### Planning of Order Schedule

The following chart shows the results of carrying out step VIII, planning a detailed schedule of items that must be completed to fulfill the order, such as dates, task, and time requirements. Once the planning process has been completed, simply follow steps IX through XV (as discussed on page 142) to complete the IRM process.

**Order No:** CM/21/87
**Program:** Cherries/Japan
**Species:** *Cydia pomonella* (lab)
**Number and Stage Required:** suborder (S.O.) 1 = 2,200 mated adults
suborder (S.O.) 2 = 220 mated adults
**Supply Dates:** S.O. 1 = 12/14/87, S.O. 2 = 12/15/87

| # Day | Date | Task | Time<br>h min | Cost<br>($8/h) |
|-------|------|------|------|------|
| 1-4 | 1987<br>October<br>17-20 | ■ Subcolony A, generation 161. Three larval rearing trays checked daily. On emergence of first adult move trays to 15°C. | 0:30 | $ 4.00 |
| 7 | 23 | ■ Prepare 6 oviposition cages with wax paper (6 min/cage). | 0:36 | $ 4.80 |

| # Day | Date | Task | Time<br>h  min | Cost<br>($8/h) |
|-------|------|------|------|------|
| 10 | 26 | ■ Subcolony A, collect and transfer 600 adults by vacuum collection to 6 gauze oviposition cages (50 pairs/cage at 10 min/cage).  Place at 25°C. | 1:00 | $  8.00 |
| 12-17 | 28-2<br>November | ■ Collect egg sheets daily from 6 cages for 5 days, a total of 30 sheets (5 min/cage/day).  Label with date.<br>■ Place eggs collected on 29-31 at 20°C.<br>■ Place eggs collected on 1-2 at 25°C. | 2:30 | $ 20.00 |
| 21-24 | 6-9 | ■ Sterilize egg sheets (30 sheets at 30 min/5 sheets). | 3:00 | $ 24.00 |
| 24-25 | 9-10 | ■ Inoculate 6 trays of diet - (20 min/tray), Label Order No. CM/21/87. | 2:00 | $ 16.00 |
| 50-52 | December<br>5-7 | ■ Check rearing trays for adult emergence.  Move trays from 25°C to 15°C as soon as the first adults emerge. | 0:30 | $  4.00 |
| 52 | 7 | ■ Prepare 32 oviposition cages (without wax paper) at 5 min/cage. | 2:40 | $ 21.33 |
| 53-56 | 8-11 | ■ S.O. 1<br>Move daily 6 rearing trays of diet from 15°C to 25°C.  Collect up to 2,600 adults (to give 1188 pairs) individually in test tubes. Place adults at 15°C. | 21:40 | $173.29 |

| # Day | Date | Task | Time h min | Cost ($8/h) |
|-------|------|------|------------|-------------|
| 56 | 11 | ■ S.O. 2<br>Collect up to 240 adults (to give 110 pairs) individually in test tubes from 3 larval trays used for S.O. 1. | 2:00 | $16.00 |
|  |  | ■ S.O. 1<br>Transfer 1,188 pairs to 27 oviposition cages without wax paper (44 pairs/cage at 15 min/cage). Place at 15°C. | 6:45 | $54.00 |
| 57 | 12 | ■ S.O. 1<br>Move cages from 15°C to 25°C. | 0:10 | $ 1.33 |
| 58 | 13 | ■ S.O. 2<br>Transfer 110 pairs to 5 oviposition cages at 22 pairs/cage (8 min/cage). Place at 25°C. | 0:40 | $ 5.33 |
| 59 | 14 | ■ S.O. 1<br>Supply 25 oviposition cages. Place egg sheets in 2 quality control cages. | 0:20 | $ 2.67 |
| 60 | 15 | ■ S.O. 1<br>Collect eggs from 2 quality control cages (10 min/cage) and place new egg sheets in cages. Place eggs at 25°C for development. | 0:20 | $ 2.67 |
|  |  | ■ S.O. 2<br>Supply 5 oviposition cages. | 0:10 | $ 1.33 |

| # Day | Date | Task | Time<br>h min | Cost<br>($8/h) |
|-------|------|------|---------------|----------------|
| 61 | 16 | ■ Collect egg sheets from 2 control cages and discard adults. | 0:10 | $ 1.33 |
|  |  | ■ Count eggs from both days collections and compare with data from colony.  Total of 4 sheets (20 min/egg sheet). | 1:20 | $ 10.67 |
| 67 | 22 | ■ Check fertility of eggs from control cages. | 1:20 | $10.67 |
| 69 | 24 | ■ Receive quality control form from the user group and assess results. | 0:15 | $ 2.00 |

The total cost of labor for this insect order can then be found by adding the subtotal from the order schedule to the cost of labor for the items in Table 1B.

**Table 1B. Cost of labor for producing the insect order**

| | | |
|---|---|---|
| **Subtotal from Order Schedule** | **47:56** | **$383.42** |
| Order preparation time, evaluation, and task sheet preparation time. | 6:00 | $ 48.00 |
| Diet preparation time for 6 larval trays at 70 min/tray. | 7:00 | $ 56.00 |
| Washing and cleaning time 6 larval trays at 15 min each. | 1:30 | $ 12.00 |
| Washing and cleaning time 32 gauze oviposition cages at 5 min each. | 2:40 | $ 21.33 |
| **Final Total** | **65:06** | **$520.80** |

The final cost of the order includes:

Labor cost at 65 h 06 min. . . .$520.80
Diet cost . . . . . . . . . . . . .  <u>$52.63</u>
Final cost of order. . . . . . . . **$573.43**

Estimate cost = $602
Difference between estimated and actual cost = $28.57

# 10

# Multiple-Species Insect Rearing in Support of Research

## E. G. King, Jr., and G. G. Hartley

### Introduction

A dependable supply of insects of known quality is required to support basic and applied research for plant protection. About 500 species of insects and mites are maintained for this purpose worldwide (Dickerson et al. 1980). Artificial diets have been developed to facilitate rearing many of these species (Singh & Moore 1985). The capability to consistently rear large numbers of insects has usually preceded pioneering efforts in entomological research, particularly in biological and genetic control (Knipling 1984).

Research at the disciplinary level and on suppression or control are served by insect rearing. For this reason, the Southern Insect Management Laboratory (SIML) developed one of the most advanced multiple-species insect rearing programs (MSIRP) in the world (Fig. 1).

#### Stoneville Multiple-Species Insect Rearing Program

**Facilities and Equipment.** The MSIRP, located on the first floor of the USDA, ARS, Jamies Whitten Delta States Research Center, Stoneville, Mississippi, occupies 324 m² of which 259 m² is "clean area" (Fig. 2) (King et al. 1985b). Heating and cooling ducts in this area are fitted with high efficiency particulate aie (HEPA) filters (99.99% effective for particles greater than 3 microns). Primary activities that occur in the clean areas include small-scale diet mixing, egg sterilization, egg implantation, pupal sexing, larval development, adult emergence and oviposition packaging of insect life stages for shipment, and laundering of laboratory clothing. The remaining area (65 m²) must be entered from the outside.

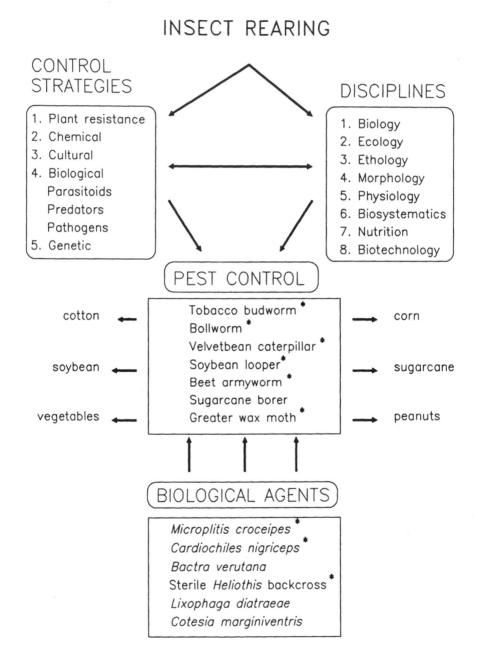

**Fig. 1. Insects produced in support of research by the multiple-species insect rearing program of the USDA, ARS Southern Insect Management Laboratory, Stoneville, Mississippi. Asterisks denote species reared at present.**

161

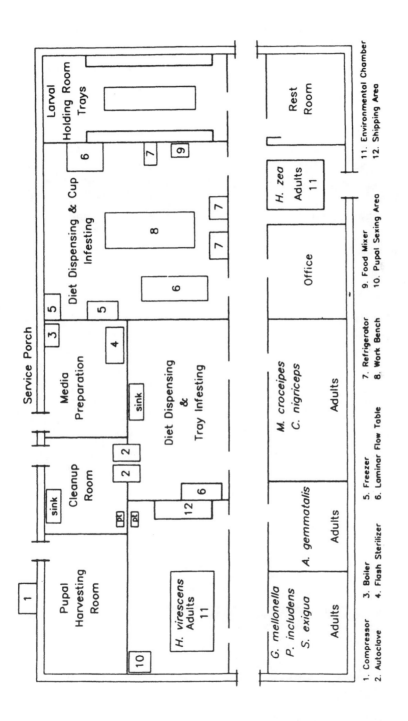

**Fig. 2. Floor plan of the multiple-species insect rearing facility, ARS Southern Insect Management Laboratory, Stoneville, MS.**

The MSIRP has two 11-m$^2$ rooms for holding adults of the velvetbean caterpillar *(Anticarsia gemmatalis)*, soybean looper *(Pseudoplusia includens)*, beet armyworm *(Spodoptera exigua)*, and greater wax moth *(Galleria mellonella)*, and a 22-m$^2$ room for holding cocoons and adults of *Microplitis croceipes* and *Cardiochiles nigriceps*. These insect holding areas are humidified with steam and centrifugal humidifiers. Electric heaters in the holding rooms supplement the center's heating system. Bollworm and tobacco budworm moths and pupae are held in environmental chambers that independently control light, temperature, and relative humidity. Larvae of all species are reared in a separate 27-m$^2$ room maintained at 30°C and 55 $\pm$ 5% relative humidity.

Clean conditions are emphasized in several other areas, including the 52-m$^2$ diet dispensing and cup infesting room (cup rearing) and the 33-m$^2$ diet dispensing and tray infesting room. In these areas, the diet is potentially exposed to contamination. Diet can be supplied to either of the two diet dispensing and egg/larval infesting rooms from a Cherry Burrell Unitherm IV spiral-tube food sterilizer, which is housed in the media preparation room outside the clean area.

Harvesting of pupae, which can be a major source of contamination, occurs outside the clean area. After the moth pupae or parasitoid cocoons have been harvested, the trays, cell inserts, and covers are washed in the cleanup room. Heat-resistant materials are transferred to the clean area through one of two double-door autoclaves. The remaining materials are passed through a 1% sodium hypochlorite solution before being re-introduced into the clean area. Other equipment in this facility includes four laminar flow hoods, four freezers, one refrigerator, an automatic diet dispenser for filling small plastic cups, a clothes washer and dryer, a commercial food mixer, three balances, three dissecting microscopes, a 3.8-liter blender, and a gas fed hot plate.

Sanitation is strictly enforced throughout the facility. All equipment entering the clean room areas are either autoclaved or chemically disinfected with sodium hypochlorite. All areas of the facility are wet-mopped daily with a 0.2% sodium hypochlorite solution. The rooms where the diet is exposed for egg and larval infesting are wet-mopped, and the tables are cleaned before and after the rearing operations. Eggs, pupae, or cocoons are disinfected with sodium hypochlorite or formaldehyde before placing them on diet or in emergence chambers. Antibiotics and antimicrobial agents are used in diets for all insect species. Movement of personnel is closely controlled. Personnel harvest insects do not work in the diet-infesting rooms. Workers wear laboratory clothing and often wear hair nets and filter masks to prevent the spread of contaminants and protect them from respiratory hazards. Laboratory clothing and tennis shoes are approved by the laboratory. The clean area is locked to prevent entry of unauthorized personnel, particularly field workers. Field-collected insects are allowed into the facility only under strictly controlled conditions.

**Diet.** The artificial diet used in the SIML-MSIRP for rearing noctuids (Table 1) is a greatly modified soybean flour-wheat germ diet first described by Vanderzant (1974). This soybean flour-wheat germ diet may be used to rear insects in small plastic cups or cotton-stoppered glass vials (King & Hartley 1985a) and in multicellular trays (King et al. 1985b). Where the trays are used for large scale production of tobacco budworm and bollworm pupae, much of the agar may be replaced with corn cob grit. The high-agar diet is required for rearing larvae in cups, as well as for production of non "Heliothis" species. Preparation of the high-agar diet is described in detail by King and Hartley (1985a), and the low-agar diet is described in King et al. (1985b).

**Larval Rearing.** Lepidopterous larvae are reared in a multicellular unit consisting of a molded, fiberglass tray (64.4 x 34.4 x 2.8 cm); a polystyrene-cell unit made by gluing together three sheets of light-diffusion louver (61 x 30 cm with 666 or 903 cells); a porous-polypropylene cover with 125-cm openings (64.4 x 34.3 cm x 9 m); a cover reinforcement made from one sheet of 0.9-cm-thick polystyrene light-diffusion louver (64.4 x 34.4 x 0.9 cm); and an (64.4 x 34.4 cm ID) made of 25-cm aluminum angle stock (1.6mm thick). The unit is strapped at each end with 9-mm wide polypropylene strapping fastened with crimp-type metal seals (Hartley et al. 1982). There are 666 or 903 cells in a unit, depending on their size. The cell size is determined by the amount of diet required for a larva to complete development to the pupal stage.

**Table 1. Soybean flour-wheat germ diet used in the Stoneville Multiple-Species Insect Rearing Program**

| Ingredient | Amount (g/liter) |
|---|---|
| Soybean flour (Nitrisoy Flour #40) | 41 |
| Wheat germ | 35 |
| Wesson salt | 10 |
| Sugar | 41 |
| Vitamin mix | 9.5 |
| Agar | 22.4 |
| Methyl paraben | 1 |
| Aureomycin (6% Soluble Power) | 1 |
| Sorbic-acid | 1 |
| Water | 930  ml |

The eggs used to infest the trays are removed from the polyester oviposition cloths by washing them in 0.2% sodium hypochlorite for 2 min followed by rinsing in tap water. These loose eggs are disinfected under OSHA approved fume hoods with formalin and rinsed again in distilled water. Then the eggs are dried and deposited in the cells using an aluminum-sheet "planter" (Raulston & Lingren 1972). Eggs of the soybean looper, velvetbean caterpillar, and beet armyworm are treated similarly to the other species, except they are deposited on paper toweling after being removed from the oviposition substrates and disinfected. The egg-laden paper towels are placed on the diet surface. After the trays are sealed, they are wrapped in a single layer of paper (18.2 g weight) to prevent microbial contamination and retard drying. The trays are held at 29 ± 1°C and 55 ± 5% relative humidity until the larvae complete development to the pupal stage. Neonate larvae are easily obtained for infesting cups by placing egg-covered cloths (after disinfection) into a sterile 2-liter Erlenmeyer flask containing sterile cotton moistened with distilled water. The flask is plugged with sterile cotton and wrapped in aluminum foil. The hatching positively phototactic larvae crawl to the top, where they are removed and transferred to the diet surface.

**Pupal Harvesting.** Bollworm, tobacco budworm, beet armyworm, and velvetbean caterpillar pupae are harvested by lifting the cell unit from the tray and bumping it against the table, causing the pupae to fall free. The pupae are separated from the diet and frass by use of a two-part separator and a high-volume blower (Hartley et. al. 1982). The air current of the blower separates the pupae from the lighter material and suspends them above the sieve, thereby cushioning them and preventing damage.

Since soybean looper pupae are retained in the cells by their silk webbing, 1% sodium hypochlorite solution must be used to remove them. The freed pupae float to the surface for removal and a tap water rinse. Pupae are disinfected and either shipped to cooperators or used for colony maintenance.

**Adult Holding.** Except for the soybean looper and velvetbean caterpillar, male and female pupae are placed in separate containers for moth emergence. A 3.8-liter cardboard bucket covered at the top with white polyester cloth is used as an emergence chamber. As the adults emerge, 40 males and 40 females are removed and placed together in cardboard containers identical to the emergence chambers. For the bollworm and tobacco budworm, an additional polyester cloth strip is extended from the top edge of the container to the bottom. Crumpled strips of wax paper serve as a streamer for the beet armyworm. Soybean looper pupae are difficult to sex because of their light coloration. Consequently, their pupae are mixed without sexing and allowed to emerge, mate, and oviposit in the original container. This cage consists of a 20-cm x 30-cm-diameter hole cut about 2.5 cm from the bottom to attach them to a scale collector similar to one reported by Hartley et al. (1977). Cardboard oviposition containers may be cleaned, autoclaved, and reused about six times. Acrylic containers last indefinitely, but they must be disinfected in a 1% sodium hypochlorite solution.

The oviposition cloths are removed each day beginning the 2nd night after mating. Maximum egg production for the beet armyworm moths occurs between the second and third nights so they are discarded after 5 days. For the other species egg production peaks between the 4th and 6th nights after mating. These moths are discarded after 10 days. Moths of all five species are held at $24 \pm 1°C$ and $80 \pm 5\%$ relative humidity with a 14:10 (L:D) photoperiod and provided with 5% sucrose solution as a food and water source. For the bollworm, tobacco budworm, and beet armyworm, a cotton pad saturated with the solution is placed on the cloth cage cover. For the other two species, the saturated pad is placed in a petri dish inside the cage. Regardless, the pad is resaturated daily.

**Parasitoid Rearing.** Large-scale rearing of *M. croceipes* and *C. nigriceps* is accomplished by exposing several hundred tobacco budworm larvae in a multicellular tray to caged parasitoids for a designated period of time (Powell & Hartley 1987). The polypropylene sheet is secured atop the cell insert after larvae are exposed to the parasitoids. Upon emergence from their moribund hosts, the parasitoid larvae spin cocoons on the polypropylene sheet and upper portions of the cell insert. Parasitoid cocoons are easily harvested using a 0.4% sodium hypochlorite wash; a subsequent 70% isopropyl alcohol wash separates the parasitoid cocoons from host larval exuviae.

## Applications for Multiple-Species Insect Rearing

More than 175 million life stages of eight insect species were produced in 1987 at the Stoneville location (Table 2). These insect life stages were shipped to about 100 scientists associated with 30 private firms, 15 universities, and 7 Agricultural Research Service (ARS) laboratories. These were distributed across all disciplines but typically in the areas of host plant resistance and chemical, genetic, and biological control (Fig. 1).

Basic information on the biology, ecology, and behavior of insects is essential to understanding and improving control measures. Often detailed observations of the insects are not possible in the field and the ability to colonize the insect in the laboratory greatly facilitates its study.

**Table 2.** **Insect production during 1987 in the Stoneville Multiple-Species Insect Rearing Program**

| Species | No. eggs produced | No. pupae produced | Purpose of production |
|---|---|---|---|
| Tobacco budworm | 70,504,000 | 406,800 | Host plant resistance, insecticide screening, parasite rearing |
| Bollworm (corn earworm) | 50,466,000 | 378,750 | Host plant resistance, insecticide screening |
| Velvetbean caterpillar | 20,266,000 | 147,200 | Host plant resistance |
| Soybean looper | 17,835,000 | 167,200 | Host plant resistance |
| Beet armyworm | 14,681,000 | 149,200 | Host plant resistance |
| Greater wax moth (larvae) | | 26,000 | Parasite production |
| *Microplitis croceipes* | | 387,419 | Biological control |
| *Cardiochiles nigriceps* | | 229,416 | Biological control |

**Host Plant Resistance.** The availability of laboratory-reared insects is essential for research on plant resistance to insects. Plants can be infested during particularly vulnerable periods and the variability in population levels can be reduced year-round.

The Stoneville-MSIRP produced tobacco budworms to select cotton lines with increased levels of resistance (Meredith et al. 1979a, 1979b). Jenkins et al. (1982) dispensed 1st instar larvae mixed in corn cob grits. Bailey and Meredith (1984) dispensed eggs suspended in xanthan gum solution into plate terminals using a syringe. Identification of resistant soybean genotypes is accomplished by a) releasing 4,000 - 5,000 pairs of beet armyworm, velvetbean caterpillar, soybean looper, or bollworm moths in 0.07 ha field cages in which soybean is growing; or b) infesting greenhouse-grown plants with 1st instar larvae. See Hartwig et al. (1984) for release of soybean germplasm line D75-10169.

**Chemical Control.** Colonization of insect species is essential for initial screening and subsequent continuous testing of chemical control methods. The availability of large numbers of insects accelerates development of new, more effective, and more economical control measures such as insecticides, pheromones, and chemosterilants.

**Biological Control.** The Stoneville-MSIRP has supplied host material and artificial diet to support a variety of biological control projects. During 1980-82, the Stoneville Research Quarantine Facility (SRQF) received 320 shipments of natural enemies from 23 foreign countries. These organisms were colonized, and 333 shipments were distributed to cooperators in 14 states and 4 foreign countries (Jones et al. 1985). Control by importation and establishment of natural enemies often required colonization in quarantine for host range studies and subsequent distribution. Conservation was supported by providing insects for detailed studies in the laboratory and field. Augmentation often required propagation and release of the natural enemy in the field.

The Stoneville-MSIRP has been essential in the production of insects for evaluating the feasibility of augmenting natural enemies. For example, the selectivity of pyrethroids for *M. croceipes* was demonstrated in the laboratory and field using laboratory-reared wasps (Powell et al. 1986, Powell & Scott 1985, Elzen et al. 1987). The presence of a sex pheromone from female wasps was demonstrated by using traps baited with lab-reared virgin females. Other studies showed that "Heliothis" larvae fed less if parasitized by *M. croceipes* (Hopper & King 1984). During the period 1973 - 1976, 4.4 million tachinids, *Lixophaga diatraeae*, were produced on natural and unnatural hosts (King et al. 1979). These parasites were used to prove, on a large scale, that it is feasible to control a row crop pest, the sugarcane borer *Diatrea saccharalis*, by augmentative releases of *L. diatraeae* (Summers et al. 1976, King et al. 1981). The moth *Bactra verutana* was mass produced for biological control studies on nutsedge. An average of 90,500 larvae per day were produced over an 87-day period (Frick et al. 1983). Finally, it was demonstrated that *Trichogramma pretiosum* can be mass produced, shipped, and applied uniformly over a large cotton-producing area. Parasitism was dramatically increased in release fields and reduced damage was reflected in greater yields. Though technical feasibility was demonstrated, economic feasibility was not (King et al. 1985c).

**Genetic Control.** During the period 1978 to 1983, over 11.2 million tobacco budworm pupae, as well as pupae from a sterile backcross, were produced (King et al. 1985b). Release of the sterile backcross insects on St.

Croix, U.S. Virgin Islands, demonstrated sterility infusion and suppression of the feral tobacco budworm population.

## Regional Multiple-Species Insect Rearing Facility

Insect rearing is essential to nearly all entomological research. This need can best be fulfilled by strategically locating a few facilities to rear several of the most important insect species. This would assure a dependable supply of insects at a cost far less than required to maintain numerous small colonies at many locations. Moreover, often small insect-rearing programs are conducted in poorly designed and equipped facilities by inadequately trained personnel. Often personnel have other duties, and insect rearing has become their responsibility through default. Periodically, programs such as these inevitably fail in their ability to continuously supply projected numbers of high-quality insects.

A regional facility could rear a range of lepidopterous larvae, selected predators and parasitoids, and pathogens that require the continuous production of hosts. Multiple-species insect rearing may be associated with a quarantine facility for importing and clearing natural enemies of pests. Availability of these organisms would facilitate and accelerate research to evaluate new chemicals for effectiveness and selectivity, to screen plant germplasm for resistance, and to introduce biological control agents.

The prototype facility consists of 1,045 m$^2$, with 613 m$^2$ of enclosed space and 432 m$^2$ of balcony or open space (Fig. 3). The enclosed area is divided into 22 rooms and the hallway. The open balcony is used for housing equipment and conducting some rearing operations. Each of the enclosed rooms (except for Harvest, Autoclave, Diet Preparation, and Diet Storage) may be accessed from the hallway through the central part of the facility.

Personnel movement through these rooms may occur only from the covered deck, but disinfected insect life stages may be passed from Harvest to Shipping for packaging. Equipment and containers may be passed from the Autoclave Room to the Cup Room and Tray Room, after being autoclaved or chemically disinfected. Sterilized diet is pumped from the Diet Preparation Room into either the Tray or Cup Rooms. Diet supplies do not enter the facility as they are passed from Storage to the Diet Preparation Room.

Clean conditions should be a goal in all enclosed rooms accessed by the inside hallway, but this is an *absolute necessity* in the two diet dispensing rooms (cup and tray). These two rooms should be equipped with absolute air filters. Moreover, any material entering these rooms should be sterilized either by passage through double door autoclaves or soaking in 1 % sodium hypochlorite.

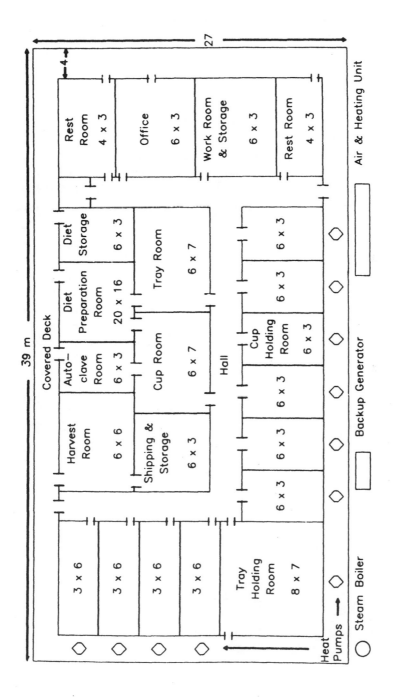

**Fig. 3. Prototype multiple-species insect rearing facility for producing insects in support of research.**

Air and heating for the building are provided by a centralized unit located outside the facility. Ten or more species of insects are to be held in the facility at any one time, and each species may be held under different conditions depending on their environmental requirements. So, the 11 holding rooms should each be equipped with a heat pump for varying temperature and independent sources of humidity and light. Thus, in the event of mechanical heating and cooling problems in one of the units, insect colonies could be temporarily moved to another room.

The inside floors of the facility should be constructed on a concrete slab covered with quarry tile. Each room should be equipped with floor drains. The interior walls of the building should be 9.3 mm (3/8 inch)-thick marine-grade plywood, and it should be painted with marine-grade enamel paint. Under very moist conditions, the walls may be covered with ceramic tile. Exterior walls of the facility can be covered with brick veneer. Energy-saving construction techniques are important. Waterproof fluorescent lamps must be used for lighting within the facility.

Suggested equipment in the facility includes four laminar flow hoods in the diet-dispensing and cup-infesting rooms, two double-door autoclaves, a diet sterilizer with mixer in the diet-preparation room, a diet dispenser for placing diet in cups, a steam boiler for humidifying rooms and operating the autoclaves, a still for provision of sterile deionized water, a backup electrical generator, and a small truck for transporting supplies. Other equipment would include balances, carts, stereoscopic microscopes, shelving, refrigerators, freezers, rearing containers, and other general laboratory supplies.

Twelve people would be sufficient to operate the facility, including a supervisor, maintenance personnel (4), and general workers (7). The movement of personnel in the facility should generally be from clean to less clean areas. Personnel could enter the facility through the restrooms, where street clothing would be exchanged for sterile laboratory garments.

The cost of constructing, equipping, and maintaining a facility such as just described will be expensive during the first year. Thereafter, the annual cost should be considerably less than the initial cost and would include personnel, utilities, supplies, repair, and maintenance. This cost may be greatly reduced by constructing a small facility following the same principles.

## References

Bailey, J. C. & W. R. Meredith, Jr. 1984. Technique for applying *Heliothis virescens (F). (Lepidoptera: Noctuidae)* eggs to cotton. J. Econ. Entomol. 77: 759-762.

Dickerson, W. A., J. D. Hoffman, E. G. King, N. C. Leppla & T. M. ODell. 1980. Arthropod species in culture in the United States and other countries. Entomological Society of America, College Park, MD.

Elzen, G. W., P. J. O'Brien, G. L. Snodgrass & J. E. Powell. 1987. Susceptibility of the parasitoid *Microplities croceipes* (Hymenoptera: Braconidae) to field rates of selected cotton insecticides. Entomophaga 32: 545-550.

Frick, K. E., G. G. Hartley & E. G. King. 1983. Large-scale production of *Bactra verutana* (Lep.: Tortricidae) for the biological control of nutsedge. Entomophaga 28: 107-116.

Hartley, G. G., C. W. Gantt, E. G. King & D. F. Martin. 1977. Equipment for mass rearing of the greater wax moth and the parasite *Lixophaga diatraeae*. ARS-S-164.

Hartley, G. G., E. G. King, F. D. Brewer & C. W. Gantt. 1982. Rearing of *Heliothis* sterile hybrid with a multicellular larval rearing container and pupal harvesting. J. Econ. Entomol. 75: 7-10.

Hartwig, E. E., S. G. Turnipseed & T. C. Kilen. 1984 Registration of soybean germplasm line D75-10169. Crop Science. 24: 214-215.

Hopper, K. R. & E. G. King. 1984. Feeding and movement on cotton *Heliiothis species* (Lepidoptera:Noctuidae) parasitized by *Microplitis croceipes* (Hymenoptera: Braconidae). Environ. Entomol. 13: 1654-1660.

Jenkins, J. N., W. T. Parrott, J. C. McCarty, Jr. & W. H. White. 1982. Breeding cotton for resistance to the tobacco budworm: Techniques to achieve uniform field infestations. Crop Science 22: 400-404.

Jones, Jr., W. A., J. E. Powell & Egar G. King. 1985. Stoneville Research Quarantine Facility: A national center for support of research on biological control of arthropod and weed pests. Bull. Entomol. Soc. Am. 31: 20-26.

King, E. G., G. G. Hartley, D. F. Martin, J. W. Smith, T. E. Summers & R. D. Jackson. 1979. Production of the tachnid, *Lixophaga diatraeae* and its natural host, the sugarcane borer and an unnatural host, the greater wax moth. USDA, SEA, Advances in Agricultural Technology, AAT-S-3.

King, E. G. & G. G. Hartley. 1985a. *Heliothis virescens*. pp. 323-328. *In* P. Singh & R. F. Moore [eds.], Handbook of insect rearing, vol. 2. Elsevier, Amsterdam.

King, E. G., G. G. Hartley, D. F. Martin & M. L. Laster. 1985b. Large scale rearing of a sterile backcross of the tobacco budworm (Lepidoptera: Noctuidae). J. Econ. Entomol. 78: 1166-1172

King, E. G., D. L. Bull, L. F. Bouse & J. R. Phillips. 1985c. Introduction: Biological control of *Heliothis* spp. in cotton by augmentative releases of Trichogramma, pp. 1-10. *In* E. G. King, D. L. Bull, L. F. Bouse & J. R. Phillips [eds.], Biological control of *Heliothis* spp. in cotton by augmentative releases of Trichogramma, vol. 8. Southwest Entomol. Suppl.

King, E. G., J. Sanford, J. W. Smith & D. F. Martin. 1981. Augmentative release of *Lixophaga diatraeae* (Dip.: Tachnidae) for suppression of early-season sugarcane borer populations in Louisiana. Entomophaga 26: 59-69.

Knipling, E. F. 1984. Foreword: What colonization of insects means to research and pest management, pp. ix-xi. *In* E. G. King & N. C. Leppla [eds.], Advances and challenges in insect rearing. Bull. U.S. Dept. of Agricultural.

Meredith, W. R., Jr., B. W. Hanny & J. C. Bailey. 1979a. Genetic variability among glandless cottons for resistance to two insects. Crop Science 19: 651-653.

Meredith, W. R. Jr., V. Meyer, B. W. Hanny & J. C. Bailey. 1979b. Influence of five *Gossypium* species cytoplasms on yield, yield components, fiber properties, and insect resistance in upland cotton. Crop Science 19: 647-650.

Powell, J. E. & William P. Scott. 1985. Effect of insecticide residues on survival of *Microplitis croceipes* adults (Hymenoptera:Braconidae) in cotton. Fla. Entomol. 68: 692-693.

Powell, J. E., E. G. King & C. S. Jany. 1986. Toxcity of insecticides to adult *Microplitis croceipes* (Hymenoptera: Braconidae) J. Econ. Entomol. 79: 1343-1346.

Powell, J. E. & G. G. Hartley. 1987. Rearing *Microplities croceipes* (Hymenoptera: Braconidae) and other parasitoids of Noctuidae with multicellular host-rearing trays. J. Econ. Entomol. 80: 968-971.

Raulston, J. R. & P. D. Lingren. 1972. Methods for large scale rearing of the tobacco budworm. U.S. Dept. Agric. Prod. Res. Rep. 45.

Sikorowski, P. P., J. G. Griffin, J. Roberson & O. H. Lindig. 1984. Boll weevil mass rearing technology. Univ. Miss. Press, Jackson, MS.

Singh, S. P. & R. F. Moore, editors. 1985. Handbook of insect rearing, vol 2. Elsevier, Amsterdam.

Summers, T. E., E. G. King, D. F. Martin & R. D. Jackson. 1976. Biological control of *Diatraea saccharalis* in Florida by periodic releases of *Lixophaga diatraeae*. Entomophaga 21: 359-366.

Vanderzant, E. S. 1974. Development, significance and application of artificial diets for insects. Annu. Rev. Entomol. 19: 139-160.

# 11

# Artificial Rearing Technique for Asian Corn Borer, *Ostrinia furnacalis* (Guenee), and Its Applications in Pest Management Research

*Zhou Darong, Ye Zhihua, and Wang Zhenying*

## Introduction

The Asian corn borer (ACB), *Ostrinia furnacalis* (Guenee), is an important insect pest in the East and Southeast Asian countries, such as Japan, Korea, China, Thailand, the Philippines, Indonesia, and Malaysia. It causes serious damage to corn, sorghum, millet, and cotton. To support research on management of this noxious pest, it is fundamental to establish artificial rearing techniques. Researchers have been developing such systems for the last 30 years.

In Japan, the first artificial diet used to rear the ACB was the same as that for the European corn borer (ECB), *O. nubilalis* (Hubner). Later, Kojima and Nakayama (1979) found that a diet for *Spodoptera litura* (Fabricius) could be used to rear the ACB. Saito and Nakayama (1981) compared the effects of three different diet formulas for ACB rearing and confirmed that, with certain ingredient modifications, the diet for *Spodoptera litura* was best. Miyahara and Saito (1982) improved the technique by using petri dishes (9 cm in diam.) for rearing young larvae, large plastic containers (20 x 14 x 8 cm) for more advanced instars, and corrugated cardboard strips as pupation devices. After rearing for three generations, the yield reached 75 to 116 adults per container without a decrease in larval survival and pupal weights.

In Thailand, research on artificial rearing of ACB was begun in the early 1960s, and preliminary success was obtained. However, Rangdang et al. (1970) reported that some problems remained when the insects were reared in large dishes. Patanakamjorn et al. (1978) modified the formula and rearing technique developed by Rangdang et al. (1971), but the results were still not fully satisfactory under large-scale rearing conditions.

In the Philippines, research on artificial rearing of ACB was begun in the early 1970s. The first recorded attempt at mass rearing of ACB was that of Rodriguez (1972). She used fresh bush sitao, *Vigna sesquipedalis*, pods to grow larvae. However, rearing the borer in a fresh plant medium was laborious and messy, and only a limited number of insects were produced at any given time. Later, Camarao (1976) introduced the use of 'Opaque-2' corn in a synthetic diet modified from the methods of Rangdang et al. (1971). This diet produced an improvement in larval and pupal weights and a corresponding increase in egg mass production. Since Opaque-2 corn was not grown locally, Ceballo and Marallo-Rejesus (1983) substituted ordinary yellow maize plus tryptophan or lysine and found that the survival, growth and development, oviposition, and vitality of the insect all appeared to be the same as that of ACB reared on Opaque-2 corn diet. Hirai and Legacion (1985) improved the diet by using mung bean, *Vigna radiata*, as a substitute for the kidney bean, *Phaseolus vulgaris*, which increased the oviposition rate by 4-fold and reduced the cost of the diet to one-tenth. Microbial contamination was a problem encountered in the laboratory. Adalla et al. (1984) found that 5% sodium hypochlorite solution was an effective surface sterilant for egg masses prior to infestation.

In China, owing to the importance of corn (ranking third in grain crops) and ACB as a corn pest (ranking first), the Chinese government has paid much attention to ACB research ever since the establishment of the People's Republic in 1949. Research on artificial rearing of ACB was begun in 1962. This effort, founded on a knowledge of ECB rearing methods based on the classic studies of Beck (1949), sought to develop a rearing technique based on equipment and ingredients available in China. A preliminary formula was developed. Unfortunately, the research work was interrupted at that time, and could not be continued until years later, when the first formula was tested and verified. Following modification and improvements, this formula was published (Corn Borer Research Group [CBRG] 1975). However, the size and weight of the pupae were reduced, so fourth instar larvae had to be transferred onto a newly prepared diet. A new formula was developed that could be used to rear the ACB from egg to pupa with normal size and weight (Zhou et al. 1980). This was the first successful formula for rearing a phytophagous insect in China.

Beginning in 1979, the rearing container was further improved through a cooperative project between the Institute for Application of Atomic Energy and the IPP of the CAAS. The container (Fig. 1), made of polyethylene, consisted

Fig. 1. Improved rearing container consisting of a dish (A) and cover (B). The assembled container (C) is shown in the background.

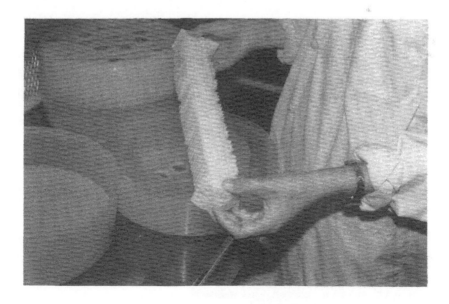

Fig. 2. Special paper strip with honeycomb-like cells used as a pupation site.

of a dish 28.5 cm in diameter and 7.5 cm in height (Fig. 1A) and a cover 27.5 cm in diameter and 7.5 cm in height (Fig. 1B). The cover had 17 ventilation holes. Those in the center were 3 cm in diameter; the rest were 2.5 cm in diameter. All the holes were sealed with 100-mesh copper screening to prevent the larvae from escaping. Fig. 1C shows the assembled container. On the inner side of the cover there was a holding device (Fig. 1B) designed to hold two specially manufactured paper strips, each measuring 54 to 56 cm (L) x 6.5 cm (W) x 3.7 cm (H) with about 600 small cells like that of a honeycomb (Fig. 2), which served as a pupation site. It was advantageous for the collection and separation of the pupae. The dish can be used to hold about 1,000 g of diet for rearing 1,000 to 12,000 larvae.

With the new diet, containers, and corresponding rearing techniques described, peak production at one time was 200 containers with more than 200,000 larvae, producing about 150,000 adults of good quality. The oldest culture maintained in the authors' lab using these techniques was in its 60th generation. The same technique was once used to rear the ECB successfully.

## Diet Formula and Rearing Techniques

### *Diet Formulation and Preparation*

A successful, practical diet for ACB rearing was developed by Zhou et al. (1980) (Table 1). Some of the ingredients, namely, the soybean meal, corn meal, and yeast powder, must be preprocessed before use, either for increasing the utilization ratio of soybean protein or for sterilization purposes. For soybean meal, treatment at 130°C in an oven for about 100 min is needed. For the other two, the temperature of the oven is set at 120°C with the same duration. The preparation procedure is as follows:

1) Mix the three ingredients of Group A.
2) Mix B-1 with B-2 and then add B-3; after dissolved, pour them into Group A. (B-4 is rather expensive in China; it is used only in summer when *Bacillus thuringiensis* might be a problem).
3) Put C-1 into C-2, stir, and allow the corn meal to fully absorb water for several minutes.
4) Put D-1 into D-4 and heat the mixture until ingredients are fully dissolved. Add group C, keep boiling, and stir for 5 min to increase the viscosity of the agar solution (otherwise more agar is needed). Add D-2 directly into the boiling agar solution with constant stirring. A larger

container may be needed to avoid overflowing. Stop heating and add D-3.

5) Cool the agar solution to 60 or 65°C, pour it into the mixture of Group A and B, and stir slowly until the larger particles stop sinking.

6) While the completed diet mixture is still liquid, pour it into rearing containers with a ladle and cool. After the diet solidifies, cut it with long-armed forceps (16.5 cm long) with its two arms separated to a width of 1.5 to 2.0 cm. Cut a cross in the center of the diet, from surface to bottom. Then dig out the cross and place on the surface of the rest of diet. The remaining pieces of diet, thus divided into four separate quarters, are moved toward the center of the container with forceps to form a cross-shaped crevice at the center and a circular crevice between the diet and the wall of the container. These crevices are important because they facilitate feeding and hiding of the larvae.

## Table 1. A practical artificial diet formula for ACB rearing in China

| Group | Ingredients | |
|---|---|---|
| A | 1. Soybean meal | 120.0 g |
| | 2. Corn meal | 120.0 g |
| | 3. Brewer's yeast powder* | 72.0 g |
| B | 1. Glucose (or sucrose) | 60.0 g |
| | 2. Ascorbic acid | 4.0 g |
| | 3. Water | 320.0 ml |
| | 4. Erythromycin | 500,000.0 I.U. |
| C | 1. Corn meal | 32.0 g |
| | 2. Water | 80.0 ml |
| D | 1. Agar | 12.0 g |
| | 2. Sorbic acid | 4.0 g |
| | 3. Formaldehyde (40%) | 1.6 ml |
| | 4. Water | 60.0 ml |

* Produced by the Beijing Brewery.

## Technique for Rearing

Two honeycomb-like pupation strips are put into the holding device of each cover (Fig. 3). There are four small holes spaced at equal distance along the inside of the wall of the holding device.

Egg masses obtained from overwintering moths are used for infestation. The moths are reared in single pairs. The egg masses from each pair of moths are kept separately and laid on wax paper. Only egg masses from those pairs that lay more eggs and without protozoan infection are used to begin a laboratory culture. Before infestation, egg masses are submerged in 2% formalin solution for 20 minutes for disinfection. Individual egg masses are cut out and strung on four insect pins (10 egg masses per pin, and about 40 eggs per mass). The pins are then stuck through four small holes and rest on the paper pupation strip so the eggs will not fall and contact the diet. This technique ensures normal hatching. The containers are kept under a constant temperature of 28°C with 60 to 85% relative humidity and 16 h of illumination for about 17 days until most of the larvae pupate (Fig. 4). The pupation strips are then taken out and put into metal-screen oviposition cages (40 x 25 x 20 cm) (Fig. 5). The cages are kept in an oviposition room at 25 to 28°C constant temperature, 90 to 100% relative humidity, and 16 h illumination for about 8 to 10 days until the adults emerge. After copulation and oviposition, a sheet of wax paper is placed on top of the oviposition cage for oviposition (Fig. 6). The sheet should be changed every day to ensure egg masses are the same age.

**Fig. 3. Two paper pupation strips are put into the holding device inside the cover of the rearing container.**

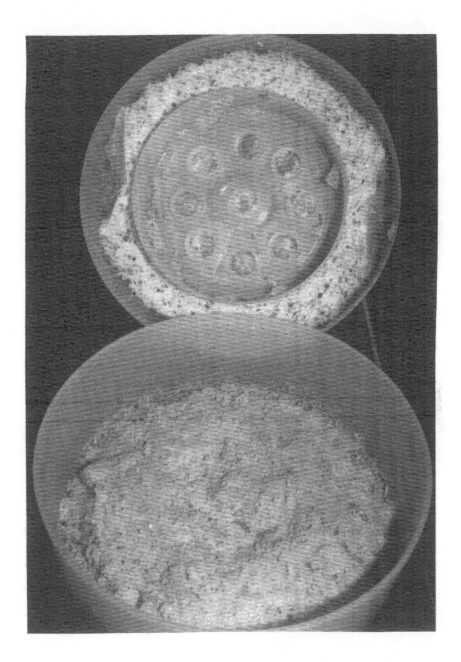

Fig. 4. Interior of container after most of the larvae have pupated.

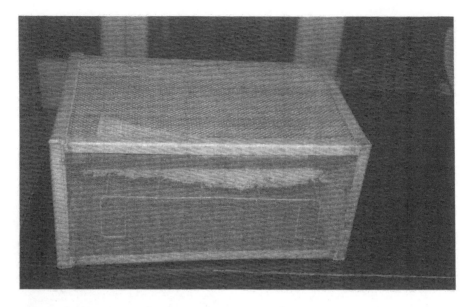

Fig. 5. Paper strips suspended in an oviposition cage.

Fig. 6. A sheet of wax paper is placed on top of the cage for oviposition.

Using these techniques, the average yield per container is about 800 to 850 pupae. The average weight for female and male pupae is about 120 mg and 80 mg, respectively. The average oviposition rate of a single female is about 8 to 10 egg masses with excellent hatchability.

## Quality Control of ACB

To determine whether there is any significant decrease in vitality of successive ACB generations reared on artificial diet, quality control experiments were carried out in 1985 and 1986.

### Flight Ability of ACB Adults

A flight ability test with flight mills showed no significant difference in the artificially reared $F_1$, $F_2$, and $F_{24}$ adults versus the wild ones. Under a 24-h test, the average accumulative flight duration of an adult exceeded 22 h with a distance exceeding 100 km. The artificially induced diapausing $F_{19}$ culture, after termination of diapause and emergence into adults, still maintained flight ability equal to that of the wild population (Table 2). Meanwhile, the $F_{11}$ culture labeled with Sudan Blue-II at a dosage of 0.01% of fresh diet weight during larval development as described by Hang et al. (1984) showed no significant difference from those unlabeled individuals of the same generation (Table 3). The results demonstrated that artificially reared, internally labeled ACB adults could be used for release-recapture experiments to determine the dispersal distance of this species under natural conditions (Zhou & Wang 1986).

### Virulence of ACB Larvae

Guthrie et al. (1971, 1974) reported that the European corn borer, *O. nubilalis*, reared artificially on a meridic diet for more than 14 generations, eventually lost its virulence on susceptible inbred lines of corn. However, this is not the case in ACB reared with our methods. Egg masses of $F_1$, $F_{13}$, $F_{15}$, and $F_{26}$ were used to infest highly susceptible inbred Zi-330, with egg masses of the wild population as a check. The leaf-feeding ratings (1 = Highly resistant; 9 = Highly susceptible) for the four artificial cultures were 8.6, 8.7, 8.8, and 8.7, respectively, while that of the check was 8.7. In addition to leaf-feeding ratings, the virulence in causing holes on the stalk of a susceptible single cross Zhong Dan No. 2 also showed no significant difference. The number of holes per plant was 1.8, 2.2, 1.9, and 1.9 for the four artificial cultures, and 2.1 for the check (Zhou & Wang 1987).

**Table 2. Twenty-four-hour flight performance of wild and artificially reared ACB adults**

| Sex | Generation | No. of tethered moths | Total flight duration (h) x ± SE | Maximum | Flight velocity (m/min) x ± SE | Maximum | Distance of flight (km) x ± SE | Maximum |
|---|---|---|---|---|---|---|---|---|
| ♀ | W | 10 | 22.48 ± 1.57 | 24 | 83.38 ± 10.67 | 134 | 112.887 ± 18.56 | 147.885 |
| | $F_1$ | 10 | 22.09 ± 1.91 | 24 | 85.70 ± 10.01 | 119 | 112.668 ± 11.98 | 131.345 |
| | $F_{11}$ | 8 | 22.09 ± 1.91 | 24 | 82.00 ± 9.04 | 126 | 109.367 ± 16.30 | 150.764 |
| | $F_{19}$* | 6 | 22.16 ± 2.53 | 24 | 83.75 ± 7.63 | 113 | 110.570 ± 17.86 | 137.753 |
| | $F_{24}$ | 7 | 23.40 ± 0.38 | 24 | 78.71 ± 11.01 | 112 | 113.060 ± 13.75 | 133.343 |
| ♂ | W | 5 | 23.32 ± 0.79 | 24 | 76.50 ± 8.58 | 107 | 108.367 ± 16.02 | 122.830 |
| | $F_{19}$* | 8 | 23.76 ± 0.91 | 24 | 75.00 ± 9.51 | 119 | 101.779 ± 9.27 | 122.175 |
| | $F_{24}$ | 7 | 23.65 ± 0.43 | 23.78 | 71.00 ± 4.05 | 97 | 102.631 ± 5.20 | 111.347 |

$F_{19}$* = Artificially induced diapausing larvae.

**Table 3. Influence of internal marking on flight ability of ACB adults**

| Treatment | No. of tethered moths | Total flight duration (h) x ± SE | Maximum | Flight velocity (m/min) x ± SE | Maximum | Distance of flight (km) x ± SE | Maximum |
|---|---|---|---|---|---|---|---|
| CK | 9 | 22.20 ± 1.99 | 24 | 71.22 ± 7.90 | 112 | 93.820 ± 9.74 | 109.736 |
| Marked | 7 | 22.06 ± 1.12 | 23.78 | 65.83 ± 4.72 | 105 | 92.073 ± 7.99 | 100.260 |

## Mating Behaviors of ACB Adults

Wang (1986) reported that under laboratory conditions no isolation in reproduction or preference in mating was observed between artificially reared and wild ACB. This is of utmost importance in conducting a sterile male technique program. Zhang et al. (1985a) reported that artificially reared ACB adults, after being irradiated with a substerile dosage in the pupal stage, performed well in competition with wild males under field conditions.

## Other Criteria

Wang (1986) compared other criteria of $F_{11}$ and $F_{26}$ with first generation offspring of a wild population. No significant differences were found in the duration of larval development, survival rate of larvae, pupation rate, emergence rate, fecundity of females, and longevity of adults.

These rearing techniques can be used to successfully rear the ACB for research purposes under either laboratory or field conditions.

## The Importance of ACB Rearing Technique in Pest Management Research in China

Owing to its economic importance, the ACB has received much attention from the Chinese government and entomologists as well. Soon after the establishment of the PRC in 1949, the first ACB research project was begun in the early 1950s. However, because of the lack of artificial rearing techniques, the wild population was the only source of test material for many years. This became a limiting factor for conducting ACB research requiring large numbers of insects. Furthermore, since ACB is an important insect pest in north China, where the growing season is comparatively short and the winter is long, it is essential that a stable and plentiful supply of test insects are available.

## Identification of Dominant Corn Borer Species in China

The dominant species of corn borer in China has long been known as the European corn borer, *Ostrinia nubilalis* (Hubner). However, Mutuura and Munroe (1970) reported that ECB is distributed only in Europe, Northwest Africa, West Asia, and North America, and that the dominant species in East Asia and Oceania is the Asian corn borer, *O. furnacalis* (Guenee). The morphological characters used to distinguish the two species are not distinct and stable, so classification of the family Pyralidae remained uncertain. Later, a sample of synthetic ECB pheromone did not show biological activity in quite a large area of China. Since 1977 the All China Corn Borer Research Group (ACCBRG) has been conducting a long-term project throughout most of the Chinese territory where corn borers occur. To overcome disadvantages of morphological identification, three other identification techniques were employed: hybridization tests, response of males to different pheromones, and identification and comparison of pheromone chemical structure.

For the hybridization test, overwintering larvae collected from specific regions were reared in the laboratory to emergence and oviposition of adults. Egg masses were placed on artificial diet and reared for one or more generations until enough adults were available. Adults from different regions were paired with each other. The mating preference of the adults, fecundity of females, and fertilization of egg masses were observed, and the parents were considered to be of different species. If not, the fertile egg masses were used to rear for more generations. Adults from this first filial generation were paired for further observation of the fertility of their egg masses. If the eggs were again of good hatchability, the two original parents were conclusively identified as the same species. However, with this method it was still impossible to know the actual species of the tested insects. For further identification, a known ACB or ECB individual was used as one of the parents. In this way, the unknown species could easily be determined.

Pheromone monitoring employed several approaches, in addition to the use of ACB and ECB synthetic pheromone for trapping local males in specific regions. Live virgin females of both species, reared on the diet, were used as lures for trapping. Crude extracts of abdominal tips of virgin females reared on the diet were analyzed for chemical identification of pheromone structures. The project lasted 8 years, involving 34 institutions and nearly 100 participants. The program conclusively demonstrated that the dominant species of corn borer in China is the Asian corn borer, *Ostrinia furnacalis* (Guenee), rather than the European corn borer, *O. nubilalis* (Hubner) as had long been thought (ACCBRG 1988).

## Studies on Corn Varietal Resistance to the ACB

In China, the utilization of corn varieties resistant to the infestation of ACB is considered one of the most important control measures in an integrated pest management program (IPM). However, this important research was not carried out until successful artificial rearing techniques provided the large numbers of egg masses necessary for manual infestation. Now, screening for resistance against ACB is also accomplished in sorghum and cotton. From 1975 to 1983, the All China Corn Borer Research Group (ACCBRG) screened for resistance a total of 1,770 corn samples, including inbreds, hybrids, open-pollinated varieties, and synthetics. Of these, 47 were highly resistant to whorl infestation,

86 were resistant, and 64 were moderately resistant (ACCBRG 1983). Wang et al. (1983) adapted the artificial infestation technique and successfully evaluated 18 inbred lines of corn for resistance against first generation ACB infestation. Li (1987) studied the mechanism and pattern of inherited resistance in corn to ACB by using artificial infestation.

After evaluation of the resistance of 20 inbreds and 44 hybrids to the first and second generation ACB, he found that the resistance of $F_1$ hybrids to the first generation ACB is influenced comparatively more by the paternal inbred than the maternal, whereas with the second generation, it is the opposite. The senior author and his colleagues, in cooperation with other entomologists and breeders of the ACCBRG, developed a highly resistant inbred named Zhi-dian 122 and crossed it with a commercial inbred Huang-zao 4, resulting in a resistant single cross called Zhi-dan No. 1. As the first single cross resistant to ACB occurring in the whorl stage of corn, Zhi-dan No. 1 has been planted by farmers in a total area about 1,000 ha to control the ACB, demonstrating high potential for commercial use (Zhor et al. 1987). Beginning in 1984, scientists for the IPP, CAAS, and IPP of the Jilin Academy of Agricultural Sciences began to screen corn with silk and tassel resistance to ACB in the pollen-shedding stage of corn. Certain resistant lines have been selected that may serve as a base for breeding corn hybrids resistant to both the first and second (in case of spring-sown corn), or second and third (in case of summer-sown corn) generation of ACB, although this work is still preliminary (Liu et al. 1987), Zhang et al. 1985).

The artificially reared ACB were also used to screen sorghum and cotton varieties for resistance to ACB. Chen et al. (1983) reported selecting for two highly resistant sorghum varieties, Red Shell and Hubocuo, from a screening of 514 local varieties in Jilin Province. The leaf damage ratings of the Red Shell and Hubocuo were 1.5 to 2.4 and 1.5 to 2.9, respectively (1=Highly resistant, 9=Highly susceptible), which were significantly different from that of the susceptible check. Based upon the screening work, a breeding program was carried out with the goal of selecting new sorghum varieties not only with good agronomic characters but also with high resistance to the ACB. He found that the content of HCN-p in seedlings and whorl tissues played an important role in resistance to the first generation ACB. After heading, the hardness of the first internode of the plants is a determining character for resistance and is significantly different among varieties.

## Sterile Male Techniques

It is well known that efficient artificial rearing is the most important prerequisite to conducting a sterile technique program, either using chemosterilants or irradiation.

**Sterilization with Chemosterilants.** Gong et al. (1980) reported that adult ACB became sterile after a 30 to 60 sec exposure to the wall of a small plastic screen cage containing 25 micrograms/mm$^2$ of Thio-tepa. Also, when pupae were dipped for 5 min in a 4% aqueous solution of Thio-tepa or for 240 min in a 2% aqueous solution of the same compound, the resulting adults also were sterile. In addition, sterility in the $F_1$ and $F_2$ generations was inherited. Moreover, Thio-tepa can also be used in combination with sex pheromone traps. Male moths, after getting into a specially designed trap, were sterilized and released automatically so as to mate with normal wild females and cause sterility.

**Sterilization with Irradiation.** Scientists of the Institute for Application of Atomic Energy, CAAS, conducted a program using gamma-ray-induced sterility to control the ACB in the early 1960s. However, since an artificial rearing technique was not available by that time, they could use only the overwintering wild ACB. Thus, their research work was limited mainly to the laboratory.

After the success in artificial rearing, an insectary was established with the potential of rearing 20,000 to 30,000 larvae at one time through a collaborative project with the IPP, CAAS. Pilot tests were carried out from 1981 to 1983 on an isolated small island called the Mo-Pan-Shan Isle in Liaoning Province. The ACB were irradiated with a substerilizing dosage of 25 to 30 kR of cobalt-60 gamma ray 1 to 2 days prior to emergence. After irradiation, the pupae were stored at 15°C for 24 h and then released. The irradiated moths mixed evenly and dispersed as well as the wild population. With a 0.64:1 ratio of irradiated to wild moths, the rate of sterile and substerile moths was found to be 44%. The density of ACB larvae per 100 plants decreased by 34.2% compared with the check field, suggesting a definite control effect using this technique, especially considering the very low ratio of irradiated to wild moths (Zhang et al. 1985a).

## Determination of Dispersal Distance of ACB Adults

Although Chinese entomologists have conducted much significant work on ACB since the early 1950s, a very important project, namely, measuring the dispersal distance of ACB adults, has never been carried out because of its technical difficulty. It is not known whether the local overwintering ACB population can be considered the only source for the first generation of next year, or whether immigration from other areas contributes significantly. An area-wide management program must understand the "Minimum Effective Area" that must be treated to produce an area-wide decrease in the ACB population.

Although it is well-known that the ACB is not a migratory insect, Zhai et al. (1989) found that under a flight-mill test conducted in 1985 the ACB adults of a wild population display a considerable potential for long distance flight. Thereafter, Wang (1986) conducted a series of detailed flight-mill tests and proved that the flight ability of artificially reared ACB adults was comparable to that of the wild ACB and thus provided a scientific basis for conducting a release-and-recapture experiment with artificially reared and internally labeled ACB adults to determine the dispersal distance of ACB adults under natural conditions. In 1987 and 1988, about 150,000 and 600,000 labeled ACB adults were reared and released, and 461 and 2,654 specimens were recaptured with pheromone traps at different distances from the releasing site, respectively. Based on the results, it can be concluded that: 1) Although the ACB adults have great potential for long distance flight, under field conditions during overwintering emergence, most of them actually disperse within an area of about 50 $km^2$ from their site of hibernation. Within this area, the moths fly without the aid of wind. 2) The local overwintering population is the main source of ACB for the next year. 3) The "Minimum Effective Area" for a population control program should be not less than 50 $km^2$ surrounding each hibernation site (Zhou et al. unpublished data).

## Rearing of ACB Parasites

Success in rearing the ACB enabled the artificial rearing of ACB parasites. For example, *Macrocentrus linearis* (Nees) is an important natural enemy of ACB larvae. It has two to three generations a year. The natural parasitization rate of ACB larvae by this wasp is about 30 to 40%, with a maximum rate as high as 70%. On artificially reared ACB larvae as the hosts, 39.12 parasitoids were produced per third instar larva. One female parasitoid adult can parasitize an average of 11.7 ACB larvae, yielding 460.05 progeny. No significant difference in biological characters was found between the artificially reared parasitoid and that of the natural population. Further, after releasing artificially reared *M. linearis*, the parasitization rate of ACB larvae in the field was increased (Feng et al. 1987). There was, however, a gradual decrease in vitality in the later generations (Lou 1986).

The Tachinid fly, *Lydella griaescens*, is another important parasite of ACB which is mass produced on artificially reared larvae (Zhao & Wang 1985).

The protozoan *Nosema furnacalis*, is one of the natural control factors of ACB populations under field conditions. Guo et al. (1989) developed a technique to propagate this protozoan under laboratory conditions. Spore propagation in the host follows a logistic pattern. The maximum spore yield ($6.3 \times 10^8$ spores/pupa) was obtained at an inoculation level of $1 \times 10^6$ spores per third instar larvae, with an optimum incubation temperature of 26°C for 19 days. The critical factor for commercial production of *N. furnacalis* is the production of low-cost ACB.

The above programs demonstrate that the use of artificially reared ACB as an intermediate host to rear its natural enemies may serve as an important tool for biological control.

## Studies on Resistance of ACB to Insecticides

Artificial rearing has been widely used to study the resistance of ACB to chemical insecticides in China.

Chemical control, although limited, is still the most effective way to control ACB infestations. In recent years, researchers have successfully utilized mass rearing technology to study resistance of the ACB to insecticides. Liu and Ji (1986) studied the resistance of ACB to fenvalerate. Toxicity tests under laboratory conditions showed that the increase in fenvalerate resistance was quite apparent among ACB after several years of application in cotton. Wei et al. (1987) evaluated the resistance of ACB to parathion and found that the sensitivity of ACB larvae collected from Shangqiu District of Henan Province, where this insecticide had been used for 12 years, was basically the same as that of a laboratory culture in the $F_{14}$ generation. Mu and Wang (1987) conducted detailed studies on the development of resistance to BHC on the ACB. This insecticide has been used in China for ACB control from the 1950s. The results showed that resistance developed in certain provinces, such as Shandong, Hebei, Henan, Anhui, Tianjin, Jiangsu, Liaoning, Shanxi, and Shaai where corn was planted twice a year and large amounts of BHC had been applied. Conversely, in San-dao-he of Hailin County, Heilongjiang Province, where the corn corp has never been treated with any insecticides, the resistance was just less than one-tenth of the above. In certain provinces where corn was planted only once a year and less BHC has been used, such as Jilin, Gansu, and some mountainous regions of South China, the resistance has not significantly developed. In addition, populations resistant to BHC do not show any cross-resistance to methyl parathion, parathion, or Phoxim, which have not been widely used in ACB control but showed very slight cross-resistance to Carbofuran and Sumicidin. Four insecticides and one mixed insecticide were tested topically on the fifth instar larvae to select for resistance. After continuous selection for up to 30 generations, one strain with 40-fold resistance to fenvalerate and another with 30-fold resistance to Furandan were obtained (Mu & Wang 1988). Obviously, it would be almost impossible to conduct such experiments without the artificial rearing technique.

## Basic Research Using ACB Artificial Rearing Technique

In addition to the pest management research mentioned above, the ACB artificial rearing technique has been used in other studies. Studies on the influence of photoperiod and temperature on the diapause of the ACB indicated that a short light signal influenced diapause (Gong et al. 1984). The critical photoperiod of the ACB population collected from Qinshue (35°N and 112°-113°E) of Shanxi Province was 14.5 h at 25°C. Xiao (1988) studied the phototactic response of the newly hatched ACB larvae and found it to be positive rather than negative as in the ECB. No difference in this positive response was found due to light intensities between 30,000 and 130 Lux or between wild or artificially reared $F_{46}$ cultures. Wang (1989) studied the effects of two major meteorological factors, temperature and humidity, on the survival and fecundity of ACB. Under constant temperatures ranging from 15 to 35°C, the hatching rate increased with increases in relative humidity. The favorable temperature range for hatching was from 18 to 32°C, 26.1°C appearing optimal. During pupation, if relative humidity is too high or low, pupation declines along with subsequent emergence and oviposition. The artificial rearing technique was especially useful for determining the influences of short-duration treatments (4 to 8 h) and high (34 and 38°C) or low (6°C) temperatures on fecundity. Chen et al. (1980, 1982) conducted preliminary studies on use of juvenoids and effect of chitin synthesis inhibitors on the larvae. Cheng and Zhu (1986) studied the effects of L-canavanine on the growth and development of ACB. Chiu et al. (1984) conducted research on growth-disruptive effects of azadirachtin on the larvae.

## A New Nonagar Diet for ACB Rearing

Agar though expensive is still widely used in artificial rearing of insects. However, a substance called *JSMD*, can be used as an agar substitute with very promising effects. A very importance advantage of the new formula is that no cooking is needed in the preparation procedure. In 1988, the new diet and improved techniques were used to rear 800 containers of ACB at one time. More than 600,000 healthy adults were produced and labeled with 0.01 % Sudan Blue-II. The new nonagar diet and improved techniques will be published soon after a patent is granted.

## Conclusion

The success of ACB artificial rearing has contributed greatly toward pest management research on the ACB. Moreover, the principal diet ingredients and related rearing techniques should be useful for rearing other species of the lepidopterous borers.

## References

Adalla, C. B., M. T. Caasi & D. M. Legacion. 1984. Refinement of mass-rearing techniques for corn borer, *Ostrinia furnacalis* (Guenee) (Pyraustidae: Lepidoptera). Phil. Agri. 67: 345-349.

All China Corn Borer Research Group, IPP. CAAS. 1975. An artificial diet for rearing the corn borer. Lun Chong Zhi Shi (4): 22.

All China Corn Borer Research Group. 1983. An evaluation of resistance of corn varieties to the Asian corn borer. Plant Protection 9(2): 41-42.

All China Corn Borer Research Group. 1988. Studies on the identification of the dominant corn borer species in China. Acta Phytophyl. Sinica 15 (3): 145-152.

Beck, S. D. 1949. Nutrition of the European corn borer, *Pyrausta nubilalis* (Hbn.) I. Development of a satisfactory purified diet for larval growth. Ann. Entomol. Soc. Amer. 42: 483.

Camarao, G. C. 1976. Population dynamics for the corn borer, *Ostrinia furnacalis* (Guenee) I. Life cycle, behavior and generation cycles. Phil. Entomol. 3: 1179-2000.

Ceballo, F. A. & B. Morallo-Rejesus. 1983. Trytophan and lysine supplemented artificial diet for corn borer (*Ostrinia furnacalis* Guenee) Phil. Entomol. 6: 531-538.

Chen, Pei, Huifen Gong, Zhihong Xia, Meili Lian, Rui Wang, Ruilin Liu & Mingshi Fu. 1980. Preliminary studies on the application of Juvenoids in rearing European corn borer, *Ostrinia nubilalis* (Hbn.) Acta Entomol. Sinica 23: 224-227.

Chen, Pei, Huifen Gong, Rui Wang & Zhihong Xia. 1982. Activity and effect of the chitin synthesis inhibitors against the larvae of the Asiatic corn borer, *Ostrinia furnacalis* (Guenee). Acta Phytophyl Sinica 9: 217-222.

Chen, Shurong, Jun Dai & Huimin Chen. 1983. Preliminary studies on resistance of sorghum varieties to Asian corn borer, *Ostrinia furnacalis* (Guenee). Jilin Agri. Sci. (2): 15-17.

Cheng, Zhenheng & Zhiqiang Zhu. 1986. L-canavanine effects on growth and development of *Ostrinia furnacalis*. Acta Entomol. Sinica 29: 143-148.

Chiu, Shin-foon, Xing Zhang, Siu-king Liu & Duanping Huang. 1984. Growth disruptive effects of azadirachtin on the larvae of the Asiatic corn borer (*Ostrinia furnacalis* Guenee). Acta Entomol. Sinica 27: 241-247.

Feng, Jianguo, Gengshen Shi, Xun Tao, Yong Zhang & Shirong Jiang. 1987. Biology of *Macrocentrus Linearis* (Nees) and its use against Asian corn borer, *Ostrinia furnacalis*. Chinese J. Biological Control 3(3): 102-105.

Gong, Huifen, Pei Chen & Rui Wang. 1980. Studies on chemosterilization of the Asian corn borer. Acta Phytophyl. Sinica 7: 239-246.

Gong, Huifen, Pei chen, Rui Wang, Meili Lian, Zhihong Xia & Yi Yang. 1984. The influence of photoperiod and temperature on the diapause of the Asian corn borer, *Osternia furnacalis* (Guenee). Acta Entomol. Sinica 27: 280-287.

Guo, Peilin, Binjun Li & Chentseng Wenn. 1989. Mass propagation of *Nosema furnacalis* (Microsporida: Nosematidae) in *Ostrinia furnacalis* (Lep. Pyralidae). Chinese J. Biological Control 5(1): 30-33.

Guthrie, W. D., W. A. Russell & G. W. Jennings. 1971. Resistance of maize to second brood European corn borer. Proc. Annual Corn and Sorghum Research Conf. 26: 165-179.

Guthrie, W. D., Y. S. Rathore, D. F. Cox & G. L. Reed. 1974. European corn borer: Virulence on corn plants of larvae reared for different generations on a meridic diet. J. Econ. Entomol. 67: 605-606.

Hang, Shuqun, Qiongru Liu, Guiying Wang, Xinzhi Huang & Shujie Wang. 1984. Studies on mass rearing and marking technique of corn borer, *Ostrinia furnacalis* (Guenee). Application of Atomic Energy in Agriculture 4: 27-33.

Hirai, Y. & D. M. Legacion. 1985. Improvement of the mass-rearing techniques for the Asiatic corn borer, *Ostrinia furnacalis* (Guenee) in the Philippines. Japan Agricultural Research Quarterly 19: 224-233.

Kojima, I. & I. Nakayana. 1979. Rearing of various lepidopterous larvae on artificial diet of identical composition. Jap. J. Appli. Entomol. Zool. 23: 261-263.

Li, Jianping. 1988. Mechanism of resistance to first generation Asian corn borer, *Ostrinia furnacalis* (Guenee), found in sorghum. M.S. thesis, Graduate School, CAAS.

Li, Xinhua. 1987. Assessment of resistance in maize to Asian corn borer and its genetic law. Journal of Shenyang Agricultural University 18(1): 7-15.

Liu, Dejun & Huo Ji. 1986. The resistance of *Ostrinia furnacalis* (Guenee) to fenvalerate. Plant Protection 12(2): 33.

Liu, Haifeng, Yungsheng Wang & Rong Zhang. 1987. Preliminary studies on resistance of corn silks to Asian cron borer, *Ostrinia furnacalis* (Guenee). Sci. Agri. Sinica 20(6): 58-60.

Lou, Juxian. 1986. A preliminary report on artificial breeding *Macrocentrus linearis* (Nees). Journal of Shenyang Agricultural University 17(4): 61-65.

Miyahara, Y. & O. Saito. 1982. Improved rearing method for *Ostrinia furnacalis* (Lep:Pyralidae) on an artificial diet. Jap. J. Appl. Entomol. Zool. 26: 160-165.

Mu, Liyi & Kaiyun Wang. 1987. Studies on the resistance of Asian corn borer to BHC and alternation insecticide in China. Acta Phytophyl. Sinica 14: 209-215.

Mu, Liyi & Kaiyun Wang. 1988. Studies on selection of resistance in Asian corn borer with insecticides in laboratory. Ibid. 15: 209-214.

Mutuura, A. & E. Munroe. 1970. Taxonomy and distribution of the European corn borer and allied species: 27 Genera: *Ostrinia*. Mem. Entomol. Soc. Canada.

Patanakamjorn, S., W. D. Guthrie & W. R. Young. 1987. Meridic diet for rearing the tropical corn borer, *Ostrinia furnacalis*. Iowa State Journal Research 52: 361-370.

Rangdang, Y., Y. Charosensom & R. Grandos. 1970. Entomological problems in corn. Thailand National Corn and Sorghum Research Center (Faculty of Agriculture, Kasetsart University, Bankok) Annual Report 1970:76-78.

Rangdang, Y., S. Jamormarn & G. Granados. 1971. Mass rearing corn stem borer, *Ostrinia furnacalis*. Ibid. Annual Report 1971: 264-266.

Rodriguez, C. P. 1972. Topical toxicity of ten insecticides to the larvae of *Ostrinia furnacalis* (Guenee). B.S. thesis, University of the Philippines.

Saito, O. & I. Nakayama. 1981. A simple rearing method for the oriental corn borer, *Ostrinia furnacalis* (Guenee) (Lepidoptera; Pyralidae). Bulletin of the Tohoku National Agri. Exp. Sta. 63: 242-247.

Wang, Yunsheng, Rong Zhang & Fenfen Feng. 1983. An evaluation of resistance of some corn inbred lines to the Asian corn borer. Acta Phytophyl. Sinica 10: 231-234.

Wang, Zhenying. 1986. Studies on quality control index in the mass-reared Asian corn borer, *Ostrinia furnacalis* (Guenee), on a practical diet. M.S. thesis, Graduate School, CAAS.

Wang, Zhongyue. 1989. Studies on effects of major meteorological factors on survival and fecundity of the Asian corn borer, *Ostrinia furnacalis* (Guenee). M. S. thesis, Graduate School, CAAS.

Wei, Chen, Ruiyun Wei, Xianlin Fan & Shaoning Huang. 1987. Tests for the resistance of Asian corn borer, *Ostrinia furnacalis* (Guenee) to Parathion. Plant Protection 13(6): 28-29.

Xiao, Liang. 1988. Studies on the behavior of the newly-hatched larvae of the Asian corn borer, *Ostrinia furnacalis* (Guenee). M.S. thesis, Graduate School, CAAS.

Zai, Baoping & Ruilu Chen. 1989. Flight capacity of Asian corn borer. Jilin Agri. Sci. (1): 40-46.

Zhang, Hequin, Caidao Zhao, Huasong Wang, Xinzhi Huang & Shujie Wang. 1985a. The effectiveness of irradiation-substerile-technique for controlling corn borer population on Mo-Pan-Shan Island. Application of Atomic Energy in Agriculture (2): 5-9.

Zhang, Rong, Yunsheng Wang & Weijiu Guan. 1985b. Preliminary studies on the resistance of the DIMBOA contents in whorl leaves and corn silk to Asian corn borer, *Ostrinia furnacalis* (Guenee). Jilin Agri. Sci. (3): 66-70.

Zhao, Ianming & Xingjian Wang. 1985. Studies on mass rearing of the tachinid fly, *Lydella grisescens* on its natural host, corn borer. ActaPhytophyl. Sinica 7: 113-122.

Zhou, Darong, Yuying Wang, Baolan Liu & Zhengli Ju. 1980. Studies on the mass rearing of corn borer I. Development of a satisfactory artificial diet for larval growth. Acta Phytophyl. Sinica 7: 113-122.

Zhou, Darong, Zhengli Ju, Ruiyun Wei, Caiceng Chen, Yunxia Gao, Liping Wen, Kanglai He, Xiuzhi Li & Chengde Liu. 1987. Utilization of corn borer resistance in maize and introduction of a resistance single cross Zhidan No. 1. Plant Protection 13(5): 16-18.

Zhou, Darong & Zhenying Wang. 1986. Comparison of flight ability between wild and mass reared Asian corn borer *Ostrainia furnacalis* (Guenee). Proc.14th Symp. Int. Working Group on *Ostrinia*, Beijing. 165-169.

Zhou, Darong & Zhenying Wang. 1987. Quality evaluation of the mass reared Asian corn borer, *Ostrinia furnacalis* (Guenee). Prod. Int. Symp. on Modern Insect Control: Nuclear Techniques and Biotechnology, 1987. Vienna, IAEA-SM-310/5: 281-284.

# 12

## Comparison of Artificial Diets
## for Rearing the Sugarcane Borer

*J. R. P. Parra and L. H. Mihsfeldt*

### Introduction

The sugarcane borer, *Diatraea saccharalis* Fabr., occurs in every agroecosystem of the western hemisphere where sugarcane (interspecific hybrids of *Saccharum officinarum* L.) is grown (Williams et al. 1969). Besides sugarcane, the borer also attacks other important crops such as corn, sorghum, rice, and wheat, as well as wild grasses (Silva et al. 1968).

This insect causes both direct and indirect losses; in Brazil, high losses are caused in the production of sugar and alcohol. Considering that Brazil is the world's largest sugar and alcohol producer, the importance of this insect cannot be overemphasized (Gallo et al. 1988). For each 1% intensity of infestation, losses occur in the order of 0.5% in sugar, 1.4% in alcohol, and 0.14% in weight. In addition, this intensity of infestation often reaches 10% in the main producing centers of this country. In the State of Sao Paulo alone, losses reached US$ 119 million in 1983 (Botelho 1984).

As a general rule, in Latin America this insect is controlled biologically; in Brazil this is done mainly through the introduction and multiplication of parasitoids, such as *Cotesia flavipes* Cameron (Botelho 1980) or through increasing native larval parasitoids, such as *Metagonistylum minense* Towns, and *Paratheresia claripalpis* Wulp. (Gallo 1952, Macedo et al. 1983). In the last few years, the possibility of utilizing natural enemies of eggs, such as *Trichogramma* spp. (Parra et al. 1987), is being investigated.

Biological control required development of a technique for rearing *Diatraea saccharalis* in the laboratory, ultimately making its mass production feasible. Natural diets were initially used for this purpose (Holloway & Loftin 1919, Holloway et al. 1928, Meadows 1938, Bergamin 1943, Taylor 1944). Since

195

1960, when research on artificial diets underwent a great advance, studies of *Diatraea saccharalis* rearing on meridic diets were greatly increased.

After the technique for rearing the insect on an artificial diet was mastered in the 1980s, other aspects of insect rearing, such as the control of microorganisms, mechanization, quality control, and lowering the cost of the diet were approached. In these studies, the diets of Adkisson et al. (1960) and Hensley and Hammond (1968), based on casein and wheat germ, were used (King & Hartley, 1985).

Fig. 1, extracted from Roe et al. (1981), shows the chronology of the studies of *Diatraea saccharalis* on natural and artificial diets.

In Brazil, studies on artificial diets for *Diatraea saccharalis* were initiated in the Department of Entomology, Escola Superior de Agricultura "Luiz de Queiroz," Piracicaba, Sao Paulo, Brazil. Here Gallo et al. (1969) utilized Hensley and Hammond's diet (1968), applying some modifications to meet Brazilian conditions. These studies were followed by the research of Sgrillo et al. (1977), Mendonca Filho (1973), Macedo et al. (1983), and Araujo et al. (1985), always with adaptations of diets utilized in other countries.

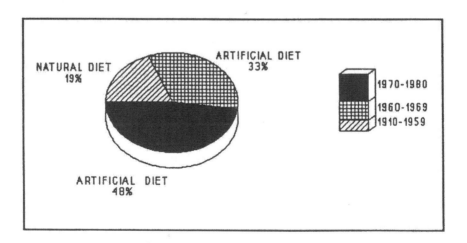

**Fig. 1. Chronology of studies on *D. saccharalis* using natural and artificial diets, according to Roe et al. (1981).**

The objective of this study was to compare the establishment and development of *Diatraea saccharalis* on different artificial diets, as compared with Hensley & Hammond's diet (1968), which is still used to a great extent in Brazil where it has become a standard in developing or adapting diets compatible with Brazilian conditions.

## Materials and Methods

The research was conducted in the biology laboratories of the Department of Entomology, Escola Superior de Agricultura "Luiz de Queiroz," Universidade de Sao Paulo, in Piracicaba, State of Sao Paulo, Brazil. Laboratories were maintained at 25 ± 1°C, 70 ± 10% relative humidity, and 14 h photophase. The *Diatraea saccharalis* were obtained from stock cultures maintained on King and Hartley's (1985) artificial diet. All of the methodology was based on Melo and Parra (1988); glass tubes 8.5 cm in length and 2.5 cm in diameter were used for rearing the larvae.

In the first phase of the study some diets containing variable protein sources (beans, soybeans, corn, wheat germ, wheat meal, and casein) were tested in four experiments:

**Experiment 1** - Bean varieties (Aeta, Aroana, Carioca, Catu, Goiano Precoce, and Rosinha G 3) and hybrid corn meal.

**Experiment 2** - Wheat germ, soybean meal (commercial product Sojinha), yeast, casein, beans (Carioca), and wheat meal.

**Experiment 3** - Soybean meal (varieties IAC 2, IAC 5, IAC 6, IAC 7, IAC 8, IAC 9, IAC 773090, and Santa Maria) and corn meal (variety Nutrimaiz).

**Experiment 4** - Corn meal (varieties Nutrimaiz, VDwx, VD-2 $br_2o_{29}$ , VD-2 $br_{29}$ $su_2o_2br_2$).

These diets were compared with two standard diets (C and D in Table 1) based on Hensley and Hammond (1968).

The best diets were selected by evaluating the following criteria: phagostimulation (beginning of the feeding), duration and viability of the larval and pupal stages; weight of pupal at 24 h of age; percent of adults with malformed wings, and percent contamination of diets by microorganisms (especially fungi).

In the first phase, two diets were selected and compared with the diets considered standard in a second phase.

In the second phase, the following parameters were studied in addition to those analyzed in the first: preoviposition period, total number of eggs per

**Table 1. Composition of the artificial diets utilized for rearing *D. saccharalis***

| Component | Diets | | | |
|---|---|---|---|---|
| | A | B | C | D |
| Corn meal | - | 28g[3] | - | - |
| Yeast | 25.5g | 7.5g | - | - |
| Wheat Germ | 40g | 7g | 108g | 54g |
| Casein | - | - | 108g | 54g |
| Cooked beans | 173g[2] | - | - | - |
| Sucrose | - | - | 180g | 90g |
| Wessons salts | - | - | 36g | 18g |
| Choline chloride | - | - | 3.6g | 1.8g |
| Vitamin solution[1] | - | 36ml | 20ml | - |
| Ascorbic acid | 2.6g | 1g | 14.4g | 7.2g |
| Benzoic acid | - | 0.25 g | - | - |
| Formaldehyde | 0.65ml | - | 1.8ml | 2ml |
| Glacial acetic acid | - | - | 2.0ml | - |
| Methyl para-hydroxibenzoate | 1.6g | 0.2g | 5.4g | 2.7g |
| Tetracycline | - | - | - | 1,000mg |
| Aureomycin | - | - | 1,000mg | - |
| Sorbic acid | 0.8 g | - | - | - |
| Water | 611ml | 200ml | 3,116ml | 1,450ml |
| Agar | 10.2g | 5g | 72g | 36g |

Note: Source of diets. A = Burton (1969);
B = Poitout & Bues (1970);
C and D = Hensley & Hammond (1968).
[1]Composition of the vitamin mixture:
    n= niacinamide, 1.0g
    calcium pantothenate, 1.0g
    thiamine, 0.5g
    pyridoxine, 0.25g
    folic acid, 0.1g
    biotin, 0.02g
    vitamin B12 (1,000g/cm3), 2.0 ml.
[2]'Carioca' variety.
[3]'Nutrimaiz' variety.

female, egg viability per laying day, incubation period, longevity of males and females, and number of larval instars (determined by measuring the head capsule width).   Life tables were prepared (Andrewartha & Birch 1954) and the following nutritional indices were determined based on Waldbauer (1968) and Slansky & Scriber (1985): AD, approximate digestibility; ECI, efficiency of conversion of ingested food; and ECD, efficiency of conversion of digested food.

Total protein on the four diets and of the larvae at maximum development was analyzed through disk electrophoretic analysis in polyacrylamide gel (Brewer & Ashworth 1969).   A cluster analysis (Sneath & Sokal 1973) was conducted to compare development of *Diatraea saccharalis* on the four diets using operational taxonomic units (Otu's) involving both biological and nutritional parameters.   Costs of the components were also taken into account for diet selection.

## Results and Discussion

After conducting the first four experiments, the following two diets were selected for comparison with the diets considered as standard: the wheat germ, Carioca bean (Parra & Carvalho 1984), and yeast (diet A); and the corn meal ("Nutrimaiz"), wheat germ, and yeast (diet B).

Diets with a total viability (i.e., egg-adult survival) below 75% were discarded, a reasonable limit for a diet to be considered nutritionally adequate for insect rearing (Singh 1983).   There were cases in which the discarded diets showed viabilities over 75%; however, they were inferior with regard to other biological aspects considered, such as phagostimulation, duration of immature phases, weight of pupae, malformation of adults, and contamination by microorganisms, especially fungi.

Larval phase duration was shorter on the standard diets (C and D) (Table 2).

Values obtained for the two diets selected for comparison are within the range suggested in some of the literature (Bowling 1967; Brewer 1981), even though many of the studies were conducted at higher temperature and duration of the larval period (King et al. 1975, Melo & Parra 1988).   Although diet B lengthened the larval phase in relation to the standard diets, it presented the highest viability and the lowest contamination by *Aspergillus* spp.   In general, contamination by fungi was considerably lower than that found by Hensley and Hammond (1968), (11% at 25 ± 1°C); by Roe et al. (1982), (16.2% at 27 ± 1°C); and by Melo and Parra (1988) (1 to 6.5% at 10 to 32°C range).   Number of instars on the four diets varied from five to six.   Diet B had a higher percentage of larvae with six instars (Table 2), and growth ratio on all diets

**Table 2.** Duration, viability of larval phase, percentage of larvae with six instars and percentage of diet contaminated by microorganisms for *Diatraea saccharalis* reared on four artificial diets

| Diet[a] | Duration (days) | Viability (%) | Larva with 6 Instars (%) | Contamination (%) |
|---------|-----------------|---------------|--------------------------|-------------------|
| A | 27.01 ± 3.04a | 89.00a | 40 | 3.00 |
| B | 24.12 ± 3.40b | 93.00a | 80 | 0.00 |
| C | 22.30 ± 2.03c | 87.00b | 30 | 5.00 |
| D | 21.45 ± 2.24d | 89.00b | 30 | 3.00 |

Means followed by the same letter do not differ statistically from each other by Tukey's test at the 5% level of probability.

[a]Diet  A = yeast, wheat germ, cooked beans;
        B = corn meal, yeast, wheat germ;
        C = wheat germ, casein;
        D = wheat germ, casein.

followed Dyar's rule (Dyar 1890), in accord with Roe et al. (1982).

Weight of pupae was higher for females than for males on all diets, a characteristic that is typical for *Diatraea saccharalis* (Santa Cruz et al. 1964, Bowling 1967, King et al. 1975, Brewer 1976, Melo & Parra 1988). Lighter weight pupae originated from only diet A. For females, the heavier pupae were obtained on diet C (Table 3). Egg-laying capacity was constant for the four diets (Table 3).

The preoviposition period was constant and equal to two days on all of the diets, and male longevity was higher than that of females.

Although Araujo et al. (1980) and Melo and Parra (1988) obtained higher egg laying under lower temperatures (20-22°C), the total number of eggs obtained at 25°C was quite high and similar to that reported by King and Hartley (1985). The 1:1 sex ratio used in this study was optimal for oviposition. Other authors varied the sex ratio inside the observation cages (Miskimen 1965, Sgrillo 1973, Guevara 1976, Araujo et al. 1980).

Pupal phase durations were very similar and mortality very low on all of the diets (Table 3).

The percentage of adults with malformed wings was 2, 1, 4, and 2%, respectively, on diets A, B, C, and D.

**Table 3.** Duration, viability of the pupal phase, weight of pupae and number of eggs per female of *Diatraea saccharlis* reared on four artificial diets

| Diet [a] | Duration (days) | Viability (%) | X | Weight (mg) | TT eggs/ female |
|------|------|------|------|------|------|
| A | 8.20a | 98.00 | 132.72b | 104.94c ± 160.52c | 749.9a |
| B | 8.61a | 94.00 | 149.45a | 118.89ab ± 180.00b | 866.9a |
| C | 8.00a | 96.00 | 154.45a | 113.26b ± 195.64ab | 738.6a |
| D | 8.24a | 99.00 | 147.52a | 122.04a ± 173.00b | 766.2a |

Means followed by the same letter do not differ statistically from each other by Tukey's test at the 5% level of probability.

[a]Diet  A = yeast, wheat germ, cooked beans;
        B = corn meal, yeast, wheat germ;
        C = wheat germ, casein;
        D = wheat germ, casein.

The egg-laying rate varied according to diet with the highest percentage on the first day (Fig. 2). On diets A and C, there were only six egg-laying days, and females began to diet from the fourth day on. Thus, on the fourth, fifth, and sixth days of egg laying, few females remained and eggs oviposited. On the sixth egg-laying day, hatching percentage was null for both diets. On diets B and D, death of females started on the fifth day, and egg viability remained high through the seventh. The incubation period varied from 6.08 days (diet B) to 6.74 days (diet C) and was statistically similar on all of the diets. Percent hatch was 89, 87, 83, and 87% for diets A, B, D, and C, respectively. The phagostimulating effect of the diets on neonate larvae was high (93-98%). The highest value was obtained on the corn diet (diet B) and the lowest value was on the C diet.

The net reproduction rate (Ro) was higher on the Nutrimaiz corn diet (diet B), where population increased 338.38 times for each generation. Values for the remaining diets were lower and similar to each other. The lowest value (261.88) was found on diet C; however, it was higher than that registered by Melo (1984). The shortest period to complete a generation (T) was found on diet C (41.19 days), a value much lower than that found by Melo and Parra (1988) (57.90 days). The longest period was shown for the Carioca bean, wheat

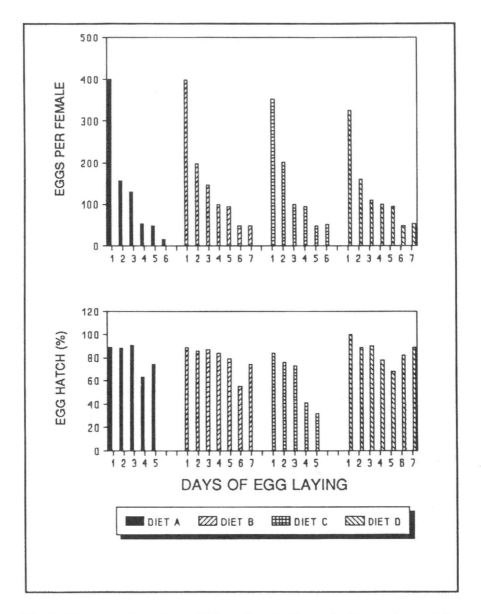

**Fig. 2.** Mean number of eggs laid per female *D. saccharlis* per day and the percentage of daily hatching of those eggs for insects reared on diets A (wheat germ, yeast, and beans), B (corn meal, yeast, and wheat germ), C (wheat germ and casein), and D (wheat germ and casein).

germ, and yeast diet (diet A). Diets B and D had intermediate values for the duration of a generation. The capacity to increase in number (rm) was lower on diet A and presented similar and higher values for the remaining diets. The finite increase ratio was similar for the standard diets and the Nutrimaiz corn diets and lower on diet A (Table 4).

The amount of food consumed (FC) was similar on the four diets (Fig. 3 and 4). However, weight gained by larvae at maximum development (WG) was higher on the diet containing Nutrimaiz corn (diet B). Greater conversion efficiency (ECI and ECD) was also shown on this diet, demonstrating its greater nutritional suitability.

The percentage of total protein was lower on diet B (17.88) and about the same for the other three diets. The value found for diet C was 24.14%, which was similar to that reported by Martins (1983). However, the protein contents of larvae at maximum development were similar (variation from 43.08 to 46.64%). This indicated that the larvae compensated for different protein percentages of the diets. Thus the corn, wheat germ, and yeast diet (diet B) was the most suitable for sugarcane borer rearing. In spite of presenting a low protein percentage, the conversion efficiency of larvae was very high on this diet, possibly due to the greater availability of the proteins. Electrophoretic analysis revealed that diet C had the higher number of protein bands; this is due probably to its number of components. The Carioca bean, wheat germ, and yeast diet (diet A) had 12 bands, while diets B and D had 11 and 9, respectively. Therefore, no correlation existed between the presence of bands and protein availability for the insect.

**Table 4. Fertility life table of *Diatraea saccharlis* reared on four artificial diets**

| Diet[a] | $r^m$ | T | Ro | λ |
|------|--------|-------|--------|--------|
| A | 0.1209 | 46.26 | 268.72 | 1.1285 |
| B | 0.1336 | 43.60 | 338.88 | 1.1429 |
| C | 0.1352 | 41.19 | 261.88 | 1.1448 |
| D | 0.1317 | 42.68 | 276.74 | 1.1408 |

[a]Diet  A = yeast, wheat germ, cooked beans;
  B = corn meal, yeast, wheat germ;
  C = wheat germ casein;
  D = wheat germ, casein.

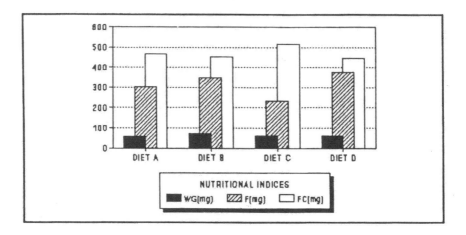

Fig. 3. *D. saccharalis* larvae: food consumed (FC), feces (F), and weight gain (WG) on four different artificial diets (A, B, C, and D).

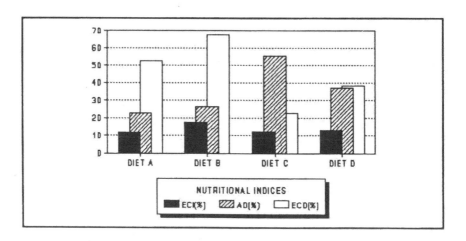

Fig. 4. Nutritional indices of *D. saccharalis* larvae on the four artificial diets under study. AD = approximate digestibility; ECI = efficiency of conversion of ingested food; ECD = efficiency of conversion of digested food.

## Conclusions

The phenogram depicts two distinct groups: one consisting of the diets taken as standard (diets C and D) plus the Carioca bean, wheat germ, and yeast diet (diet A), and the other consisting of the Nutrimaiz corn, yeast, and wheat germ diet (diet B) (Fig. 5).

This analysis shows that diet B may be used for *Diatraea saccharalis* rearing, especially under Brazilian conditions, due to the availability of ingredients and their low cost. The production cost of this diet is 2.70 times lower than that of diet C and 2.54 times lower than that of diet D (same diet with additional anticontaminants). In addition, diet B produced the highest total viability, had no contamination by fungi, and resulted in a low percentage of malformed adults and higher egg-laying capacity. There was, however, a slight lengthening of the larval phase as compared with the standard diets (C and D). This lengthening of the larval phase led to a higher percentage of larvae with 6 instars, although, according to Parra et al. (1977) and Slansky and Rodriguez (1987), a higher number of instars is a consequence of nutritional unsuitability.

If 75% total viability is established as the threshold level for an artificial diet to be used in mass rearing, the Nutrimaiz corn diet (diet B) may be considered as the most suitable for sugarcane borer rearing.

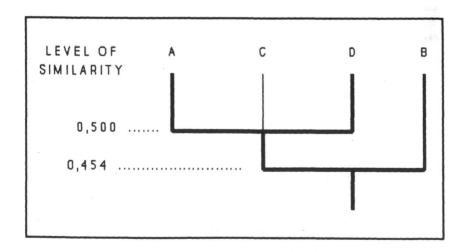

**Fig. 5. Phenogram obtained from cluster analysis of four artificial diets (A, B, C, and D) selected according to their biological and nutritional parameters.**

Diets taken as standard, especially that of Hensley and Hammond (1968) were below that value. The importance of corn in this diet was evident, since the components of diet A are the same, except that beans were substituted for corn. The Nutrimaiz corn has an amino acid profile with increased amount of lysine (4.35%) and tryptophan (1.01%), and a reduction in leucine (7.47%), (Silva et al. 1978). This is different from ordinary corn. In addition, the VD-2 $br_2o_2$ variety of corn (opaque corn) has an increase in lysine due to a reduction of zein and, consequently, increased albumine, globuline, and gluteline (Tosello 1980). This variety and the mutant V-475 (Maya-opaque-2-$o_2$Su) may be utilized in the diet for rearing *Diatraea saccharalis* (Mihsfeldt 1985, Parra et al. 1986). The utilization of the corn diet has permitted *Diatraea saccharalis* rearing in our laboratory for successive generations. The nutritional suitability of this diet, the nutritional indices, was also demonstrated by the production of vigorous *Cotesia flavipes* Cameron, comparable to the parasitoids reared on other diets for *Diatraea saccharalis* (Padua 1986).

## References

Adkisson, P. L., E. S. Vanderzant, D. L. Bull & W. E. Allison. 1960. A wheat germ medium for rearing the pink bollworm. J. Econ. Entomol. 53: 759-762.

Andrewartha, H. G. & L. C. Birch. 1954. The distribution and abundance of animals. Univ. of Chicago Press, Chicago.

Araujo, J. R., S. M. S. Araujo, P. S. M. Bothelho & N. Degaspari. 1980. Obtencao de posturas de *Diatraea sacharalis* em condicoes de laboratorio. Brasil Acucareiro. 93(3): 67-73.

Araujo, J. R., P. S. M. Bothelho, S. M. S. Araujo, L. C. Almeida & N. Degaspari. 1985. Nova dieta artificial para criacao de *D. saccharalis*. Saccharum A.P.C. 8(36): 45-48.

Bergamin, J. 1943. Metodos de laboratorio para observacao e criacao de *Diatraea saccharalis* (Fabricius, 1794), a broca de cana. Arq. do Inst. Biol. Sao Paulo. 14: 351-355.

Botelho. P. S. M. 1980. *Apanteles flavipes* performance in the control South region of Brazil. Entomol. Newsletter 8: 15.

Botelho, P. S. M. 1984. Cana de acucar:tudo sob controle. A Granja. 40(442): 70-74.

Bowling, C. C. 1967. Rearing of two lepidopterous pests of rice on a common artificial diet. Ann. Entomol. Soc. Am. 60: 1215-1216.

Brewer, F. D. 1976. Development of the sugarcane borer on various artificial diets. USDA - ARS 5-116.

Brewer, F. D. 1981. Development of *Heliothis virescens* and *Diatraea saccharalis* on a soyflour-corn oil diet. Ann. Entomol. Soc. Am. 74(3): 320-323.

Brewer, F. D. & D. F. Martin. 1976. Substitutes for agar in a wheat germ used to rear the corn earworm and the sugarcane borer. Ann. Entomol. Soc. Am. 69(2): 255-256.

Brewer, J. M. & R. B. Ashworth. 1969. Disc electrophoresis. J. of Chem. Educ. 46: 41-45.

Burton, R. L. 1969. Mass rearing of the corn earworm in the laboratory. USDA Paper ARS 33-134.

Dyar, H. G. 1890. The number of molts of lepidopterous larvae. Psyche. 5: 420-422.

Gallo, D. 1952. Contribuicao para o controle biologico da broca da cana de acucar. An. da Escola Superior de Agricultura "Luiz de Queiroz." 9: 135-142.

Gallo, D., F. M. Wiendle, R. N. Willians & E. Berti Filho. 1969. Metodo de criacao artificial da broca de cana de acucar para emprego no seu controle, pp. 4-5. *In* Resumos da 2a. Reuniao Anual de Entomologia, 1969. Recife.

Gallo, D., O. Nakano, S. Silveira Neto, R. P. L. Carvalho, G. C. de Batista, E. Berti Filho, J. R. P. Parra, R. A. Zucchi, S. B. Alves & J. D. Vendramim. 1988. Manual da Entomologia Agricola. Agronomica Ceres, Sao Paulo.

Guevara, L. A. C. 1976. Aspectos da biologia em condicoes naturais e frequencia de acasalamento da *Diatraea saccharalis* (Fabr., 1794) (Lepidoptera:Crambidae) a broca da cana-de-acuar. Mestrado, Escola Superior de Agricultura "Luiz de Queiroz"/USP, Piracicaba.

Hensley, S. D. & A. M. Hammond Jr., 1968. Laboratory techniques for rearing the sugarcane borer on an artificial diet. J. Econ. Entomol., 61: 1742-1743.

Holloway, T. E. & U. C. Loftin. 1919. The sugarcane moth borer. USDA Bulletin 746.

Holloway, T. E., N. E. Haley, U. C. Loftin & C. Heinrich. 1928. The sugarcane moth borer in the United States. USDA Technical Bulletin 41.

King, E. G. & G. G. Hartley. 1985. *D. saccharalis*, pp. 265-70. *In* P. Singh & R. F. Moore [eds.], Handbook of insect rearing, vol.2. Elsevier, New York.

King, E. G., F. D. Brewer & D. F. Martin 1975. Development of *Diatraea saccharlis* (lep:Pyralidae) at constant temperatures. Entomophaga. 20(3): 301-306.

Macedo, N., P. S. M. Botelho, N. Degaspari, L. C. Almeida, J. R. Araujo & E. A. Margrini. 1983. Controle biologico da broca de cana de acucar. IAA/Planalsucar, Piracicaba.

Martins, J. F. S. 1983. Resistencia de variedades de arroz a *Diatraea saccharalis* (Fabricius, 1794) (Lepdoptera:Pyralidae) e sua associacao com caracteristicas biofisicas e bioquimicas das plantas. Doutorado, Escola

Superior de Agricultura "Luiz de Queiroz"/USP, Piracicaba.

Meadows, C. M. 1938. The biology of the sugarcane borer *Diatraea saccharalis* (F). M. S. thesis, Louisiana State Univ., Baton Rouge.

Melo, A. B. P. 1984. Biologia de *Diatraea saccharalis* (Fabricius, 1794) (Lepidoptera:Pyralidae) em diferentes temperaturas para determinacao das exigencias termicas. Mestrado, Escola Superior de Agricultura "Luiz de Queiroz"/USP, Piracicaba.

Melo, A. B. P. & J. R. P. Parra 1988. Biologia de *Diatraea saccharalis* (Fabricius, 1794) (Lepidoptera:Pyralidae) em diferentes temperaturas. Pesq. Agropec. Bras. 23(7): 663-680.

Mendonca Filho, F. A. 1973. Criacao artificial em laboratorio dos parasitos da broca de cana de acucar *(Diatraea* spp.)(Lep:Crambidae). Bras. Acuc. 81(4): 47-80.

Mihsfeldt, L. H. 1985. Comparacao de dietas artificiais para a criacao de *Diatraea saccharalis* (Fabricus, 1794)(Ledpdoptera:Pyralidae). Mestrado, Escola Superior de Agricultura "Luiz de Queiroz"/USP, Piracicaba.

Miskimen, G. W. 1965. Non aseptic laboratory rearing of the sugarcane borer *D. saccharalis*. Ann. Entomol. Soc. Am. 18: 820-823.

Padua, L. E. M. 1986. Influencia da nutricao, temperatura e umidade relativa do ar na relacao *Apanteles flavipes* (Cameron, 1891) -*Diatraea saccharalis* (Fabricius, 1794). Doutorado, Escola Superior de Agricultura "Luiz de Queiroz"/USP, Puracicaba.

Parra, J. R. P. & S. M. Carvalho. 1984. Biologia e nutricao quantitativa de *Spodoptera frugiperda* (J.E. Smith, 1797) em meios artificiaia compostos de differentes variedades de feijao. An. Soc. Entomol. Bras. 13(2): 305-319.

Parra, J. R. P., A. A. C. M. Precetti & P. Kasten Jr. 1977. Aspectos biologicos de *Spodoptera eridania* (Cramer, 1782) (Lepidoptera:Noctuidae) em soja e algodoeiro. An. Soc. Entomol. Bras. 6(2): 147-155.

Parra, J. R. P., R. A. Zucchi & S. Silveira Neto. 1987. Biological control of pests through egg parasitoids of the genera Trichogramma and/or Trichogrammatoidea. Mem. Inst. Oswaldo Cruz. 82(Suppl. 3): 153-160.

Parra, J. R. P., L. C. Almeida, W. J. Silva & P. S. M. Botelho. 1986. Aspectos biologicos de *Diatraea saccharalis* (Fabr., 1794) em dietas artificiais contendo diferentes mutantes de milho, pp. 30. *In* Resumos do 10o. Congresso Brasileiro de Entomologia, 1986. Rio de Janeiro.

Poitout, S. & R. Bues. 1970. Elevage de plusieurs especes de lepidopteres Noctuidae sur milieu artificiel riche, et sur milieu artificiel simplifie. Ann. Zool. Ecol. Anim. 2: 79-91.

Roe, R. M., A. M. Hammond Jr. & T. C. Sparks. 1982. Growth of larval *Diatraea saccharalis* (Lep.:Pyralidae) on an artificial diet and synchronization of the last larval stadium. Ann. Entomol., Soc. Am. 75: 421-429.

Roe, R. M., A. M. Hammond Jr., T. E. Reagan & S. D. Hensley 1981. A bibliography of the sugarcane borer, *Diatraea saccharalis* (Fabricius), 1887-1980. USDA-ARM -S-20.

Santa Cruz, J. M. S., C. S. Moss, G. G. Raynaud & C. G. Montavalo. 1964. Cria artidicial de *Diatraea saccharalis* Fabr, (Lepidoptera:Pyralidae) y su aplicacion en la evolucion de resistencia em maiz. Agrocienc. 18:3-13.

Sgrillo, R. B. 1973. Criacao em laboratorio da broca da cana de acucar *Diatraea saccharalis* (Fabricius, 1794), visando seu controle. Mestrado, Escola Superior de Agricultura "Luiz de Queiroz"/USP, Piracicaba.

Sgrillo, R. B., J. M. M. Walder & F. M. Wiendl. 1977. Progressos na criacao de broca de cana de acucar, *Diatraea saccharalis* (F). realizados no Centro de Energia Nuclear na Agricultura-CENA. O. Solo. 69(1): 58-60.

Silva, W. J., J. P. F. Teixeira, P. Arruda & M. B. Lovato. 1978. Nutrimaiz, a tropical sweet maize cultivar of high nutritional value. Maydica. 23: 129-136.

Silva, A. G. D. A., C. R. Goncalves, D. M. Galvao, A. J. L. Goncalves, J. Gomes, M. N. Silva & L. Simon. 1968. Quarto catalogo dso insetos que vivem nas plantas do Brasil, seus parasitos e predadores, t.1, pt2. Insetos, hospedeiros e inimigos naturais. Ministerio da Agricultura/Departamento de Defesa e Inspecao Agropecuaria, Rio de Janeiro.

Singh, P. 1983. A general purpose laboratory diet mixture for rearing insects. Insect Sci. Appl. 5(4): 357-362.

Slansky Jr., F. & J. M. Scriber. 1985. Food consumption and utilization, pp. 87-163. *In* G. A. Kerkut & L. I. Gilbert [eds.], Comprehensive insect physiology biochemistry and pharmacology, vol. 3. Pergamon Press, Elmsford.

Slansky, Jr., F. & J. G. Rodriguez, 1987. Nutritional ecology of insects, mites, spiders and related invertebrates: an overview, pp. 1-69. *In* F. Slansky Jr. & J. G. Rodriguez [eds.], Nutritional ecology of insects, mites, spiders and related invertebrates. Wiley, New York.

Sneath, P. H. A. & R. R. Sokal. 1973. Numerical taxonomy. W. F. Freeman, San Francisco.

Taylor, D. J. 1944. Life history studies of the sugarcane moth borer, *Diatraea saccharalis* Linn. Fla. Entomol. 27(1): 10-13.

Tosello, G. A. 1980. Milhos especiais e seu valor nutritivo. In E. Paterniani (coord.) Melhoramento e producao de milho no Brasil. Esalq. Piracicaba.

Waldbauer, G. P. 1968. The consumption and utilization of food by insects. Adv. in Insect Physiol. 5: 229-288.

Williams. J. R., J. R. Metcalfe, R. W. Mungomery & R. Mathes, editors. 1969. Pests of sugarcane. Elsevier, New York.

# 13

# Rearing Lepidoptera for Plant Resistance Research

## F. M. Davis and W. D. Guthrie

### Introduction

Rearing lepidoptera for plant resistance research has become an integral part of many programs aimed at developing crop plants resistant to attack by these insects. In this chapter, we discuss the development of a model rearing program to furnish a lepidopteran for a highly successful plant resistance program; other artificial rearing of lepidoptera for plant resistance research; the importance of artificial rearing capabilities for the development of resistant plants; and the future of plant resistance and artificial rearing to support it.

### European Corn Borer: A Model Rearing System for Support of Plant Resistance Research

The first lepidopteran for which entomologists in the United States developed suitable diets and rearing procedures was the European corn borer, *Ostrinia nubilalis* (Hubner). Bottger (1942) developed the first synthetic medium for rearing the European corn borer (ECB) to compare the effect of individual nutritive constituents on the physiological development of the insect and its relationship to borer-resistant strains of maize. In succeeding years, entomologists at the University of Wisconsin continued to modify and improve diets for this insect and developed procedures for small-scale laboratory rearing (Beck et al. 1949, Beck 1950, Beck & Stauffer 1950, Beck & Smissman 1960). Diets, procedures, and equipment for large-scale rearing of ECB, again primarily for plant resistance research, were developed by researchers located at the U.S. Department of Agriculture, Agricultural Research Service

211

(USDA-ARS) laboratory at Ankeny, Iowa (Becton et al. 1962, Guthrie et al. 1965, 1971, Guthrie 1989). Through the years, research has been conducted at Ankeny to improve each phase of ECB rearing. These efforts have resulted in a highly reliable, cost effective rearing program with the capability of supplying sufficient ECB for plant resistance programs at Ankeny and other USDA-ARS locations.

The plant resistance program at Ankeny-Ames has been highly successful in identifying sources of maize with ECB resistance, developing these sources into usable forms for public release, and determining the genetics, mechanisms, and causal factors of resistance. This success is largely due to the rearing program; without it, progress in identifying and developing resistant maize would have been impossible (Guthrie 1989).

Studies of maize resistance to ECB by ARS researchers in Iowa were conducted using natural infestations from 1927 to 1931. Since 1932, all studies have been conducted with artificial infestations. Eggs obtained from field-collected borers were used to infest plants until the early 1970s when they were provided by artificial rearing (Guthrie et al. 1971). Research to develop ECB-rearing capability at Ankeny-Ames began in 1960, and by the spring of 1963 an easy-to-prepare and relatively inexpensive artificial diet was available (Guthrie et al. 1965). However, the addition of maize whorl-leaf powder to the diet was required for the insects to develop normally. This was costly and inconvenient because the powder was obtained by growing a susceptible maize hybrid in the field, cutting the plants at the desired growth stage, and heat drying the tissue. Fortunately, wheatgerm was found to be a suitable substitute for whorl-leaf powder (Lewis & Raun 1966, Lewis & Lynch 1969).

Early in the program, ECB were reared on artificial diet plugs placed into 3-dram vials. Each of the vials was infested with two neonate larvae using an artist's brush. Cotton wads were used to close the vials. The vial method was highly successful but too inefficient for producing the number of moths needed for egg production in plant resistance research. In 1964, efforts were made to rear ECB, a noncannibalistic larva, in plastic dishes (30.5 cm in diameter by 10.2 cm deep) containing artificial diet. The cover for the dish was a snap-on friction lid with a "screen window" in the middle for ventilation. Each dish was infested by placing egg masses (200-300 eggs) attached to wax paper disks on the diet. The dishes were then placed in an incubator operated at 28°C, 85% relative humidity, and continuous light for development to the pupal stage.

Several problems had to be solved for dish rearing to be successful. Incidence of the disease, *Nosema pyraustae,* and diet contamination caused by molds and bacteria increased substantially when using dishes. These problems were solved by adding the antibiotic Fumidil B to the diet for control of *Nosema* (Lewis & Lynch 1970, Lynch & Lewis 1971) and by using an assortment of antimicrobial agents (e.g., methyl p hydroxybenzoate, propionic acid, formaldehyde, and aureomycin) (Guthrie et al. 1971).

Also, the neonate larvae tended to crawl extensively over the diet and on the sides of the dish before feeding. Scarifying the surface of the diet with a fork prior to infestation was found to enhance initiation of feeding (Guthrie et al. 1971).

Another problem with dish rearing was inefficient pupal harvest. Many larvae pupated in large quantities of silk matted over the diet. Initially, these silken mats containing pupae were removed from the dishes and placed in a 1.5% sodium hypochlorite solution to dissolve the silk. The free-floating pupae were skimmed off the surface with a strainer (Guthrie et al. 1965). Several years later, a more efficient method of harvesting pupae was developed. Strips of corrugated paper (2.5 cm wide, treated with hot wax containing sorbic acid as a mold inhibitor) were formed into a ring inside each dish for larvae to crawl into and pupate. Significantly more larvae were found to leave the diet to pupate in the corrugated paper if the lid and upper half of the dish were painted black (Reed et al. 1972). Pupae were harvested by removing the corrugated paper ring from each dish. Pupae and larvae remaining in the dish were discarded.

Initially, the rings containing pupae were hung in a room held at 21°C and 80-85% relative humidity for moth emergence. Adults were collected daily from the walls of the room using an adapted dog dryer. About 100 pairs of moths were collected in a cone-shaped container mounted on the end of the dryer. The moths within the container were then shaken into an oviposition cage (Guthrie et al. 1971). At present, corrugated paper strips containing pupae are placed directly into the oviposition cage, eliminating the need for moth collection (Guthrie 1989).

Rectangular cages (60 cm long by 31 cm wide by 31 cm high) have been used for many years for oviposition by 200-300 pairs of moths (Guthrie 1989). The cages consist of a wooden frame covered on the sides, ends, and bottom by screen with openings between the wires too small for moths to escape. Wooden surfaces within the cages are covered with screen because moths oviposit on smooth surfaces but not on screen. Hardware cloth (4 mesh) is placed over the cage top so that the moths can oviposit between the wires onto strips of wax paper or plastic placed on the cage top and held in position by a rubber pad. Cages are sprinkled with water daily to enhance oviposition and moth survival (Kira et al. 1969). Occasionally, the oviposition room and cages become contaminated with disease organisms (e.g., the fungus *Beauveria bassiana*). Disinfecting the oviposition room and cages with 5% Roccal before restocking cages with moths minimized the problem (Guthrie 1989).

The room containing the oviposition cages is maintained at a temperature of 27°C for 16 h each day (0500 to 2100) and 18-20°C for the remaining 8 h. The cycling temperature is required to ensure adequate mating (Sparks 1963). The lights are on from 0600 until 2000. Relative humidity is maintained at 80-85%. A fan is used to prevent layering of air (Guthrie 1989).

ECB reared for more than 14 generations on artificial diet survive poorly on field-grown susceptible maize (Guthrie et al. 1974). To provide ECB equivalent to the wild type in ability to establish on and damage maize, a new culture is started each year by collecting overwintering larvae from infested maize plants in the fall, surface sterilizing them in a phenylmercuric nitrate solution to prevent infection by diseases such as *B. bassiana,* and then storing them at 4°C until needed in the spring or by collecting moths from light traps during the summer and placing them directly into the rearing program. To minimize the incidence of the disease caused by *Nosema pyraustae* obtained from field-collected ECB, larvae of the first two to three generations in the laboratory are fed a diet containing three times the normal rate of Fumidil B (Guthrie 1989).

Many public research institutions and about eight maize seed companies in the U.S. and several laboratories in foreign countries (e.g., in France and Canada) have modeled their ECB rearing programs after the one developed at Ankeny. Most U.S. maize seed companies have developed centralized rearing programs to furnish ECB eggs via mail to plant breeders located at field stations across the U.S. and, in some cases, in other countries. For shipment, the eggs are removed from the wax paper or plastic sheets (Mihm 1983a, Guthrie 1989) and placed in small containers such as styrofoam drink cups with snap-on lids. Holes are punched in the lids for ventilation before the cups are placed in insulated mailing boxes. To avoid egg hatch in transit, the boxes must be delivered within 2 days.

In 1986, public and private researchers produced about 1.5 billion ECB eggs for plant resistance studies in maize (Guthrie 1989). This clearly demonstrates the importance placed on artificial rearing for the development of ECB-resistant maize.

### Other Artificial Rearing of Lepidoptera for Support of Plant Resistance Research

Research on the development of crops resistant to lepidoptera pests increased during the 1960s. This increase was partially due to the emphasis on developing economically acceptable pest management tactics that can be used as alternatives to chemical insecticides and to the advent of practical artificial diets for rearing lepidoptera. Most of the increase in plant resistance programs in the

U.S. during this period occurred within USDA-ARS; for example, the tobacco budworm, *Heliothis virescens* (Fab.), on cotton at Brownsville, Texas; the corn earworm, *H. zea* (Boddie), and fall armyworm, *Spodoptera frugiperda* (J. E. Smith), on maize and sorghum at Tifton, Georgia; and the southwestern corn borer, *Diatraea grandiosella* Dyar, on maize at Mississippi State, Mississippi. The rearing programs at Brownsville and Tifton were developed as large, centralized operations that provided insects to scientists involved in many types of research, whereas a medium-sized rearing operation for furnishing the insects at Mississippi State had to be developed by the entomologist on the plant resistance team.

One of the early international programs for developing insect resistance in crop plants utilizing a rearing system was developed in India. In 1964, the U.S. Agency for International Development (USAID) funded a project to develop a host plant resistance program in maize to lepidopterous stalk borers. An artificial rearing program was developed for supplying *Chilo partellus* (Swinhoe) (Chatterji et al. 1968).

Technology for rearing lepidoptera in the laboratory on artificial diets improved significantly during the late 1960s and 1970s. Many of the plant resistance programs using artificially reared insects were able to demonstrate that reliable and economical rearing was possible and that consistent progress could be made in identifying and developing resistant plants using these insects.

Plant resistance programs for developing crop cultivars resistant to lepidoptera increased significantly during the 1970s and 1980s in the U.S. and abroad, including institutions that are public (state and national), private (especially maize seed companies in the U.S.), and international (e.g., the International Maize and Wheat Improvement Centre in Mexico; the International Institute of Tropical Agriculture in Nigeria; the International Crops Research Institute for Semi-Arid Tropics in India; the International Rice Research Institute in the Philippines; and the International Center of Insect Physiology and Ecology in Kenya).

Most institutions engaged in lepidopterous plant resistance research have established internal rearing operations or contracted with other laboratories or commercial insectaries (e.g., French Agricultural Research Inc., Lamberton, Minnesota or private maize seed companies) to supply their insects. The rearing operations that furnish lepidoptera vary in age, size, and state of technological development (Table 1). Many have evolved into highly reliable and efficient systems capable of producing enough insects for the research programs. This evolutionary process involved solving problems dealing with facilities, personnel, diets, rearing procedures, diet contamination, diseases, and genetic quality of laboratory colonies. To solve these problems, the researchers and rearing staff have had to learn the biological needs of the insect, its behavioral characteristics, and how abiotic and biotic factors influence its survival, growth

**Table 1. Some advantages and disadvantages of using different sources of insects for plant resistance research**

| Insect source | Advantages | Disadvantages |
|---|---|---|
| Natural | Requires low investment; utilizes wild pest population. | Often fails to separate resistance from susceptible plant genotypes because of variations in population densities and spatial distribution of pests; allows for no control over time of infestation in relation to plant growth or selection pressure (density/plant) of pest; severely restricts scope of research program. |
| Field collected | Aids in providing uniform pest infestations when screening for resistance. | Affects consistency of progress because of year-to-year variations in pest populations; requires extensive labor, storage technology, and facilities; restricts number of plant genotypes that can be screened and scope of total research program. |
| Laboratory reared on host plants | Aids in providing uniform pest infestations when screening for resistance; utilizes insects reared on natural host. | Requires facilities to grow plants and feed insects; requires extensive labor; and restricts number of plant genotypes that can be screened and scope of total research program. |

**Table 1, Continued**

| Insect source | Advantages | Disadvantages |
|---|---|---|
| Laboratory reared on artificial diet | Provides a reliable source of insects to produce uniform infestations when desired plant growth stages occur and with the desired number of insects per plant when screening for resistance; allows for complete plant resistance program which includes screening large number of plant genotypes for resistant sources, developing identified sources into usable forms (e.g. inbreds, populations, varieties), determining genetics of resistance, determining effectiveness of resistance, and determining mechanisms and bases of resistance; results in consistent progress toward program goals. | Requires substantial investment in facilities, equipment, supplies, and trained personnel; laboratory reared insects may differ from wild-type behaviorally or physiologically. |

and reproduction. In addition, they have had to keep abreast of literature that addresses problems associated with laboratory rearing of insects. Two important books that have recently become available are *Advances and Challenges in Insect Rearing*, edited by E. G. King and N. C. Leppla and published by USDA-ARS in 1984, and Volumes 1 and 2 of *The Handbook of Insect Rearing*, edited by P. Singh and R. F. Moore and published by Elsevier Publishing Co. (Amsterdam, Holland) in 1985.

The following observations concerning problems associated with laboratory rearing of lepidoptera and suggested solutions are based on the authors' experiences. Rearing facilities should be functional and practical. Rooms or chambers must be available where temperature, humidity, and photoperiod can be controlled. Unreliable environmental control can result in severe adverse effects on the insects and increases in microbial contamination of diet. Equipment to maintain these environmental conditions should be simple, reliable, and, if possible, obtained from local sources so that repairs can be made quickly. Also, space should be available so that clean tasks (e.g., preparation, dispensing, and infesting of diet) can be separated from dirty tasks (e.g., harvesting pupae and maintaining moth colonies). Separation of clean and dirty areas is an excellent way to help manage microbial contamination of diet and diseases.

Personnel are a key ingredient to successful rearing. They must carefully and consistently follow procedural guidelines, actively participate in the development of a more reliable and efficient rearing system, and be dedicated to rearing large numbers of high-quality insects. Also, at least one member of the rearing group should be capable of fabricating and maintaining equipment.

Many of us are fortunate that the lepidoptera that we rear can be easily grown on available diets or on these diets with slight modifications. The advent of using wheat germ as a primary nutritive source for formulating diets introduced by Adkisson et al. (1960) for rearing the pink bollworm, *Pectinophora gossypiella* (Saunders), opened the door to rearing many lepidopterous species. However, there are some lepidopterous species for which suitable diets have not yet been developed. Extensive research may be required, and in some cases an insect dietetic specialist may need to be consulted before a suitable diet is obtained. The ideal diet should be simple, economical, and suitable for the development of functional adults. If possible, diet ingredients should be easily accessible from local sources. It may be necessary to find substitutes for ingredients that are difficult to obtain or are too costly. When testing diets for suitability, the following biological characters should be considered: survival of all life stages, larval and pupal developmental times, pupal size, and adult longevity and fecundity. For recent reviews of insect diets, see Singh (1984 & 1985).

The development of procedures that are practical, economical, and most of all biologically acceptable to the reared insect are critical to successful rearing. Procedures must be carefully worked out for each rearing task (e.g., preparation and dispensing of diet into diet containers, infestation of diet-filled containers with eggs or larvae, time and method of pupal harvest, and handling of adult colony and eggs). A problem that seems to confront researchers constantly, especially in developed countries, is the cost of labor, diet, and other essential rearing materials. To solve this dilemma in part, researchers have devised many labor-saving procedures, including the development of specialized equipment. For example, at the USDA-ARS laboratory at Mississippi State, Mississippi, the rearing program for the southwestern corn borer and fall armyworm has gone through three distinct developmental periods. From 1968 to 1976, the rearing program was small and centered around the use of 30-ml plastic cups and paperboard caps as rearing containers. These containers were selected primarily because of the highly cannibalistic behavior of the southwestern corn borer. Procedures were mostly manual (Davis 1976). During this period, the output of the rearing program was a limiting factor to the number of maize genotypes that could be screened for resistance. In 1976, there was an increased emphasis on developing maize resistant to the southwestern corn borer (Scott & Davis 1976). The new effort required that the rearing program furnish several times the number of insects produced previously. To meet this challenge, the researchers developed a system based on equipment to semiautomate cup and cap rearing. Equipment was fabricated to meter diet into cups accurately (Davis et al. 1978), to infest diet-filled cups with larvae mixed in sterilized corn cob grits and cap them in a continuous process (Davis 1980), and to remove pupae from cups mechanically (Davis 1982a). Also, during this period large moth cages were fabricated with special features for handling pupae, feeding adults, removing and replacing oviposition substrates, and removing dead moths to replace the labor-intensive small cage procedure (Davis 1982b, Davis et al. 1985). By the mid-1980s, the cost of cups and caps plus the labor required to process them had increased so much that an effort was made to devise a new rearing container. An existing multicellular tray (Sparks & Harrell 1976) was modified for rearing the lepidopterans. It consists of a 32-cell clear plastic tray with a heat sealable, preperforated mylar lid. The only new items of equipment required for tray rearing were a hand-operated lidder (custom made by private industry) that sealed the lid onto the top of the tray and racks to hold rearing trays while the insects develop. The multicellular rearing system has proven to be a highly reliable and cost effective way of furnishing the insect needs of the present plant resistance program (Davis et al. 1990).

Microbial contamination of artificial diets is a universal problem associated with laboratory rearing. Some common methods that we and others use to prevent contamination are: 1) separation of clean from dirty task areas and directed traffic from clean to dirty areas only; 2) development of strict sanitary procedures for personnel, facilities, utensils, and equipment; 3) addition of antimicrobial agents (e.g., sorbic acid, methylparaben, and antibiotics) into the diet; and 4) use of air-purifying equipment (e.g., clean air hoods). Before rates of antimicrobial agents are increased or new agents are added to the diet, preliminary tests should be done to determine their effects on the insect. For recent reviews of contamination problems and solutions, see Sikorowski (1984), Shapiro (1984), and Sikorowski and Goodwin (1985).

Diseases caused by microorganisms (e.g., bacteria, fungi, viruses, and microsporidia) can be a serious threat to rearing lepidoptera. Their effects can be acute or chronic on the insects' survival, development, and fecundity. The researcher and rearing personnel must be able to recognize symptoms and signs of diseased insects, to determine causative organisms (which may require the assistance of an insect pathologist), and to take the proper procedures to eliminate the disease. The laboratory system to rear the insects should be developed in such a manner as to minimize stressing the insects because stresses (e.g., overcrowding) can cause diseases directly or increase susceptibility to diseases (Goodwin 1984). The measures mentioned above for managing diet contamination organisms are compatible with disease control. In addition, insect stages, especially eggs, can be surface sterilized with chemicals (e.g., sodium hypochlorite) to kill disease organisms (e.g., viruses) that occur on their surfaces. To avoid introduction of diseases that can be transmitted within the eggs, wild males are used as the only donor when infusing wild genes into the laboratory colony. If a new colony is started from the wild, a method for avoiding diseases is to separate larvae during development on diet and retain only healthy ones for colony increase. See Goodwin (1984), Shapiro (1984), and Sikorowski and Goodwin (1985) for procedures to deal with disease problems within rearing laboratories.

The plant resistance researcher needs to rear not only a healthy insect but one that is also physiologically and behaviorally equivalent to its wild counterparts. As discussed previously, continuous rearing of European corn borers in the laboratory results in an insect that can no longer establish on or inflict damage to maize plants equally with the wild type. This insect should not be used in screening maize for resistance. Quisenberry and Whitford (1988) reported similar results for laboratory-reared fall armyworms used to evaluate bermudagrass accessions for resistance. Until recently, monitoring to see if the laboratory-reared insect is equivalent to the wild type for screening plants for resistance required comparing the two on the host plant or comparing expected to actual degree of damage that occurs on susceptible plant hosts. Mason et al. (1987) suggest that starch gel electrophoresis be used to monitor changes in genetic heterozygosity between laboratory reared and wild insects. Using this procedure, the researcher can readily detect changes in genetic structure by examining protein products of the DNA that have been repeatedly demonstrated to have a genetic basis. Researchers rearing lepidoptera for plant resistance research should closely monitor the quality of their laboratory colonies. When the colony is no longer equivalent to the wild, new genes from the wild should be infused into the laboratory colony or the colony should be replaced with one from the wild. Some plant resistance researchers routinely infuse or start new colonies each year to ensure a reared insect equivalent to the wild type.

In recent years, commercial industry has offered insect rearers a variety of items, such as artificial diets, special equipment, and rearing containers. Lists of rearing supplies and equipment available commercially are published periodically in *Frass*, a newsletter of the Insect Rearing Group. Also, in describing their rearing program, many researchers include a list of supplies and vendors (e.g., Guthrie 1989, Davis 1989, 1990). Private industries' involvement in providing essential supplies and equipment has greatly helped insect rearers in solving many of their problems.

## Importance of Artificial Rearing to Plant Resistance Research

The degree of susceptibility or resistance to an insect pest is established by comparisons to standard susceptible and/or resistant plant genotypes. The comparisons are based on estimations of either the effect of the insect pest on the plant (i.e., leaf, fruit, or stem damage or yield loss) or the effect of the plant on the insect (i.e., reduced survival, slower growth, or increased developmental time). Therefore, in screening a plant for resistance to an insect, it is imperative that insect and plant be brought together in a way that the resulting interaction can be evaluated. Thus, the researcher creates a situation in which each plant has the same opportunity of being equally damaged as its neighbors. Normally, thousands of plants must be evaluated for response to insect attack before resistant sources are identified and developed into useful cultivars.

Options for obtaining insects for evaluating host plants include: 1) natural populations, 2) field-collected insects, 3) insects reared on host plants in the laboratory or insectary, and 4) insects reared on an artificial diet in the laboratory. Each option has its advantages and disadvantages (Table 2). Most researchers agree that little, if any, progress can be made in identifying and developing resistant plant genotypes using only natural populations of lepidoptera as the pest source. For species in which suitable artificial diets have yet to be developed, or where funding prevents the development of an artificial rearing program, field-collected insects or those reared on host plants in the laboratory or insectary should be used for resistance screening. The most efficient and consistent way to identify and develop resistant plants and to allow for a complete plant resistance program is through the use of insects reared on artificial diet in the laboratory. Therefore, development of an artificial rearing

**Table 2.  A partial list of lepidopterous species and crops in which laboratory reared insects are used in plant resistance studies along with respective rearing references**

| Insect Species | Crops | References |
|---|---|---|
| *Anticarsia gemmatalis* (Hubner) | Soybeans | Hartley 1989 |
| *Chilo partellus* (Swinhoe) | Maize, sorghum | Ochieng et al. 1985; Taneja & Leuschner 1985; Taneja & Nwanze 1988 |
| *Chilo suppressalis* (Walker) | Rice | Heinrichs et al. 1985 |
| *Cochylis hospes* Walsingham | Sunflower | Barker 1988 |
| *Diaphania hyalinata* (L.) | Cucurbits | Elsey et al. 1984 |
| *Diaphania nitidalis* (Stoll) | Cucurbits | Elsey & McFadden 1981; Elsey et al. 1984 |
| *Diatraea grandiosella* Dyar | Maize | Mihm 1983a, 1989; Davis 1989; Davis et al. 1989 |
| *Diatraea saccharalis* (Fab.) | Maize | Mihm 1983a, 1989 |
| *Elasmopalus lignosellus* Zeller | Southern peas and snap beans | Chalfant 1975 |

**Table 2, Continued**

| Insect Species | Crops | References |
|---|---|---|
| *Eldana saccharalis* Walker | Maize | Jackai & Raulston 1982; Bosque-Perez & Dabrowski 1989 |
| *Heliothis virescens* (Fab.) | Cotton | Raulston & Lingren 1969; King & Hartley 1985; Davis et al. 1989 |
| *Heliothis zea* (Boddie) | Maize, sorghum | Mihm 1982, 1989; Burton & Perkins 1989 |
| *Maruca stulalis* Geyer | Cowpeas | Jackai & Raulston 1982; Ochieng & Bungi 1983 |
| *Ostrinia furnacalis* (Guenee) | Maize | Hirai & Legacion 1985 |
| *Ostrinia nabilalis* (Hübner) | Maize | Guthrie 1989 |
| *Pectinophora gossypiella* (Saunders) | Cotton | Bartlett & Wolf 1985 |
| *Pseudoplusia includens* (Walker) | Soybeans | Hartley 1989 |
| *Sesamia calamistis* Hmps. | Maize | Jackai & Raulston 1982; Bosque-Perez & Dabrowski 1989 |
| *Spodoptera exigua* (Hubner) | Soybeans | Hartley 1989 |
| *Spodoptera frugiperda* (J.E. Smith) | Maize, sorghum | Mihm 1983b, 1989; Burton & Perkins 1989; Davis 1989; Davis et al. 1989 |

system must be considered an important prerequisite to having a successful plant resistance program for lepidopterous pests.

## Future of Plant Resistance and Artificial Rearing Support

The role of plant resistance in helping manage lepidopterous pests should increase significantly in the future because of the urgent need for effective alternatives to chemical control in developed and Third World countries. Resistant cultivars or varieties offer an alternative control tactic that is economically, environmentally, and sociologically acceptable. They can be used as the principal management tactic or in conjunction with other tactics to form an integrated approach. Ideally, the resistant cultivar should form the foundation on which other tactics are brought into play as needed. Such an approach to the management of insect pests is especially important as we move toward low-input sustainable agriculture.

Until recently, traditional entomological and plant-breeding techniques have been relied upon for development of plants resistant to lepidoptera. While traditional techniques have proven to be successful for developing resistant cultivars, several to many years are usually required to develop cultivars ready for grower use. Genetically engineered resistant plants may require less time to develop; however, the high cost of biotechnology research could limit its usefulness. Regardless of how researchers develop resistant plants, a reliable source of insects equivalent to the wild type will be needed to evaluate the pest-plant interactions in the field.

At present, there is a real need for artificial rearing capabilities by many national research programs in developing countries. Genes for resistance to a number of lepidopterous pests are available; however, these genes must often be transferred from unadapted plant sources to adapted cultivars before farmers in various regions can use the resistance. Successful transfer of genes requires laboratory insect rearing capabilities to furnish sufficient quantities to infest plants uniformly for identification of resistant segregates. Information on rearing lepidoptera on artificial diets in the laboratory is available. Some international research centers, such as the International Maize and Wheat Improvement Center (CIMMYT) in Mexico, offer training and consulting in rearing.

Rearing of lepidoptera on artificial diets in the laboratory will continue to be an essential support service for plant resistance research. Its use will increase as entomologists develop suitable diets for some of the other major lepidopterous pests and as research is initiated on new pest species.

## References

Adkisson, P. L., E. S. Vanderzant, D. L. Bull & W. E. Allison. 1960. A Wheat germ medium for rearing the pink bollworm. J. Econ. Entomol.53: 759-62.

Beck, S. D. 1950. Nutrition of the European corn borer, *Pyrausta nubilalis* (HBN). II. Some effects of diet on larval growth characteristics. Physiol. Zool. 23: 353-361.

Beck, S. D. & E. E. Smissman. 1960. The European corn borer, *Pyrausta nubilalis*, and its principal host plant. VIII. Laboratory evaluation of host resistance to larval growth and survival. Ann. Entomol. Soc. Amer. 53: 755-762.

Beck, S. D. & J. F. Stauffer. 1950. An aseptic method for rearing European corn borer larvae. J. Econ. Entomol. 43: 4-6.

Beck, S. D., J. H. Lilly & J. F. Stauffer. 1949. Nutrition of the European corn borer, *Pyrausta nubilalis* (HBN.). I. Development of a satisfactory purified diet for larval growth. Ann. Entomol. Soc. Amer. 42: 483-496.

Becton, A. J., B. W. George & T. A. Brindley. 1962. Continuous rearing of European corn borer larvae on artificial medium. Iowa State J. Sci. 37: 163-172.

Bottger, G. T. 1942. Development of synthetic food media for use in nutritional studies of the European corn borer. J. Agric. Res. 65: 493-500.

Chatterji, S. M., K. H. Siddiqui, V. P. S. Panwar, G. C. Sharma & W. R. Young. 1968. Rearing of maize stem borer, *Chilo partellus* on artificial diet. Indian J. Entomology 30: 8-12.

Chatterji, S. M., G. C. Sharma, K. H. Siddiqui, V. P. S. Panwar & W. R. Young. 1969. Laboratory rearing of pink borer, *Sesamia inferens* Walker on artificial diet. Indian J. Entomology 31: 75-77.

Davis, F. M. 1976. Production and handling of eggs of southwestern corn borer for host plant resistance studies. Mississippi Agric. and Forestry Exp. Stn. Tech. Bull. 74.

Davis, F. M. 1980. A larval dispenser-capper machine for mass rearing southwestern corn borer. J. Econ. Entomol. 73: 692-693.

Davis, F. M. 1982a. Mechanically removing southwestern corn borer pupae from plastic rearing cups. J. Econ. Entomol. 75: 393-395.

Davis, F. M. 1982b. Southwestern corn borer: Oviposition cage for mass production. J. Econ. Entomol. 75: 61-63.

Davis, F. M. 1989. Rearing southwestern corn borers and fall armyworms at Mississippi State, Mississippi, pp. 27-36. *In* Toward insect resistant maize for the Third World: Proceedings of the International Symposium on Methodologies for Developing Host Plant Resistance to Maize Insects. Mexico, D.F.: CIMMYT.

Davis, F. M. & T. G. Oswalt. 1979. Hand inoculator for dispensing lepidopterous larvae. USDA, SEA. AAT-S-9.

Davis, F. M., T. G. Oswalt & J. C. Boykin. 1978. Insect diet dispenser for a medium-sized rearing program. USDA, ARS. ARS-S-182.

Davis, F. M., T. G. Oswalt & S. S. Ng. 1985. Improved oviposition and egg collection system for the fall armyworm (Lepidoptera: Noctuidae). J. Econ. Entomol. 78: 725-729.

Davis, F. M., S. Malone, T. G. Oswalt & W.C. Jordan. 1990. A medium-sized lepidopterous rearing system using multicellular rearing trays. J. Econ. Entomol. 83: (in press).

Goodwin, R. N. 1984. Recognition and diagnosis of diseases in insectaries and the effects of disease agents on insect biology, pp. 96-129. *In* E. G. King & N. C. Leppla [eds.], Advances and challenges in insect rearing. U.S. Dept. of Agric., Agric. Res. Serv.

Guthrie, W. D. 1989. Advances in rearing the European corn borer on a meridic diet, pp. 46-59. *In* Toward insect resistant maize for the Third World: Proceedings of the International Symposium on Methodologies for Developing Host Plant Resistance to Maize Insects. Mexico, D.F.: CIMMYT.

Guthrie, W. D., W. A. Russell & C. W. Jennings. 1971. Resistance of maize to second-brood European corn borers. Proc. Annual Corn and Sorghum Res. Conf. 26: 165-179.

Guthrie, W. D., Y. S. Rathore, D. F. Cox & G. L. Reed. 1974. European corn borer: Virulence on corn plants of larvae reared for different generations on a meridic diet. J. Econ. Entomol. 67: 605-606.

Guthrie, W. D., E. S. Raun, F. F. Dicke, G. R. Pesho & S. W. Carter. 1965. Laboratory production of European corn borer egg masses. Iowa State J. Sci. 40: 65-83.

Kira, M. T., W. D. Guthrie & J. L. Huggans. 1969. Effect of drinking water on production of eggs by the European corn borer. J. Econ. Entomol. 62: 1366-1368.

Lewis, L. C. & R. E. Lynch. 1969. Rearing the European corn borer, *Ostrinia nubilalis* (Hubner), on diets containing corn leaf and wheatgerm. Iowa State J. Sci. 44: 9-14.

Lewis, L. C. & R. E. Lynch. 1970. Treatment of *Ostrinia nubilalis* larvae with *Fumidil B* to control infections caused by *Perezia pyraustae*. J. Invert. Pathol. 15: 43-48.

Lewis, L. C. & E. S. Raun. 1966. Consumption and utilization of laboratory diets by European corn borer. Iowa State J. Sci. 41: 173-180.

Lynch, R. E. & L. C. Lewis. 1971. Reoccurrence of the microsporidan, *Perezia pyraustae* in the European corn borer, *Ostrinia nubilalis*, reared on diet containing *Fumidil B*. J. Invert. Pathol. 17: 243-246.

Mason, L. J., D. P. Pashley & S. J. Johnson. 1987. The laboratory as an altered habitat: Phenotypic and genetic consequences of colonization. The Florida Entomologist 70: 49-58.

Mihm, J. A. 1982. Techniques for efficient mass rearing and infestation in screening for host plant resistance to corn earworm, *Heliothis zea*. Centro Internacional de Mejoramiento de Maiz y Trigo. El Batan, Mexico.

Mihm, J. A. 1983a. Efficient mass rearing and infestation techniques to screen for host plant resistance to maize stem borers, *Diatraea sp*. Centro Internacional de Mejoramiento de Maiz y Trigo. El Batan, Mexico.

Mihm, J. A. 1983b. Efficient mass rearing and infestation techniques to screen for host plant resistance to fall armyworm, *Spodoptera frugiperda*. Centro Internacional de Mejoramiento de Maiz y Trigo. El Batan, Mexico.

Mihm, J. A. 1989. Mass rearing armyworms, borers, and earworms at CIMMYT, pp. 5-21. *In* Toward insect resistant maize for the Third World: Proceedings of the International Symposium on Methodologies for Developing Host Plant Resistance to Maize Insects. Mexico, D.F.: CIMMYT.

Quisenberry, S. S. & F. Whitford. 1988. Evaluation of bermudagrass resistance to fall armyworm (Lepidoptera: Noctuidae): Influence of host strain and dietary conditioning. J. Econ. Entomol. 81: 1463-1468.

Reed, G. L., W. B. Showers, J. L. Huggans & S. W. Carter. (1972). Improved procedure for mass rearing the European corn borer. J. Econ. Entomol. 65: 1472-1476.

Scott, G. E. & F. M. Davis. 1976. Breeding for resistance to the southwestern corn borer. Proc. Annual Corn and Sorghum Res. Conf. 31: 118-128.

Shapiro, M. 1984. Microorganisms as contaminants and pathogens in insect rearing, pp. 130-142. *In* E. G. King & N. C. Leppla [eds.], Advances and challenges in insect rearing. U.S. Dept. Agric., Agric. Res. Serv.

Sikorowski, P. P. 1984. Microbial contamination in insectaries: occurrence, prevention, and control, pp. 143-153. *In* E. G. King & N. C. Leppla [eds.], Advances and challenges in insect rearing. U.S. Dept. of Agric., Agric. Res. Serv.

Sikorowski, P. P. & R. H. Goodwin. 1985. Contaminant control and disease recognition in laboratory colonies, pp. 85-105. *In* P. Singh & R. F. Moore [eds.], Handbook of insect rearing, vol. 1. Elsevier, Amsterdam.

Singh, P. 1984. Insect diets: Historical development, recent advances, and future prospects, pp. 32-50. *In* E. G. King & N. C. Leppla [eds.], Advances and challenges in insect rearing. U.S. Dept. of Agric., Agric. Res. Serv.

Singh, P. 1985. Multiple-species rearing diets, pp. 19-66. *In* P. Singh & R. F. Moore [eds.], Handbook of insect rearing, vol. 1. Elsevier, Amsterdam.

Sparks, A. N.  1963.  Preliminary studies of factors influencing mating of the European corn borer.  Proc. North Central Branch Entomol. Soc. Amer. 23: 95.

Sparks, A. N. & E. A. Harrell.  1976.  Corn earworm rearing mechanization. USDA, ARS Tech. Bull. 1554.

# 14

## Influence of Artificial Diet on Southern Corn Rootworm Life History and Susceptibility to Insecticidal Compounds

*Pamela Marrone, Terry B. Stone, and Steven R. Sims*

### Introduction

Because of their economic importance, corn rootworms (*Diabrotica* spp.) are currently reared extensively in laboratories throughout the United States. Most industrial labs use Diabrotica in insecticide discovery screening operations. Insects of high quality, genetic heterozygosity, and uniformity are critical to the success of the screening operation. In Monsanto's Insect Control Discovery Program, corn rootworms reared on their natural diet (corn) (Marrone et al. 1985) proved to be of acceptable quality. However, the rearing methods were time consuming and labor intensive, so we developed an artificial diet procedure that was more efficient than rearing on corn (Marrone et al. 1985). We reared southern corn rootworms (SCR) for over 25 generations on the artificial diet. To determine if the diet colony was as vigorous as the corn colony, we compared several life history parameters, such as adult and larval survival, fecundity, and susceptibility to insecticidal compounds.

### Materials and Methods

The SCR colony was established in 1983 with eggs obtained from the Northern Grain Insects U.S. Department of Agriculture Research Laboratory in Brookings, South Dakota and Colorado Insectary, Durango, Colorado. The SCR were reared on artificial diet according to procedures described by Marrone et al. (1985). Larvae were reared on this diet until egg production dropped too low to continue colony maintenance (about 26 generations).

The techniques described by Marrone et al. (1985), with some modifications, were used to rear the corn colony. Instead of plastic sweater boxes, uncovered metal planter boxes (43 x 18.5 x 13 cm) were used as larval rearing pans. After one week, new corn seedlings were added to the bottom of each planter box. Water was added to the boxes as needed (about every 2 days). Ten days after the addition of newly hatched larvae to the boxes, the larvae that reached the pupation threshold weight of 18 mg were removed from the box and transferred to round plastic dishes (24 cm in diameter by 9 cm high) filled with autoclaved Peat-Lite potting mix. These procedures were also used for rearing a "mixed" colony, which consisted of larvae from the 25th generation of diet-reared adults fed corn instead of artificial diet.

For the diet-reared colony, 20 trays of larvae, each containing about 250 neonate larvae, were set up per generation. The number of neonate larvae added to each tray and the number of larvae reaching a pupation weight of 18 mg in 10 days were recorded. Two hundred larvae were placed in each of the pupation containers. Pupal survival was estimated by recording the number of adults that emerged from each pupation container.

One thousand adults of a single generation were placed in two cages each holding about 250 males and 250 females. For each generation, the number of adults that died and the number of eggs laid in each cage were recorded every 1-3 days until egg production dropped to zero. The mean number of eggs per female versus adult age and the mean oviposition period (adult survival during the reproductive period) were calculated. An analysis of variance was performed and means compared using Duncan's multiple range test (corn versus diet versus mixed). The procedures described previously for the diet-reared colony were used for the corn-reared colony. Larval pans were used instead of larval trays.

To compare the susceptibility of the two colonies to various insecticides, the $LD_{50}$ for carbofuran and a gougerotin producing *Streptomyces griseolus* were determined. Aliquots of technical carbofuran formulated in a solution of acetone (20%) and water (80%), or the microbial culture broth were incorporated into ungelled artificial diet. This liquid mixture was poured into wells of 96-well polystyrene trays and allowed to gel (about 20 min). Forty-eight neonate larvae were tested against each of five doses of the two insecticides. Larvae were added individually to each well with a camel hair brush. Trays were sealed with Mylar, a small hole was added for air circulation, and they were incubated in a dark room at 27 ±1°C, 55% relative humidity. Six days later, the number of dead larvae exposed to each dosage was recorded. The experiment was repeated on three consecutive days. A probit analysis (SAS) was used to compare $LD_{50}$'s between the diet and corn larvae.

## Results

The number of eggs laid per female by diet-reared adults was consistently less than the number laid by corn-reared adults (Table 1). The $\bar{x}$ ova/female of the diet colony decreased dramatically at generations 4 and 6 and again at generation 22 (Table 1). At generation 22, larval survival on diet also decreased, from 50.7 to 31.0%, and pupal survival gradually dropped from generation 1 (56.2%) to generation 22 (18.7%) (Table 2). The corn colony had a similar but much smaller drop in egg production at generation 10 (Table 1). No corresponding decline in larval or pupal survival of the corn colony was observed (Table 2).

The decrease in egg production for the 6th generation reared on diet corresponds to a decrease in mean days of adult survival during the oviposition period (Table 1). The preoviposition period of diet-reared adults decreased from 7-8 days at generation 1 to 4-5 days at generation 6, although this difference is statistically significant only in generations 12-14 and generation 20. This same decrease did not occur in the corn-reared adults (Table 1).

Although the oviposition period of the mixed colony was similar to that of corn-reared adults (Table 1), peak egg production was similar to egg production of the diet-reared colony during generations 6-20 (Table 1). Larvae reared on diet (generation 24) had low survivorship on corn (Table 2). Adult emergence (pupal survival) of the mixed colony was only 20.9%. Survival and egg production increased with each subsequent generation on corn. Larval and pupal survival in the mixed colony were equivalent to the corn-reared colony after six continuous generations. Egg production reached only half that of the production of the corn colony in this time (Table 1).

The $LD_{50}$ of carbofuran for corn-reared larvae, 0.357 ppm (95% fiducial limits, 0.291-0.438), was similar to the $LD_{50}$ for diet-reared larvae—0.326 (0.235-0.438). In contrast, diet-reared larvae (0.984, 0.625-1.420) were almost twice as susceptible to an insecticidal Streptomycete as corn-reared larvae (1.756, 1.106-2.755).

## Discussion

Lower fecundity of the $F_1$ diet-reared generation compared with corn-reared $F_1$ suggests that nutritional value of artificial diet may be inferior to corn. It is also possible that mold inhibitors and antibiotics in the diet caused the reduction in egg production and adult survival. Antimicrobial agents have been reported to cause numerous detrimental effects such as reduced fecundity, fertility, larval size, and prolonged larval development (Singh 1977, Shapiro 1984). Unfortunately, the antimicrobial compounds in the rootworm diet are at minimum levels necessary for prevention of excessive bacterial contamination (Marrone et al. 1985).

232

Table 1. Egg production, oviposition period, and pre-oviposition period of SCR reared on corn and artificial diet[1]

| Gener- ation | Eggs/♀ | | | Preoviposition Period (Days) | | | Oviposition Period (Days) | | |
|---|---|---|---|---|---|---|---|---|---|
| | Corn | Diet | Mixed[1] | Corn | Diet | Mixed | Corn | Diet | Mixed |
| | 325±10.5a | 212±3.5b | 65±5.2c | 7.0±0.5a | 7.0±0.9a | 8.0±0.4a | 32.5±1.8a | 31.1±1.1a | 36.9±1.8a |
| 2 | — | 213±4.2a | 86±7.7b | — | 9.0±0.4 | | 25.5±1.9a | 27.5±1.9a | 30.2±1.4a |
| 4 | 306±15.0a | 143±6.3b | 145±6.8b | 8.0±1.0a | 7.5±0.9a | | 25.0±1.4a | 35.5±1.9a | 28.6±1.9a |
| 6 | 412±22.0a | 59±5.1b | 222±8.6c | 9.0±0.5a | 5.5±0.5a | | 27.5±1.9a | 10.5±0.4b | 26.7±1.4a |
| 8 | 328±14.5a | 59±2.5b | | 6.5±0.3a | 3.0±0.0b | | 37.5±0.2a | 13.5±0.4a | |
| 10 | 195±9.2a | 45±3.1b | | 7.5±0.3a | 3.0±0.0b | | 19.5±0.4a | 12.0±0.7a | |
| 12 | 198±9.2a | 39±5.6b | | 7.0±0.4a | 2.5±0.7b | | 11.0±0.4a | 17.5±0.9a | |
| 14 | 186±10.1a | 32±4.3b | | 8.0±0.2a | 3.0±0.3b | | 18.0±1.4a | 13.5±0.5a | |
| 16 | 192±12.6a | 36±3.2b | | 8.5±0.9a | 4.0±0.7b | | 14.5±0.4a | 11.5±0.7a | |
| 18 | 196±5.2a | 38±4.1b | | 7.0±0.7a | 5.0±0.9a | | 14.0±0.7a | 12.5±0.3a | |
| 20 | 182±3.1a | 36±4.0b | | 7.0±0.7a | 4.0±0.7b | | 10.5±0.9a | 12.4±0.7a | |
| 22 | 185±8.2a | 22±3.5b | | 6.0±0.0a | 5.0±0.7a | | 12.0±0.9a | 7.5±0.7b | |
| 24 | 179±4.5a | 20±2.6b | | 6.0±0.0a | 6.0±1.4a | | 12.5±0.7a | 5.5±0.7b | |

Note: Means compared among corn, diet, and mixed colonies within equivalent generations or corn and diet colonies where not mixed colony data was taken. Means with a different letter indicate significant differences at $P \leq 0.05$.

[1]Mean (±S.D.) of 2 cages, 1,000 adults (500 ♂'s, 500 ♀'s) per cage.

[2]Mixed colony raised on diet for 25 generations and on corn for 1 generation.

**Table 2.** Comparison of mean larval and pupal survival percentage of southern corn rootworm reared on artificial diet and corn

| Gen. | Corn | Larval Diet | Mixed | Corn | Pupal Diet | Mixed |
|---|---|---|---|---|---|---|
| 1 | 82.1±7.8a | 64.1±20.3b | 20.4±19.4c | 76.2±16.2a | 56.2±15.1b | 20.9±20.7c |
| 2 | 81.2±6.4a | 62.3±17.6b | 36.2±10.4c | 74.5±15.4a | 52.1±14.2b | 48.6±10.1b |
| 4 | 78.5±10.2 | 55.2±12.4b | 49.6±12.1b | 73.1±12.1a | 53.6±12.2b | 55.2±3.2b |
| 6 | 76.4±7.6a | 50.7±19.3b | 69.4±16.4a | 72.1±12.6a | 54.1±16.5b | 69.1±12.1a |
| 8 | 78.4±8.2a | 54.2±10.2b | | 73.4±17.2a | 48.9±13.6b | |
| 10 | 69.7±10.4a | 48.1±7.6b | | 71.2±16.1a | 47.6±14.2b | |
| 12 | 76.7±6.7a | 56.2±8.6b | | 76.1±15.2a | 42.5±15.3b | |
| 14 | 74.2±7.6a | 52.2±5.4b | | 73.2±12.3a | 43.6±12.1b | |
| 16 | 73.1±4.8b | 48.6±6.7b | | 69.9±13.4a | 33.5±8.7b | |
| 18 | 76.2±8.7a | 46.2±4.5b | | 74.2±11.6a | 31.2±6.5b | |
| 20 | 71.0±10.4a | 41.4±2.3b | | 68.6±12.3a | 26.5±4.3b | |
| 22 | 68.7±8.6a | 31.0±1.0b | | 71.5±9.4a | 18.7±0.6b | |

Note: Means (±S.D.) compared among corn, diet and mixed colonies or corn and diet colonies where no mixed colony data was taken. Means with a different letter indicate significant differences at P≤0.05.

The dramatic drops in egg production and adult survival that occurred after the 5th and 21st generations may be a result of inbreeding depression caused by directional selection or, more specifically, truncation selection (Joslyn 1984). Only those individuals that developed to a pupation weight of 18 mg in 10 days (the truncation point) were saved for pupation, thus potentially reducing the heterozygosity of the remaining population. This directional selection acts on polygenic traits such as fecundity, dietary preference, and longevity (Joslyn 1984), which are traits affected in our corn rootworm colony.

Because a reduction in adult survival and egg production also occurred in the corn-reared colony (after generation 9), nutritional factors are less likely to be the cause. Nearly 100% of the corn-reared larvae develop to 19 mg in 10 days (Marrone et al. 1985); therefore, a deliberate selection of the fastest-developing individuals is not necessary. The heterozygosity of the corn colony is maintained better under these conditions. Directional selection may still occur under these circumstances as the colony becomes adapted to a static laboratory environment (constant temperature, humidity, etc.). Raulston (1975), for example, reported a change in mating and oviposition pattern after the 6th or 7th generation in a wild strain of tobacco budworm that had become adapted to the laboratory environment.

There are several ways to maintain the heterozygosity of a laboratory colony among which are gene infusion, variable laboratory environment, and adequate effective population size (Joslyn 1984). We increased the vigor of our 25th generation diet colony by mass mating diet-reared adults with corn-reared adults. In addition, at least 25% of the eggs set up each week for the diet colony were from the corn colony. The cutoff point for selection of pupation weight larvae on diet was increased from 10 days to 12 days to slow the loss of genetic variability among insects maintained in the colony. Larvae from eggs obtained from outside insectaries have since been added to both the corn and diet colonies at regular intervals. We also mated $F_1$ adults from a field-collected colony with adults from our corn colony. Adults collected from Missouri corn fields were not mated directly with our colony adults because of the presence of tachinid parasites, the fungus *Beauveria bassiana,* and an unidentified microsporidium in wild adults. With these procedures, possible dramatic drops in colony production were prevented.

Although colony vigor is essential to any rearing operation, it is also critical to an insecticide screening operation. Ideally, reared insects should be similar to natural populations in insecticide susceptibility. The difference in $LD_{50}$ values between corn and diet-reared larvae for the microbe may be the result of differential degradation in the insect gut. Bacteria-free boll weevils were more susceptible to methyl parathion than weevils with normal gut flora (Hurej et al. 1982). The difference was attributed to differential rates of degradation in the weevil gut. Tran and Marrone (1988) found large differences in the bacterial species composition and abundance of SCR larvae reared on artificial diet and corn. The total bacterial population in diet-reared larvae was about 100 times less than those in corn-reared larvae. This difference provides one possible explanation for the differential susceptibility of corn and diet insects to the microbial pesticide, but not to the carbofuran. Clearly, more information is needed about the physiology in the gut and mode of action of the microbial to fully explain the results.

Because of the convenience of artificial diets for rearing and bioassay, artificial diets will continue to be widely used in industry testing programs. The results of this study suggests some caution should be used when basing decisions solely on tests from diet-reared larvae.

## References

Hurej, M., P. P. Sikorowski & H. W. Chambers. 1982. Effects of bacterial contamination on insecticide treated boll weevils (Coleoptera: Curculionidae). J. Econ. Entomol. 75: 651-654.

Joslyn, D. J. 1984. Maintenance of genetic variability in reared insects, pp. 20-29. *In* E. G. King & N. C. Leppla [eds.], Advances and challenges in insect rearing. USDA-ARS.

Marrone, P. G., F. D. Ferri, T. R. Mosley & L. J. Meinke. 1985. Improvements in laboratory rearing of the southern corn rootworm, *Diabrotica undecimpuncata howardi* Barber (Coleoptera: Chrysomelidae), on an artificial diet and corn. J. Econ. Entomol. 78: 290-293.

Raulston, J. R. 1975. Tobacco budworm: observations on the laboratory adaptation of a wild strain. Ann. Entomol. Soc. Amer. 68: 139-142.

Shapiro, M. 1984. Micro-organisms as contaminants and pathogens in insect rearing, pp. 130-142. *In* E. G. King & N. C. Leppla [eds.], Advances and challenges in insect rearing. USDA-ARS.

Singh, P. 1977. Artificial diets for insects, mites and spiders. Plenum Press, New York.

Tran, M. T. & P. G. Marrone. 1988. Bacteria isolated from southern corn rootworms, *Diabrotica undecimpunctata howardi* (Coleoptera: Chrysomelidae), reared on artificial diet and corn. Environ. Entomol. 17: 832-835.

# 15

## Importance of Host Plant or Diet on the Rearing of Insects and Mites

*M. J. Berlinger*

### Introduction

Andrewartha (1965), when he set out to compile a list of, "everything that might influence an animal's chance to survive and multiply," found that the factors fell into four major categories: 1) food, 2) weather, 3) other organisms, and 4) a place in which to live. The host plant may affect one or more of these categories, directly or indirectly.

Food ingestion and digestion by the insect must fulfill, first of all, its nutritional requirements for normal growth and development. A quick glance at the dietary requirements of an insect reveals the following: Carbohydrates are a common source of energy and, although not always essential, are usually necessary for normal growth. Some 10 amino acids are essential for tissue and enzyme production, but fats are usually essential only in very small quantities. A direct source of sterols is necessary for all insects, since they are unable to synthesize these compounds. Various vitamins are essential in the diet, and a source of inorganic salts is also necessary. In the absence or imbalance of certain requirements, growth or molting may not occur or may be impaired (Chapman 1969). Host plant morphology may also affect the insect's microclimate. The shape of the cotton leaf produces changes in host suitability for *Bemisia tabaci* (Sippell et al. 1983); leaf pubescence may interfere with the activity of natural enemies; citrus mealybug may find shelter from predators by hiding beneath fruit sepals (Berlinger & Gol'berg 1978).

Quite often attempts are made to rear insects for research, education, or for suppression of pest populations by releasing sterilized insects or natural enemies, etc. Sometimes, insects are reared for economic purposes (e.g., the silkworm and the honey bee): not only for production of food and fiber but also pollination. Insect rearing is also important in plant breeding programs to gain insights into producing plant cultivars that are less suitable hosts.

This paper demonstrates the great variety of host plant effects on insects, specifically spider mites and their natural enemies. The effects of plant chemistry, plant morphology, and endophytic microorganisms are considered.

## The Diversity of Insect/Host Plant Relations

### *The Effect of Plant Chemistry*

Fertilization of host plants can strongly influence growth rates, reproduction rates, etc. of insects in culture. A shorter generation time may thus dramatically reduce the work needed to rear insects or mites for mass release purposes.

**Reproduction and Survival.** An increase in nitrogen (N) fertilization is usually followed by an increase in populations of phytophagous insects. This is probably a result of an increase in insect fertility due to a higher N content of the leaves, as well as better protection because of a denser plant canopy (Harbaugh et al. 1983, Jackson et al. 1973).

Harrewijn (1970) measured the reproduction rate of caged *Myzus persicae* on potato plants growing in nutrient solution. He found that the levels of total and soluble N in the leaves and the reproduction rate of the aphid were positively correlated with the amount of N in the nutrient solution.

Wermelinger et al. (1985) studied the effect of the N content in apple tree leaves on the two-spotted spider mite, *Tetranychus urticae*. N-deficiency increased pre-imaginal development and preoviposition periods and decreased female weight, fecundity, and oviposition rate of the mites. Nitrogen, water, amino acid, and sugar content were positively correlated with development time and preoviposition period. Whereas negative correlations were found with total phenol content of leaves, reduction of leaf N by 50% resulted in a 10-fold decline in fecundity. The stress affected oviposition rate and to a lesser extent oviposition period. Net reproductive rate (avg. no. female offspring per female, $R_0$), mean length of generation time (T), and the innate capacity for increase ($r_m$) were $R_0 = 40.3$, $T = 17.1$, and $r_m = 0.22$ for the standard N concentration, and 4.7, 25.0, and 0.06 for strong N-deficiency, respectively.

Depletion of soil nitrogen frequently leads to reduced concentrations of soluble amino acids and amides, whereas nitrogen excess tends to limit proteolysis, which also can lead to depleted levels of sap nitrogen. Potassium deficiencies, on the other hand, have been associated with accumulation of soluble nitrogen and carbohydrates, owing to inhibition of protein synthesis and increased rates of proteolysis. Phosphorus deficiency also tends to increase soluble nitrogen levels through inhibition of protein metabolism. Thus a common influence of variation in nitrogen, phosphorus, and potassium fertility lies with changes in levels of soluble nitrogen. Although the complex interactions associated with nitrogen, phosphorus, and potassium nutrition limit the value of generalizations, many arthropods tend to suffer from quantitative and qualitative depletion of soluble amino and amide nitrogen (Tingey & Singh 1980).

Harbaugh et al. (1983) noticed that an increase in N fertilization influences leafminer *(Liriomyza trifolii)* damage to chrysanthemums. The number of leafmines per plant, and thus the damage, increased linearly as leaf nitrogen content increased from 2.2 to 4.0%. The number of leafminers also rose with an increase in percent of leaf K, but no relationship was found between percent of leaf P and number of leafmines. This indicates that N was the most critical factor correlated with leafminer damage.

**Egg Production.** It is well known that for some insect species (house flies, fruit flies, etc.) the adults require food containing protein before eggs can develop and oviposition can take place. McCaffery (1975) found that oocyte development is not initiated when female *Locusta migratoria migratorioides* are fed on poor, low protein *Agropyron repens*. Survival on this diet is improved by the provision of water and small quantities of lush *A. repens*. When maturing female locusts (with developing oocytes), previously fed on lush grass, are provided with poor quality grass, the rate of egg pod production is reduced and terminal oocyte resorption is increased. The final percent resorption and the possibility of oviposition are determined by the overall quality of food during vitellogenesis. In locusts fed on poor grass, the levels of ingestion and utilization are low, suggesting that quantitative factors are likely to be critical.

Vanderzant (1963) obtained good oviposition by the boll weevil *(Anthonomus grandis)* on protein-free diets only when amino acids and dextrin were substituted for casein. No eggs were produced when arginine, histidine, isoleucine, lysine, threonine, tryptophan, valine, methionine, or phenylalanine were omitted one at a time from the diet. Eggs were laid by females fed diets containing the 10 indispensable amino acids, glutamic acid, and glycine as the only sources of nitrogen for protein formation.

A more appropriate N-fertilizer application could contribute to keeping pest population densities below the economic injury level. In The Netherlands, N dressings have been substantially reduced in the last couple of years as a result of the studies of the nutritional relationships between host plants and phytophagous mites (Storms 1980).

**Duration of Development.** The time required by *Bemisia tabaci* to complete its development from egg to adult at 26.7 ± 1°C is influenced by the host with which it is confined. Development was completed in 30% less time on lettuce, cucumber, eggplant, and squash than on broccoli or carrot (Coudriet et al. 1985).

**Changes in Plant Suitability.** Cuttings from resistant alfalfa clone 4956 and susceptible clone 4959 were watered, for 9-14 weeks, with solutions that contained an excess, a medium amount, or a deficient amount of Ca, Mg, N, K, P, or S before they were infested with *Therioaphis maculata* (Kindler & Staple 1970). None of the treatments rendered the susceptible clone more resistant. Resistance was decreased significantly but not eliminated when the resistant clone was treated with deficient levels of Ca or K or excess levels of Mg or N, but resistance was increased significantly in plants receiving deficient levels of N. Sulfur did not affect resistance.

**Host Plant Attraction.** Some insects are attracted to their host plants by allelochemicals that are elucidated by the plants. Often a diet must have the proper chemical cues to induce feeding, although frequently these cues have nothing to do with the nutritional value of the diet. In some cases a diet must have the proper nonnutritional chemical cues in order to enhance or induce feeding.

**Leaf Odor Attraction and Oviposition.** The sweetclover weevil, *Sitona cylindricollis*, is a severe pest of sweet clover. Thorsteinson (1960) reported a preliminary observation that this weevil is attracted by the host odor of coumarin, a characteristic constituent of its principal food plant. It is reasonable to assume that the dispersal flights of this weevil, observed in the field during May and June, are associated with the initial infestation of these fields. Knowledge of the environmental factors most favorable to the initiation of dispersal flights would be useful for the timing of control measures. Confirmation of the role of plant odor (coumarin) would facilitate surveys of weevil populations in the fields. The attractiveness of color and odor to flying weevils was investigated with baited traps. The weevils did not show any reaction to colors during their flight. However, it is clear that coumarin odor is a significant food-plant stimulus that induced termination of the dispersal flight (Hans et al. 1961).

Females of *A. bipunctata* and *Coccinella septempunctata* deposited more eggs on leaves of *Berberis vulgaris* than on the leaves of many other plants (e.g., *Malus domesticus, Prunus avium, P. cerasus, Lonicera per-iclymenum, Cotoneaster tomentosa*). Leaf extract from *B. vulgaris* enhanced ovi-position in both coccinellid species when sprayed on the leaves of *P. cerasus* (Shah 1983).

**Host Repellency.** Gibson and Pickett (1983) reported that the aphid *M. persicae* (Sulzer) is repelled, at a distance of 1-3 mm, from leaves of the wild potato *Solanum berthaultii* Hawkes. In addition, air from above the foliage induced rapid dispersal of settled aphid colonies, behavior similar to that of aphids exposed to the aphid alarm pheromone, (E)-ß-farnesen. The presence of substantial quantities of (E)-ß-farnesen in the air around *S. berthaultii* foliage and in ethanolic washings from it was demonstrated by gas chromatography (GC) and mass spectrometry (MS); the (E)-ß-farnesen is probably exuded from type B glandular hairs on the foliage. Gibson and Pickett (1983) believe that this represents the first time, certainly for major food crops, that a plant has been shown to use an aphid alarm pheromone as an allomone. If unknowingly a plant is chosen as host to rear aphids, and this plant has the ability to produce an aphid alarm pheromone, the whole rearing will fail. With the discovery of a plant that produces an aphid alarm pheromone, such a feature may possibly be incorporated into cultivated plants, providing some protection against aphids and hence against aphid-borne viruses.

**Sex Pheromone Inhibition.** The influence of fresh plant material (about 25 g per trial) on the mating behavior and pheromone release activity of female *Spodoptera littoralis* was investigated by Klingauf and Aboul Ela (1982). Various plant species were exposed in cages (40 x 50 x 60 cm) for a period of 7 h and the mating behavior of moths was recorded. In trials with 15 pairs (5 pairs/cage, three replicates; observations at short intervals) four plant species showed no significant effect (11-14 mating pairs), and five showed slightly reduced mating activity (6-9 pairs). Leaves of walnut, celery, and cabbage had a strong inhibitory effect (4, 1, and 0 mating pairs, respectively). In the presence of effective plant species, pheromone release was completely inhibited (cabbage) or the release activity was reduced (late activity start, reduced release frequency, shortened duration). Essential oils from caraway and parsley also lowered mating frequency. It was therefore concluded that the volatile compounds from certain plants influence pheromone release and consequently the reproductive behavior of *S. littoralis*. While rearing insects, especially for pheromone trials, the possible effects of the diet on pheromone production and release must be considered.

**Juvenile Hormone Mimics.** Insect growth regulators, such as juvenile hormone mimics, may occur in plants. This phenomenon is illustrated in the case of *Spodoptera littoralis* reared on a wild tomato. Second and third instar larvae were fed in the laboratory, with *Lycopersicon pennellii* leaves. The larvae were affected by a leaf substance that acted like a juvenile hormone, causing high larval mortality, morphological deformations of wings, and sterility in those few adults that succeeded to develop. These symptoms were not found in the control when the larvae fed on cultivated tomato foliage (Berlinger & Tamim, unpublished data). The effect of a juvenile hormone mimic is not necessarily expressed by high mortality and pronounced symptoms of deformations. Sometimes the effect may be "milder" and expressed only by reduced fertility caused by an "unknown" factor.

**Plant Growth Hormones.** Although insects periodically ravage vegetation in their habitats, the host plants and the insects remain extant. Thus, both the insects and plants that they feed on are adapted, not only to the physical factors in their environment, but also to each other. Because plant growth hormones (PGHs) change in kind and concentration with changes in temperature, moisture, and day length, as well as with aging of the plant, it seems possible that changing concentrations of these substances in host plants might be related to the physiological and reproductive changes that result in insect outbreaks.

The addition of gibberellic acid ($GA_3$) and abscissic acid (ABA) to leaves of *Agropyron smithii* fed to pairs of *Aulocara elliotti* significantly inhibited the production of viable eggs. When indoleacetic acid (IAA), $GA_3$, and kinetin (a synthetic cytokinin) were fed with this grass, highly significant increases in their reproductive processes and longevity were observed. However, since PGHs regulate plant metabolism, it was uncertain whether these results reflected direct effects on the physiology of the insects or indirect effects brought about by an altered nutrient status of the host plant. To resolve this uncertainty, experiments were undertaken in which nymphs and adults of *A. elliotti* were fed PGHs at three concentrations in a defined diet devoid of unrefined plant materials. In addition, several physiological and morphological traits of these variously treated insects were compared. Nymphal survival and days to adult ecdysis are generally reduced by $GA_3$, which reduces the number of fertile females and the mean number of eggs per female but not the mean days to first eggs. The addition of other PGHs to the defined diet also affected nymphal development and survival (Visscher 1987).

**Diapause.** One of the most important aspects of host plant relations in oligophagous or monophagous insects is their coincidence in the course of the year. This implies synchronization of the life cycle of the insect with the vegetative season of the host plant. Diapause, and especially avoiding its onset, is critical to successful rearing throughout the year. Onset of diapause can potentially shut down a colony.

Some signals that trigger the process of preparing the insects for diapause are of particular importance, especially in annuals with a relatively short growth period, and especially toward the end of the growth season. de Wilde et al. (1969) found in Colorado beetle females fed with physiologically aged potato leaves an inhibition or even a standstill of reproduction, followed by diapause.

*Bruchidius atrolineatus* is a tropical beetle. It appears in the field at the end of the rainy season, mid-August, when its host plant, *Vigna unguiculata*, begins to flower, and reproduces on the young pods as soon as they are formed. Two generations occur, one during and one after ripening of cowpea pods. Most adults of the latter generation are in reproductive diapause that is induced by the prevailing thermoperiodic conditions. These diapausing insects probably take refuge in protected sites during the dry season (5 to 6 months) until the flowering of the host plant. When 90-day-old diapausing males and females were placed in the presence of *V. unguiculata* inflorescence or green young pods, development of the reproductive organs was observed as early as the 10th day in most of the insects tested. Nevertheless, the leaves and the dry pods still showed no influence even after 20 days (Huignard et al. 1987).

**Resistance to Insecticides.** Many plants are thought to contain toxic chemicals for defense, yet almost all species of wild and cultivated plants are attacked by some insects. To enable them to feed and survive toxic diets, many phytophagous insects possess a wide range of potent detoxifying enzymes. Of these, the microsomal cytochrome-P450-defendent monooxygenase system is considered to play a vital role in the oxidative metabolism of a huge variety of xenobiotics. Likewise, pesticides may be metabolized to more polar compounds by a variety of detoxifying enzymes. Resistance to insecticides is frequently associated with a rapid induction of these enzymes. It seems likely that allelochemicals in host plants may induce detoxifying enzymes in insects grazing on them, and this may in turn render the insects less susceptible to subsequently applied insecticides. When larvae of *Heliothis armigera* and other lepidopteran larvae are fed on a variety of diets they are subsequently found to have different levels of susceptibility to insecticides. The effects seen here are clearly dependent on the host plant (or allelochemical) and insecticide combination. If certain crops are more potent inducers of detoxifying enzymes, then it is clear that greater quantities of certain insecticides will be required (McCaffery et al. 1987).

Tolerance to parathion in phytophagous insects is significantly influenced by the host plant. This fact was demonstrated for *Ostrinia nubilalis*, *Pieris rapae*, and *Hylemya brassicae* (Wieb & Radcliffe 1973).

Resistance to insecticides is a very important phenomena with practical implications. A knowledge of how resistance may be influenced by diet is critical to anyone measuring the response of laboratory or wild populations to insecticides.

**Webbing.** The type of host may influence the ease with which a colony may be handled and maintained. Too much webbing may be problematic when collecting or observing mites without damaging them or their host plants.

Silk production by the carmine spider mite, *Tetranychus cinnabarinus*, was compared on seven host plants (bean, rose, sweet potato, hibiscus, castor bean, cotton, and Algerian ivy). The most silk was spun on beans, and the least on Algerian ivy. A very good correlation was obtained between the amounts of web produced and the number of eggs deposited on each host plant. A good correlation was found between the intrinsic rate of increase (Ro), the feeding, and the webbing on the seven host plants (Gerson & Aronowitz 1981).

**Wing Formation.** Various ecological factors affect wing formation in aphids. Sometimes wing formation is convenient for experimental purposes but at other times it may be unwanted when aphids are to be mass reared. Among the factors affecting wing formation are day length, temperature, and crowding. The contents and quality of the diet on which aphids feed also exert a significant effect on wing determination. Nutritional effects have been studied by rearing the aphids either on host plants or synthetic media. On unsuitable plants, mainly alatae are produced. The stage of plant growth may also influence wing formation. Whereas mature leaves produced more alatae, seedlings caused the formation of more apterae. Higher proportions of aptera were also obtained when the aphids were grown on synthetic diets suboptimal for growth. When the sucrose-amino acid ratio deviated from the median of 10% sucrose to 2.4% amino acids, more alatae *M. persicae* were produced. Individual vitamin deficiencies had no apparent effect on wing determination. Lack of iron or zinc in diets increased the percentage of alatae, whereas larvae deprived of phosphate developed mainly to apterae. When copper, manganese, or magnesium were omitted, no effect on wing formation was found. Attention is drawn to the apparently similar effect on wing determination of imbalanced synthetic diets and waning host plants (Raccah et al. 1971).

**Parasitoid Egg's Encapsulation.** Egg encapsulation is a phenomenon that may strongly reduce parasitoid population growth. The Florida wax scale female is one of many insects that is able to prevent the development of the parasitoid *Tetrastichus ceroplastae* within its body by encapsulation of the deposited egg. This process begins within 1-3 days after parasitization. A highly sclerotized capsule is formed, within which the egg remains. A host scale that has encapsulated a parasitoid egg subsequently lays normal eggs. The parasitoid rearer must be aware of this phenomenon since the mean incidence of encapsulation is host plant related. It reached 90% when the scale was reared on sweet lime leaves compared with 27% on English ivy leaves (Ben-Dov 1972).

## The Effect of Plant Morphology

**Leaf Shape.** Cotton leaf shape influences the surrounding microclimate. Both day and night relative humidity were considerably higher (54 and 95%

relative humidity) in a normal-leaf cultivar than in a super okra (43 and 76% relative humidity) possessing very narrow-lobed leaves. The temperatures were slightly lower (27.9 and 14.7°C) in the normal type than in the super okra type (29.1 and 15.1°C). Lower relative humidity and higher temperature are known to affect *B. tabaci* development and survival (Sippell et al. 1983).

**The Effect of Light.** Sometimes climatic conditions affect the morphology of plants. Various wild Lycopersicon accessions are resistant to the tobacco whitefly. It was noticed that some accessions are resistant in summer but not in winter (Berlinger et al. 1983, Kennedy et al. 1981, Snyder & Hyatt 1984). A laboratory test showed that one of the *L. hirsutum v. glabratum* accessions was remarkably more susceptible when grown under lower light intensity, independent of day length, whereas in *L. pennellii* accessions only a combination of low light intensity with a short-day photoperiod resulted in susceptibility (Berlinger et al. 1983). The effect of light intensity on leaf morphology is a common phenomenon and must be considered while choosing the climatic conditions under which the host plants are grown, especially in breeding programs concerning plant resistance to pests.

**Plant Tissue Hardness.** The development of *Ostrinia nubilalis* larvae on different maize cultivars was tested. A highly significant negative correlation (r = -0.967 to -0.969) was found between the tissue hardness, determined by a penetrometer, and the relative growth rate of the larvae. The results indicate that selection for high tissue hardness in maize could reduce damage by *O. nubilalis* (Viereck 1983).

**Plant Morphology and Thigmotropism.** Many insects exhibit the tendency to settle or hide in narrow places such as crevices or underneath bark, where they are in maximum contact with their surroundings (thigmotropism). To rear these species effectively, it is important to fulfill this requirement, to provide them with "a place in which to live."

The citrus mealybug, *Planococcus citri*, is a severe citrus pest in Israel. Among the various citrus fruits, grapefruit is the most severely infested species. After a period of wandering, the newly hatched crawlers settle down in shaded spots, particularly where they obtain maximum contact with their surroundings. Therefore, they tend to concentrate on citrus fruits, beneath the sepals. The fruit sepals were removed by means of a scalpel and replaced by a paper cone, which was stuck onto the fruit (artificial sepals). Ten days after the artificial infestation the percent of normal fruits infested was grapefruit, 70; orange, 24; lemon, 18; and Troyer, 18 with averages of 4.5, 1.1, 0.2, and 0.4 mealybugs per fruit, respectively. Replacing the plant's sepals with artificial sepals resulted in a very similar rate of infestation of all four citrus species (30-32% of infested fruits) and very similar average numbers of mealybugs per fruit (1.5-1.6). Thus the natural sepals of grapefruits provided the best "place in which to live" (Andrewartha 1965). Although the mealybug population of fruits without sepals was 0.1, and on normal fruits 4.0 per fruit, total parasitism was relatively high on fruits without sepals (45%), compared with about 5% on the normal fruits (Berlinger & Gol'berg 1978).

The hall scale *Nilotaspis halli* (Diaspididae) is found on peach trees exclusively beneath buds, hidden between the buds and the twig, on 1- or 2-year-old branches. On the peach cv. 'Hermosa' the greatest scale populations were found (73 scales per branch), on cv. 'Summerset' the lowest (12 scales per branch), and on cv. 'Suwannee' intermediate numbers (53 scales per branch). Thus, cv. Hermosa seems to be a better host for this scale. A closer look revealed that more scales were found in closely attached buds (4.9 scales per bud) than in protruding buds (1.5 scales per bud). A high correlation was also found when the number of scales per branch was compared with the distribution of "percent attached buds" on the three peach cultivars (Berlinger et al. 1987).

The importance of shelters for thigmotrophic insects is so high that it can be used to trap them in order to monitor the pest or its natural enemies, e.g., by attracting the citrus mealybug to "artificial sepals" made of paper caps (Berlinger & Gol'berg 1978) or to twig traps that consist of a strip of black cloth (15 mm wide) that was wrapped around a peach branch and caused the scales to settled on the branch under the traps (Berlinger et al., 1987). When the traps (shelters) were removed, the scales were fully exposed to their natural enemies, and both the pests and their natural enemies could be investigated easily.

**Search Capacity of Natural Enemies.** Contrary to the situation with tomatoes, sufficient control of the greenhouse whitefly by the parasitoid *Encarsia formosa* is rarely achieved in greenhouse cucumbers, despite repeated large introductions of the wasp. The wasp searches for its prey mainly by walking, not flying. The failure of control was caused by cucumber leaf pubescence and the honeydew that stuck to them. These factors reduce the mobility of the wasp, and thus its parasitizing efficiency. It was found that on glabrous leaves the wasp was no longer hampered by the hairs and it covered, per unit of time, a distance three to four times greater than on the hairy leaves (de Ponti 1980).

The behavior of 1st and 4th instar larvae of the predatory coccinellid *Adalia bipunctata* was observed on leaves with different surface characters. Searching success was high, where single scattered hairs forced larvae to change their direction often and to cross the interveinal areas (e.g., Chinese cabbage and radish). On glabrous leaves with a thick, slippery wax, layer (e.g., kohlrabi or Brussels sprouts), larvae could move only along the edge of narrow protruding veins that they could clasp with their legs. Larvae were unable to search leaves with dense upright or hook-shaped hairs (e.g., tomato, tobacco, and bush bean leaves) (Shah 1982).

## The Effect of Microorganisms

**Endophytic Fungi.** Fungi occurring occasionally in host plants may alter the biology of the insects which thrive on them. Graminaceus plants harboring endophytic fungi (*Acremonium* spp.) deter 18 insect species of a wide taxonomic range that include Coleoptera, Lepidoptera, Hemiptera, and Orthoptera. In chewing insects, acute toxicity is associated with the consumption of plant leaf sheaths, where the fungal level is highest. There is evidence for a high polar antifeedant compound in *Lolium perenne* that suppresses feeding of *Listronotus bonariensis* adults. The antibiosis of sap-sucking Hemiptera that feed primarily on leaf blades suggests that some toxic or antifeedant allelochemical is translocated from the endophyte-infected area to other parts of the plant. The suspected substances, of either fungal or plant origin, are ergot alkaloids and loline-type pyrrolizidine alkaloids in *Festuca arundinacea* and a neurotoxic indole called lolitrem B and an antifeedant of unresolved structure in *L. perenne* (Ahmad & Funk 1987).

Fusariotoxin, a preparation from the microfungus *Fusarium sporotrichilla*, proved to be highly toxic to all developmental stages of *Tetranychus telarius* mites at concentrations of 0.01, 0.02, and 0.03% and also induced sterility in the treated females. It is apparent that the endophytic fungi are important sources of insect resistance that plant breeders can incorporate into new crop varieties (Chhabra 1973).

**Mycorrhiza.** Mycorrhiza is the association between a fungus (Gr.- Mykes) and the root (Gr.- rhiza) of a plant. Four soybean (*Glycine max* L. Merr) cultivars were either inoculated with a vesicular-arbuscular mycorrhizal (VAM) fungus, *Glomus fasciculatum*, or were left uninoculated but fertilized with 0.2 mM P (Pacovsky et al. 1985). Leaves were detached from the plants and fed to neonate lepidopteran larvae of *Heliothis zea* and *Spodoptera frugiperda*. The average larval weights for *H. zea* and *S. frugiperda* fed leaves from VAM plants were reduced by more than 40% compared with larvae fed P-fertilized plant foliage. When fed VAM plants, the larvae of both species took longer to pupate; the average pupal weight was 17% lower and the mortality of *H. zea* was 15% greater than that of the P-fertilized controls. Growth reduction for larvae fed VAM leaves did not correlate with leaf-N amino acid, carbohydrate, micronutrient, or phenolic content. Dry weights of P-fertilized plants were 25% greater than those of VAM plants, but they contained 50% more P. However, the decrease in insect growth was not well correlated with the lower P content of VAM plants and it is unlikely that this effect was due to a P deficiency. The VAM-induced decrease in insect growth was evident in all four cultivars. The VAM-induced decrease in insect growth may be specific for larvae that are foliar feeders, since no difference was observed by Pacovsky et al. in the feeding or reproductive behavior of a phloem-feeding aphid (*Schizaphis graminum*) tested on VAM-colonized or P-fertilized sorghum.

**Plant Viruses.** The biology of an insect can be affected by the presence of a virus in the host plant; an insect can become adapted to a new host plant, previously resistant to it because of infection by a plant virus. Similarly, certain aphids are able to colonize yellowed aster plants, while noninfected asters appeared to be fairly resistant to aphids. In the course of an epidemiological study of the maize rough dwarf virus (MRDV) in Israel, it was found that *Laodelphax striatellus* (Homoptera: Delphacidae), which is the principal vector of MRDV, cannot survive on noninfected Bermuda grass (*Cynodon dactylon* ) plants for more than 4 days. On the other hand, when this plant becomes infected with the virus, the vector can develop and breed on it at a rate equal to that on wheat, oats, and barley, which are its natural host plants. In fact, the presence of MRDV in its body is enough to render the vector capable of completely adapting itself to Bermuda grass (Klein & Harpaz 1969). Furthermore, virus-infection may change the outcome of a screening test, since TYLCV-infected tomatoes seem to be less attractive to the tobacco whitefly than healthy tomato plants (Berlinger, unpublished data).

## Concluding Remarks

Interest in insect rearing may be purely academic, but quite often it is essential for the solution of economically important problems. Therefore, the impact that the host plant may have on the biology, behavior, and performance of phytophagous insects and on their natural enemies must be considered. The host plant may be involved in all of the four categories (Andrewartha 1965) that affect population growth: 1) Food—The host plant must fulfill all the nutritional and palatability requirements and not interfere with the normal development of both phytophages and entomophages; 2) Weather—In many cases the host plant provides a favorable microclimate for the herbivores and their natural enemies; 3) Other organisms—The host plant may attract competitors, as well as parasitoids and predators, or alter their activity and performance; 4) A place in which to live—To leaf miners, stem borers, and to many insect and mite species that dwell beneath bark, buds, or fruit sepals, the host plant also provides shelter.

This paper describes the crucial role that a plant plays in the life of an insect beyond the well-known effects of attraction, repellency, or poisoning, and illustrates the great diversity of the insect-plant relationship. The information included here will help rearing specialists understand the implications in type and quality of host or diet and will assist them in finding suitable solutions to specific problems.

## References

Ahmad, S. & C. R. Funk. 1987. Role of endophytic fungi in enhancing host plant resistance to herbivores, p. 364. *In* V. Labeyrie, G. Fabres & D. Lachaise [eds.], Insect-Plants. Junk, Dordrecht.

Andrewartha, H. G. 1965. Introduction to the study of animal populations. Phoenix Science Series, third impression. University of Chicago Press, Chicago & London.

Ben-Dov, Y. 1972. Life history of *Tetrastichus ceroplastae* (Girault) (Hymenoptera: Eulophidae), a parasite of the Florida wax scale, *Ceroplastes floridensis* Comstock (Homoptera: Coccidae), in Israel. J. Entomol. Soc. South Africa 35: 17-34.

Berlinger, M. J. & A. M. Gol'berg. 1978. The effect of the fruit sepals on the citrus mealybug population and on its parasite. Entomol. Exp. Appl. 24: 238-243.

Berlinger, M. J., R. Dahan & Esther Shevach-Urkin. 1983. The effect of light on the resistance of wild species of solanaceae to *Bemisia tabaci*. Phytoparasitica 11: 63.

Berlinger, M. J., Ch. Fallek, R. Dahan & S. Mordechi. 1987. The relationship between the Hall scale, *Nilotaspis halli*, (Diaspididae) and its host plant: the effect of the plant on the scale, pp. 371-372. *In* V. Labeyrie, G. Fabres & D. Lachaise [eds.], Insect-Plants. Junk, Dordrecht.

Chapman, R. F. 1969. The insects—structure and function. English University Press, London.

Chhabra, K. S. 1973. Influence of certain phytocides and antibiotics on the reproduction and development of Tetranychus mites, pp. 645-647. *In* Proceedings 3rd Int. Congr. of Acarology, 1971 Prague, Pague. Junk, The Hague.

Coudriet, D. L., N. Prabhaker, A. N. Kishaba & D. E. Meyerdirk. 1985. Variation in developmental rate on different hosts and overwintering of the sweetpotato whitefly, *Bemisia tabaci* (Homoptera: Aleyrodidae). Environ. Entomol. 14: 516-519.

Gerson, U. & A. Aronowitz. 1981. Spider mite webbing. V. The effect of various host plants. Acarologia 22: 277-281.

Gibson, R. W. & J. A. Pickett. 1983. Wild tomato repels aphids by release of aphid alarm pheromone. Nature 302: 608-609.

Hans, H., D. Peschken & A. J. Thorsteinson. 1961. The influence of physical factors and host plant odour on the induction and termination of dispersal flights in *Sitona cylindricollis* Fahr. Entomol. Exp. Appl. 4: 165-177.

Harbaugh, B. K., J. F. Price & C. D. Stanley. 1983. Influence of leaf nitrogen on leafminer damage and yield of spray chrysanthemum. HortScience 18: 880-881.

Harrewijn, P. 1970. Reproduction of the aphid Myzus persicae related to the mineral nutrition of potato plants. Entomol. Exp. Appl. 13: 307-319.

Huignard, J., J. F. Germain & J. P. Monge. 1987. Influence of the inflorescence and pods of *Vigna unguiculata* Walp (Phaseolinae) on the termination of the reproduction diapause of Bruchidius atrolineatus (Pic) Coleoptera Bruchidae, p. 183. *In* V. Labeyrie, G. Fabres & D. Lachaise [eds.], Insect-Plants. Junk Publishers, Dordrecht.

Jackson, J. E., H. O. Burhan & H. M. Hassan. 1973. Effects of season, sowing date, nitrogenous fertilizer and insecticide spraying on the incidence of insect pest on cotton in the Sudan Gezira. J. Agric. Sci. 81: 491-505.

Kennedy, G. G., R. T. Yamamoto, M. B. Dimock, W. G. Williams & J. Bordner. 1981. Effect of day length and light intensity on 2-tridecanone levels and resistance in *Lycopersicon hirsutum f. glabratum* to Manduca sexta. J. Chem. Ecol. 7: 707-716.

Kindler, D. & Staple, R. 1970. Nutrients and the reaction of two alfalfa clones to the spotted alfalfa aphid. J. Econ. Entomol. 63: 938-940.

Klein, M. & I. Harpaz. 1969. Changes in resistance of graminaceous plants to delphacid planthoppers induced by maize rough dwarf virus (MRDV). Z. angew. Entmol. 64: 39-43.

Klingauf, F. & A. Aboul Ela. 1982. Volatile plant constituents as pheromone inhibitors in cotton leaf worm, *Spodoptera littoralis* Boisd. (Lepid.: Noctuidae). Med. Fac. Landbouww. Rijksuniv. Gent 47/2, 473-480 (German with English summary).

McCaffery, A. R. 1975. Food quality and quantity in relation to egg production in *Locusta migratoria migratorioides*. J. Insect Physiol. 21: 1551-1558.

McCaffery, A. R., A. J. Walker & S. M. Lindfield. 1987. Effect of host plant on susceptibility of lepidopteran larvae to insecticides, p. 53-58. *In* V. Labeyrie, G. Fabres & D. Lachaise [eds.], Insect-Plants. Junk, Dordrecht.

Pacovsky, R. S., L. B. Rabin, C. B. Montllor & A. C. Waiss, Jr. 1985. Host-plant resistance to insect pests altered by *Glomus fasciculatum* colonization. *In* Proceedings 6th North American Conf. on Mycorrhizae. Bend, OR, USA 1984.

Ponti, O. M. B. de. 1980. Contributions of plant breeding to integrated pest control illustrated on the glasshouse cucumber and its pests *Tetranychus urticae* and *Trialeurodes vaporariorum*. EPPO Bull. 10: 357-363.

Raccah, B., A. S. Tahori & S. W. Applebaum. 1971. Effect of nutritional factors in synthetic diets on increase of alatae forms in Myzus persicae. J. Insect Physiol. 17: 1385-1390.

Shah, M. A. 1982. The influence of plant surface on the searching behaviour of Coccinellid larvae. Entomologia Exp. Appl. 31: 377-380.

Shah, M. A. 1983. A stimulant in *Berberis vulgaris* inducing oviposition in coccinellids. Entomologia Exp. Appl. 33: 119-120.

Sippell, D. W., O. S. Bindra & H. Khalifa. 1983. Resistance in cotton to whitefly *(Bemisia tabaci)*. Proc. 10th Int. Congr. Plant Protection, vol. 2. Brighton, England.

Snyder, J. C. & J. P. Hyatt. 1984. Influence of daylength on trichome densities and leaf volatiles of Lycopersicon species. Plt. Sci. Letters 37: 177-181.

Storms, J. J. H. 1980. Nutritional relationship between host plants and phytophagous mites, pp. 155-158. *In* A. K. Minks & P. Gruys [eds.], Integrated control of insect pests in the netherlands. PUDOC, Wageningen.

Thorsteinson, A. J. 1960. Host selection in phytophagous insects. Ann. Rev. Entomol. 5: 193-218.

Tingey, W. M. & S. R. Singh. 1980. Environmental factors influencing the magnitude and expression of resistance, pp. 87-113. *In* F. G. Maxwell & P. R. Jennings [eds.], Breeding plant resistance to insects. Wiley, New York.

Vanderzant, E. S. 1963. Nutrition of the adult boll weevil: oviposition on defined diets and amino acid requirements. J. Insect Physiol. 9: 683-691.

Viereck, A. 1983. Der Einfluss der Gewebehrte auf die Resistenz von Maisgenotypen gegen den Maiszuensler, *Ostrinia nubilalis* Hbn. Z. PflZuecht. 90: 75-84 (German, with English summary).

Visscher, S. N. 1987. Plant growth hormones: Their physiological effects on a rangeland grasshopper *(Aulocara elliotti)*, pp. 37-41. *In* V. Labeyrie, G. Fabres & D. Lachaise [eds.], Insect-Plants. Junk, Dordrecht.

Wermelinger, B., J. J. Oertli & V. Delucchi. 1985. Effect of host plant nitrogen fertilization on the biology of the two-spotted spider mite, *Tetranychus urticae* Entomol. Exp. Appl. 38: 23-28.

Wieb, J. & E. B. Radcliffe. 1973. Tolerance to parathion of phytophagous insects influenced by host plant. Envir. Entomol. 2: 537-540.

Wilde J. de, W. Bongers & H. Schooneveld. 1969. Effect of host/plant age on phytophagous insects. Entomol. Exp. Appl. 12: 714-720.

# 16

## Seasonal and Nutritional Influences on the Toxicological Response of the First Instar Larvae of *Spodoptera littoralis* (Boisduval) in a Mass Rearing Culture

*V. Flueck, F. Bourgeois, and P. Stoecklin*

### Introduction

The mass rearing of insects for the industrial screening of insect control products requires insects of consistent quality to ensure good reproduction of test results. However, the economic aspects of materials, methods, and labor are also very important.

Until about 10 years ago we preferred to rear all our insect species on plants. It was felt that with this method insects from a laboratory colony conformed more closely to individuals from the field in their reactions to feeding on the host plant.

This approach is justified to some extent, particularly when stomach insecticides are being tested on the larval instars. The insects' digestion suffers no change when the food plant and test plant are the same.

However, various researchers demonstrated that the action of an insecticide may also be affected by the nature of the host plant. One such case was reported by El Sayed (1975) using Sevin against *Agrotis ipsilon*. Sevin was 10 times more effective against insects reared on lettuce than against those reared on ricinus (castor-oil plant). In contrast, the use of spinach or cabbage produced no essential differences in larval mortality.

Zoebelein (1977) indicates particularly consistent and comparable results for toxicological experiments with *Spodoptera littoralis* when castor-oil leaves or cotton leaves are used as food and test plant.

Our problem in the rearing of *Spodoptera littoralis* was not so much the choice of suitable host plants but rather whether or not a synthetic substrate should be used instead of plants. There were several reasons for considering these alternatives:

1) When we fed our insects on plants grown for test purposes, the variation in the quality of host plants with season affected the quality of insecticide trials.
2) As experimental capacity expanded, it became increasingly difficult to obtain sufficient host plants, especially in the winter months. Increasing use had to be made of expensive greenhouse space to grow the plants.
3) The use of plants as insect feed was also excessively time consuming. The feed had to be renewed daily and any excrement removed.

In view of these disadvantages for quality, economy, and working method, we considered the use of artificial nutrients for breeding insects, particularly Lepidoptera species such as *Spodoptera littoralis*. Experience in the preparation and use of artificial diets is now quite extensive. Literature summaries on this subject have been published by House (1967), Vanderzant (1974), King and Leppla (1984), and Singh and Moore (1985). Various papers have also been published on artificial nutrients for *S. littoralis* in particular (Poitou et al. 1972, Isa & Khadr 1974, Navon 1976, Cabello-Garcia et al. 1984, & Rahbe et al. 1986).

Important differences in the composition of published nutrient media stem chiefly from the sources of proteins and carbohydrates. Sources mentioned include dried powders of various beans, such as broad beans, snap beans, and kidney beans. Mention is also made of alfalfa meal and, as being highly successful, corn meal.

However, since Adkisson et al. (1960) discovered that wheat germ is a particularly good nutrient for pink bollworm, wheat germ has also been successfully used for *Spodoptera*. Wheat germ is especially rich in amino acids, protein, sugar, triglycerides, phospholipids (including choline and inositol), the B vitamins, tocopherolene, carotins, minerals, and more than 50 enzymes (MacMasters et al. 1971).

Breeding on an artificial diet offers economic advantages and reduces the work load. Before deciding in favor of this method, however, we needed to know whether it would alter the biological quality of insects that had been reared for many years on castor-oil leaves or cotton leaves. We also examined, as an intermediate form, insects that were reared first on artificial nutrients over a long period and then on cotton leaves for 3-5 generations.

In further studies we concentrated on the significance of the biological quality of the insects fed by different methods and used for toxicological trials. Special attention was given to the question of seasonal influences on the toxicological response by using insects bred on artificial nutrients.

## Materials and Method

### Experimental Insects

In the present investigation the insects were obtained from a laboratory colony continuously reared for 10 years on a modified Vanderzant diet (Vanderzant et al. 1962). In the following discussion, we call this strain the *D strain*.

From the same strain we developed the *D strain*, which was reared on diet and then fed for 3-5 generations on cotton leaves only.

In our experiments we compared the D and DC strains with a further strain, the *C strain*. This consisted of insects reared exclusively on castor-oil leaves for about 10 years and then fed for 3-5 generations on cotton leaves only.

All the strains used originally came from Egypt.

### Artificial Diet

The composition of the artificial diet used is given in Table 1.

### Preparation of the Diet

The diet ingredients, with the exception of the vitamins, antimicrobials, and choline chloride, were dissolved in the amount of water required for blending and then left to swell for 1 h. The ingredients were then mixed in a kneading machine for 20 min, and choline chloride and formaldehyde were added. The agar was dissolved separately in the remaining water at 100°C and added to the other ingredients. This mass was then cooled to about 60°C. The vitamins, antimicrobials, aureomycin, and CGA 39896 were added and the diet was ready for use.

A major problem in the use of artificial diet is the elimination of bacteria and fungi. However, antimicrobials such as benzoic acid, methyl-para-hydroxybenzoate (Nipagin), formaldehyde, and sorbic acid have proved adequate in overcoming this problem.

Wyniger (1974) has provided an even better solution to the problem of aspergilli and penicillia using CGA 39896 (earlier GS 36896)—a sulfonyl thiadiazole—as fungistat.

**Caution.** The CGA 39896 must be added in an efficient fume cupboard. Gloves must also be worn to protect sensitive persons against an allergic reaction. This substance must be used at the experimenter's own risk, since toxicology of the product is not fully know.

**Table 1. Composition of the artificial diet for rearing the cotton leaf-worm *Spodoptera littoralis***

| Ingredients | Quantity |
| --- | --- |
| Casein (alkali soluble) | 378 g |
| Wheat Germ | 324 g |
| Glucose | 18 g |
| Ascorbic acid | 42 g |
| Vanderzant Vitamin Mix | 18 ml |
| Wesson's Salt | 108 g |
| Choline Chloride (10% aqueous solution) | 108 g |
| Formaldehyde 38% | 9 ml |
| Potassium Hydroxide (22.5% aqueous solution) | 54 ml |
| Aureomycin (9 capsules of 250 mg AI each) | 2.25 g AI |
| CGA 39896 (dissolved in acetone/ water 1:2) | 1.2 g AI (Fig. 1) |
| Sodium Alginate (to improve consistency) | 54 g |
| Agar | 270 g in 7,000 ml water |
| Water | 3,000 ml |

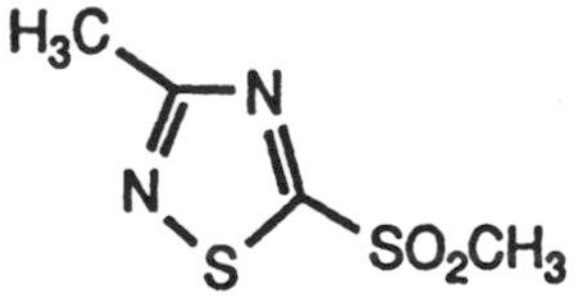

Acute oral $LD_{50}$ for rats: 55 mg/kg

**Fig. 1. CGA 39896, a very effective fungistat in diets; not available commercially.**

## Rearing Procedure on Artificial
## Diets and on Young Cotton Leaves

All stages of the cotton leafworm were cultured at room temperature, 25°C ± 1°C, and 60% relative humidity.

Generally, the rearing of *S. littoralis* on artificial diet in our insectary (D strain) served two purposes:

1) Maintenance of the strain. For this purpose the eggs were disinfected in 2.5% formaldehyde and then placed on artificial nutrients in petri dishes 5 cm in diameter. After reaching the 3rd instar, one, two, or three individuals were transferred to a new dish until pupation. The collected pupae were also disinfected with formaldehyde and stored over sawdust in large petri dishes. About 600 hatched adults were introduced into 5-liter plastic jar, lined with filter papers, for egg deposition. The filter paper with the eggs was removed every day.

2) Production of larvae for screening purposes only. For this purpose the eggs were not disinfected. They were introduced into a large petri dish containing a smaller dish filled with artificial nutrients as food for the hatched larvae. The larvae not used within the first to third instar were discarded.

The larvae of the strain fed with young cotton leaves (C Strain) were kept in large petri dishes containing several layers of filter paper. Cotton leaves were added every day and the excrement removed. The last larval instars were transferred for pupation to another dish containing sawdust, partly covered first with a filter paper and then cotton leaves. Filter paper is added to reduce humidity from excrement.

## Determination of the Reproductive Biology of *S.* littoralis

The biology of the strains was compared by determination of: 1) the average larval period, in days, and the pupation rate and 2) the pupal weight and hatching rate of adults.

## Results and Discussion of the
## Reproductive Biology of *S. littoralis*

The results for the average larval period, pupation rate and hatching rate for adults are given in Table 2. The shortest larval period and the highest pupation rate are shown by the D strain. Although a slightly lower hatch of adults occurs in the D strain, it may be concluded that this is compensated by a higher pupation rate (Table 2) and a greater pupal weight (Fig. 2). Therefore, we conclude that the use of an artificial diet is most favorable not only for the reproductive biology of *S. littoralis* but also in allowing a reduced generation time, notably from a shorter larval period.

We paid particular attention to the weight of the pupae. Measuring pupal weight is not only a relatively simple means of checking quality; it is also a good indicator of progress when breeding Lepidoptera. There should, in fact, be a direct relationship between breeding capacity and pupal weight, since it is known that the heavier the adult female, the more eggs she is likely to lay.

**Table 2. Mean larval periods, pupation rates, and percentage of hatched adults for strains of *S. littoralis* reared on different diets**

| Strains | Larval Period (days) | Pupation Rate (days) | Hatched Adults (%) |
|---|---|---|---|
| **D Strain** (diet only) | 19.7 B* | 97.9 A | 92.5 B |
| **DC Strain** (diet, then 3 generations on cotton leaves) | 28.1 A | 71.3 C | 87.8 C |
| **C Strain** (castor oil plant leaves, then 3 generations on cotton leaves) | 28.5 A | 88.9 B | 97.2 A |

* Values with different letters are significantly different from each other: P= 0.05 (Duncan's multiple range test).

The results plotted in the graph in Fig. 2 show that there is little difference in pupal weight between insects of the DC and C strains fed on cotton. However, both differ significantly from pupae reared on the artificial diet. With one larva per feeding dish, the pupal weight was more than twice that of insects reared on cotton leaves; with three larvae per dish, the pupal weight was almost twice as high.

As already indicated, we consider pupal weight an important indicator of breeding quality. Returning to the question posed in the introduction of whether insects reared on different diets show biological differences, we found that insects reared on an artificial diet are biologically superior with respect to reproduction. This finding is supported by the shorter larval periods for insects reared on an artificial diet, as shown in Table 2.

Although the highest pupal weight is achieved by feeding the larvae individually, we find that three larvae per dish is satisfactory; this method still gives good results and is more efficient.

As the number of generations increases, the DC strain shows a trend toward increasing pupal weight compared with the C stain (Fig. 3). This trend could indicate that the insect suffered shock from the change of feed, but were adapting increasingly to the cotton leaves. However, further trials would be necessary to prove this point.

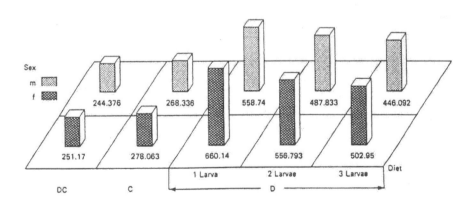

**Fig. 2. Mean weight of male and female pupae of strains (D, DC, and C) of S. *littoralis* reared on different diets.**

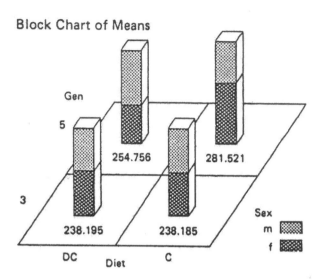

**Fig. 3. Mean pupal weight of the 3rd and 5th generations of the DC and C strains of *S. littoralis*.**

## Toxicological Trials Using 1st Instar
## Larvae of *S. littoralis* Bred on Various Diets

As in the biological reproductive trials, the strains used here were again the D, DC, and C strains.

### Method of Testing Insecticides

Discs 20 mm in diameter were punched from cotton leaves and sprayed by the Potter spray tower method (Potter 1941). The tower nozzle was calibrated to release 120 liter spray volume per hectare.

The leaf discs sprayed in the Potter tower were placed on a thin layer of agar in small petri dishes. The agar prevents the leaves from drying out quickly. One of the larvae selected by means of the fitness ramp (Brassel, pers. comm.) was then placed on each leaf disc. For each test 30 replicates were used. Mortality was evaluated 6 days after infestation. The trials were conducted with 4 to 5 replicates on various days using 5 to 7 concentrations. Mortality was evaluated by Probit analysis (FINNEY).

## Insect Control Products Used in the Experiments

**CGA 184699, a Benzolphenyl Urea.** CGA 184699 is an insect growth regulating compound that interferes with chitin synthesis. It is active against the larvae of Lepidoptera and Coleoptera, and it acts more by ingestion than by contact.

Structural formula:

Not yet released.
Acute oral $LC_{50}$ for rats: > 2.000 mg/kg

CGA 15324 was discovered and introduced onto the market by CIBA-GEIGY Limited, Basle, Switzerland, under the trade names Curacron and Selecron. It is a broad-spectrum insecticide with contact and stomach action for use in plant protection.

Structural formula:

Acute oral $LC_{50}$ for rats: 358 mg/kg techn. a.i.

**The Fitness Ramp.** Before being used to infest the treated leaf discs, the first instar larvae had to pass the "fitness ramp."

The fitness ramp selects insects of consistent quality, thereby assuring low scattering in the subsequent infestation trials. Pieces of filter paper holding larvae that had emerged from the eggs were introduced into a plastic beaker. The latter was connected to a similar beaker placed at a higher level by a 20-cm length of tubing. The first beaker and the tubing were shaded, but the second beaker at the top end of the tube was exposed to light. Exploiting the positive phototaxis of larvae, the insects used for infestation were those that passed within 15 min through the sloping tube into the beaker exposed to light.

## Results of the Toxicological Trials

Before considering the toxicological responses of the various strains, it is interesting to demonstrate the effect of CGA 184699 on 1st instar larvae before and after they had passed the fitness ramp. From the results in Table 3, it is evident that this method selects less sensitive larvae in comparison to those removed from the breeding dish or beaker on the bottom of the ramp. This technique produces more homogeneous results in toxicological trials. The sensitivity of these larvae is more comparable to that of feral individuals. The effects of nutrition on the toxicological response of *S. littoralis* to CGA 184699 and CGA 15324 are summarized in Table 4. In the toxicological trials, a steady decrease in sensitivity to both test substances was observed as follows:

D strain > DC strain > C strain

However, the difference in sensitivity to benzoylphenylurea, CGA 184699, was extremely slight. The sensitivity ratio for the $LC_{50}$ ($SR_1$) and $LC_{95}$ ($SR_2$) values between the C and D strains was only 1.27 - 1.3 (Table 4).

**Table 3. Effect of selection by fitness ramp on the sensitivity of *S. littoralis* L-1 to the Potter test (insects reared on artificial diet: host plant = cotton)**

| Point of removal of L-1 larvae from the breeding dish (A) or from the fitness ramp (B) | Mortality percentage of larvae 6 days after placement on leaf disks treated with CGA 184699 | |
|---|---|---|
| | 0.8 ppm AI | 0.4 ppm AI |
| A | 100 | 67 |
| B | 100 | 97 |
| B | 93 | 33 |

Table 4. Effects of nutrition on the toxicological response of the 1st instar larvae of *S. littoralis* to CGA 184699 and CGA 15324 (Curacron)

| Product | Strain | LC$_{50}$ ppm AI | Conf. limits | LC$_{95}$ ppm AI | Conf. limits | β (slope) | SR$_1$ | SR$_2$ |
|---|---|---|---|---|---|---|---|---|
| CGA 184699 | Diet (D) | .79 | .76 - .82 | 1.4 | 1.29 - 1.52 | 6.5 | 1.27 | 1.3 |
| | Diet/cotton (DC) (3-5 gener. on cotton) | .88 | .85 - .92 | 1.65 | 1.52 - 1.80 | 6.0 | 1.14 | 1.10 |
| | Cotton (C) | 1.0 | .96 - 1.05 | 1.82 | 1.68 - 1.98 | 6.3 | 1.0 | 1.0 |
| CGA 15324 (Curacron) | Diet (D) | 24.6 | 23.3 - 25.8 | 47.4 | 44.0 - 51.0 | 5.8 | 1.98 | 2.75 |
| | Diet/cotton (DC) (3-5 gener. on cotton) | 29.1 | 27.9 - 30.64 | 54.9 | 50.9 - 59.1 | 6.0 | 1.67 | 2.37 |
| | Cotton (C) | 48.7 | 45.1 - 52.6 | 130.3 | 100.1 - 169.7 | 3.9 | 1.0 | 1.0 |

SR$_1$ = sensitivity ratio LC$_{50}$ C/LC$_{50}$ D, DC
SR$_2$ = sensitivity ratio LC$_{95}$ C/LC$_{95}$ D, DC

There was also only a very slight difference in sensitivity to the second test substance, the OP ester (CGA 15324), between the D strain and the DC strain. The C strain, however, proved less sensitive to this substance; the sensitivity ratio relative to the D strain was in this case 1.98 for the $LC_{50}$ and approximately 2.75 for the $LC_{95}$.

In view of the difference in the basic diet of strains D and C over a period of years, the difference in sensitivity is amazingly low.

For screening, however, it is important that the results are independent of season. We used the compound CGA 189699 to test the seasonal independence of results from the Potter test method for insects reared on artificial diet. The "breeding strain" used for these tests was slightly more sensitive than the one (D strain) used in the trials described in the present paper.

The results of trials repeated several times over a period of 1 year showed surprisingly little variation (Table 5).

The toxicological experiments from February 1986 to January 1987 demonstrate a mean mortality of 91% at a dosage of 0.4 ppm AI. The standard deviation of 9.5 is very low. Inferior dosages such as 0.2 ppm AI, tending towards the "dosis tolerata," give a higher standard deviation of 31.6.

## Discussion and Conclusion

After establishing the superior quality of insects reared on artificial diet in breeding trials on different substrates, we examined the suitability of such insects for toxicological trials as required for the screening of our products. In this context, quality means uniformly reproducible results that are independent of season.

The toxicological trial results are statistically very consistent (Table 4) and give very narrow confidence limits.

In the present trials, adaptation of the test insects to plant material (DC strain) prior to the trial was found to have no effect on toxicological response. Since this result was obtained for two compounds that are chemically very different, it is very possible that this finding is generally valid.

The most important conclusion to be drawn from our trials is that noctuid species such as *S. littoralis*, reared on artificial diet, consistently produce representative and seasonally constant results for the selection of products (Table 5). The economic and methodological advantages of rearing insects on artificial nutrients are in no way gained at the expense of the relevance of screening results.

**Table 5.** Seasonal toxicological response of 1st instar larvae of *S. littoralis* reared on artificial diet to the benzoylphenyl urea CGA 189699 (Potter Tower test method)

| Date of experiment (mo/day/yr) | *Spodoptera littoralis* first instar mortality % | | |
|---|---|---|---|
| | 0.8 ppm AI | 0.4 ppm AI | 0.2 ppm AI |
| 2/14/86 | 100 | 87 | - |
| 2/27/86 | 100 | 100 | - |
| 3/5/86 | 100 | 100 | - |
| 5/23/86 | 100 | 73 | - |
| 6/27/86 | 100 | 87 | - |
| 7/4/86 | 100 | 93 | - |
| 8/23/86 | - | 100 | - |
| 9/5/86 | - | 97 | 93 |
| 9/12/86 | - | 100 | 13 |
| 9/17/86 | - | 80 | 13 |
| 9/24/86 | - | 100 | 83 |
| 10/16/86 | - | 90 | 53 |
| 10/29/86 | - | 100 | 87 |
| 11/14/86 | - | 80 | - |
| 12/12/86 | - | 80 | 30 |
| 1/16/87 | - | 80 | 30 |
| 1/30/87 | - | 100 | 63 |
| Number of trials | | 17 | 9 |
| Mean | | 91 | 51.7 |
| Standard deviation | | 9.5 | 31.6 |

## References

Adkisson, P. L., E. S. Vanderzant, D. L. Bull & W. E. Allison. 1960. A wheat germ medium for rearing the pink bollworm. J. Econ. Entomol. 53: 759-762.

Cabello-Garcia, T., H. Rodriguez-Menendez & P. Vargas-Piqueras. 1984. Development longvity and fecundity of *Spodoptera littoralis* (Boisd.) (Lep. Noctuidae) reared on eight artificial diets. Appl. Entomology 97(5): 425-536.

El Guindy, M. A., M. M. El Sayed & Y. H. Issa. 1979. Biological and toxicological studies on the cotton leaf worm *Spodoptera littoralis* (Boids) reared on natural and artificial diets, Zeitschr.f. Pflanzenkrankenheiten und Pflanzenschutz 86: 180-189.

El Sayed, E. I. 1975. Effect of different host plants on larval development and on response of the black cutworm *Agrotis ipsilon* (Hufnagel) to Sevin, Z. Angew, Entomol. 79: 365-369.

House, H. L. 1967. Artificial diets for insects: a compilation of references with abstracts, Inform. Bull, Can. Agric. Res. Inst. Belleville. 5: 156.

Isa, A. L. & G. D. Khadr. 1974. Rearing of *Spodoptera littoralis* larvae on a simple artificial diet, Agric. Res. Review 52(1): 9-14.

King, E. G. & N. C. Leppla, editors. 1984. Advances and challenges in insect rearing. Agric. Res. Service, USDA.

MacMasters, M. M., J. J. C. Hinton & D. Bradbury. 1971. Microscopic structure and composition of the wheat kernel, Chap. 3: 51-113. *In* Y. Pomeranz [ed.], Wheat chemistry and technology. Am. Assoc. Cereal Chemists.

Navon, A. 1976. The role of dietary studies in developing biological control methods with special reference to *Spodoptera littoralie* (Boisd) and *Zeuzera pirina L.*, Med. Fac. Landbouw, Rijksuniv. Gent. 41: 2.

Poitou, S. Bues, R. & C. Lerumeur. 1972. Elevage sur miliue artificial simple de deux noctuelles parasites du coton, Earias insulana et *Spodoptera littoralis*, Entomol. Exp. et Appl. 15: 341-350.

Potter, C. 1941. Original Potter Spray Tower, Ann. Appl. Biol. 28: 142.

Rahbe, Y., M. J. Brochon & P. Nardon. 1986. Use of a new gel in a conditioned artificial diet for rearing of the cotton leaf worm *Spodoptera littoralis*, Insect Sci. Applic. 7(6): 803-808.

Singh, P. & R. F. Moore. 1985. Handbook of insect rearing, vol. 2, Elsevier, Amsterdam.

Vanderzandt, E. S., C. D. Richardson & S. W. Fort. 1962. Rearing of the boll worm on artificial diet, J. of Econ. Entomol. 55(1): 140.

Vanderzandt, E. S. 1974. Development, significance and application of artificial diets for insects, Ann. Rev. Entomol. 19: 139-160.

Wyniger, R. 1974. Insektenzucht, Edit. Eugen Ulmer, Stuttgart.

Zoebelein, G. 1977. Practical experiences gained during twelve years of crop protection trials work in the Middle East and North Africa, Fourth report. The influence of different host plants on the development of cotton leafworm *Spodoptera littoralis* (Boisd.), and on its response to insecticides, Pflanzenschutz-Nachrichten Bayer 1977, recd. 1979, 30(2): 164-212.

# 17

# Evaluating the Role of Genetic Change in Insect Colonies Maintained for Pest Management

*Robert L. Mangan*

A large volume of literature has shown that the selection and breeding processes in mass rearing and laboratory environments have significant, measurable effects on insect populations. For insects including Lepidoptera (Roush 1986), mosquitos (Haeger & O'Meara 1970), several fruit fly species (Tephritidae) and the screwworm (*Cochliomyia hominivorax*, Calliphoridae) (Baumhover 1965), important behavioral and physiological traits have been shown to change over time in artificial rearing conditions. These include fecundity and preoviposition period (reviewed in Leppla et al. 1976, de Souza et al. 1988, Hightower et al. 1972), courtship songs (Kanmiya et al. 1987 a,b), flight (Sharp et al. 1983), oviposition (Greany & Szentzi 1979, Fletcher et al. 1973), rate of development, production of pheromones (Mazomenos & Pantazy-Mazomeny 1989, Hammack 1986, Sower et al. 1973), response to pheromones (Hammack 1987), eye morphology (Davis et al. 1983), visual sensitivity (Goodenough et al. 1977), metabolic rate (Leppla et al. 1976), and resistance to stress (Vargas & Hofmann 1983). For programs requiring appropriate behavior for population control by sterile insect release, the importance of genetic change in mass rearing programs has been reviewed by Boller (1972), Mackauer (1976), and Huettel (1976).

Evidence of change in colonized insect populations, including the above studies, has suggested or demonstrated some degree of genetic control over these traits. Usually "wild" (newly colonized) strains were reared under the same conditions as "colony" strains. Differences in the traits between populations were attributed to genetic differences. Other factors, such as maternal transfers of trace nutrients (Kircher 1978, Whitten 1985) may confound these interpretations. But in general, follow-up studies using crossing schemes or selection tests have verified these interpretations for several of the traits (de Souza et al. 1988, Mangan 1988, Sharp et al. 1983, Fletcher et al. 1968).

## Significance

The importance of laboratory or mass rearing adaptation in sterile insect release programs for population eradication is not well documented, even for species known to undergo genetic changes under mass rearing conditions. For the Tephritidae and the screwworm, traits typically shown to be "laboratory adapted" such as development rate, oviposition, and reduced susceptibility to temperature stress greatly reduce the cost of producing the insects. The relationship of traits that improve efficiency of mass production to those involved in the essential activity of the insect (such as successful mating with target populations) should determine the degree of laboratory adaptation acceptable for the produced insects. This relationship is evaluated for the screwworm by comparisons of rearing traits between "old" and "new" strains, and field testing the new strain for satisfactory mating with the wild insects under natural conditions. The two most important factors are 1) that the new strain can be reared at an acceptable efficiency and 2) that the new strain can achieve an acceptable level of mating in a field test.

There is conflicting evidence concerning the importance of laboratory adaptation for field performance of reared insects. Williamson et al. (1983) tested field performance of tsetse flies (*Glossina moristans* Westwood) for a 15 month period and found no diminution of male effectiveness. The colony management system selected for earliest emerging (mostly female) flies for the rearing colony while later emerging flies (mostly male) were sterilized for release. This process imposed strong selection for rapid development. In addition, a backup colony of flies, which was maintained by in vitro methods (and presumably experienced different selective pressures), was equally successful in comparison with the less colony-adapted strain in this field test.

A genetically manipulated (females susceptible to propoxur, males resistant) strain of *Anopheles albimanus*, derived from lines collected in 1975 (Kaiser et al. 1978), replaced a locally collected El Salvador strain in a field test carried out in 1978 to 1979. Improved handling and distribution were possible (because sexes did not have to be separated) and superior results were achieved with the manipulated insects (Dame et al. 1980), though considerable hybridization and selection occurred during the formulation of this strain. The 1978 to 1979 tests showed that an intensely genetically manipulated strain, that had undergone laboratory adaptation for 2 to 3 years longer than a wild strain, performed better in the field than the wild strain. The genetic sexing process apparently induced less stress on the mosquitoes than the previously used mechanical system.

The results reported by Dame et al. (1980), contrast with those of Reisen et al. (1980), for a similar test using a genetically manipulated *Culex tritaeniorhynchus* strain. Despite attempts to "induce hybrid vigor and a wild-type genetic background" (Reisen et al. 1980), released sterile males did not successfully compete for wild females, though they were successful in mating with their siblings.

Direct field observation of copulation to assess mating competitiveness of colonized males with native females in the field is generally not possible for screwworms, mosquitoes, tsetse flies, or Tephritidae. Observations were possible in tobacco budworm experiments conducted with backcrosses of hybrid (*Heliothis subflexa* females X *H. virescens* males) males to *H. virescens* females, which produce sterile males. Collection of sufficient data from observed mating pairs was made to identify components of mating behavior that affect mating success of released males. Raulston et al. (1976), used a strain derived from colony-adapted *H. virescens* to show that, while wild males and backcross males were successful in mating with backcross females, backcross males were not as successful as wild males in mating with wild females. In a later experiment, Proshold et al. (1983) used backcross males having the last 3 generations of *H. virescens* parents from a newly colonized strain. Released males in this test were fully competitive with the wild males with both backcross females and wild females. Both studies measured mating competitiveness by collecting mating pairs through the night. Backcross males and females derived from colony-adapted *H. virescens* mated earlier than the wild moths, causing the backcross males to miss the wild females.

Another well documented study attributing poor field performance to colony adaptation was carried out during eradication of the melon fly, *Dacus cucurbitae*, from the Japanese island of Kume. A nearby island (Kudaka) served as a control (Iwahashi et al. 1983). Sterile males and females were marked, released, and then recaptured along with wild flies. Competitiveness of sterile males was calculated over a 10 month period of eradication, corresponding to generations 5 to 18 in the rearing colony. Competitiveness decreased from generations 10 to 18 in the field. Competitiveness indices (calculated from laboratory data) indicated that, at the same time the released males were decreasing in competitiveness in the field, these males were highly competitive (even superior to wild males) in small cages. A regression model relating release rate and percent sterility showed a significant decline in competitiveness when an outlier date near the end of the test was omitted. This outlying competitive index was more than 3 times greater than any other competitive index calculated. The authors did not explain this outlier or justify its deletion.

Later investigations (Hibino & Iwashi 1989) found that colony-adapted (about 20 generations) males attempted to directly mount females more frequently than did wild males. The rate of success for wild and colony-adapted males was about the same for 10- to 12-day-old females. For the younger females (9-day-old), however, the direct mounting approach was less successful. In the wild, the more elaborate pre-mounting courtship (mainly by wild males) may have a significant advantage, since most of the females will be mated at the younger age.

Both the *Heliothis* experiments and the melon fly data suggest deleterious effects of colony adaptation on male performance under field conditions. However, both tests are weakened because the ecological communities differed for tests of the wild-type and colony-adapted flies. The backcross *Heliothis* males were tested near Brownsville, Texas, while the wild-crossed strain was tested in St. Croix, Virgin Islands. Both newly colonized and colony-adapted melon flies were tested on the same island, but newly colonized melon flies were tested when the native population was at much higher density. Moreover, the native population presumably had a different distribution than occurred later in the eradication period when the colony-adapted strain had allegedly deteriorated. The strain apparently did change during the test, but the ability of released males to find wild females may have been reduced by a lower density and different distribution of the wild population since, at this time, the native population was nearly eradicated.

Reviews and discussions of the genetics of insects held in mass rearing colonies have been unanimous in supporting periodic changes or rejuvenation of genetic material. The degree of difficulty in strain change varies greatly. Strain development for screwworms, for example, is simple. Diet and handling improvements (Taylor & Mangan 1987), and improved knowledge for collecting material (Thomas & Mangan 1989) has made the construction of strains with minimal apparent selection or inbreeding into a routine procedure. Other insects, however, require complex genetic manipulation for control of sex ratio (Kaiser et al. 1978, Reisen et al. 1988). Medflies must undergo considerable adaptation in order to be mass produced. For these species, the process of preparing new strains is more complex and laborious.

The change of strains in a mass rearing plant also involves considerable effort to maintain production and avoid contamination. For the screwworm (the only mass produced insect that, to my knowledge, undergoes periodic, scheduled strain change), a separate colony must be maintained in isolation from the main colony. Pupae are passed from the "new" colony to the adult holding section of the main production area. Emerging "new" flies provide eggs for the main production. Pupae produced in the main colony are sterilized for release. Fertile pupae are passed continuously from the "new" colony to the main colony for several generations until the main colony has been completely replaced. If production of pupae from the more stress-sensitive new strain is sufficient and emerging adults oviposit, total production may not suffer. Reports for the Mexican American commission for Screwworm Eradication show, however, that a month-long decrease in production of 30% has been experienced during strain change.

Other costs of strain change include the risk that diseases or parasites may be introduced with the new strain or that the strain may have undesirable recessive mutations. The cost of collecting and constructing the new strain is not a major factor (the last three strains that I constructed from the Yucatan peninsula, Belize, and Guatemala each cost about $5,000 in the field and $5,000 in technicians salaries in the laboratory). However, the strain must be expanded, reared, and then tested to determine quality control and rearing characteristics, and evaluated for field performance (see Hofmann [1985] for a description of these processes and Mangan [1985] for summary of past field tests). These tests and making the strain change cost millions of dollars in labor, facilities maintenance, transportation and air dispersal operations, and technician per diem.

## Applications to the Screwworm Eradication Program

Alleged strain deterioration in mass-reared screwworms is one of the most widely accepted examples in laboratory colonization (Bush et al. 1976, Bush 1978, Bush 1979). However, the strain used to eradicate screwworms from the southeastern U.S. and Texas was used in the Florida and Texas production colonies from 1955 until the strain was changed in 1966. The eradication of screwworms in Texas (1962 to 1964) was carried out using a sub-colony of the strain used in the southeastern U.S. (Bushland 1975). Several reports (Curtis 1985, Whitten 1985, and Krafsur 1985) have challenged the allegation that strain deterioration was related to periodic program failures.

In addition to actual costs, strain change also includes the perception of considerable risk that the new strain will be inferior (Marroquin 1985). At present, no unequivocal method exists to compare performance factors for two (or more) mass-reared strains in the field. In addition to the problem of replication field sites, it is impossible to assure that environmental conditions under which two strains are being reared are identical. Rather drastic differences in rearing techniques, temperatures, larval handling, pupation, and adult maintenance existed (between strains in the replacement colony and the main production colony) during all field tests assessing the strain changes (as discussed by Marroquin 1985 and Hofmann 1985).

## Genetic Control of Field Performance

While little is known about variation in mating performance among strains, even less is known about the relationship of genetic control of traits that improve sterile fly production and genes that control successful performance in the field. For the screwworm, the alleged effects of adaptation to mass rearing are avoided by regular strain changes. Introduction of wild genotypes into colonies (rejuvenation) might also decrease detrimental effects if suitable methods were devised. Understanding the mode of genetic control, especially the degree of additive genetic variance, for the traits under consideration is essential if program managers wish to plan the frequencies of such changes or rejuvenation.

Understanding the mode and degree of genetic control of traits that are important for performance of reared insects has immediate use in determining what to evaluate in quality control programs. This knowledge will aid in decisions concerning the construction of strains and the rearing environmental factors that should receive attention for avoiding undesirable genetic changes. The following section discusses a series of experimental crosses and tests designed to elucidate the mode and degree of genetic control of mating behavior in screwworms. No attempt was made to determine "heritability" in either the broad sense (degree of genetic control) or the narrow sense (percent of phenotypic variance attributable to additive genetic variance). Heritability is discussed in such texts as Falconer (1981) and a range of examples of applied investigations of genetic variability are presented in Weir et al. (1989).

### Trials Studied

The mating behavior of screwworm males consists of perching on vegetation, especially flowering plants, and "striking" females that enter a defended zone (Krafsur 1978, Guillot et al. 1978). Courtship behavior in "wild" males after a strike is normally elicited by a contact pheromone found in the cuticle of young females. In cages or confined areas, males will touch the female, but if a lab-adapted female is more than 6 to 8 days old, wild males will not court them (Hammack 1987). That study also showed that a major difference between females of wild and laboratory-adapted strains is in the rate of loss of the attractiveness of these lipids, laboratory adapted females lose this attractiveness within 48 h of emergence, for wild females, the attractiveness peaks at 3 to 6 days. Hammack (1987) also suggested that selection pressure from male aggressiveness in caged populations leads to the loss or modification of these lipids because harassed females are killed or do not oviposit.

Mangan (1988) studied the genetic control of both mating aggressiveness and cuticular pheromone selectivity in male screwworms by a series of interstrain crosses and back-crosses. Males of wild strains and laboratory-adapted strains were compared in crosses to females of the two strain types.

Hybrid males were then tested with females of the two parental strains. In the series of crosses, differences were noted between wild-type and lab-adapted males interacting with the two types of females. Activity of hybrid males (wild-type x lab-adapted) was also compared when paired with females from wild-type and lab-adapted lines. Hybrid males, when paired with wild-type females, were as aggressive as lab-adapted males. When paired with lab-adapted females, however, hybrid males showed much less mating activity. They had the same behavior as the wild-type males. Those data suggested that aggressiveness and the selectivity requirement for female pheromones are controlled by separate genetic systems and that the genetic systems for aggressiveness and selectivity were both dominant in the hybrid males.

These crossing schemes demonstrated the underlying genetic control of two aspects of differences between colony-adapted and wild screwworm populations. Alone, this information has little predictive value for determining when a mass-reared strain may be expected to have diverged from the wild-type behavior so that the strain is nonfunctional in the sterile release program. The experiments do predict that infusion of wild males into screwworm colonies will have little effect unless they are hybridized with colony females under controlled conditions. Both wild males and their hybrid male offspring are not attracted to colony females. Genetic complexes for rejuvenation would only become established in the colony through the hybrid females (wild female x colony male). Other characteristics of wild females, such as slow maturation and reduced oviposition rate (Hightower et al. 1972), would further reduce the probability of survival of wild genotypes in a mixed colony. Another method would be to develop new strains by combining the traits allowing easy rearing (prompt oviposition and uniform development rates) with those controlling wild-type mating behavior. This requires a knowledge of the genetic control of both rearing traits and mating behavior. A set of genetic markers must be available to allow the correct selection of chromosomes or their segments for construction of the composite strain. Progress has been made in developing marker strains (Taylor, in press) but, thus far, markers have been recessive mutations, which are difficult to trace in hybridization programs, or electrophoretic variants, which require complex technical training and equipment for their detection.

## Genetic Architecture of Mating Behavior

Understanding the underlying mode of genetic control for behavioral traits in sexually reproducing organisms involves determining the degree of additive genetic variance for the trait in the population of interest. This allows a prediction concerning the expected response to selection for such applications as breeding programs or provides explanations for evolutionary and sociobiology studies. The major reference for interpretation of genetic control and expression of quantitative traits is Falconer (1981). Weir et al. (1988) have presented a collection summarizing a wide variety of recent advances in applied and theoretical quantitative genetics. Discussions of genetic control of animal behavior have been presented in Ehrman and Parsons (1976). An extensive review of insect mating behavior have been presented in Ehrman and Parsons (1976). An extensive review of insect mating behavior, including some treatment of the genetic control and evolution of a variety of reproductive patters, is found in Thornhill and Alcock (1985).

Typically, control of fitness and behavior traits of interest for selection programs are classified by their mode of control as either polygenic, under the control of more than one allele at more than one locus, or under the control of a major gene with alleles at one or a few loci controlling a significant amount of the variation. For studies of laboratory adaptation and mass production of insects, it is desirable to understand the mode of control of the essential behaviors in the field and their relationship to traits important for efficient rearing.

## Diallel Crosses

The expression of traits from different strains in a series of hybrids can be analyzed by the method of diallel crosses. Literature on the use of this method deals mainly with plant breeding, but Parsons (1964), Fulker (1966, 1972) used diallel analysis to analyze the mating behavior of *Drosophila* and control of several behaviors in rats. A thorough discussion of the Parsons and Fulker studies, and comparisons of diallel results with other approaches to genetic analysis of mating behavior in *D. melanogaster*, can be found in Spiess (1970).

I used the method of diallel crosses to assess the relative importance of additive and dominant genetic variance in comparison with environmentally associated variance for screwworm male mating capacity. The complete results of this analysis is fully described in Mangan (1991). Here, I present results that are significant in understanding the genetic control of mating behavior relative to methods for developing strains and evaluating possible problems in performance of mass reared insects. Table 1 presents data comparing performance of males from inbred lines (males from parents of the same line) to that of male hybrids with parents from each of the strains tested. In 10 of the 11 comparisons, the hybrids had greater mating percentages than did the inbred parental lines.

In the overall analysis of variance from the diallel crosses (Table 2), dominance effects were the greatest followed by effects of individual crosses. Differences were not significant among inbred lines for the three replicates in the first experiment or the four replicates in the second experiment. The strong

**Table 1. Comparison of hybrid and inbred males' mating activity**

| Experiment 1 | | | | | Experiment 2 | | | |
|---|---|---|---|---|---|---|---|---|
| Line — | Hybrid Paternal | Hybrid Maternal | Inbred Line | | Line — | Hybrid Paternal | Hybrid Maternal | Inbred Line |
| G3101 | 77.7 | 74.2 | 69.8 | | G3101 | 15.0 | 18.3 | 8.3 |
| 009 | 85.6 | 79.8 | 75.7 | | 009 | 27.5 | 14.2 | 16.7 |
| Sinaloa | 82.2 | 86.0 | 72.8 | | Sinaloa | 13.3 | 24.2 | 6.7 |
| B2709 | 82.0 | 79.6 | 73.0 | | CIH34 | 20.0 | 17.5 | 6.7 |
| V81 | 75.8 | 83.8 | 66.6 | | CH85 | 19.2 | 25.8 | 12.5 |
| | | | | | OW87 | 18.3 | 20.8 | 15.0 |

Note: Percentage of females mated (experiment 1) and number of females mated (experiment 2) are listed for males by parental lines. Hybrid-paternal group consists of hybrid males with a common paternal parent. The hybrid maternal group consists of hybrid males with a common maternal parent. The inbred line consists of males with both parents from given line.

**Table 2.** Anova tables for diallel cross experiments 1 and 2

**Experiment 1**

| Source | df | SS | F | P |
|---|---|---|---|---|
| Genotypes | 24 | 6582.16 | 1.72 | 0.055 |
| Crosses | 19 | 4758.69 | 1.57 | 0.104 |
| Selfs | 4 | 302.79 | 0.46 | 0.754 |
| Cross vs Self | 1 | 1520.68 | 9.54 | 0.003 |
| Replicates | 2 | 2546.44 | 7.99 | 0.001 |
| Error | 48 | 7650.25 | - | - |
| Total | 74 | 16778.84 | - | - |
| Reciprocals | 10 | 2521.31 | 1.58 | 0.140 |

**Experiment 2**

| Source | df | SS | F | P |
|---|---|---|---|---|
| Genotypes | 35 | 142.69 | 1.94 | 0.005 |
| Crosses | 29 | 117.97 | 1.93 | 0.008 |
| Selfs | 5 | 5.71 | 0.54 | 0.744 |
| Cross vs Self | 1 | 19.01 | 9.02 | 0.003 |
| Replicates | 3 | 55.52 | 8.78 | <0.001 |
| Error | 105 | 221.23 | - | - |
| Total | 143 | 419.44 | - | - |
| Reciprocals | 15 | 73.75 | 2.33 | 0.006 |

over-dominance effect (hybrid males having greater mating success than inbred males) suggests that investigations of intraspecific variation in mating behavior should be performed with flies collected from the wild or with outbred strains rather than isofemale lines.

The Analysis of Variance (ANOVA) tables for other genetic and environmental effects from the diallel crosses are given in Table 2. Complete descriptions of the methods are presented in Mangan (1991). Experiment 2 was performed with more strains (6) than experiment 1 and with better control of environmental factors (rearing and adult handling). The two experiments were nearly identical in the ranking of the importance of the 3 genetic factors, with crosses versus self (heterotic effects) highly significant, selfs (effects of inbred lines) not significant, and crosses (additive genetic effects) significant in experiment 2, but not experiment 1. The marginal importance of additive genetic effects was further shown in an artificial selection for mating aggressiveness test using wild strains from Guatemala and Belize. Divergence of selected and nonselected lines occurred, but no increase in activity occurred for lines selected for high mating activity (Mangan 1991).

Comparisons of mating attempts and matings for outbred and inbred strains recently collected (F2-F4) from Guatemala were reported in Mangan (1988). In those tests, reciprocal crosses with 4 different old lines were made. No differences were detected between inbred and outbred males, as neither type mated and only about 2% of the males attempted to mate with females from the old lines. This suggests that selectivity (the requirement for appropriate female pheromones) can preclude aggressiveness in newly colonized males.


## Applications of Genetics to Sterile Insect Technique

The importance of appropriate mating behavior has long been appreciated for sterile insect release programs and some progress has been made in elucidating the mode of genetic control of this behavior. However, actual application of this knowledge to construction and testing of various strains has not been carried out. For screwworms, the discovery that mating aggressiveness is very weak, led us to revise our strain construction methods to emphasize inter-line crosses and to avoid bottlenecks in the development process. We made less effect to avoid crowding or other artificial conditions if they were necessary to maintain a large population. While the strain resulting from these techniques (OW87) has apparently been successful in completing the eradication of the screwworm from Mexico, no tests or comparisons have demonstrated that, under field conditions, this strain is superior to any other.

For programs requiring genetic manipulation to eliminate females, several tests were carried out to determine whether the constructed strains were adequate or whether colony-adapted material was superior to wild material for parental lines (as in the *Heliothis* tests). I know of no field tests designed to determine the best way to construct strains.

There are many practical problems involved in designing tests of mating behavior. In the screwworm program, the method for determining strain suitability has been to challenge native populations with sterile insects from a candidate strain at the density used by the eradication program. As reviewed in Mangan (1985), past tests involved the whole technology of the program on a reduced scale. Such factors as rearing conditions, transport, release pattern, and native population density all interacted to determine whether the candidate strain as well as eradication methodology was suitable for program success. Attempts to run tests under more controlled conditions (e.g., Brenner 1984) have been made, but even those tests could not control or measure native fly density, migration, and rearing quality with sufficient precision to compare strain behaviors.

## Sterile Insect Technique in the Insect Community

Most reviews of laboratory or mass rearing adaptation and its importance for field performance of sterile insects have assumed that mating behavior of the wild population is the ideal behavior for the released insects. Unfortunately, the process of adapting the insects for mass rearing incurs a change in the genetic control of courtship behavior of the production strain. This lab-adapted behavior is different from that observed in the field or observed for newly collected strains observed in the laboratory. According to these reviews, mating success rate is much lower for the released, sterile males than for the wild males.

The changes in mating behavior in colony-adapted strains may be direct responses to selection for the most successful males in the mass rearing environment. However, other selective factors such as prompt oviposition, rapid or uniform development time, or resistance to adult or pupal stress (disease, temperature, or humidity) may also be involved in altering mating behavior. Other factors, such as population bottlenecks or reduction in heterozygosity associated with long-term rearing of small populations, have also been suggested as forces that may change mating behavior.

Recent studies of several Diptera species have shown, however, that mating behavior in wild populations frequently involves more than one type of 'mating strategy.' In several species, the males may conditionally utilize a strategy very similar to that described as 'colony adapted.' Notable examples of environmental sensitive alternative strategies occur in dung flies (Borgia 1982), cactus flies (Mangan 1979), and *Drosophila* (Hoffmann & Cacoyianni 1990). In those studies, courting males switched from territorial defense behavior to an alternative strategy, such as searching for and directly mounting females, when environmental conditions, such as high population density or abundant substrates, made those tactics advantageous.

Reviews of the role of appropriate male courtship behavior in sterile insect technique (SIT) programs have not considered the role of mixed strategies. For the screwworm, only fragmentary observations exist of mating behavior in the wild and no observations exist for alternative strategies in the broad array of habitats occupied by this species. The existence of lek and non-lek strategies has been described for several fruit flies including the Caribbean fruit fly, *Anastrepha suspense* (Loew) (Hendrichs 1986); the Mediterranean fruit fly, *Ceratitis capitata* (Wiedemann) (Prokopy & Hendrichs 1979); and the Mexican fruit fly, *Anastrepha ludens* (Loew) (Robacker et al. 1991). Most studies comparing the success rates of lek and non-lek behaviors for male fruit flies have shown that the males defending leks are most successful under simulated (cages in laboratory) or confined to cages in natural habitats. My observations of diptera in which males defend a food or ovipositional resource (cactus flies on necrotic spots, Sepsidae on dung, screwworms on flowers) suggest that leks and resource territory defense both involve defending and satellite males. There appear to be conditions under which nonterritorial strategies are profitable for males. The genetic potential for the loss of territorial behavior seems to be widespread among species undergoing mass-rearing adaptation. This suggests that the genetic system for nonterritorial behavior is widespread and may displace the mixed behavior under crowded conditions when no defendable substrate is provided.

A small cage advantage of colony-adapted melon fly *Dacus cucurbitae* Coquillet, described by Soemori et al. (1980), may serve as an example of the alternate (high density) strategy displacing the behavior shown by wild or recently colonized males at lower densities. The density of released plus native insects may prove to be as important as the overflooding (sterile:native) ratio for determining SIT success. This would also suggest that there is no constant mating advantage for wild males over colony-adapted males, rather their relative competitive abilities would be dependant on the total population density. Rossler (1975b) found, in contrast to the melon fly data, that laboratory-adapted medfly males were inferior to wild males in inseminating both wild and laboratory-adapted females in small, high density cages. This apparent discrepancy may be due to species differences between medflies and melon flies, however the 'field strain' in Rossler (1975b) had been in the laboratory for less than a year while the 'laboratory strain' had been maintained for 13 years. The 'laboratory' strain described by Soemori et al. (1980) had been mass reared for 33-41 generations and was compared to 'wild' flies reared from natural substrates or taken as adults from traps. The melon fly laboratory strain was similar to the medfly field strain in numbers of generations in colony (about 3 years and 1 year, respectively). The melon fly field strain, however, differed from the medfly field strain in both numbers of generations of colony rearing (0 generations versus approximately 10 generations) and the rearing environment (nature versus laboratory) of the adults tested. If most laboratory adaptation of medflies occurs during the first few generations in artificial rearing, Rossler's test may have been a comparison of a new (but still somewhat colony-adapted) strain with an old strain. The melon fly strains compared by Soemori et al. (1980), on the other hand, were wild versus old.

## Recommendations

The effect of laboratory or colony adaptation on the success of SIT programs has been interpreted as detrimental (for SIT), but for many pest species, some degree of adaptation is required for efficient production of large numbers of insects. This is true for insects such as mosquitoes, which require genetic manipulation of their sex ratio, but it also has been an inevitable consequence of screwworm and fruit fly production. At present, there is widespread agreement among behavioral geneticists and ecologists that a proper behavioral repertoire is essential for the success of SIT programs. It is suggested here that the detrimental effect of genetic changes must be evaluated for each species in the habitat to be treated. The approximately year-long program of field testing carried out by the screwworm eradication program serves as a model for other programs in which performance of mass-reared strains are a concern.

The stereotyped pattern of behavioral changes, which have been demonstrated for many mass-reared or laboratory-colonized species suggest an underlying, alternative behavioral strategy that may be advantageous under certain conditions. The possibility of inducing conditions (by 10- to 100-fold increases in population density) under which nonterritorial behavior or forced copulation courtship is advantageous exists for species that have been shown to have density-dependent alternative mating strategies. Other possible interactions such as female harassment by nonterritorial or aggressive males have been proposed to explain reduction in oviposition rate for wild females immediately following sterile male release (Mangan 1985) and may be effective under sufficiently high densities of released insects.

Future research is needed to determine whether alternative mating strategies exist for targeted pests and whether the laboratory-adapted behavior is one of the alternatives. Field tests might then be carried out to determine the economic as well as biological feasibility of releasing sufficient sterile insects to create densities of released and native populations that are high enough for the alternative behavior to be advantageous.

## Conclusions

Rearing insects in laboratory or mass production environments subjects the population to selective regimes that alter their physiology and behavior. This adaptation has been shown for a wide array of taxa reared for sterile insect release programs. The cost of replacing strains (in money and in risk) must be balanced with the degree of loss of effectiveness as a strain adapts. Even in species for which laboratory-adaptation is well documented, such as the screwworm, loss of effectiveness in the field has not been demonstrated in replicated field comparisons of strains.

Investigations of the genetic control of traits, such as mating aggressiveness and mate selectivity in screwworm males, have shown that attempts to infuse wild-type traits into mass-reared populations are likely to succeed only if asymmetry in mate selection and dominance of mating traits in hybrids are considered. When the mass reared populations are genetically manipulated to alter sex ratio (as in mosquitos) or to produce sterility (as in *Heliothis*), problems of strain replacement or genetic infusion become more complex.

In determining the optimum behavior for mass reared, released insects, I propose that the behaviors identified as 'laboratory-adapted' are actually alternate behaviors that are advantageous under certain conditions, such as high population density. Field tests need to be designed to determine the relative importance of absolute (sterile plus native) population density and sterile:native ratio to determine the success of the alternate strategies. This may allow an interaction of release methods and knowledge about the predominant strategy of the mass reared strain to determine optimal community conditions for mating effectiveness.

## References

Baumhover, A. H. 1965. Sexual aggressiveness of male screwworm flies measured by effect on female mortality. J. Econ. Entomol. 58: 544-545.

Boller, E. 1972. Behavioral aspects of mass-rearing of insects. Entomophaga 17: 9-25.

Borgia, G. 1982. Experimental changes in resource structure and male density: Size-related differences in mating success among male *Scatophaga stercoraria*. Evolution 36: 307-3115.

Brenner, R. J. 1984. Dispersal, mating, and oviposition of the screwworm (Dipera: Calliphoridae) in southern Mexico. Ann. Entomol. Soc. Am. 77: 779-788.

Bush, G. L. 1978. Planning a rational quality control program for the screwworm fly, pp.37-34. *In* R. H. Richardson [ed.], The screwworm problem: Evolution of resistance to biological control. University of Texas Press, Austin.

Bush, G. L. 1979. Ecological genetics and quality control, pp.145-152. *In* M. Hoy & J. J. McKelvey [eds.], Working papers: Genetics in relation to insect Management. The Rockefeller Foundation.

Bush, G. L., R. Neck & G. B. Kitto. 1976. Screwworm eradication: Inadvertent selection for non-competitive ecotypes during mass rearing. Science 193: 491-493.

Bushland, R. C. 1979. Screwworm research and eradication. Bull. Entomol. Soc. Am. 21: 23-26.

Curtis, C. F. 1985. Genetic control of pests: Growth industry or lead balloon? Biol. J. Linnean Soc. 26: 359-374.

Dame, D. A., R. A. Loe & D. L. Williamson. 1980. Assessment of release sterile *Anapheles albimanus* and *Glossina morisitans*. Proc. XVI Int'l. cong. Entomol. Symp. on Genetics and Insect Control. Kyoto. 1980.

Davis, J. C., H. R. Agee & E. A. Ellis. 1983. Comparative ultrastructure of the compound eye of the wild, laboratory reared and irradiated Mediterranean fruit fly, *Ceratitis capitata* (Diptera: Tephritidae). Ann. Entomol. Soc. Am. 76: 322-332.

De Souza, H. M. L., S. R. Matioli & W. M. Naciemento. 1988. The adaptation process of *Ceratitis Capitata* to the laboratory analysis of life-history traits. Entomol. Exp. Appl. 49: 195-201.

Ehrman, L. & P. A. Parsons. 1976. The genetics of behavior. Sinauer Assoc. Sunderland, Mass.

Falconer, D. S. 1981. Introduction to quantitative genetics, 2nd ed. Longman, London.

Fletcher, L. W., H. V. Clabron, J. P. Turner & E. Lopez. 1968. Difference in response of two strains of screwworm flies to the male pheromone. J. Econ. Entomol. 61: 1386-1388.

Fletcher, L. W., J. P. Turner & C. N. Husman. 1973. Surface temperature on selection of ovipositional sites by three strains of the screwworm. J. Econ. Entomol. 66: 422-423.

Fulker, D. W. 1966. Mating speed in male *Drosophila Melanogaster*: A psychogenetic analysis. Science 153: 203-205.

Fulker, D. W. 1972. Applications of a simplified triple-test cross. Behav. Genetics 2: 185-198.

Goodenough, J. L., D. D. Wilson & H. R. Agee. 1977. Electroretinographic measurements for comparison of visual sensitivity of wild and mass-reared screwworm flies, *Cochliomyia hominivorax* (Diptera: Calliphoridae). J. Med. Entomol. 14: 309-312.

Greany, P. D. & A. Szentesi. 1979. Oviposition behavior of laboratory-reared and wild caribbean fruit flies *Anastrepha suspense* (Diptera: Tephritidae): II. Selected Physical Influences. Ent. Exp. and Appl. 26: 239-244.

Guillot, F. S., H. E. Brown & A. B. Broce. 1978. Behavior of sexually active screwworm flies. Ann. Entomol. Soc. Am. 71: 199-201.

Haeger, J. S. & G. F. O'Meara. 1970. Rapid incorporation of wild genotypes of *Culex nigripalpus* (Diptera: Culicidae) into laboratory-adapted strains. Ann. Entomol. Soc. Amer. 63: 1390-1391.

Hammack, L. 1986. Pheromone-mediated copulatory responses of the screwworm fly, *Cochliomyia hominivorax*. J. Chem. Ecol. 12: 1623-1631.

Hammack, L. 1987. Chemical basis for asymmetric mating isolation between strains of the screwworm fly *Culex nigripalpus* (Diptera: Culicidae) into laboratory-adapted strains. Ann. Entomol. Soc. Amer. 63: 1390-1391.

Hendrichs, J. 1986. Sexual selection in wild and sterile Caribbean fruit flies, *Anastrepha suspensa* (Loew) (Diptera:Tephritidae). M. S. Thesis, Univ. Florida, Gainesville.

Hibino, Y. & O. Iwahashi. 1988. Mating receptivity of wild type females for wild type males and mass-reared males in the melon fly, *Dacus cucurbitae* Coquillet (Diptera:Tephritidae). Appl. Ent. Zool. 24: 152-154.

Hightower, B. G., J. J. O'Grady, Jr. & J. J. Garcia. 1972. Ovipositional behavior of wild-type and laboratory-adapted strains of screwworm flies. Environ. Entomol. 1: 227-229.

Hoffmann, A. A. & Z. Cacoyianni. 1990. Territoriality in *Drosophila melanogaster* as a conditional strategy. Anim. Behav. 40: 526-537.

Hofmann, H. C. 1985. Methods development activities in Mexico. *In* O. H. Graham [ed.], Symposium on Eradication of the Screwworm From the United States and Mexico. Misc. Publ. Entomol. Soc. Am. 62: 41-48.

Huettel, M. D. 1976. Monitoring the quality of laboratory reared insects: A biological and behavioral perspective. Environ. Entomol. 5: 807-814.

Iwahashi, O. , Y. Ito & M. Shiyomi. 1983. A field evaluation of the sexual competitiveness of sterile melon flies, *Dacus (Zeugodacus) Cucurbitae*. Ecol. Entomol. 8: 43-48.

Kaiser, P. E., J. A. Seawright, D. A. Dame & D. J. Joslyn. 1978. Development of a genetic sexing system for *Anapheles Albimanus*. J. Econ. Entomol. 71: 766-771.

Kanmiya, K., K. Nakagawa, A. Tanaka & H. Kamiwada. 1987a. Comparison of acoustic properties of tethered flight sounds for wild, mass-reared and irradiated melon flies, *Dacus Cucurbitae* Coquillet (Dipera:Tephritidae). Ann. Ent. Zool. 22: 85-97.

Kanmiya, K., A. Tanaka, H. Kamiwada, K. Nakagawa & T. Nishioka. 1987b. Time-domain analysis of the male courtship songs produced by wild, mass-reared, and by irradiated melon flies, *Dacus cucurbitae* Coquillet (Diptera:Tephritidae). Appl. Ent. Zool. 22: 181-194.

Kircher, H. W. & M. A. Gray. 1978. Cholestanol-cholesterol utilization by axenic *Drosophila melanogaster*. J. Insect Physiol. 24: 555-559.

Krafsur, E. S. 1978. Aggregations of male screwworm flies, *Cochliomyia hominivorax* (Coq.), in south Texas (Diptera:Calliphoridae). Proc. Entomol. Soc. Wash. 80: 164-170.

Krafsur, E. S. 1985. Screwworm flies (Diptera:Calliphoridae): Analysis of sterile mating frequencies and covariates. Bull. Entomol. Soc. Amer. 31: 36-44.

Leppla, N. C., M. D. Huettel, D. L. Chambers & W. K. Turner. 1976. Comparative life history and respiratory activity of "wild" and colonized Caribbean fruit flies (Diptera:Tephritidae). Entomophaga 21: 353-357.

Mackauer, M. 1976. Genetic problems in the production of biological control agents. Ann. Rev. Entomol. 21: 369-385.

Mangan, R. L. 1979. Reproductive behavior of the cactus fly, *Odontoloxosus longicornis*, male territoriality and female guarding as adaptive strategies. Behav. Ecol. and Sociobio. 4: 265-278.

Mangan, R. L. 1985. Population ecology and genetics research on Mexican screwworms. *In* O. H. Graham [ed.], Symposium on Eradication of the Screwworm from the United States and Mexico. Misc. Publ. Entomol. Soc. Am. 62: 56-66.

Mangan, R. L. 1988. Pedigree and heritability influences on mate selectivity and mating aggressiveness in the screwworm, *Cochliomyia hominivorax* (Diptera:Calliphoridae) Ann. Entomol. Soc. Am. 81: 649-656.

Mangan, R. L. 1991. Analysis of genetic control of mating behavior in screwworm (Diptera:Calliphoridae) males through deallet crosses and artificial selection. Theor. Appl. Genet. (in press).

Marroquin, R. 1985. Mass production of screwworm in Mexico. *In* O. H. Graham [ed.], Symposium on Eradication of the Screwworm from the United States and Mexico. Misc Publication Entomol. Soc. Am. 62: 31-40.

Mazomenos, B. E. & A. Pantazi-Mazomenoy. 1989. Olive fruit fly *Dacus Oleae* pheromone: Comparison of the quantity produced between a laboratory strain and a wild strain, pp. 151-158. *In* Fruit Flies of Economic Importance 87. Proc. of the CEC/IOBC International Symp. Rome/Italy. Balkema, Rotterdam.

Parsons, P. A. 1964. A diallel cross for mating speeds in *Drosophila melanogaster*. Genetica 35: 141-151.

Prokopy, R. J. & J. Hendrichs. 1979. Mating behavior of *Ceratitis capitata* on a field-caged host tree. Ann. Entomol. Soc. Am. 72: 642-648.

Proshold, F. I., J. R. Raulston, D. F. Martin & M. L. Laster. 1983. Release of backcross insects on St. Croix to suppress the tobacco budworm (Lepidoptera:Noctuidae): Behavior and interaction with native insects. J. Econ. Entomol. 76: 626-631.

Raulston, J. R., H. M. Graham, P. D. Lingren & J. W. Snow. 1976. Mating interaction of native and laboratory-reared tobacco budworms released in the field. Environ. Entomol 5: 195-198.

Reisen, W. K., R. K. Sakai, R. H. Baker, H. R. Rathor, K. Raana, K. Azra & S. Niaz. 1980. Field competitiveness of *Culex tritaeniorhynchus* (Giles) males carrying a complex chromosomal: A second experiment. Ann. Entomol. Soc. Am. 73: 479-484.

Robacker, D. C., R. L. Mangan, D. C. Moreno & A. M. Tarshis Moreno. 1991. Mating behavior and male mating success in wild *Anastrepha ludens* (Diptera:Tephritidae) on a field caged host tree. J. Insect Behav. (in press).

Rossler, Y. 1975. The ability of inseminate: A comparison between laboratory reared and field populations of the Mediterranean fruit fly. Ent. Exp. & Appl. 18: 255-260.

Roush, R. T. 1986. Inbreeding depression and laboratory adaptation in *Heliothis virescens* (Lepidoptera:Noctuidae). Ann. Entomol. Soc. Am. 79: 583-587.

Sharp, J. L., E. F. Boller & D. L. Chambers. 1983. Selection for flight propensity of laboratory and wild strains of *Anastrepha suspensa* and *Ceratitis capitata* (Diptera:Tephritidae) J. Econ. Entomol. 76: 302-305.

Sower, L. L., D. W. Hagstrum & J. S. Long. 1973. Comparison of the female pheromones of a wild and laboratory strain of *Cadra cautella* and male responsiveness to the pheromone extracts. Ann. Entomol. Soc. Am. 66: 584-585.

Soemori, H., S. Tsukaguchi & H. Nakamori. 1980. Comparison of mating ability and mating competitiveness between mass-reared and wild strains of the melon fly *Dacus cucurbitae* Coquillet (Diptera:Tephritidae). (English summary) Jap. Jour. Appl. Entomol. and Zool. 24: 246-250.

Speiss, E. B. 1970. Mating propensity and its genetic basis in *Drosophila*. *In* Essays in evolution and genetics in honor of Theodosius Dobzhansky. Appleton-Century-Crofts, New York.

Taylor, D. B. 1989. Genetics of screwworm: New genetic markers and preliminary linkage. Map. Jour. Heredity (in press).

Taylor, D. B. & R. L. Mangan. 1987. Comparison of gelled and meat diets for rearing screwworm larvae, *Cochliomyia hominivorax* (Diptera:Calliphoridae). J. Econ. Entomol. 80: 427-432.

Thomas, D. B. & R. L. Mangan. 1989. Oviposition and wound-visiting behavior of the screwworm fly, *Cochliomyia hominivorax* (Diptera:Calliphoridae). Ann. Entomol. Soc. Am. 82: 526-534.

Thornhill, R. & J. Alcock. 1983  The evolution of insect mating systems. Harvard Univ., Cambridge, Mass.

Vargas, M. & H. C. Hofmann.  1983.  Evaluacion de la Variedad A-82 de Moscas Esteriles del Gusano Barrenador del Ganado *Cochliomyia hominivorax* (Coquerel).  Rpt. Mex. Amer. Comm. Erad. Screwworm, Mexico City, Mexico.

Weir, B. S., E. J. Eisen, M. M. Goodman & G. Namkoong.  1988.  Proc. of the Second International Conference on Quantitative Genetics.  Sinauer Assoc., Sunderland, Mass.

Whitten, M. J.  1985.  The conceptual basis for genetic control, pp. 465-528. *In* G. A. Kergut & L. I. Gilbert [eds.], Comprehensive insect physiology, biochemistry, and pharmacology, vol. 12.  Pergamon, New York.

Williamson, D. L., D. A. Dame, D. B. Gates, P. E. Cobb, B. Bakuli & P. V. Warner.  1983.  Integration of insect sterility and insecticides for control of *Glossina morsitans* Westwood (Diptera:Glossinidae) in Tanzania.  The impact of sequential releases of sterilized tsetse flies.  Bull. Ent. Res. 73: 391-404.

# 18

## Problems with Entomopathogens in Insect Rearing

*George G. Soares, Jr.*

### Introduction

The need to produce healthy insects in large numbers goes back to the very beginnings of civilization, in the development of the ancient practices of sericulture and apiculture. Insect diseases, although they were not recognized as such in ancient times, were responsible for premature mortality in silkworm cultures and appear to have been recognized by early sericulturists. Steinhaus (1975) notes that a number of maladies of silkworm were described in 18th and 19th century Chinese and Japanese accounts but that these were often attributed to overcrowding, poor ventilation, food quality, or other factors.

It was not until the nineteenth century that the origins of these diseases were studied and some practical methods for their control were developed. Agostino Bassi (1835) was the first to elucidate the cause of muscardine, a disease afflicting silkworm in Europe in the early 19th century. In a careful series of experiments he identified the fungus, *Beauveria bassiana*, as the cause of this disease, showed that "seeds" from infected larvae could infect healthy larvae and that warm humid conditions favored the disease. He also described methods for preventing the disease (Steinhaus 1949, 1975). Beyond its importance to sericulture, Bassi's work was particularly noteworthy as the first postulation and experimental confirmation of the germ theory of disease.

Beginning in 1849, epizootics of pebrine, a protozoan disease caused by the microporidian, *Nosema bombycis*, had a devastating effect on the silkworm industry in France, reducing silk production by over 80% (Hukahara 1987). In a classical study, Louis Pasteur (1870) investigated this disease and after 5 years was able to develop a practical method for its control. He first determined that infected moths contained spores in the hemolymph and that these infected individuals could transmit the disease via trasovarian transmission. His method was based on identifying infected females after oviposition and destroying all of their egg masses. Only the eggs of healthy uninfected individuals were then used for production stock. This basic method is still used today. Diseases of the silkworm and honeybee remain a concern and have dictated many of the practices and regulations governing sericulture and apiculture in different countries (Hukahara 1987).

With the advent of the field of entomology came the need for rearing large numbers of healthy insects for research purposes. There is little doubt that the consistent production of high quality, healthy insects has played a key role in almost every aspect of entomology. As our ability to culture various species has grown, we have been able to learn more and more about many aspects of insect biology including insect behavior, ecology and biological control, physiology, and toxicology.

However, in rearing insects under artificial conditions and in large numbers, we are often placing them under stress due to crowding and suboptimal diets, temperatures and relative humidities. Under these conditions pathogens and contaminants are much more likely to cause disease problems.

As in sericulture and apiculture, controlling insect diseases is critically important in rearing insects for research. Our ability to prevent or at least minimize occurrence of these diseases influences how rapidly progress can be made. There is little doubt that our ability to overcome some of these disease problems has helped the field of entomology to progress rapidly in many areas.

This review covers 1) the various pathogens generally encountered in insect rearing and 2) addresses some of the approaches that one can take in preventing problems and dealing with them should they arise. Entomopathogens will be defined rather broadly to include the true primary pathogens as well as a range of other microorganisms that we might call "facultative pathogens," this latter group including microorganisms that through contamination and growth on insect feeding substrates can cause mortality or morbidity.


## Pathogens and Contaminants Associated with Rearing

One of the chief obstacles to rearing insects has been disease (Helms & Raun 1971). Anyone who has experienced an epizootic of a baculovirus in one of their colonies knows the devastating and costly effects insect disease can have on a rearing program. The insidious and debilitating effects of microsporidia or cytoplasmic polyhedrosis viruses can be even more frustrating since they often go undetected.

The development of insect diets has substantially reduced the incidence of certain diseases in insect rearing. The baculoviruses, for instance, have generally been less of a problem in rearing than in the past, since contaminated host plant material was often a source of virus. The nuclear polyhedrosis virus of cabbage looper, for example, was found to naturally occur on field grown cabbage at the rate of 4.36 x $10^3$ to 3.3 x $10^7$ polyhedra per gram of heads harvested in Maryland (Thomas et al. 1972).

However, with diets, which can serve as rich nutrient sources for microorganisms, there can be greater problems with contaminants that can act as pathogens. So regardless of how one might be rearing a given species, entomopathogens are a problem that must be anticipated.

## *Identification of the Problem*

The primary interest of those involved in insect rearing, of course, is to produce healthy insects of high and consistent quality in the numbers required. This makes the prevention of disease a key objective. For this reason, the insectary worker should be familiar with the potential disease problems that may arise in their colonies and should be able to recognize them when they occur. When a disease problem arises in a rearing program, it is imperative that the causal agent be identified so that the appropriate steps can be taken. Often it may be difficult to determine at first if the problem is related to the diet or other rearing conditions or whether a pathogen is involved. Burges (1981), Cantwell (1974), Goodwin (1984), and Poinar and Thomas (1984) are valuable references in this regard, providing identification keys and photographs of infected insects and pathogens. The general key from Goodwin (1984) is shown in Key 1. In many cases a specialist in insect pathology should be consulted to make a positive identification of the causal agent and provide advice on how to eliminate it.

## *Entomopathogens of Importance in Insect Rearing*

Insects are infected by a diverse group of microorganisms that include viruses, bacteria, rickettsia, fungi, protozoans, and microsporidians. These pathogens can be quite specific in their host range, as with certain microsporidia and baculoviruses. However, some pathogens can infect a broad range of species, as are certain fungi. They can be relatively easy to identify as with most fungi, which sporulate on the surface of the infected insect. Obligate intracellular pathogens, such as the viruses and rickettsia, are often much more difficult to identify. These pathogens cannot be subcultured in vitro, and their signs and symptoms are often more difficult to interpret. Frequently, only certain tissues are infected or effects on the host are subacute. In these cases a positive identification can be made only by examining infected tissues (midgut epithelium, fat body, etc.)

**Key 1. Symptoms and signs of insect disease as seen with the unaided eye or dissecting microscope below 80X**

---

1. Stunted larvae, abnormally whitened . . . . . . . . . . . . . . . . . . . . 2
   Stunted larvae, no discoloration . . . . . . . . . . . . . . . . . . . . . . 3

   Normal-sized insects showing nervous
     activity and trembling followed by lack
     of coordination and paralysis . . . . . . . . . . . . . . Picornaviridae
   Normal-sized insects showing discolorations
     (yellow, pink, red, purple, brown, green, blue) . . . . . . . . . . 5
   Insect fails to pupate or molt normally . . . . . . . . . . . . . . . . . . . 6

2. Light-colored stunted larvae (open dissection):
     With midgut abnormally whitened . . . . Reoviridae or microsporidia
     With fat body abnormal in appearance
       (distended, fragile, fragmented,
       or discolored) . . . . . . . . . . . *Entomopoxvirus*, microsporidia,
                                        Neogregarinida, or Baculoviridae

3. Stunted larvae (no other signs or symptoms) . . . . . . . . . . *Chlamydia*
   Stunted larvae showing both regurgitation and
     anal discharge (sometimes with violent
     muscular contractions) . . . . . . . . . . . . . . *Enterella* (rickettsia)
   Stunted larvae showing white fecal exudate . . . . . . . . microsporidia

4. Sawfly larvae showing white regurgitate (open
     dissection reveals abnormally whitened gut) . . . . . . . Baculoviridae
   Insects showing both regurgitation and copious
     anal discharge (death may occur in hours after
     symptom appearance) . . . . . . . . . . . . . . . . . . . Picornaviridae

5. Pronounced or subtle color changes followed by
     solidification (mummification) at time of death
     (aquatic insects do not solidify) . . . . . . . . . . . . . . . . . . . fungi
   Subtle color changes followed by liquefaction of
     larval contents after death . . . . . . . . . . . . . . . . Baculoviridae
                                              (nuclear polyhedrosis)
   Color changes followed by rotting of cadaver with
     sweet or putrid odors . . . . . . . . . . . . . . . . . . . . . . bacteria

Color changes persisting after death
(decomposition may be delayed) . . . . . . . . . . . . Rickettsiella or
Baculoviridae (granulosis)
Larvae with transparent cuticles may show
progressive translucence of normally white
internal organs, finally becoming
a dirty, mottled brown . . . . . . . . . . . . . . . Coccidia (protozoa)

6. Insect becomes abnormally swollen-
elongate (often lighter colored
than normal) . . . . . . . . . . . . . . . *Entomopoxvirus, Rickettsiella,*
Baculoviridae, or microsporidia
Insect dies while molting or pupating
or emerges as a deformed
adult (chalky appearance) . . . . . . . . . Neogregarinidia (protozoa)
Insect dies while molting or pupating, then
darkens and rots (sweet or putrid odor) . . . . . . . . . . . bacteria

---

**Viruses.** Viruses are defined as submicroscopic, intracellular, obligate pathogens that affect their hosts by causing host cells to replicate virus genome. They consist of protein and nucleic acid (DNA or RNA depending on the type of virus) and require the living host cell to reproduce. They are frequently more resistant to certain environmental factors than some other pathogen groups. They can, for example, survive freezing and desiccation very well.

The insect viruses are classified as DNA or RNA viruses according to which nucleic acid they contain. They are further separated according to those in which the virions are embedded in a protein matrix or inclusion body (occluded viruses) and those in which the virions occur free in the infected cells. The major families of insect viruses are:

- Baculoviridae—ds DNA; occluded; nuclear polyhedrosis viruses (NPV) and granulosis viruses
- Reoviridae—ds RNA; occluded; cytoplasmic polyhedrosis viruses (CPV)
- Picornoviridae—ss RNA; nonoccluded; *Enterovirus*
- Poxviridae—ds DNA; occluded; *Entomopoxvirus*
- Parvoviridae—ss DNA; nonoccluded; *Densovirus*
- Iridoviridae—ds DNA; nonoccluded; *Iridovirus*

The baculoviruses contain the majority of insect viruses of importance to insect rearing and as group are know to infect Lepidoptera, Hymenoptera, Diptera, Neuroptera, Coleoptera, Trichoptera, Crustacea and mites (Payne & Kelly 1981). The NPVs produce inclusion bodies within the cell nucleus that are 0.5 - 15 μm and readily visible in the light microscope. Virions are enclosed by an outer membrane and may be embedded singly in the inclusion (single embedded virus or SEV) or embedded in packets of varying numbers of virions (multiple embedded virus or MEV). Many different lepidoptera are infected by NPVs, which tend to be fairly species specific.

The insect viruses are infectious via ingestion. The infective viral inclusion bodies or nonoccluded virions enter the host through the mouth during feeding. In the case of occluded viruses, virions are released as the inclusions are dissolved within the midgut. Virions attach to and infect the midgut epithelium. After invasion and replication within the gut epithelium, certain viruses, such as the NPVs spread to many different tissues in the host. Often this leads to the complete liquefaction of the internal tissues of the host larva at death. Some viruses, such as the GVs and the CPVs, are restricted to certain tissues, such as the midgut or the midgut and fat body.

The following list describes some of the symptoms and signs of viral infection:

- Behavioral—NPV infected lepidoptera larvae may crawl to the tops of plants and cease feeding
- Affected tissues may change color—for example, to milky white for some granulosis viruses or blue for iridoviruses
- Growth retardation
- Sluggishness, cessation of feeding
- Dysentery
- Disintegration of tissues and liquefaction of infected dead larvae—nuclear polyhedrosis viruses (NPV)
- Paralysis.

For example, larvae of the cabbage looper, *Trichoplusia ni*, infected with the *T. ni* single-embedded nuclear polyhedrosis virus begin showing faint white mottling around the 5th and 6th abdominal segments within 2 to 3 days of ingesting a high dose of virus. By 4 to 5 days this symptom is visible in almost all larvae. The mottling becomes more extensive until the entire larva takes on a creamy, whitish coloration. Larvae at this stage are very lethargic, cease feeding, and appear swollen when compared to normal, healthy larvae. Dark patches then begin to appear on the integument, which becomes shiny. Infected larvae at this stage climb to high places and upon death become dark brown to black in color, often hanging limply from their last perch. At this point, all tissues have been infected and liquified within the thin sac of the integument (Drake & McEwen 1959; Vail & Hall 1969).

Once a virus has been introduced into an insectary it can be transmitted on dust particles, aerosols, air currents, and adult wing scales and setae. Strict sanitation and high standards of cleanliness are critically important in preventing viral infections and in eliminating them when they do occur. Regular use of disinfectants to sterilize plastic ware and work surfaces, the use of ultraviolet radiation and HEPA filtration of the air supply can all help prevent contamination with viruses and eliminate aerial transmission. Specific techniques are discussed later. For identification of viruses see Payne and Kelly (1981).

**Bacteria.** The bacteria are ubiquitous prokaryotic organisms that generally range in size from 0.5 to 5 microns. Bacteria are classified according to differences in cell wall composition as reflected in the gram stain. Cell morphology is also used in classification, with the four main groups being the rods or bacilli, the spherical cocci, the spiral or comma-shaped bacteria, and the filamentous branching cells of the Actinomycetes.

Bacteria are widely distributed in nature and are naturally present on insects and in the insectary environment. Some of these are constituents of the normal gut microflora, some are symbiotes with insects, some are primary pathogens, and many of these can be serious contaminants, causing mortality or morbidity in insect colonies, particularly under crowded and stressed conditions. For this reason precautions must be taken to minimize bacterial contamination.

Bacterial contaminants can negatively affect insects in a multitude of ways. Typically bacteria cause disease when they are able to penetrate the gut epithelium or exoskeleton and enter into the hemocoel. Once in the hemolymph many species are able to multiply until they eventually kill the host. Under stressful conditions insect immune responses may be compromised and the host is less able to combat infection effectively.

Disease can also result from the multiplication of certain species within the insect gut lumen, as in the case of *Clostridium brevifaciens* and *C. malacosomae*, pathogens of tent caterpillars (Bucher 1981).

Bacteria can also cause problems in a number of other ways. Sikorowski (1984a) observed that fast-growing *Leuconostoc* spp. could quickly overgrow diet and smother any boll weevil larvae present. Sikorowski and Thompson (1980) showed that feeding *Heliothis virescens* on diet contaminated with *Staphylococcus epidermis*, *Escherichia coli*, and *Bacillus subtilis* resulted in decreased pupation, lower pupal weights, and lower egg hatch of progeny of exposed larvae. In the case of *B. subtilis*, which severely retarded pupation rate, adult emergence was also significantly lower than it was in the controls.

Nonsporulating bacteria comprise a large and diverse group of microorganisms. Their interactions with insects has been reviewed (Bucher 1963, 1981; Krieg 1987) and an overview of diagnostic methods including isolation and identification provided (Bucher 1981).

*Serratia* and *Pseudomonas* spp. are often associated with laboratory reared insects. In most cases they act as occasional pathogens, causing disease under stressful conditions such as high humidity or temperature (Goodwin 1968, Greany et al. 1977, Habib 1978) or when the host is injured, causing breaks in the integument or gut epithelium. Generally, disease results when the bacteria breach the gut epithelial barrier and enter the hemocoel. *Pseudomonas* spp. can be common contaminants of artificial diets and under the proper conditions can cause disease in insects. *Pseudomonas aeruginosa*, for example, can cause disease outbreaks in laboratory colonies of grasshoppers, with the young nymphs being particularly sensitive, and rapid mortality often reaching epizootic levels (Bucher 1963). *P. aeruginosa* was also described as the most frequent cause of death for larvae of the armyworm, *Pseudeletia unipuncta*. Infection and mortality increased at higher rearing temperatures (McLaughlin, 1962; Goodwin, 1968). *Pseudomonas fluorescens* has been found to infect grasshoppers in western Canada but causes little mortality (Bucher 1963). *P. fluorescens* can also be an occasional contaminant on artificial diet used in rearing *Diabrotica* spp., leading to some larval mortality.

*Serratia marcescens* is commonly associated with insects and seems able to invade the hemocoel in certain hosts and in those cases is generally considered to be a facultative pathogen. This gram-negative, motile, short, rod-shaped organism is known to cause disease in boll weevil (Sikorowski 1984b). *S. marcescens* caused extensive disease outbreaks in boll weevils shipped from Mexico to Mississippi (McLaughlin & Keller 1964). Presumably, the stress of crowding and suboptimal conditions involved in shipment contributed to the disease outbreak in this case. Typically, *S. marcescens* produces red colonies on agar, but many achromogenic strains are also known (Faust 1974). In insect larvae infected with the red strains, the body takes on a red or rosy color at death.

Recently, a new species, *Serratia entomophila*, was found causing honey disease in field populations of the grass grub, *Costelytra zealandica*. Epizootics of this disease have been frequently observed in New Zealand. Infection rates can exceed 80% and lead to the collapse of the population in the field (Trought et al. 1982, Stucki 1984, Jackson et al. 1986). Infected grubs take on a honey-like color.

*Micrococcus* spp. have also been implicated in insect disease. For example, *Micrococcus pseudoflacidoflex* was isolated from gypsy moth, *Lymantria dispar*, and found to cause high mortality when fed to gypsy moth larvae and tent caterpillar larvae (*Malacosoma americanum*) on foilage treated with a $10^9$ cell/ml suspension (Concannon & Wadewitz 1983).

Sporulating bacteria are important pathogens of insects. Many of the *Bacillus* species can contaminate insect colonies from time to time. *Bacillus thuringiensis*, for example, will occasionally cause mortality in diet-reared lepidoptera, particularly in the more susceptible species. In silkworm rearing in Japan, it has occasionally caused disease outbreaks resulting in serious losses. In fact, this species was first described from diseased silkworm larvae in Japan (Ishiwata 1911). *B. thuringiensis* produces a parasporal body composed of selective toxins called delta-endotoxins that are highly active against a number of susceptible insect species (Heimpel & Angus 1963). Many varieties of *B. thuringiensis* have delta-endotoxins that are highly active against lepidopteran larvae, while other varieties are active against certain diptera and coleoptera. The lepidopteran-active varieties can occasionally cause problems in rearing programs for certain moths, particularly if sanitation procedures and other conditions favor propagation of this toxigenic saprophyte in the insectary (for example, see Stewart et al. 1976). *B. thuringiensis* is the active ingredient in many biopesticide products used in the control of various caterpillars in agriculture. As a result, spores may sometimes be introduced into the insectary on treated foliage or soil.

Other *Bacillus* spp. that are known pathogens of insects include *Bacillus popilliae*, causing milky disease of whit grubs Dutky (1963), *Bacillus sphaericus*, infecting mosquito larvae, and *Bacillus larvae*, the casual agent for American foulbrood of honeybees. *B. popilliae*, an obligate pathogen, infects though the gut and propagates and sporulates in the hemolymph, turning it from a clear to milky fluid as the septicemia progresses.

The following list provides some of the symptoms and signs of bacterial infection in insects:

- Feeding drops off or stops
- Sluggishness, generally reduced levels of activity and feeding
- Septecemia; presence of bacterial cells in the hemolymph, often very high numbers
- Hemolymph changes color as septicemia progresses; milky to dark brown
- Infected larvae become darker as other tissues are broken down
- Dysentery
- Flacherie; body begins to "wilt" or becomes flaccid as body tissues are destroyed.

**Rickettsia and Chlamydia.** The rickettsia are a primitive group of small coccoid to rod-shaped or kidney-shaped prokaryotes that are primarily pathogens of vertebrates. Those associated with insects are found in the tribe Wolbachieae and can infect a broad taxonomic range of insects. Some of the *Wolbachia* species may also be symbiotes or commensals with insects, and some species in this group will also infect vertebrates.

*Ricketsiella* and *Enterella* are also in this group. The few *Rickettsiella* spp. that have been studied thus far are known to infect a broad range of insect hosts, including certain flies, beetles, grasshoppers, cockroaches and crickets.

Infection occurs through the midgut epithelium and develops primarily within the fat body. Other tissues may also be infected, depending on the species. Infected individuals become sluggish as the disease progresses. They may also take on a chalky white appearance as the tissues become filled with rickettsiae. In scarab, carabid, tenebrionid, acridid, and gryllid hosts this massive development of rickettsiae within the host fat body also causes the abdomen to swell.

The *Enterella* spp. generally infect and grow within the gut epithelium of the host. Relatively few species have been described. The effects on the host can range from no effect to destruction of the gut epithelium and death of the host due to a severe dysentery.

There are also several chlamydia resembling both *Enterella* and *Rickettsiella* infections that cause stunting in a number of hosts, including *Trichoplusia ni, Heliothis zea, Diatraea saccharalis, Ostrinia nubilalis,* and *Blattella germanica.* (Jean Adams, personal communication).

The following list describes some of the symptoms and signs of rickettsia and chlamydia:

- Sluggishness
- Dysentery
- Swelling of the abdomen due to fat body infections *Ricketsiella*
- Discoloration of infected larvae; often a chalky white.

The rickettsia are particularly difficult to identify, and the associated symptoms of stunting and dysentery resemble those of other diseases. Consequently, all other possible pathogens should be ruled out before a rickettsial agent can be suspected. Conclusive identification can often be made only by careful microscopic examination of infected tissues, usually with a transmission electron microscope.

**Protozoa.** The protozoa are unicellular animals that are associated with insects in many different ways. Some protozoa are symbiotes with insects, as in the classical case of the ciliates associated with termites. Others, such as the coccidia and neogregarines, are primary pathogens. Comprehensive reviews of protozoan pathogens of insects include Steinhaus (1949), Aoki (1957), Lipa (1963, 1967), Weiser (1963, 1966), and Brooks (1988, 1974).

The microsporidia are the most important of the protozoan pathogens of insects and the most likely protozoans to be encountered in insect rearing. In fact, the microsporidia are among the most commonly encountered insect pathogens in insectaries. However, because of the debilitating, sublethal effects on the host, they are also among the most difficult to detect. This makes the microsporidia among the most difficult pathogens to deal with.

The protozoa and microsporidia are generally present at low, enzootic levels in the field, although certain microsporidia often cause epizootics in host populations. *Nosema fumiferanae*, for example, may reach epizootic levels in field populations of the spruce budworm, *Choristoneura fumiferanae*. In an insectary environment, infection can spread rapidly, particularly for those species that transmit the disease transovarially to their progeny, as in the case of some microsporidia such as *N. bombycis*, the cause of pebrine in silkworm. The microsporida can be particularly troublesome in rearing a number of species such as the European corn borer, *Ostrinia nubilalis*; the spruce budworm; and the corn earworm, *Heliothis zea Nosema* spp. have also been serious problems on occasion in laboratory colonies of alfalfa weevil, *Hypera postica*, and the boll weevil, *Anthonomus grandis*. Recently, a microsporidia was described from a laboratory colony of *Diabrotica undecempuntata* (J. C. Pershing, personal communication).

The following list describes some of the symptoms and signs of infection of microsporidia:

- Milky white appearance in transparent insects
- Whitish appearance of gut, malpighian tubules, fat body, or other tissues
- Melanized black spots on the integument
- Sluggishness, lack of coordination, abnormal movement
- Reduced feeding
- White fecal exudate
- Presence of typical elliptical spores in infected tissues
- Pressure on cover slip of slide prep causes extrusion of polar filament from spore.

It is important to know if the particular species being reared is susceptible to a microsporidian disease. This will make it possible to better anticipate and prevent any problems. Should infections appear, thorough sterilization of all cages, surfaces, and equipment and selection of clean stock (as is done with *Nosema bombycis*) or new stock from an outside source can eliminate the problem. Other strategies are discussed in the section on insectary design and traffic patterns.

**Fungi.** The fungi are very widely distributed in nature and are common contaminants of insect diet and the insectary environment. Fungal species cause problems in rearing both as true pathogens and as contaminants (see Table 1). A number of species, such as the deuteromycetous fungi, *Beauveria bassiana* and *Metarhizium anisopliae*, are primary pathogens that can infect a broad range of insect species, producing white and olive green conidia, respectively, on the outside surface of infected and killed individuals. Fungal entomopathogens generally infect the host via direct penetration of the insect cuticle. Once the host has been colonized and killed with its tissues converted to fungal biomass, hyphae penetrate back out through the cuticle. Sporulation then takes place on these emerging hyphae.

The fungi produce spores that are readily transmitted to diet, insects, and other surfaces on air currents, the integument of contaminated insects, and dust particles. Certain fungi are readily transferred in aerosols or in water.

**Table 1. Fungi often associated with insect rearing**

**Pathogens**

| Deuteromycotina | Zygomycotina | Mastigomycotina |
|---|---|---|
| *Beauveria bassiana* | Entomophthorales | *Lagenidium gigantium,* |
| *Matarhizium anisopliae* | (*Entomophthora,* | *Saprolegnia* spp., |
| *Nomuraea rileyi* | *Neozygites,* | *Leptolegnia* spp. |
| *Paecilomyces fumoso-roseus* | *Entomophaga,* etc.) | |

**Contaminants and Facultative Pathogens**

| Deuteromycotina | Zygomycotina | Ascomycotina |
|---|---|---|
| *Aspergillus niger,* | *Rhizopus* spp. | Yeasts |
| *A. ochraceous,* etc. | *Mucor* spp. | |
| *Penicillium* spp. | | |

The fungi are most often problems as contaminants in insect rearing, particularly on artificial diets. A number of *Aspergillus* spp., for example, are the most commonly encountered fungal species in insect rearing since they are among the most common contaminants of insect diets. When contamination occurs during larval development it will often determine the extent of the effect on the larvae. Contamination by *Aspergillus* early in larval development, for example, prevented pink bollworm larvae from feeding (Ouye 1962). Early contamination of boll weevil diet by *Aspergillus favus* severely reduced the number of larvae completing development. Later contamination when the larvae were older had little or no effect on the number of insects produced per rearing dish (Sikorowski 1984b).

Primary pathogens are often a problem under conditions of high humidity, particularly for plant reared species or species in contact with soil. *Beauveria bassiana* can be an occasional problem in rearing a number of species including *Diabrotica* species and Coloarado potato beetle, *Leptinotarsa decemlineata*. In these situations, the rearing conditions that favor infection and rearing conditions that stress the insects can significantly contribute to the problem.

Reviews of fungal entomopathogens that contain information useful in diagnosing the specific causal agents include Samson (1981) on the Deuteromycetes, King and Humber (1981) on the Entomophthorales, Bland et al. (1980) on the major fungal pathogen groups affecting aquatic insects, as well as the general reviews of Poinar and Thomas (1984) and Goodwin (1984).

Larvae that are suspected of being infected by a fungus can be placed in a sterile petri dish with moistened filter paper and incubated for 4 to 7 days at 20-25°C. The insect can be surface sterilized with a 1% NaOCl solution and rinsed three times in sterile water prior to transfer to the moist chamber. After sporulation has occurred, the fungus can then be identified by examining the spores and fruiting structures produced. A phase contrast microscope with 400-1000X magnification is most useful.

The following list describes some of the symptoms and signs of fungal infection:

- Sluggishness, decreased growth rate
- Generalized color change; pink to red for some *Beauveria* spp. strains, yellow to brown in most cases
- Presence of yeast-like cells or hyphal bodies in the hemolymph, going to mycelium as infection progresses
- Cheese-like consistency of larvae gradually changing to leathery consistency and eventually becoming hard mummy under drier conditions
- Mycelium growing out of cadaver, sporulation on surface of integument; Fungus is easiest to identify at this stage.

Fungi can generally sporulate within 2 to 5 days after death on infected insects and generate large quantities of infectious spores that can infect other individuals. For this reason, for species and rearing systems known to have problems with fungal entomophathogens, rearing cages should be checked daily and infected individuals removed and properly disposed of. Warm, humid conditions favor infections, as do conditions that stress the insect host.

Similar considerations apply to fungal contaminants of artificial diets. Diet containers should be opened only in an area far removed from diet preparation and infestation areas, and contaminated cups or trays should be properly and carefully disposed of. Clear containers have the advantage of showing whether or not fungal contaminants are present without opening the container and releasing spores into the air. In addition, bioassays using artificial diet that are occasionally contaminated with fungi should be read and opened in a separate area removed from where assays are being infested.

## Methods of Preventing Contamination

Sikorowski and Goodwin (1985), Sikorowski (1984a), and Shapiro (1984) have reviewed in detail the methods commonly used in reducing or eliminating contamination in a rearing program. Table 2 summarizes the various methods that can be utilized. All of these practices have as their ultimate goal to minimize the potential for pathogens or contaminants to come in contact with the susceptible stages of the insects being reared.

Prevention is the key. Once an outbreak of disease caused by a highly infectious pathogen occurs, a therapeutic approach is rarely the best option; at that point the colony must usually be replaced.

### Disease-free Stock

In establishing a colony, one should always strive to use stock that is known to be free of disease. Generally, this involves obtaining stock from an established colony that has a good history and is known to be free of pathogens. This is usually not a problem for those species that are commonly reared in many different labs, such as many of the noctuid moths.

### Quarantine of Field-collected Stock

For those species that are not commonly available, one must often resort to field collected rearing stock. This has the obvious problem of potentially introducing pathogens into the rearing system. When the use of field collected insects to start a colony is unavoidable, certain precautions should be taken. A knowledge of the diseases commonly associated with the species would be an

**Table 2. Methods of preventing contamination and disease problems in insect rearing**

---

1. Use of disease-free stock
2. Quarantine of field collected stock
3. Use of appropriate insectary design and traffic patterns
4. Sterilization of eggs
5. Sterilization of pupae
6. Sterilization of diet
7. Prevention of bacterial and fungal growth on diet
8. Use of insectary sanitation practices
9. Use of air handling systems
10. Monitoring microbial contamination
11. Monitoring insect quality
12. Quarantining field-collected stock

---

important starting point. Field collected insects can be set up in quarantine and observed for signs of disease. This can be done with different cohorts of insects, ideally based on single females, and only healthy cohorts used as rearing stock to initiate the colony.

## Insectary Design and Traffic Patterns

The design of an insectary and the layout of various functions is critical in preventing contamination. One of the most important considerations for minimizing pathogen contamination is to lay out the rooms so that diet preparation and egg infestation areas, where contamination of insects or diet is most likely to occur, are physically removed from potential sources of contamination, such as pupal harvest areas. Leppla and Fisher (1985) described a multispecies rearing facility in which there were periodic outbreaks of virus infections and fungal contamination of diet. This problem was not solved until the diet preparation area and the pupal harvest room were physically separated and direct access between these areas eliminated. A UV lamp pass-through into the diet preparation room also helped. Fisher (1984a,b) and Leppla and Fisher (1985) provide a good overview of insectary design concepts and issues related to contamination control.

Insectary designs have been discussed by Tanaka (1978) for three species of tephritid fruit flies; by Debolt and Petterson (1978) for cabbage looper, *Trichoplusia ni*, and beet armyworm, *Spodoptera exigua*; by Leppla et al. (1978)

for several different noctuids; by Gantt et al. (1978) for sugarcane borer, *Diatraea saccharalis*, and one of its parasites, *Lixophaga diatraea*; by Brewer et al. (1978) for corn earworm, *Heliothis zea* and *D. saccharalis*; by Griffin and Lindig (1978), Griffin (1984a,b) and Griffin and Lindig (1974) for boll weevil, *Anthonomus grandis*. Fisher (1984a) discussed the layout and traffic patterns of a multiple species industry insectary.

## Egg Sterilization

A key method for reducing contamination of insect diet with pathogens or microbial contaminants is to infest with eggs or larvae that are free of contamination. This is often best accomplished by surface sterilizing eggs. Without egg sterilization, the levels of diet contamination with bacteria and/or fungi can be extremely high, even with the use of antimicrobial compounds in the diet. For this reason, most insectaries using artificial diet routinely surface sterilize eggs.

Formaldehyde and sodium hypochlorite are the two most widely used surface sterilants for treatment of insect eggs. These disinfectants are used to treat eggs or egg papers. For any chemical agent used in sterilizing eggs, the specific concentration and contact time must be determined for each species since egg sensitivity to a given disinfectant will vary from species to species.

Sterilization is ideally carried out in a way that ensures direct contact with the sterilant. Agitation of the bath during surface sterilization is important for this reason. Treatment of individual egg sheets is also helpful, and Fisher (1984a) described a simple rack that suspends egg sheets individually in the sterilant.

After sterilization and thorough rinsing in water to remove the chemical sterilant, eggs or egg papers are dried. At this point it is important to ensure that drying takes place in a clean environment that is free of potential contaminants. An effective way to dry eggs or egg papers while minimizing contamination is to dry in a HEPA filtered laminar airflow hood. If a sterile airflow is used in drying, care should be taken to ensure that eggs are not exposed for so long that desiccation becomes a problem. This is particularly important for eggs that have been sterilized in sodium hypochlorite, which will partially dechorionate the eggs, making them more susceptible to desiccation.

**Formaldehyde.** Formaldehyde is bacteriocidal, sporicidal, and fungicidal. It also deactivates viruses, often more effectively than sodium hypochlorite. Formalin, a 37% solution of gaseous formaldehyde, is an effective sterilant against all microorganisms, including sporeformers and viruses. Thompson and Steinhaus (1950) described a method using a 10% formaldehyde solution to surface sterilize *Colias eurytheme* eggs with a 90-min exposure. Vail et al. (1968) described methods for sterilizing eggs of the cabbage looper, *Trichoplusia ni*, using formaldehyde and NaOCl as a means of preventing baculovirus infections.

Formalin has been used for egg sterilization of many different species. Among these are boll weevil, *Anthonomus grandis*, using a 3.1% AI. (formaldehyde) solution for 20 min (Roberson 1984, Sikorowski et al. 1977); cabbage looper, *T. ni*, using a 3.7% AI solution for 45 min (Leppla et al. 1984); pink bollworm, *Pectinophora gossypiella* (Bullock et al. 1969); spruce budworm, *Choristineura fumiferana*, using a 10% AI solution for 15 min (Robertson 1984); and *Heliothis virescens* using a 3% AI solution for 10 min (King & Hartley 1985). After sterilization it is important to remove residual formaldehyde by thoroughly rinsing eggs in tap water (5 to 45 min).

Formaldehyde is a known carcinogen (Sun 1981) and must be handled with care. For this reason, egg sterilizations with formaldehyde should always be conducted in a fume hood and eggs should be thoroughly rinsed afterwards.

**Sodium Hypochlorite.** Sodium hypochlorite (NaOCl), most often in the form of common laundry bleach, is a strong oxidizing agent that is used by many rearing facilities to sterilize eggs. NaOCl is bacteriocidal, sporicidal, fungicidal, and viricidal. Ignoffo and Dutky (1963) stated that NaOCl was an excellent multipurpose disinfectant, citing its good water solubility, activity on a broad range of microbes, low toxicity to human beings and insects, stability in water, availability, and low cost. The disadvantages associated with NaOCl are its corrosive properties, irritating odor, and skin irritation that can result from prolonged contact. As mentioned above, NaOCl can also partially dechorionate insect eggs, increasing their sensitivity to mechanical damage and desiccation. This latter effect, however, can sometimes be used to advantage in removing eggs from an oviposition substrate or in separating eggs in an egg mass.

Species that have been reared using NaOCl for egg sterilization include *Heliothis zea* and *Heliothis virescens* using a 0.2% AI (NaOCl) solution for 8 min (Ignoffo 1966); T.ni, using a 0.3% AI solution for 5 min with a rinse in 10% sodium thiosulfate (Leppla et al. 1984); *Agyrotaenia velutinana* using a 0.26% AI as a quick rinse; the cutworms, *Euxoa scandens* and *Euxoa messoria*, using a 0.2% AI solution for 10 min (Belloncik et al. 1985); *Spodoptera frugiperda* using a 0.16% AI solution for 10 min (Davis 1989). Concentrations used are generally 0.2% - 1% NaOCl with agitation for 5-15 min. Again, thorough rinsing with water is necessary to remove NaOCl after sterilization. Since use of NaOCl may result in rapid desiccation, treated eggs, should be stored in a suitably moist environment prior to hatch.

**Other Methods of Egg Sterilization.** Quaternary ammonium compounds have also been used to sterilize insect eggs. Brust and Fraenkel (1955) used a 0.1% aqueous solution of Zephiran chloride to sterilize blowfly eggs using a 25-min immersion. Martingnoni and Milstead (1960) evaluated Hyamine 10-X (methylbenzothonum chloride) and Zephiran chloride (benzalkonium chloride) for disinfection of eggs of the variegated cutworm, *Peridroma saucia* (Hübner). They found both had good bactericidal activity, low mammalian toxicity, and good wetting properties, but only Hyamine 10-X provided adequate egg disinfection.

Combinations of sterilants used in sequence can sometimes provide better results than a single agent alone. Surface sterilization of pink bollworm eggs was more effective in controlling a cytoplasmic polyhedrosis virus when both a 1% sodium hypochlorite and a 9% formaldehyde solution were used sequentially than when only sodium hypochlorite was used (Stewart 1984, Stewart et al. 1976).

Although the eggs of many moths will withstand treatment with formalin or sodium hypochlorite quite well, many insect species that inhabit more sheltered environments than leaf surfaces have eggs with thinner chorions. The eggs of these species can be considerably moe sensitive to chemical sterilants. *Diabrotica* spp. for example, do not tolerate either formalin or bleach very well. Marrone et al. (1985) found that peracetic acid, when used in conjunction with sodium hypochlorite, provided excellent surface sterilization and caused very little reduction in egg hatch of *Diabrotica undecimpunctata howardi*. This method involved treating eggs with a low rate of 0.0525% sodium hypochlorite (1% bleach) for 5 min followed by a 2-min treatment in 0.25% peracetic acid. This combination greatly reduced bacterial contamination with little effect on egg viability.

## Sterilization of Pupae

Pupal sterilization can substantially reduce the levels of microbial contamination passed from the larval diet to the adult moths. This technique ensures that any viral infections, for example, that may occur in a larval rearing container are less likely to be transmitted to the next generation. NaOCl is most often used for pupal sterilization and has the added advantage of dissolving the silk, making it easy to separate and clean pupae. However, it should be noted that for pathogens transmitted on the outside of the egg, an effective and consistently applied method of egg sterilization often precludes the need for pupal sterilization.

## Sterilization of Diet

Generally, good sanitation and the addition of microbial inhibitors to artificial diet will minimize the potential for contamination and growth of microorganisms and preclude the need to sterilize diet. Heat sterilization can destroy key nutrients, such as the B vitamins, and is generally not used. Nonetheless, in some instances it may be necessary to utilize some method of sterilization because of the sensitivity of a particular species to microbial inhibitors. In those cases, several methods can be used.

**Autoclaving.** Autoclaving is rarely used to sterilize insect diet. If it is used, certain key heat-labile components, such as the vitamin mix, may be sterilized by membrane filtration and added afterwards.

**Flash Sterilization.** Flash sterilizers have been used in the mass production of the boll weevil (Griffin et al. 1974, Griffin 1984a,b) and gypsy moth (Charles P. Schwalbe, personal communication). Flash sterilizers process medium very quickly by passing the molten diet through a heat exchanger. Resident time of diet at a high temperature of 145-160°C is only seconds but is sufficient to sterilize the diet and preclude the necessity for antimicrobials.

**Chemical Sterilization.** Formaldehyde is often added to a number of media as a general microbial sterilant (Singh 1977). Raulston and Lindegren (1972), for example, added 0.4 ml AI (formaldehyde) or 1.2 ml formalin per liter of diet. Many other chemicals are added to insect diets to kill or suppress microbial contaminants; these will be discussed later.

## Prevention of Bacterial and Fungal Growth on Diet

A key approach to preventing microbial contamination of artificial insect diets has been to incorporate antimicrobial compounds and preservatives into the diets. Most of these compounds, which are generally added to artificial diets during preparation, are also used in foods to prevent microbial contamination and spoilage (Chicester & Tanner 1968). Most commonly used in insect diets are the parabens, fatty acids and their derivatives, antibotics, fungicides, and other compounds with high antimicrobial activity and relatively low activity against insects (Singh 1977).

For most of these preservatives, such as the parabens and fatty acids, the antimicrobial activity is based on lipophilic properties that allow these compounds to bind to cell membranes. At the same time, these compounds must be soluble enough in water to move freely about in the aqueous phase of the diet in order to maintain the antimicrobial effect. At high concentrations, many of these preservatives can kill cells by damaging cell membranes. At lower concentrations, microbial growth is suppressed by interfering with active transport. Parabens, for example, were found to suppress uptake of amino acids in *Bacillus subtilis* (Freese et al. 1973). The lipopolysaccharide layer produced by gram-negative bacteria such as *Escherichia coli* appears to be responsible for making these bacteria more resistant to many preservatives than the gram-positive bacteria, which lack such a layer (Scheu & Freese 1973).

**Sorbates.** Sorbic acid and its potassium salt, collectively referred to as sorbates, are widely used in foods, pharmaceuticals, cosmetics, and insect diets to inhibit the growth of various microorganisms. Use rates generally vary from 0.01 to 0.3% sorbic acid (Sofos & Busta 1981, Singh 1977). Sorbates effectively inhibit the growth of bacteria, yeasts, and filamentous fungi (molds). While not so broadly effective against bacteria, the effect of sorbates against fungi is quite comprehensive with good inhibition of a diversity of molds and yeasts, including many that are common contaminants of insect diets.

Sorbates are also most effective at pH < 6.5, making it important to ensure that diet pH is one the acid side. The broader pH of the sorbates generally makes them more effective than benzoates and propionates, which have lower pH thresholds for antimicrobial activity of 5.0 to 5.5. In addition, sorbates are generally less toxic to animals than benzoates.

**Parabens.** The parabens are produced by esterification of the carboxyl group of p-hydroxbenzoic acid. In particular, the methyl ester, methyl p-hydroxybenzoate, has been widely used as a preservative in insect diets (Singh & Moore 1985, Singh 1977). The parabens are active against both bacteria and fungi but are generally most active against fungi, which require doses from 2 to 1250 times lower than those effective for bacteria (Davidson & Branden 1981, Aalto et al. 1953). The parabens are active over a broader range of pH than the straight acids and for this reason can be used in diets with pH 7 or slightly higher. Methylparaben is often used in combination with sorbic acid.

**Antibiotics.** Many different antibiotics, including the tetracyclines, streptomycins, and erythromycins, have been used in insect diets as bacteriocidal and bacteriostatic agents (Singh & Moore 1985, Singh 1977). Antibiotics can be particularly effective in controlling certain gram-negative bacteria that are not well inhibited by such preservatives as the sorbates or parabens. Chlorotetracycline is perhaps the most commonly used antibiotic diets, where it is used to control bacteria and yeasts. Ignoffo (1963) found it to be effective against *B. thuringiensis*, and it has since been used in a number of diets for many insects, including *C. occidentalis* (Robertson 1985), *H. zea*, and *H. virescens* (Patana 1985).

Shapiro (1984) tested chlorotetracycline, tetracycline, and streptomycin on gypsy moth diet against *B. cereus, P. vulgaris, E. coli, E. aerogenes,* and *S. aureus.* He found that all three antibiotics were effective against all five bacteria at concentrations of 0.05 % to 0.1 %. The only exception was chlortetracycline, which proved to be ineffective against *B. cereus.*

Antibiotics have the advantage of being more selectively toxic than many of the antimicrobial agents. For example, Sikorowski et al. (1980) found that a number of antibiotics successfully prevented microbial contamination of artificial diet in mass rearing of boll weevil, *A. grandis,* without any detectable adverse effects on larval development, egg production, hatch, and male pheromone production. Sikorowski and Goodwin (1985) showed that concentrations up to 0.05 % (w:w) of neomycin sulfate, streptomycin sulfate, chloramphenicol, kanamycin erythromycin, penicillin V, or ampicillin mixed into artificial diet did not adversely affect *Heliothis virescens* larvae.

In addition to use in diets, antibiotics have also been used in rearing insects on host plant material. Riddiford (1985), for example, sprayed leaves used to rear the wild silk moth, *Hyalophora cecropia,* with an aqueous solution of Aureomycin (chlortetracycline HCL) and Kanamycin sulfate at 0.28 and 0.016%, respectively. Eggs or larvae were then transferred to treated leaves after they had dried.

**Other Antimicrobial Agents.** Fumagillin has been described as an effective agent in suppressing *Nosema apis* in the honeybee (Katznelson & Jaimeson 1952, Bailey 1953, Moffett et al. 1969), *N. pyraustae* in the European corn borer, *Ostrinia nubilalis,* (Lewis & Lynch 1969, 1970; Lynch & Lewis 1971) and *N. fumiferanae* in the spruce budworm, *C. fumiferanae* (Wilson 1974).

Benomyl is also effective in controlling microsporidia (Shinholster 1974, Armstrong 1976, Hsiao & Hsiao 1973). Hsiao and Hsiao (1973), for example, found that benomyl at 250 ppm completely eliminated a *Nosema* infection of the alfalfa weevil, *Hypera postica,* when incorporated into an artificial diet. In these tests larvae were devoid of spores after 3 days of feeding on treated diet. Fumagillin at 10 ppm did not prevent infection and at 20 ppm inhibited larval feeding and growth. Benomyl, a systemic funcigide, is believed to affect sporogenesis in susceptible fungal species. Hsiao and Hsiao (1973) postulated that the mode of action on the microsporidia may be similar, inhibiting sporogenesis in the insect host.

Another fungicide, Phaltan (folpet), was shown to be very effective in suppressing contamination of boll weevil diet with *Aspergillus niger* van Tiegham. In combination with tetracycline and neomycin sulfate, Phaltan provided excellent control of bacteria and fungi with no detectable adverse effects on weevils (Sikorowski et al. 1980). A number of other studies have evaluated the use of fungicides to control molds on insect diets, including Ludeman and Funke (1978), Ludeman et al. (1979), Bathon (1977) and Menten et al. (1979).

**General Considerations.** For a given insect species and rearing system, the concentration and combination of antimicrobials should be optimized. These compounds are active on insects as well as on microbes. Consequently, combinations and rates of antimicrobials must be established that will provide good control of contaminants while having little or no effect on the insects. Singh and Bucher (1971) and Singh and House (1970) define a safe level of antimicrobials as the concentration that does not increase the period of development by more than 25%. Pupal weights and larval vigor may also be affected, and acceptable effects must be considered for each species and the particular rearing objectives.

## *Insectary Sanitation Practices*

General sanitation in the daily activities of an insectary are critically important to minimizing the possibility of contamination and disease in insect rearing. Insectary sanitation programs have been described by Steinhaus (1953), Davis (1976), Martignoni and Iwai (1977), and Bell et al. (1981). A key to the success of any program is the commitment and training of personnel. Staff should be carefully selected and trained and should be assigned specific responsibilities in maintaining a clean insectary environment. Monitoring and enforcing insectary sanitation rules are also important.

The objective of any insectary sanitation program is to minimize microbial contamination. In large measure this involves removing dust and microorganisms that are regularly introduced into the insectary and settle on surfaces. These particles must be removed on a daily basis, particularly in the clean areas of the insectary. Specific activities directed at maintaining a clean environment are discussed next.

**Personal Hygiene: Use of Clean Lab Coats or Coveralls, Gloves, and Masks.** Personnel should change to clean lab coats or uniforms when entering clean areas and should wear disposable gloves. In some cases, they should wear booties, hats, and masks when engaged in certain activities in these areas, particularly when preparing and infesting diet. Hands should be washed regularly.

**Surface Disinfection and General Housekeeping.** Floors and counter tops must be washed down with a disinfectant on a regular basis, usually daily to twice weekly in conventional areas and once to twice a day in clean areas. Wet mopping of floors is preferred over sweeping or dry mopping because the latter methods tend to put dust into the air. Vacuum cleaners can also be very useful in insectaries in maintaining a clean environment, provided that the exhaust air is properly vented. In-wall vacuum cleaner units that vent to the outside of the insectary or HEPA filtered units are ideal. Litsky and Litsky (1968) and others have demonstrated the effectiveness of flooding and wet vacuuming in cleaning and disinfecting floors. Disinfectants commonly used include, sodium hypochlorite, iodophors, quartenary ammonium products (Roccal, Mikro-quat, etc), phenolic compounds such as Amphyl, formalin, and various soaps an detergents.

Exspore, a chlorine dioxide surface disinfectant, is a relatively new product that is being used in pharmaceutical and dairy industries. Exspore is prepared by mixing two components (an activator and a base) with water (1:14) and spraying or wiping the solution on surfaces. The resulting chlorine dioxide has virtually the same sterilizing effect as sodium hypochlorite but is less corrosive and is safer to use since it tends to be less of a skin contact sensitizer. It also performs better as a sterilant under high organic load. This product may also have some uses in sterilization of insect eggs, and this use would be worth exploring.

**Ultraviolet Sterilization.** Ultraviolet radiation (U.V.) can also be useful in sterilizing surfaces in the insectary. U.V. radiation kills microorganisms by damaging DNA. Wavelengths in the range of 240-280 nm are the most effective. One of the limitations of U.V. is its poor penetration. U.V. can be blocked by dust on the surface that is to be treated and on the lamp itself. Germicidal U.V. lamps must be positioned directly over surfaces to be irradiated and must not be operated when people are present in order to prevent damage to the eyes. In addition, lamps must be checked periodically to ensure that they are still active. One way to check this is to test the sterilizing effect on bacterial suspensions by plating these before and after exposure. Another, simpler alternative is to use U.V. germicidal indicators like the Uvicide germicidal lamp monitors. These are simple cards that use a chemical indicator sensitive to U.V. to show if germicidal doses are being delivered to surfaces.

Although U.V. lamps are generally used over counters and floors and inside laminar airflow hoods and pass-throughs, a small portable cart with U.V. lamps mounted on three sides can be used to sterilize surfaces that are normally difficult to access, such as the floor space under counters, etc. (P. Singh, personal communication).

**Waste Disposal.** Trash should not be allowed to accumulate and should be sealed in plastic bags and discarded at the end of each day.

**Traffic Patterns.** Regulating the movement of staff within the insectary is important in minimizing contamination of clean areas. Only authorized personnel should be allowed into the most contamination-sensitive areas, such as the diet preparation area. In other areas, unnecessary traffic should be avoided. Much of this should be incorporated into the design of the facility, as described earlier.

**Air Handling System.** Insectary heating, ventilation, and air conditioning (HVAC) systems must address a number of requirements above and beyond those associated with most facilities. In addition to providing heating, cooling, and ventilation that will provide a comfortable working environment for staff, the needs of the particular insect colonies with regard to temperature, ventilation, and relative humidity must likewise be addressed. Air handling systems can also be a source of contamination by moving dust and microorganisms into sensitive clean areas. In moth rearing facilities, adult wing scales are both a health hazard to personnel and a source of contamination. Proper filtration of the room air in adult moth holding areas can go a long way toward solving both problems.

By incorporating these various considerations into the original design of the facility, future problems can be avoided. Goodenough and Parnell (1985) note that HVAC design must take into account 1) personnel health, safety, and comfort; 2) insect physiological and environment needs; and 3) engineering considerations relating to weather, economics, management/maintenance, insect production needs, etc.

Even a well-designed, well-laid out insectary with good traffic patterns can have problems if the heating and air conditioning system is not carefully designed and constructed. Clean areas, for example, should have slightly higher room pressure than surrounding areas to minimize introduction of dust and contaminants whenever doors are opened. In some situations, particularly larger operations, it is often advisable to have an HVAC system for clean areas that is separate from the one used for conventional areas in the insectary. Air locks into clean areas can also help minimize introduction of contaminants via air currents.

Air filtration is a vital tool in minimizing contamination in the insectary environment. Filters in the HVAC system can help keep particulates in the room environment at low levels. Dust particles greater than 4 microns in diameter are generally the most likely to be carrying bacteria. Individual bacterial and fungal spores can also be present in the air free of dust particles. High efficiency particulate air (HEPA) filters remove 99.97 % of particles down to 0.3 microns, effectively sterilizing room supply air, making HEPA filtration one of the best ways to keep contamination levels low. HEPA filters are also available in a number of laminar airflow devices, including vertical airflow ceiling units, horizontal flow units, curvilinear airflow devices (airflow changes direction during operation), and laminar airflow hoods and cabinets. These devices provide a clean environment ideal for diet preparation and infestation with very low probability of contamination. Laminar airflow hoods are one of the most valuable pieces of equipment in this regard, particularly for smaller operations where the cost of a clean room with full HEPA filtration may be prohibitive. These cabinets come in a variety of sizes and designs and produce a sterile airflow ideal for pouring, drying, and infesting diet, as well as other activities requiring a contaminant-free environment.

Design and operational aspects of insectary air handling systems have been reviewed by Griffin (1984a,b) and Goodenough and Parnell (1985). For more information on laminar airflow principles and uses, see Phillips and Runkle (1973).

## *Monitoring Microbial Contamination*

One means of anticipating and preventing a problem with contaminants is to monitor the levels of contamination in an insectary on a regular basis. Deviations from normal levels can then serve as an advance warning that there is a problem developing. Typically, such a program would sample insects, eggs, air, surfaces (countertops, floors, etc.), equipment, and artificial diet. The appropriate culturing techniques and media would then be employed to determine levels of contamination. For example, diet samples are merely incubated in sterile petri dishes, eggs are plated on agar medium, and surfaces are wiped with sterile swabs that are rinsed in sterile diluent with the diluent then being further diluted and plated on agar medium. For details on one of the earliest and most comprehensive of such monitoring programs, see Sikorowski (1984a).

## *Monitoring Insect Quality*

Monitoring insect quality is an important component of any insect rearing program. The feedback from quality control testing is essential to maintaining high insect quality over time. Quality characteristics that might be assessed would vary according to the overall objectives but typically would include pupal weights, development time, percent pupation, yield of pupae, yield of adults, fecundity, percent hatch, and, for some studies, specialized parameters such as pheromone production or certain behavioral responses. Once the key quality characteristics are selected, the specifications established, and the appropriate protocols and recording forms implemented, the data obtained by evaluating samples form each cohort of insects over time can be very useful in identifying disease problems. Some of the less conspicuous pathogens that cause more chronic and debilitating disease, such as the CPVs, rickettsia, and microsporidia, can be very difficult to detect. A quality control program that shows an increase in development or reduced fecundity or percentage pupation can be the first indication of a problem with one of these pathogens.

For a detailed discussion of this important topic see Boller and Chambers (1977a,b), Chamber and Ashley (1984), and Moore et al. (1985).

## Summary

The objective of any insect rearing program is to produce the highest quality insects at the required times and in the required numbers and stages to satisfy the needs of the project or program the rearing effort is supporting. To accomplish this it is important to recognize the principal types of problems that may arise with pathogens or contaminants for the particular species and rearing method being used. In this way the rearing manager can anticipate and implement methods to prevent the major disease problems associated with a given rearing system. Prevention is always the key. Good sanitation, good planning and design of the facility and its operation, well-thought-out and implemented protocols and practices, and conscientious and well-trained personnel will go a long way toward preventing major problems.

It is hoped that this review has provided some insights into the problems that can arise and the various strategies that can be used in preventing them.

## References

Aalto, T. R., M. C. Firman & N. E. Rigler. 1953. -pHydroxybenzoic acid esters as preservatives I. Uses, antibacterial and antifungal studies properties and determination. J. Amer. Pharm. Assoc. 42: 449-457.

Aoki, K. 1957. Konchu Byorigaku, Gihodu, Tokyo.

Armstrong, E. 1976. Fumidil B and benomyl: chemical control of *Nosema kingi* in *Drosophila willistoni*. J. Invertebra. Pathol. 27: 363-366.

Bailey, L. 1953. Effect of fumagillin upon *Nosema apis* (Zander). Nature (London) 171: 212.

Bassi, A. 1835. Del mal del segno calcinaccio o moscardino ma latti che affligge i bachi da seta. Parte 1. Teorica Tip. Orcesi, Lodi.

Bathon, H. 1977. Fungicides for mold control in semi synthetic diets for insect rearing. Z. Angew. Entomol. 82(3): 247-251.

Bell, S. D., E. G. King & R. J. Hamelle. 1981. Some microbial contaminants and control agents in a diet and larvae of Heliothis spp. J. Invertebr. Pathol. 37: 243-248.

Belloncik, S., C. Lavallee & I. Quevillon. 1985. *Euxoa scandens* and *Euxoa messoria*, pp. 293-299. *In* P. Singh & R. F. Moore [eds.], Handbook of insect rearing, vol II. Elsevier, Amsterdam.

Bland, C. E., J. N. Couch & S. Y. Newell. 1980. Identification of *Coelomomyces*, Saprolegniales and Lagenidiales, pp. 129-162. *In* H. D. Burges [ed.], Microbial control of pests and plant diseases 1970-1980. Academic Press, London.

Boller, E. F. & D. L. Chambers, editors. 1977a. Quality control: An idea book for fruit fly workers. Bull. SROP, WPRS.

Boller, E. F. & D. L. Chambers. 1977b. Quality aspects of mass-reared insects. *In* R. L. Ridgway & S. B. Vinson [eds.], Biological control of augmentation of natural enemies. Plenum, New York.

Brewer, F. D., C. W. Gantt & D. F. Martin. 1978. Media-preparation and brood-colony facility, pp. 72-75. *In* N. C. Leppla & T. R. Ashley [eds.], Facilities for insect research and production. USDA-ARS, Technical Bulletin No. 1576. Washington, D.C.

Brooks, W. M. 1988. Entomogenous Protozoa. *In* C. M. Ignoffo & N. Bhusan Mandava [eds.], Handbook of natural pesticides, vol. 5, Part A. CRC Press, Boca Raton, Florida.

Brooks, W. M. 1974. Protozoan infections. *In* G. E. Cantwell [ed.], Insect disease, vol. 1. Marcel Dekker, New York.

Brust, M. & G. Fraenkel. 1955. The nutritional requirements of the larvae of a blowfly, *Phormia regina* (Meig.) Physiol. Zool. 28: 186-204.

Bucher, G. E. 1981. Identification of bacteria found in insects, pp. 7-34. *In* H. D. Burges [ed.], Microbial control of pests and plant diseases, 1970-1980. Academic Press, London.

Bucher, G. E. 1963. Nonsporulating bacterial pathogens, pp. 117-147. *In* E. W. Steinhaus [ed.], Insect pathology: an advanced treatise, vol. 2. Academic Press, New York.

Bullock, H. R., C. L. Mangum & A. A. Guerra. 1969. Treatment of eggs of the pink bollworm, *Pectinophora gossypiella*, with formaldehyde to revent infection with a cytoplasmic polyhedrosis virus. J. Invertbr. Pathol. 14: 271-273.

Burges, H. D. (editor). 1981. Microbial control of pests and plant diseases, 1970-1980. Academic Press, London.

Cantwell, G. E. (editor). 1974. Insect diseases, vol. 1. Marcel Dekker, New York.

Chambers, D. L. & T. R. Ashley. 1984. Putting the control in quality control in insect rearing, pp. 256-260. *In* E. G. King & N. C. Leppla [eds.], Advances and challenges in insect rearing. USDA - ARS, Washington, D.C.

Chicester, D. F. & F. W. Tanner, Jr. 1968. Antimicrobial food additives, pp. 137-207. *In* T. E. Furia [ed.], Handbook of food additives. Chemical Rubber Co., Cleveland, Ohio.

Concannon, J. N. & A. G. Wadewitz. 1983. Studies with an insect pathogen: *Micrococcus pseudoflaccidifex*. *In* Developments in industrial microbiology, vol. 24: 451-455. Soc. Industr. Micro.

Davis, F. M. 1989. Rearing the southwest corn borer and fall armyworm at Mississippi State, p. 27-36. *In* Toward insect resistant maize for the Third World: Proceedings of the International Symposium on Methodologies for Developing Host Plant Resistance to Maize Insects. Mexico, D.F. : CIMMYTI.

Davis, F. M. 1976. Production and handling of eggs of southwest corn borer for host plant resistance studies. Miss. Agric. and For. Exp. Sta. Tech. Bull. 74.

Davidson, P. M. & A. L. Branden. 1981. Antimicrobial activity of non-halogenated phenolic compounds. J. Food Prot. 44 (8): 623-632.

Debolt, J. W. & M. A. Peterson. 1978. Rearing facility for vegetable and sugarbeet insects, pp. 64-66. *In* N. C. Leppla & T. R. Ashley [eds.], Facilities for insect research and production. USDA Technical Bulletin 1576.

Drake, E. L. & F. L. McEwen. 1959. Pathology of a nuclear polyhidrosis virus of the cabbage looper, *Trichoplusia ni* (Hubner). J. Invertbr. Pathol. 1: 281-293.

Dutky, S. R. 1963. The milky diseases, pp. 117-147. *In* E. W. Steinhaus [ed.], Insect pathology: an advanced treatise. Academic Press, New York.

Faust, R. M. 1974. Bacterial diseases. *In* G. E. Cantwell [ed.], Insect diseases, vol. 1. Marcel Dekker, New York.

Fisher, W. R. 1984a. The Insectary Manager, pp. 295-299. *In* E. G. King & N. C. Leppla [eds.], Advances and challenges in insect rearing. USDA-ARS, Washington, D.C.

Fisher, W. R. 1984b. Production of insects for industry, pp. 234-239. *In* E. G. King & N. C. Leppla [eds.], Advances and challenges in insect rearing. USDA - ARS, Washington, D.C.

Freese, E. C. W. Scheu & E. Galliers. 1973. Function of lipophilic acids as antimicrobial food additives. Nature 241: 321-325.

Gantt, C. W., F. D. Brewer & D. F. Martin. 1978. Modified facility for host and parasitoid rearing, pp. 70-72. *In* N. C. Leppla & T. R. Ashley [eds.], Facilities for insect research and production. USDA - ARS, Technical Bulletin No. 1576. Washington, D.C. pp.

Goodenough, J. L. & C. B. Parnell. 1985. Basic engineering design requirements for ventilation, heating, cooling and humidification of insect rearing facilities, pp. 137-155. *In* P. Singh & R. F. Moore [eds.], Handbook of insect rearing, vol. II. Elsevier, Amsterdam.

Goodwin, R. H. 1968. Nonsporeforming bacteria in the armyworm *Pseudeletia unipuncta*, under gnotobiotic conditions. J. Invertebr. Pathol. 11: 358-370.

Goodwin, R. H. 1984. Recognition and diagnosis of diseases in insectaries and the effects of disease agents on insect biology, pp. 96-129. *In* E. G. King & N. C. Leppla [eds.], Advances and challenges in insect rearing. USDA - ARS, Washington, D.C.

Greany, P. D., G. E. Allen, J. C. Webb, J. L. Sharp & D. L. Chambers. 1977. Stress induced septicemia as an impediment to laboratory rearing of the fruit fly parasitoid, *Biosteres (Opius) longicaudatus* (Hymenoptera:Braconidae) and the Caribbean fruit fly, *Anastrepha suspensa* (Diptera: Tephritidae). J. Invertebr. Pathol. 29: 153-161.

Griffin, J. G. 1984a. Facility and production equipment, pp 11-52. *In* P. P. Sikowrowski, J. G. Griffin, J. Roberson & O. H. Lindig [eds.], Boll weevil mass rearing technology. University Press, Mississippi.

Griffin, J. G. 1984b. General requirements for facilities that mass-rear insects, pp. 70-73. *In* E. G. King & N. C. Leppla [eds.], Advances and challenges in insect rearing. USDA - ARS, Washington, D.C.

Griffin, J. G. & O. H. Lindig. 1974. Mechanized production of boll weevil diet pellets. Trans. ASAE. 17(1): 15-16 and 19.

Griffin, J. G., O. H. Lindig & R. E. McLaughlin. 1974. Flash sterilizers: sterilizing artifical diets for insects. J. Econ. Entomol. 67(5): 689.

Griffin, J. G. & O. H. Lindig. 1978. Pilot facility for mass rearing of boll weevils, pp. 75-78. *In* N. C. Leppla & T. R. Ashley [eds.], Facilities for insect research and production. USDA-ARS, Technical Bulletin No. 1576. Washington, D.C.

Habib, M. E. 1978. A bacterial disease of the American cotton leaf worm *Alabama argillacea* (Hubner) (Lep:Noctuidae), with notes on its histopathological effects. Z. Angew. Entomol. 85: 76-81.

Heimple, A. M. & T. A. Angus. 1963. Diseases caused by sporeforming bacteria, pp. 21-73. *In* E. W. Steinhaus [ed.], Insect pathology: An advanced treatise, vol. 2. Academic Press, New York.

Helms, T. J. & E. S. Raun. 1971. Perennial laboratory culture of disease-free insects, pp. 639-654. *In* H. D. Burgess & N. W. Hussey [eds.], Microbial control of insects and mites. Academic Press, New York.

Hsiao, T. H. & C. Hsiao. 1973. Benomyl: A novel drug for controlling a microspoidian disease of the alfalfa weevil. J. Invertebr. Pathol. 22: 303-304.

Hukahara, T. 1987. Epizootiiology: Prevention of insect diseases, pp. 497-512. In J. R. Fuxa & Y. Tanada [eds.], Epizootiology of insect disease. Wiley, New York.

Ignoffo, C. M. 1966. Insect viruses, pp. 501-530. *In* C. N. Smith [ed.], Insect colonization and mass production. Academic Press, New York.

Ignoffo, C. M. 1965. The nuclear-polyhidrosis virus of *Heliothis zea* (Boddie) and *Heliothis verscens* (Fabricius) II. Biology and propagation of diet-reared *Heliothis*. J. Invertebr. Pathol. 7: 217-226.

Ignoffo, C. M. & S. R. Dutky. 1963. The effect of sodium hypochlorite on the viability and infectivity of *Bacillus* and *Beauveria* spores and cabbage looper nuclear polyhidrosis virus. J. Invertebr. Pathol. 5: 422-426.

Ishiwata, S. 1911. (On a type of Flacherie disease - sotto disease) In Japanese, Dainihon Sanshi Kaiho 9,1.

Jackson, T. A., J. F. Pearson & G. Stucki. 1986. Control of the grass grub, *Costelytra zealandica* (White) (Coleoptera:Scarabaeidae) by application of the bacteria, Serratia spp., causing honey disease. Bull. Ent. Res. 76.

King, D. S. & R. A. Humber. 1981. Identification of the Entomophthorales, pp. 107-127. *In* H. D. Burges [ed.], Microbial control of pests and plant diseases 1970-1980. Academic Press, London.

King, E. G. & G. G. Hartley. 1985. *Heliothis virescens*, pp.323-334. *In* P. Singh & R. F. Moore [eds.], Handbook of insect rearing,vol II. Elsevier, Amsterdam.

Katznelson, H. & C. A. Jamieson. 1952. Control of Nosema disease of honey bees with fumagillin. Science 115: 70-71.

Krieg, A. 1987. Diseases caused by bacteria and other prokaryotes, pp. 323-325. In J. R. Fuxa & Y. Tanada [eds.], Epizootiiology of insect disease. Wiley, New York.

Leppla, N. C. & W. R. Fischer. 1985. Insectary design and operation, pp. 167-183. *In* P. Singh & R. F. Moore [eds.], Handbook of insect rearing, vol. I. Elsevier, Amsterdam.

Leppla, N. C., P. V. Vail & J. R. Rye. 1984. Mass rearing the cabbage looper, *Trichoplusia ni.*, pp. 248-253. *In* E. G. King & N. C. Leppla [eds.], Advances and challenges in insect rearing. USDA-ARS, Washington, D.C.

Leppla, N. C., S. L. Carlyle, C. W. Green & W. J. Pons. 1978. pp. 63-64. *In* N. C. Leppla & T. R. Ashley [eds.], Facilities for insect research and production. USDA - ARS, Technical Bulletin No. 1576. Washington, D.C.

Lewis, L. C. & R. E. Lynch. 1969. Use of drugs to reduce Perezia pyraustae infections in the European corn borer. Proc. North Centr. Branch Entomol. Soc. of Am. 24: 84-87.

Lewis, L. C. & R. E. Lynch. 1970 Treatment of *Ostrinia nubilalis* larvae with Fumidil B to control infections caused by *Perezia pyraustae*. J. Invertbr. Pathol. 15: 443-48.

Lipa, J. J. 1963. Infections caused by protozoa other than sporozoa, pp. 335-361. *In* E. A. Steinhaus [ed.], Insect pathology, an advanced treatise, vol. 2. Academic Press, New York.

Lipa, J. J. 1967. Zarys Patologii Owadow, PWRiL, Warsawa, Poland.

Litsky, B. Y. & W. Litsky. 1968. Investigations on decontamination of hospital surfaces by the use of disinfectants and detergents. Am. J. Public Health 58: 534-543.

Ludemann, L. R. & B. R. Funke. 1978. Mold control in insect rearing media. Proc. N. Dakota Acad. of Sci. 32(1): 32.

Ludemann, L. R., B. R. Funke & C. E. Goodpasture. 1979. Mold control in insect rearing media survey of agricultural fungicides and evaluation of the use humectants. J. Econ. Entomol. 72(4): 579-582.

McLaughlin, R. E. 1962. The effect of temperature upon larval mortality of the armyworm, *Pseudoletia unipuncta* (Haworth). J. Insect Pathol. 4: 279-283.

McLaughlin, R. E. & J. C. Keller. 1964. Antibiotic control of an epizootic caused by *Serratia marcesscens Bizio* in the boll weevil, *Anthonomus grandis Boheman*. J. Insect Pathol. 6: 481-485.

Marrone, P. G., F. D. Ferri, T. R. Mosley & L. J. Meinke. 1985. Improvements in laboratory rearing of the southern corn rootworm, *Diabrotica undecimpunctata howardi* Barber (Coleoptera:Chrysomelidae), on an artificial diet and corn. J. Econ. Entomol. 78: 290-293.

Martignoni, M. E. & P. J. Iwai. 1977. Sanitation program for diet preparation. Standard operating procedures, RWU-2203, Forestry Sciences Laboratory, Corvallis, Oregon.

Martignoni, M. E. & J. E. Milstead. 1960. Quartenary ammonium compounds for the surface sterilization of insects. J. Insect Pathol. 2: 124-133.

Menten, J. O., O. M. Silva, C. R. Sousa & L. A. S. Menten. 1979. Solo. 71(2): 47-51.

Moffett, J. O., J. J. Lackett & J. D. Hitchcock. 1969. Compounds tested for control of *Nosema* in honey bees. J. Econ. Entomol. 62: 886-888.

Moore, R. F., T. M. O'Dell & C. O. Calkins. 1985. Quality assessment in laboratory-reared insects, pp. 107-135. *In* P. Singh & R. F. Moore [eds.], Handbook of insect rearing, vol I. Elsevier, Amsterdam.

Ouye, M. T. 1962. Effects of antimicrobial agents on microorganisms and pink bollworm development. J. Econ. Entomol. 55: 854-857.

Pasteur, I. Etudes sur la Maladies des Vers a Soie. 1870. Gauthier-Vilars, Paris.

Patana, R. 1985. *Heliothis zea / Heliothis virescens*, pp. 329-334. In P. Singh & R. F. Moore [eds.], Handbook of insect rearing, vol II. Elsevier, Amsterdam.

Payne, C.C . & D. C. Kelly. 1981. Identification of insect and mite Viruses, pp. 61-92. *In* H. D. Burges [ed.], Microbial control of pests and plant diseases, 1970-1980. Academic Press, London.

Phillips, G. B. & R. S. Runkle. 1973. Biomedical applications of laminar airflow. CRC Press, Cleveland, Ohio.

Poinar, G. O. Jr. & G. M. Thomas. 1984. Laboratory guide to insect pathogens and parasites. Plenum Press, New York.

Raulston, J. R. & P. D. Lindegren. 1972. Methods for large-scale rearing of the tobacco budworm. U. S. Dep. Agric. Prod. Res. Rep. No. 145.

Riddiford, L. M. 1985. *Hyalophora cercropia,* pp. 335-343. *In* P. Singh & R. F. Moore [eds.], Handbook of insect rearing, vol. II. Elsevier, Amsterdam.

Roberson, J. 1984. Laboratory rearing, pp. 85-113. *In* P. P. Sikowrowski, J. G. Griffin, J. Roberson & O. H. Lindig [eds.], Boll weevil mass rearing technology. pp. 85-113. University Press, Mississippi.

Robertson, J. 1985. *Choristoneura occidentalis* and *Choristoneura fumiferana,* pp.227-236. *In* P. Singh & R. F. Moore [eds.], Handbook of insect Rearing. Elsevier, Amsterdam.

Samson, R. A. 1981. Identification: Entomopathogenic Deuteromyctes, pp 93-106. In H. D. Burges [ed.], Microbial Control of Pests and Plant Diseases 1970-1980. Academic Press, London.

Scheu, C. W. & E. Freese. 1973. Lipopolysaccharide layer protection of gram-negative bacteria against inhibition by long-chain fatty acids. J. Bacteriol. 115: 869-875.

Shapiro, M. 1984. Microorganisms as contaminants and pathogens in insect rearing, pp. 130-142. *In* E. G. King & N. C. Leppla [eds.], Advances and challenges in insect rearing. USDA - ARS, Washington, D. C.

Shinholster, D. L. 1974. The effects of X-radiation and chemotherapy on the host-parasite relationships between *Tribolium castaneum* (Herbst) and twoprotozoan parasites, *Nosema whitei* Weiser and *Adelina tribolii* Heese. PhD. dissertation, Cornell University, Ithaca, New York.

Sikorowski, P. P. 1984a. Microbial contamination of insectaries:occurence, prevention and control, pp. 143-153. *In* E. G. King & N. C. Leppla [eds.], Advances and challenges in insect rearing. USDA - ARS, Washington, D.C.

Sikorowski, P. P. 1984b. Pathogens and microbiological contaminants: their occurrence and control, pp. 115-169. *In* P. P. Sikowrowski, J. G. Griffin, J. Roberson & O. H. Lindig [eds.], Boll weevil mass rearing technology. University Press, Mississippi.

Sikorowski, P. P., J. M. Wyatt & O. H. Lindig. 1977. Method of surface-sterilization of boll weevil eggs and the determination of microbial contamination of adults. The Southwestern Entomologist 2: 32-36.

Sikorowski, P. P. & A. C. Thompson. 1980. Effects of internal bacterial contamination on mass-reared boll weevils. *(Anthonomus grandis* Boheman). Miss. Agric. and For. Exp. Sta. Tech. Bull. 103.

Sikorowski, P. P., A. D. Kent, O. H. Lindig, G. Wiygul & J. Roberson. 1980. Laboratory and insectary studies on the use of antibiotics and antimicrobial agents in mass-rearing of boll weevils. J. Econ. Entomol. 73: 106-110.

Sikorowski, P. O. & R. H. Goodwin. 1985. Contamination control and disease recognition in laboratory colonies, pp. 85-105. *In* P. Singh & R. F. Moore [eds.], Handbook of insect rearing, vol. I. Elsevier, Amsterdam.

Singh, P. 1977. Artificial diets for insects, mites and spiders. Plenum Press, New York.

Singh, P. & H. L. House. 1970. Antimicrobials: "safe" levels in a synthetic diet of an insect, *Agria affinis*. J. Insect Physiol. 16: 1769-1782.

Singh, P. & G. E. Bucher. 1971. Efficacy of safe levels of antimicrobial food additives to control microbial contaminants in a synthetic diet for *Agria affinis* larvae. Entomol. Exp. Appl. 14: 297-309.

Singh, P. & R. F. Moore (editors). 1985. Handbook of Insect rearing. Elsevier, Amsterdam.

Sofos, J. N. & F. F. Busta. 1981. Antimicrobial activity of sorbate. J. Food Prot. 44(8): 614-622.

Steinhaus, E. A. 1949. Principles of insect pathology. McGraw-Hill, New York.

Steinhaus, E. A. 1953. Diseases of insects reared in the laboratory or insectary. Calif. Agric. Expt. Stn. Ext. Serv. Leaflet 9: 1-26.

Steinhaus, E. A. 1975. Disease in a minor chord. Ohio State University Press, Columbus, Ohio.

Stewart, F. D. 1984. Mass rearing the pink bollworm, *Pectinophora gossypiella*. *In* E. G. King & N. C. Leppla [eds.], Advances and challenges in insect rearing. USDA - ARS, Washington, D.C.

Stewart, F. D., R. B. Bell, A. J. Martinez, J. J. Robertson & A. M. Lowe. 1976. The surface sterilization of pink bollworm eggs and the spread of cytoplasmic polyhidrosis virus in rearing containers. U.S. Anim. Plant Health Insp. Service. Rep. 81-27.

Stucki, G., T. A. Jackson & M. J. Noona. 1984. Isolation and characterization of Serratia strains pathogenic for larvae of the New Zealand grass grub, *Costelytra zealandica*. N. Z. J. Sci. 27: 255-260.

Sun, M. 1981. Study shows formaldehyde is carcinogenic. Science 213: 12-32.

Tanaka, N. 1978. Facility for large-scale rearing of Tephritid fruit flies, pp. 63-64. *In* N. C. Leppla & T. R. Ashley [eds.], Facilities for insect research and production. USDA - ARS, Technical Bulletin No. 1576. Washington, D.C.

Thomas, E. D., C. F. Reichelderfer & A. M. Heimpel. 1972. Accumulation and persistence of a nuclear polydedrosis virus of the cabbage looper in the field. J. Invertebr. Pathol. 20: 157-164.

Thompson, C. G. & E. A. Steinhaus. 1950. Further tests using a polyhedrosos virus to control the alfalfa caterpillar. Hilgardia 19: 411-445.

Trought, T. E. T., T. A. Jackson & R. A. French. 1982. Incidence and transmission of a disease of grass grub. *(costelytra zealandica)* in Canterbury. N. Z. J. Exp. Agric. 10: 79-82.

Vail, P. V., Henneberry, T. F., Kishaba, A. N. & K. Y. Arakawa. 1968. Sodium hypochlorite and formalin as antiviral agents against nuclear-polyhedrosis virus in larvae of the cabbage looper. J. Invertebr. Pathol. 10: 84-93.

Vail, P. V. & I. M. Hall. 1969. The histopathology of a nuclear polyhidrosis virus or the cabbage looper, *Trichoplusia ni*, related to symptoms and mortality. J. Invertebr. Pathol. 13: 188-198.

Weiser, J. 1966. Nemoci Hmyzo. Academia, Prague, Czechoslovakia.

Weiser, J. 1963. Sporozoan infections, pp. 291-334. *In* E. A. Steinhaus [ed.], Insect pathology, an advanced treatise, vol. 2. Academic Press, New York.

Wilson, G. G. 1974. The use of Fumidil B to suppress the microsoiruduan *Nosema fumiferanae* in stock cultures of the spruce budworm, *Choristoneura fumiferana* (Lepidoptera: Tortricidae). Can. Entomol. 106: 995-996.

## PART FOUR

# Insect Rearing for Pest Management

# 19

## Straggling in Gypsy Moth Production Strains: A Problem Analysis for Developing Research Priorities

### T. M. ODell

### Introduction

Laboratory-reared gypsy moths produced at the U.S. Department of Agriculture, Forest Service (FS) Insect Rearing Facility, Hamden, Conn., and the Animal and Plant Health Inspection Service's (APHIS) Methods Development Center, Otis Air National Guard Base (ANGB), Mass., account for approximately 95% of the gypsy moths used for research, development, and application in the United States.

The New Jersey Standard Strain (NJSS), produced by both facilities and now in its 34th laboratory generation (G34), has become the "white rat" for scientists investigating methods for controlling this forest pest. Until 1985, the NJSS was an excellent production insect; it was relatively disease-free, developmentally predictable, and available year-round. The laboratory-produced insects provided competitive adults for the sterile insect program (Mastro et al. 1989), large larvae for virus production (Shapiro et al. 1981) and developmentally predictable larvae for many diverse scientific investigations. However, an apparent growth abnormality, straggling, characterized by delayed and unpredictable growth of larvae, has significantly reduced the availability of the NJSS, increased production costs, and affected the performance of the insect produced. For example, since 1987, irregular larval development has, for the most part, stopped further development of the F1-sterile male technique and significantly increased costs of producing acre equivalents of nuclear polyhedrosis virus ($40-100/acre compared with $19/acre), which depends on a relatively large, uniform larval size for cost effective production (J. D.

Podgwaite, personal communication). Unpredictable growth has also necessitated a 55% increase in the number of egg masses required for research production and, more importantly, reduced the confidence of scientists who utilize NJSS in their research.

Because of 1) the importance of having a developmentally predictable gypsy moth strain available year round for research and mass production and 2) the long-term complexity of establishing a new strain, it is imperative to provide a definition of the factors that *may* or *may not* be affecting straggling and potentially related growth and viability problems and to develop a comprehensive and systematic approach to solving the problem.

This problem analysis summarizes the historical and biological events that shaped the present gypsy moth rearing system, reviews the recorded occurrence and investigation of straggling, and evaluates rearing protocols in the context of what is known about performance of gypsy moth in natural environments. The analysis will serve as a foundation for discussion and development of an integrated research effort to reestablish the NJSS as a viable research and production strain and/or to develop and maintain a new strain.

## NJSS Rearing History

The NJSS was established in the APHIS Methods Development Center, Otis ANGB, Mass., in 1967, primarily for bioassay of insecticides. Its history through 1980 is summarized by ODell et al. (1984). Major technological changes occurred between 1975 and 1982 when an interagency team, with support from the U.S. Department of Agriculture's Expanded Gypsy Moth Program, began investigating methods to improve mass-rearing technology for gypsy moth (Bell et al. 1981b). In 1980, the FS established an insect rearing facility at the Northeastern Forest Experiment Station's Center for Biological Control of Northeastern Forest Insects and Diseases. Its primary purpose was to produce gypsy moth for research. In the first year, egg masses from the APHIS NJSS G20, 21, 22 (G=generation) were used to establish a production colony. Since then, NJSS-FS colonies have been reared continuously without introduction of wild or other laboratory strain genes, and the rearing technique has been consistent (Moore et al. 1985).

## Straggling: History and Description

Tanner and Weeks (1980) first used the word "straggling" to describe poor larval development. They characterized straggling as "the failure of many larvae to develop past the first or second instar," and reported that between

November 26, 1979, and February 27, 1980, the proportion of NJSS-G20 that failed to pupate within the normal 35-day development period often reached 40-50%. As a result, methods for colony rearing were modified; instead of using newly hatched neonates for weekly colony production, only larvae that had molted to the 2nd instar within 7 days were used. Straggling ceased to be a problem after February 27, 1980, so neonates again were used for colony production.

Straggling apparently was not a problem again until 1985 when Tanner and Baker (1985) reported that "straggling" or "stunting" was a major problem in the APHIS rearing program, with reduction in pupal yields of NJSS-G28 "at times" exceeding 40%. They characterized straggling as the "lack of growth" in some newly hatched larvae and classified egg masses into five groups based on the distribution of larvae in the first four instars 11 days after egg incubation (after eggs were placed on diet). A mean larval stage (MLS) was calculated for a sample from each egg mass. Ten to 15 eggs/egg mass were placed on diet and incubated for 11 days. On day 11, the instar of each larva was estimated (1, 2, 3, etc.). These were summed and divided by the number of larvae in the sample (to give the MLS). Egg masses were classified into 5 MLS groups: 1-1.25, 1.25-1.75, 1.75-2.25, and 2.25-3.00. Samples in the first group (1-1.25) had an abnormal amount of silking, and larvae showed relatively little sign of feeding or growth. Larvae in all other groups had fed (frass visible) and excessive silking was not apparent. Using these characteristics, a selection procedure was developed to eliminate straggling families (egg masses) from colony production. As a result of selection in G27-G30, the proportion of colony egg masses that produced straggling larvae dropped from 8% in generation G28 to 1% in G29 and 0.3% in G30. However, in generation 31, 9% of the colony egg masses had almost 100% straggling larvae. Tanner et al. (1987) concluded that while this selection method cannot select for a "non-straggling" strain, it can be used to select for egg masses that will produce normally developing larvae in the current generation.

Straggling was first detected in the FS NJSS-G32 colony in January 1988 when production of 24-h-old 2nd instars was significantly reduced by the failure of many first-stage larvae to molt within the normal 5-day development period. Subsequently, FS colony performance records were checked to see if significant deviations in larval development could be detected. Each NJSS-FS generation has 37 separately maintained weekly subcolonies. When subcolony pupae were harvested, all remaining larvae were counted and recorded. Examination of the records showed that within each generation the mean subcolony proportion of "larvae remaining" was G24 - 1.4, G25 - 1.2, G26 - 0.9, G27 - 3.6, G28 - 4.2, G29 - 2.5, G30 - 1.9, and G31 - 7.4. The general increase in number of larvae that did not pupate within the normal 34-d development period, beginning in G27, is indicative of an increase in asynchronous development, suggesting that straggling probably was occurring in the FS colony at approximately the same time as in the APHIS colony. By 1988, straggling had reduced production of NJSS-FS-G31 female pupae (=egg masses) by 38%. Production was maintained at normal levels by increasing the number of larvae used for each generation.

The Agricultural Research Service (ARS) established a colony of APHIS-NJSS in Beltsville, Md., in 1984. Irregular development of first-stage larvae has persisted in this colony despite rigorous selection for synchronous development. Egg masses are presampled by hatching one-half of the mass, transferring 100 neonates to diet, and counting the number of first instars 6 days after hatch. Only egg masses from which all larvae have molted to the 2nd instar by day 6 are used to produce the next generation. Nevertheless, asynchrony persists in 15-20% of the egg masses from generation to generation (R. A. Bell, personal communication).

Unpredictable, asynchronous development of first stage larvae is not confined to the NJSS. It has been a problem in production of the Hoy nondiapause strain (Hoy 1977) currently maintained by ARS and FS and in two close-to-wild strains: VIR-FS, produced monthly and now in the 4th laboratory generation, and DH-FS, recently dropped from production. DH-FS was colonized from relatively small egg masses collected from Devil's Hopyard State Park, Haddam, Conn., in February 1982; in 1981 deciduous trees in the area had been completely defoliated. Vir-FS was colonized from large egg masses collected from Matthews Arm Campground, Shenandoah National Park, Virginia, in January 1986; egg-mass density was estimated at 4,000-6,000/acre, but trees were only partially defoliated in 1985. The fact that close-to-wild strains have development problems similar to those of the NJSS suggests that the problem is not the result of long-term laboratory colonization, but may be related to the rearing procedures used uniformly for all strains that appear to be chronically plagued by straggling.

DH-FS appears to have had symptoms of chronic straggling early in the colonization process. Although a relatively large variation in development time from hatch to pupation appears to be characteristic of newly colonized laboratory populations, the variation usually decreases as the population adapts to laboratory conditions; DH-FS development asynchrony increased. In the first year (P1 generation), the average proportion of wild larvae that did not pupate within 36 days (standard harvest date for new colonies) was 8.6%. In subsequent generations the mean proportion of 12 monthly colonies/generation for the variable "larvae remaining" was 16 (G1), 23 (G2), 53 (G3), 48 (G4), 43 (G5), 24 (G6), and 40 (G7). In 1985-86 a study was conducted to determine if straggling—i.e., delayed development of 1st instars—was causing DH's asynchronous development. Neonates from 10 egg masses (DH-FS5) were transferred to 100- by 150-mm plastic petri dishes with diet, 15 larvae/dish, 5 dishes/egg mass. NJSS-FS-G28 neonates from two egg masses were used as controls. The MLS for each egg mass was determined 7 days after hatch and placement on diet. Five DH-FS egg masses had symptoms of straggling. Their mean larval stage was 1.2, 1.4, 1.2, 1.7, and 1.9. Larvae from all other DH-FS and NJSS-FS egg masses were either in the 2nd or 3rd instar. Further studies suggested that the asynchronous development of DH-FS was similar to straggling in NJSS. DH-FS was maintained for two more generations because it consistently produced stragglers and could be valuable in further investigating the problem. However, in 1987 when straggling became relatively consistent in NJSS, DH-FS was discarded.

## Straggling: Quantitative Assessment

Tanner and Baker's (1985) evaluation of NJSS egg mass straggling was based on a group MLS mean measured from date of egg incubation or neonate infest. Although this method provides a means for assessing straggling in production, a quantitative assessment of individual larval performance was required to evaluate treatments for improving colony performance. In 1987, in cooperation with APHIS, the FS initiated studies to quantitatively describe straggling in the context of viability and life history traits, to identify the etiology of straggling and to develop treatment(s). A rearing technique was developed using 59-ml clear plastic cups which allowed daily observation of all stages of individually reared gypsy moth without significantly changing development characteristics associated with normal rearing procedures. While research is still in progress, the following preliminary observations were made for NJSS reared under conditions outlined in the Appendix:

1. "Normal" development time for NJSS, from hatch to 2nd instar was 4.6 ± .5 days for males and 4.4 ± .6 days for females. First instars that take longer than 6 days to molt will not pupate within the normal 34-day production period.
2. The proportion of 1st instars that exceed normal development time is egg-mass specific and appears to be constant over the hatch period.
3. The severity of straggling within an egg mass might be measured by calculating the mean and standard deviation for the number of days from hatch to second instar, or hatch to pupation. For example, in a severely straggling family the number of days from hatch to 2nd instar/pupation was 10.0 ± 1.9/51 ± 3.2 for males and 8.7 ± 1.6/53.6 ± 9.1 for females.
4. Within 48 h after hatch nonstragglers constructed a feeding mat on the side of the container. Stragglers usually did not construct a feeding mat, but wandered around the cup leaving irregular silk trails, often resting on the top of the cup.

5. The pupal weight of stragglers was significantly less than that of nonstragglers.

## Examination and Analysis of
## Rearing Procedures and Natural Life History Traits

The terms "stunting," "prolonged development," "development delay," "slow growth," and "growth inhibition" all have been used to describe effects of environment (Leonard 1966, ODell et al. 1984), diet ingredients (Moore 1985, Singh 1977), microbial contamination (Raulston & King 1984), and genetics (Riddle et al. 1985), on larval development. These examples suggest that the etiology of straggling and related deficiencies may be linked to one or a combination of any of these broad areas and indicate a need to identify specific potential causes and develop a systematic approach for investigating them.

Rearing protocols for sustaining gypsy moth production on a year-round basis have remained relatively constant since 1980 (Bell et al. 1981b). In the following sections, these protocols are examined as they relate to environment, diet, microbial control, and strain genetics, and they are also analyzed relative to their potential effect on straggling and related strain deficiencies.

### *Environment*

For this section, the rearing process has been divided into four units: egg incubation and hatch, larval development and pupation, adult eclosion, mating, oviposition and egg embryonation, and egg diapause.

**Egg Incubation and Hatch.** At the FS Insect Rearing Facility, egg masses that have completed diapause are washed in 10% formalin (see Pathology Section), placed individually in plastic petri dishes, and held in a covered, clear plastic box at 25°C and 16:8 (L:D). Relative humidity in the box is maintained at 90-95% by keeping water in the bottom; petri dishes with egg masses are placed on a screen above the water. Plastic bags containing moist paper towels and cabinets maintained at a high relative humidity also have been used effectively. A similar procedure has been used by APHIS.

In 1982, responding to a need to improve efficiency in production, APHIS began automated infest of dehaired eggs directly onto the diet. Tanner and Weeks (1981b) reported no significant reduction in hatch of NJSS-G21 eggs in contact with the diet (75.9 ± 3.3% versus 82.0 ± 3.6%) and there was no apparent effect on larval development. The egg-infest technique was utilized for colony production for the next four generations, G24 (1982)-G27 (1985). In

1985, straggling made it necessary (selection for nonstragglers) to return to neonate infest.

While APHIS has changed infest methods (and thus the egg incubation and hatch environment several times), the FS technique has remained the same—i.e., eggs are incubated in water boxes at 25°C and neonates are transferred to diet 3 days after initiation of hatch. Since the incidence of straggling in NJSS-G32 appears similar at the APHIS and FS facilities, it does not appear that the method of infest, eggs or neonates, affects the generation-to-generation occurrence of straggling.

Capinera et al. (1977) found that incubation of wild eggs at 10°C and 75% relative humidity for 6 weeks, following diapause under natural conditions, caused depletion of yolk and resulted in only 10% hatch. When these results were compared to eggs held at -10°C and 5°C, Capinera et al. (1977) suggested that exposure to abnormally high temperatures preceding hatch might affect the development and behavior of surviving larvae. NJSS eggs are held at 7-8°C through the 170-180 day diapause period. When removed from diapause conditions and incubated at 25°C and 90-95% relative humidity, 70-80% eclosion occurs within 5 days. While this period is less than 1/6 of the 6 week incubation period reported by Capinera et al. (1977), the degree-days for 6 weeks before NJSS eclosion is approximately the same. Using 5°C as a developmental threshold (see Mithat, 1933; *Lymantria dispar* eggs do not develop at temperatures below 6°), a 6-week, 10°C incubation period would have 210 degree days (5°x 42 days). In a similar 6-week period, NJSS is exposed to 8°C for 37 days (diapause conditions) and 25°C for 5 days (incubation period) which equals 211 degree-days (3°x 37 days + 20°x 5 days).

In the Northeast, incubation of wild gypsy moth eggs often occurs when mean daily temperatures are fluctuating between 5°C and 15°C (for example, in Connecticut, 1988). This indicates that while diurnal temperatures may be between 20°C and 25°C for short periods of time, nocturnal temperatures may be below the developmental threshold. Results of Capinera et al. (1977) suggest the departure from natural incubation temperatures may induce depletion of yolk content of eggs and adversely effect larval viability, development, and behavior.

**Larval Development and Pupation.** Review of environmental records indicates that since 1980 temperature (25°C), relative humidity (50-60%), photoperiod (16:8 L:D), and chamber airspeed (3-5 cfm) have remained relatively stable in APHIS and FS rearing chambers, suggesting that short-term environmental affects probably are not directly effecting straggling. Potential long-term effects caused by static microenvironment have not been investigated, although observed straggling in two close-to-wild strains would seem to rule out this possibility.

Small changes in average temperature or humidity over a 24-h period can significantly affect rearing schedules by changing molting synchrony (Moore et al. 1985), but when these changes occur, NJSS larvae usually respond uniformly within each respective container. Straggling is characterized by erratic growth of siblings within the same container.

The microclimate conditions within the container used to rear insects can affect their growth, development, and survival (Burton & Perkins 1984). Bell et al. (1977) found that the type of rearing container and age of diet seemed to influence incidence of malformed pupae. Generally, the effects of containers on gypsy moth growth and viability are associated with moisture retention. Excessive moisture increases microbial growth; excessive desiccation results in an unacceptably dry diet. For established insect colonies these effects usually are container specific.

In 1987, siblings from 22 NJSS egg masses were reared in three types of containers for 11 days to compare incidence of straggling between containers and egg masses. Six of the 22 egg masses produced stragglers. The mean larval stage of insects from each respective egg mass was approximately the same for all three container types indicating that, within a generation, straggling was related to parentage and not directly influenced by the container (Table 1).

Parental experience relative to container type or multiple versus individual rearing may have some effect on sibling performance, although the effect probably would be diet mediated rather than direct. For example, moisture loss from the 59-ml containers used to rear larvae individually is less than loss from the standard 177.4-ml fluted cup used for colony and mass production. Drying is effected by higher moisture loss through the paper lid on the 177.4-ml cup, as compared to the punctures in the plastic cap of the 59-ml cup. The effect of moisture on larval growth and viability has not been specifically studied, but the generation-to-generation effects of rearing in 59-ml versus 177.4-ml containers is currently being investigated.

Pupation within containers usually occurs on the underside of the cap; the last instar constructs a silk harness in which it pupates. Pupation may also occur on the diet, or pupae may drop onto the diet. At normal rearing conditions, the time from pupation to adult emergence is about 10-12 days. When pupae are harvested from rearing containers, adult eclosion may have started so that within 5 days most adults have emerged. During metamorphosis there are significant changes in weight, lipid, and water content of holometabolous insects that do not feed in the imaginal stage. Gere (1964) addressed some aspects of these changes in gypsy moth relative to chemical and energetic changes during metamorphosis. Rate of respiration, as influenced by the microclimate of the container and larval density, could influence these changes and the availability of fat reserves during egg maturation. If the etiology of straggling is related to the quantity or quality of egg proteins available to the developing embryo, the effect of container on respiration rate during metamorphosis may be significant.

Adult Eclosion, Mating, Oviposition, and Egg Embryonation. When colony pupae are harvested they are maintained at normal rearing conditions in manila mating containers lined with paper (oviposition medium). Adult eclosion, mating, oviposition, and egg embryonation occur virtually uninterrupted over the next 40 days, at which time eggs are harvested and transferred to diapause conditions. NJSS-FS-colony pupae are held in .95-liter containers, 10 males and 10 females per container. APHIS uses 3.8-liter containers, 25 males and 25 females per container. No specific studies have been conducted to determine if the eclosion and mating environment affect the mating behavior or reproductive biology, and whether or not there is an effect on egg and progeny viability, growth, and behavior. In flight cages, NJSS and wild males mated up to 8 times a day but averaged 3.4 and 2.8, respectively.

Table 1. Mean larval stage 11 days after hatch for siblings reared in three container types

| Egg mass[1] | Petri dish[2] | Fluted cup[3] (177.4 ml) | Clear cup[4] (59 ml) |
|---|---|---|---|
| 1 | 1.56 ± 0.10 | 1.38 ± 0.06 | 1.45 ± 0.07 |
| 2 | 1.66 ± 0.08 | 1.74 ± 0.09 | 1.66 ± 0.05 |
| 3 | 2.12 ± 0.05 | 2.23 ± 0.10 | 2.02 ± 0.07 |
| 4 | 1.29 ± 0.12 | 1.11 ± 0.06 | 1.25 ± 0.10 |
| 5 | 2.10 ± 0.04 | 2.02 ± 0.11 | 1.98 ± 0.09 |
| 6 | 2.92 ± 0.09 | 2.58 ± 0.08 | 2.78 ± 0.08 |
| 7 | 3.82 ± 0.04 | 3.62 ± 0.04 | 4.02 ± 0.03 |
| 8 | 3.33 ± 0.05 | 3.10 ± 0.06 | 3.22 ± 0.04 |
| 9 | 3.25 ± 0.07 | 3.31 ± 0.08 | 3.12 ± 0.06 |
| 10 | 4.17 ± 0.08 | 3.97 ± 0.05 | 3.85 ± 0.08 |

[1] Cohorts from egg masses 1-6 still had 1st instars (stragglers) 10 days after hatch. Larvae from egg masses 7-10 developed normally.

[2] 100 x 150 mm plastic petri dish, 50 larvae/dish, 3 dishes/egg mass. Each dish contained 4 cubes of standard diet. Second instars were transferred to another petri dish with fresh diet, 25 seconds/dish.

[3] 177.4-ml fluted plastic cup, 8 larvae/cup, 5 cups/egg mass. Each contained 80 ml of standard diet. Cups were not opened until day 11.

[4] 59-ml clear plastic cup, 1 larva/cup, 30 cups/egg mass. Each cup contained 15 ml of standard diet. The stage of each insect was recorded daily.

All egg masses successfully embryonated, but their viability in the next generation was not determined (T.M.O., manuscript in preparation). In colony mating containers, multiple mating is probably a common occurrence, its incidence mediated in part by the synchrony of eclosion of males and females. The effects of multiple mating by either male or female gypsy moth on progeny performance is unknown.

Adults derived from straggling larvae often exhibit morphological and behavioral abnormalities. Newly emerged adults are sometimes smaller than normal and have been observed with deformed wings and nonfunctioning tarsi that have made mating difficult or impossible; small and sometimes scattered egg masses and low numbers of hatching larvae indicate other unidentified problems in the reproductive process. But, because little is known of what constitutes normal internal reproductive morphology and sexual biology of the gypsy moth, it is difficult to recognize important departures from the norm or to characterize or assess the cause of any of the abnormalities. This, in turn, makes it difficult to postulate the potential contribution of reproductive difficulties to the occurrence of straggling.

After courtship, the act of procreation among insects is divided into four distinct stages: copulation, insemination, fertilization, and oviposition. All deal with the production and movement of gametes and associated reproductive substances. An inability to generate or properly move reproductive material at any point in the four stages, or the presence of abnormal reproductive structures or behavior, can lead to a breakdown in the production of normal offspring. It has been observed for Lepidoptera that the more intricate the process of spermatophore transfer becomes, or the more complex the genitalia, the greater the number of unsuccessful matings that will occur (Callahan & Chapman 1960).

To identify possible points of disruption in the reproductive process that might impact on neonate production and affect larval growth, it is necessary to have knowledge of both the basic sexual morphology and the reproductive biology of the insect. Development of the internal reproductive system and the anatomy of both gypsy moth sexes has been described (Goldschmidt 1934, Behrenz 1952, Hollander et al. 1982), and the movement and maintenance of gametes has been studied to some extent in males (Rule et al. 1965, Maksimovic 1971, Giebultowicz 1988). Studies comparing the reproductive morphology and the movement of male and female reproductive material through the internal system in wild and laboratory-reared females is needed to assess questions raised in studies of the straggling phenomenon.

Most of the eclosion, mating, and egg laying by NJSS occurs within 7 days of pupal harvest. This means that the majority of eggs are held at 25°C and 50-60% relative humidity for approximately 35 days. This standard treatment of the embryonating eggs is different from the optimum conditions described by others and could reduce egg viability and perhaps affect larval development.

In their summary, Giese and Cittadino (1977) stated that it takes 2-4 weeks for embryonation to take place. Tanner and Kennedy (1975) found that 90% embryonation was observed after 21 days at 25°C with 70% hatch after 150 days at 4.4°C. Forrester and Bell (1978) showed that the optimum prediapause period for NJSS under laboratory conditions (26°C, 50-60% relative humidity) was approximately 21 days. Their study showed that after this time there was a significant increase in weight loss and an associated decrease in percent hatch if eggs were not transferred to the diapause chamber (7-8°C). The life history of the Hoy nondiapause strain also indicates that embryonation is, for the most part, completed in 21 days. Egg hatch begins approximately 21 d after oviposition and continues for approximately 2 weeks (R. A. Bell, personal communication).

Tanner et al. (1988) recently found that there was no significant difference in incidence of straggling (number of egg masses with stragglers) between cohorts of NJSS-APHIS-G33 egg masses held at 25°C and 50-60% relative humidity for 21, 28, 35, 42 and 49 days after oviposition and then transferred to a diapause holding chamber (7-8°C) for 170 days. Percent hatch of these egg masses varied positively with incidence of straggling and not with prediapause conditions (Tanner et al. 1988).

Temperature and humidity can affect the viability of eggs during the embryonation period. Tanner et al. (1987) found that a relative humidity of 80% increased the percent hatch of F1-sterile eggs and extended the period of time eggs could be maintained at 25°C. Capinera et al. (1977) found that embryonating wild gypsy moth eggs at 27°C, as compared with 15°C (at 75% relative humidity), significantly reduced the egg yolk content over the 6-week period following oviposition. In the northeastern United States, wild eggs are deposited from July to August in relatively moist microhabitats and are exposed to fluctuating temperatures that seldom exceed a daily mean of 20°C. The present embryonation condition for the NJSS and other laboratory strains is in direct conflict with natural conditions and it appears from previous studies that the consequent reduction in egg quality (yolk content) during this period may predispose the larva to early death (does not hatch) or abnormal growth and development after hatch.

**Egg Diapause.** Since 1980, NJSS (APHIS and FS) eggs have been held in diapause for 170-180 days at 7-8°C or 5°C (Odell et al. 1985, Tanner & Weeks 1980). The 180-day period resulted in quicker, more uniform hatch than the previous 120-day chill period (Tanner & Weeks 1980). At the APHIS facility, eggs are held either in walk-in coolers maintained at 7-8°C or 5°C. Records through 1983 indicate no difference in percent hatch of NJSS eggs chilled for 120, 150, or 180 days in the 7-8°C chamber (92.9 ± 5%, 92.0 ± 3.7%, 91.5 ± 4.6%, respectively), whereas in the 5°C chamber hatch decreased (from 90.9

± 7.2% hatch of 120-day eggs to 69.9 ± 13.6% in 180-day eggs) (from Tanner & Weeks 1981a). Tests by Tanner et al. (1985) and Tanner (personal communication) indicate that percent hatch of NJSS-APHIS-G33 eggs declined when eggs were chilled more than 150 days at 7-8°C. In their review of gypsy moth diapause literature, Giese and Cittadino (1977) indicate that percent hatch is greatest when eggs are held at constant temperatures within the 0-5°C range for 120-150 days. Tanner and Kennedy (1976a), using NJSS-G12 egg masses, reported 85% hatch after 98 days of chill with a maximum of 92% after 155 days. Hatch declined after 168 days of chill. The results of these studies indicate that the standard method (170-180 days 7-8°C) may contribute to the decline in neonate vigor and initial feeding success (Cothran & Gyrisco 1966) by promoting excessive metabolic use of egg proteins, particularly during the post-diapause phase—i.e., after 90-120 days (see Giese & Cittadino 1977).

The gypsy moth diapauses as a mature embryo. The quantity of yolk does not change during diapause, but after the required diapause period the yolk within the gut is absorbed when temperature is appropriately elevated. When the gut contents are absorbed the reserve egg yolk is consumed (Capinera et al. 1977). Capinera et al. (1977) found that yolk content did not change significantly during the diapause period when eggs were held at 27°C or 15°C, but during the pre-eclosion period, after diapause, these elevated temperatures significantly reduced egg yolk content. The present diapause holding conditions (7-8°C, 170 days) could reduce egg quality and adversely effect the performance of the NJSS larvae; Capinera et al. (1977) state, "We suspect that high temperature may convert large, yolk-rich eggs to the equivalent of small, yolk-deficient eggs. This could have dramatic effects on the development and behavior of larvae."

### Diet

Parameters for development of optimal insect diets include preparation, chemical, and physical factors that stimulate diet acceptance, nutritional adequacy, gelling and bulking agents, and microbial inhibitors. The usual criteria for evaluating these factors are the rate of development to a particular stage, size (weight) attained, fecundity, survival, and behavior (Brewer & Lindig 1984, House 1960, Moore 1985, Singh 1977). Bell et al. (1981b), Odell et al. (1984), and Moore et al. (1985) discuss gypsy moth dietetics and provide some guidelines for achieving optimal growth and survival. However, as with most discussions on insect diets, optimal growth and survival are measured on a generation basis and not generation to generation. Long-term effects of specific diet ingredients generally are not available.

Changes in diet ingredients or in diet preparation could, conceivably, affect larval viability and development. For example, Tanner et al. (1983) showed that allowing the diet to dry 2-3 h after pouring instead of 1 h significantly reduced larval development, measured as mean larval stage on days 7, 14, and 21 after egg infest. Straggling is now common in ARS, APHIS, and FS rearing facilities, each having quite different diet preparation equipment and schedules, thus diet preparation does not appear to be a significant factor.

Investigating potential for change in ingredient quality and evaluation of diet quality relative to straggling is more complex. The majority of gypsy moths produced in the United States for research and application are reared on the high wheat germ diet developed by Bell et al. (1981b). While there are two commercial suppliers of premixed gypsy moth diets (BioServ and ICN[1]), the FS buys ingredients in bulk and stores them for a maximum of 6 months in a freezer at -10°C. In 1989 APHIS changed from this technique to storage at 5°C. Although prepared diet is regularly incubated in rearing chambers for detection of microbial contamination, responsibility for ingredient quality is left to the supplier. Salts (Wesson), casein, and vitamins are purchased as pharmaceutical-grade products. The quality of the wheat germ is a special problem because it is relatively unrefined (Stewart 1984). It is the most variable ingredient in the gypsy moth diet. Variation in protein and oil content is caused by weather conditions and soil type where the wheat is grown and by the milling process. Wheat starch and bran in the wheat germ proportionately decrease the concentration of usable nutrients. Instability in the wheat germ caused by prolonged storage produces even more variability in its quality (Karmas 1975). The FS and APHIS buy wheat germ directly from Mennel Milling Company, Fostoria, Ohio. Their quality control department tests for moisture, total protein and fat, and purity to maintain specifications within particular limits. Wheat germ is a by-product of the milling process for production of flour, so no additives are required. Mennel Milling Company consistently has purchased their wheat germ from the same grain elevators that store wheat grown in northwestern Ohio. In the last 3-4 years, new harvesting and storage techniques have been implemented to reduce the use of chemicals to control insects and pathogens. This has reduced the ethylene dibromide and carbontetrachloride residue on wheat germ from 300-400 ppm to barely detectable levels. This improvement in wheat germ purity occurred at approximately the same time that straggling became a problem.

Agar is another highly variable ingredient used in relatively large quantities and could affect larval growth and survival. The APHIS Pink Bollworm Rearing Facility, Phoenix, Ariz., developed specific requirements for purchasing agar. The facility requires agar of high gel strength, low viscosity at 65°C, and a high water-holding capacity after shredding. NJSS has been reared on diet containing agar with the same specifications since 1977. Since 1987, the FS has purchased four lots of agar from Moorehead and Company, Calif. the APHIS supplier. The first lot, made from seaweed of the genus *Gelidium*, had a 10-15% lower gel strength than the standard APHIS agar (from the genus

*Gracileria*). The gypsy moth colonies reared on diet made with Gelidium failed
to pupate primarily because a thick gel crust prevented feeding. The last three
agar lots purchased were made from *Gracileria* that, according to the supplier,
came from a European manufacturer. Recent tests by T. Forrester (unpublished
data, on file at APHIS Methods Development Center, Otis ANGB, Mass.)
suggest that *Gracileria* agar (Moorehead), at the rate generally used (1% dry
weight), still provides optimal development for gypsy moth as measured by time
to pupation and pupal weight.

Finally, the age and/or deterioration of diet during the development period
needs to be considered. Straggling in NJSS-APHIS apparently was not a
problem (at least not a detectable one) until the 20th laboratory generation. In
1975 (NJSS-APHIS-G14, 15) the diet formula was changed from a modification
of the ODell and Rollinson (1966) diet to a high wheat germ diet (Bell et al.
1981b). As in the past, larvae were provided fresh food when needed; when the
high wheat germ diet was introduced, larvae had to be moved to fresh diet only
when they molted to the 4th instar, approximately 21 days after hatch. In 1978
(NJSS-APHIS-G18, 19), changes in environment, microbial control, diet
preparation, and containers allowed larvae to remain on the same diet (no
change) without significantly affecting development, survival, and reproduction.
Tanner and Buck (1979) found that females reared on high wheat germ diet with
no transfer to fresh diet were significantly heavier and laid more eggs than those
transferred to modified hornworm diet after 21 days. The no-transfer, high
wheat germ diet is now the standard for rearing NJSS and other colonized gypsy
moth strains ( Bell et al. 1981b, Odell et al. 1984).

In 1979-80, shortly after initiation of the 35-day diet schedule, straggling
became a problem in the NJSS-APHIS-G20. It only remained a problem for a
few months and was not detected again until 1985 (NJSS-APHIS-G28). While
the occurrence of straggling has remained unpredictable, it has been chronic in
FS, APHIS, and ARS colonies since 1985. Diet deterioration during the larval
development period could reduce the nutrients available to the female during
periods when regulation of yolk protein synthesis is critical (Bownes 1986).
Reduction in normal levels of egg proteins or other nutrients could theoretically
reduce hatch, larval viability and growth, and alter larval behavior in subsequent
generations.

## Pathology

Routinely, steps are taken to combat pathogens. Eggs that have completed
diapause are soaked *en masse* in 10% formalin for 1 h, rinsed for 1 h in running
water, and dried in a biological safety cabinet for 2-3 h. Formalin is a
commonly used egg disinfectant for controlling microbial contaminants and is
particularly effective against virus (Shapiro 1984), but reduced hatch after

formaldehyde treatment of insect eggs has been reported (David et al. 1972, Howell 1970). Certain concentrations of formalin in diet can also increase the duration of larval and pupal stages (Ouye 1962). These adverse effects of formalin suggest that formalin treatment of gypsy moth egg masses reduces egg viability and effects change in larval development if formalin is ingested during hatch (i.e., residue on chorion). Tanner et al. (1982) reported that formalin treatment significantly reduced hatch of NJSS-G23 egg masses. However, recent tests of split egg masses--i.e., half of each egg mass treated with formalin, the other half only washed--showed no significant differences in egg hatch and larval development (M. A. Keena, personal communication). Long-term effects have not been investigated.

Microbial inhibitors are essential ingredients of larval diets if high-quality insects are to be produced, but the toxicity of the inhibitors can be detrimental, thereby making insect production numbers unpredictable (Kishaba et al. 1968). Bell et al. (1981a) tested four antimicrobial agents that are routinely incorporated into lepidopterous larval diets for bacterial and fungal inhibition. The agents tested in all possible combinations were benomyl, chlortetracycline, sorbic acid, and methyl P-hydroxybenzoate. Combinations that contained benomyl and sorbic acid were most effective and did not inhibit insect development. Shapiro (1984) reviewed and discussed methods for controlling microbial contamination including antimicrobial treatment of diet. In tests of several antimicrobials for control of bacteria and molds in gypsy moth artificial diets, methyl-paraben (methyl P-hydroxybenzoate) and sorbic acid were the most effective antifungal compounds; when used together they prevented fungal growth (Shapiro 1984).

The standard high wheat germ diet used by FS, APHIS, and ARS (Bell et al. 1981a) has 1% sorbic acid and 2% methyl-paraben by dry weight. While these are considered relatively safe levels—i.e., above these levels insect growth is significantly retarded—below these levels diets are likely to become contaminated. Singh and Bucher (1971) state that concentrations that are innocuous to insects will not suppress mold growth. This suggests that the tolerance of the insect to levels that suppress microbials may effect long-term changes in the life cycle. For example, at certain concentrations, methylparaben and sorbic acid may be ovicidal (Singh & Bucher 1971). Since diet investigations usually are confined to within-generation effects, there are no studies to indicate effects of methylparaben and sorbic acid on progeny of parents reared on diet containing various concentrations of these compounds.

Early diets for gypsy moth included antibiotics such as aureomycin (ODell & Rollinson 1966), chlorotetracycline, and streptomycin (T.M.O., unpublished data, on file at the U.S. Forest Service Laboratory, Hamden, Conn.). Antibiotics by themselves do not appear to be particularly effective in preventing diet and insect contamination by molds and fungi (Bell et al. 1981a). Antibiotics are effective inhibitors of intracellular microorganisms

(Rickettsia-like organisms [RLO's]) in insects (Weiss 1973), tetracycline (Keller et al. 1981, Yen & Barr 1977).

Before 1975 (NJSS-G15), aureomycin was the only antibiotic used to control bacterial infections in NJSS diet (Bell et al. 1976, 1981a; ODell & Rollinson 1966). Bell et al. (1981b) developed a less complex gypsy moth diet that used only sorbic acid and methyl paraben to control microbial contamination. While aureomycin was not intended to control intracellular microorganisms, it inadvertently may have suppressed them. Conversely, the absence of an antibiotic may allow their pathogenic expression--e.g., severe stunting (Henry et al. 1986 Keller et al. 1972).

In the last 5 years, an intracellular pathogen in the order Ricketsiales has been tentatively identified in gypsy moth larvae sampled from rearing cultures, including NJSS-G27, 28, and 29 from the APHIS and the FS, and a close-to-wild strain, DH-FS-G5. This RLO has also been found in samples from at least six wild populations, including samples from outbreak (Maryland), low-density (Vermont), and moderate-increasing (Connecticut) gypsy moth populations (all samples sent to Dr. Jean Adams, ARS, by T.M. ODell in 1986 and 1987). Adams et al. (1982) described RLO's (=chlamydial-like organisms) from stunted cabbage loopers (*T. ni*). and cotton budworm (*Heliothis virescens*) and noted that inoculation of 1-day-old larvae with RLO's caused severe stunting and larvae failed to reach pupation. However, these observations have not been verified. Although there is a great interest in elucidating the importance of the RLO, few studies have been initiated because of the fastidious nature of the pathogen, chronic rather than acute infections, and a general lack of information regarding RLO pathogens in insects.

The organisms most closely related to the gypsy moth RLO, the Wolbachieae, are described as either pathogenic (*Rickettsiella*) or as not obviously pathogenic or beneficial to their hosts, although in most cases they interfere with reproduction (*Wolbachia*). Some *Wolbachia* have been found in eggs, sperm, ovaries, and associated epithelium of Lepidoptera and Diptera but rarely in other tissues (Suitor 1964). *Wolbachia* strains may differ with the geographical range of their hosts and can render their hosts sexually incompatible when the *Wolbachia* strains are different. Incompatibility is eliminated when the insects are rendered aposymbiotic by treating their diet with tetracycline (Keller et al. 1981, Yen & Barr 1974).

The described *Ricketsiella* species are more obvious pathogens capable of causing severe stunting or death of inoculated insects (Federici 1980, Henry et al. 1986, Keller et al. 1972). Some of the Ricketsiella are highly infectious (Dutky 1959). Like other *Ricketsiales, Wolbachia* and *Ricketsiella* species are inhibited by antibiotics, particularly tetracycline, erythromycin, penicillin, and p-aminobenzoic acid (Weiss 1973).

Degeneration of intracellular microorganisms can be accomplished by dietary treatments with 17-50 µg/ml oxytetracycline or by rearing larvae at 32-33°C for 5-7 days (Weiss et al. 1984). Recent tests show that a concentration of 32 µg/ml oxytetracycline in gypsy moth diet exhibited wide zones of bacterial inhibition throughout the 34-day insect development period but had minimal effect on NJSS-FS larval development and survival. However, when this same amended diet was fed to gypsy moth larvae hatching from wild egg masses, 48% died in the 1st instar. Observation by light microscopy of homogenates prepared from these dead larvae showed no evidence for the presence of bacteria or polyhedral viruses. Culture studies and analysis of homogenates of dead larvae and frass showed evidence of bacterial contamination in only 3 dead larvae out of 94 (J. Rho & T. M. O., unpublished data). The difference in survival between NJSS-FS and wild neonates to oxytetracycline-amended diets indicates a tolerance that is significantly different and suggests that further tests be conducted to clarify cause and effect (lack of feeding, inhibition of protein synthesis, and mortality of potential stragglers).

In 1988, larvae from predetermined straggling and nonstraggling families were reared individually (see Appendix) on the standard diet or the standard diet amended with 32 µg/ml oxytetracycline. Bioassays that evaluated treatments are presently being analyzed. If larvae from straggling families reared on antibiotic-amended diet do not straggle but siblings reared on standard diet do, intracellular microorganisms are probably the cause, although that will not be positively confirmed. Further investigations to confirm results could be carried out more rapidly with the recently established Hoy-FS non-diapause strain (Hoy 1977).

## Genetics

While gypsy moth has been in culture for many years, very little genetic research has been conducted. The early work of Goldschmidt (1934) still represents the state of knowledge of basic gypsy moth genetics. Hoy (1977) was able to select for a nondiapausing strain in eight generations, which demonstrates that selection for desired traits is feasible. The electrophoretic data of Harrison et al. (1983) demonstrated that genetic variation in field populations of gypsy moth in the United States was extremely low for the enzymes tested, but ODell and Harrison (unpublished data) found differences in developmental rate and egg hatch phenology in North American populations. Significant differences were found between several different strains indicating that an analysis for genetic variability in egg hatch phenology between geographic populations has a high probability of detecting such variability.

Good guidelines are available on techniques to establish and monitor genetic diversity in insect colonies (King & Leppla 1984, Singh & Moore 1985). ODell et al. (1984) have described rearing techniques specific to gypsy moth and

Moore et al. (1985) have described quality control techniques for measuring colony performance.

The use of laboratory reared gypsy moth for research and application requires insects that are uniform and predictable relative to production requirements but have enough genetic variability to: 1) be useful predictors in bioassays 2) be appropriate hosts for production of natural enemies (virus, parasitoids, and predators) and 3) be competitive for sterile insect release. Genetic evaluation of wild and laboratory strains for traits important to these objectives is an essential part of most mass rearing but has never been incorporated into the gypsy moth program. The present rearing program does not evaluate criteria for any of these objectives.

Inbreeding depression and selection for laboratory ecotypes are frequent concerns among researchers who maintain insect colonies in the laboratory (Mackauer 1976). Bartlett (1984, 1985), Collins (1984), and Joslyn (1984) reviewed and discussed genetic considerations for establishment and maintenance of insect colonies in order to provide methodology for minimizing these concerns. However, they do not address rearing effects on specific life-history traits, such as instar duration. Instead, they suggest that, unless rearing techniques are developed to maintain genetic variability, inadvertent selection may result in deleterious characteristics.

NJSS has been maintained in continuous culture since 1967 and is now in its 34th generation. There has been no planned strategy for maintaining genetic variability, and behavioral and development changes have inadvertently occurred. Some of these changes include altered response to natural host defensive mechanisms, females that pupate after five instars (90% of laboratory *versus* 2-3% of wild populations), and reduction in generation time (at 25°C, laboratory colonies of males take 43 d, females take 53; wild males take 46, females 54; T.M.O., unpublished data, on file at the U.S. Forest Service Laboratory, Hamden, Conn.).

Scientists in all three USDA rearing facilities (APHIS, ARS, FS) select normally developing larvae for production of colony egg masses with the objective of producing normal progeny in the next generation. These attempts to reduce or eliminate straggling and related strain deficiencies in subsequent generations have been unsuccessful. In addition, as noted previously, similar abnormal larval growth has been documented in two established close-to-wild strains and a non-diapause strain, effectively eliminating immediate replacement of the standard strain.

In 1989, FS scientists began a research program to address the problem of strain improvement for specific objectives (i.e., research, sterile male, and NPV), and the development of technology to maintain these strains. These efforts should establish the genetic etiology of rearing problems and information to correct them.

## Conclusions

The environmental conditions under which NJSS and other strains are currently reared could effect qualitative changes, which are expressed as straggling and reduced hatch. The evidence presented suggests that temperature and humidity conditions during embryonation, diapause, and incubation accelerate depletion of egg nutrient reserves that may negatively affect egg and larval viability and/or induce qualitative shifts in larval growth and behavior.

Leonard (1970) and Capinera et al. (1977) have discussed the potential for qualitative shifts in gypsy moth populations due to environmental and dietary stress. They found that the reduction in nutrient reserves available in the egg is associated with both environmental and dietary stress, noting that both events occur during population outbreaks. The two close-to-wild laboratory strains which have expressed straggling came from outbreak populations; the DH-FS from a year-after-defoliation population in Connecticut, the VIR-FS from a 4,000 to 6,000 egg mass/acre population in Virginia (Shenandoah National Park). It is probable that eggs from both populations were "undernourished" relative to nutrient reserves, and that laboratory colonization using the standard techniques may have accelerated expression of an on-going qualitative change. Because the adult female gypsy moth does not feed, the food available to the female larva controls the nutritional supply for oocytes and embryo development. Darkening of the high wheat germ diet is noticeable within 10 days after it is made. The change in diet color is particularly accentuated 3-4 weeks after the gypsy moth has been feeding on it, or about the time the female molts to the last instar. The darkening of the diet is the result of oxidation, probably through the action of enzymes. It is conceivable that the diet available to the female, particularly during the last instar, is nutritionally deficient for optimum egg maturation.

Intracellular microorganisms—e.g., RLO's—have been implicated as affecting the expression of straggling in laboratory reared colonies of gypsy moth. However, to date, there is no conclusive evidence to support this. RLO's have been found in all wild and laboratory reared insects, but there is no correlation of organism density with straggling or nonstraggling insects. On-going experiments to eliminate or reduce the expression of intracellular microorganisms using antibiotics will help clarify the possible importance of RLO's.

The present system for mating laboratory reared populations likely contributes to an overall decline in genetic variability and population quality. The contribution of mating environment, multiple matings, and mating disruption on gypsy moth reproductive biology is completely unknown but, based on research with other insects, it could significantly affect the viability and long term quality of gypsy moth laboratory cultures as well as F1-sterile egg production.

Routine quality-control monitoring of gypsy moth life-history traits has been an on-going process in both FS and APHIS rearing facilities. These measurements are used to discern trends in production quantity and quality and should provide an early warning system for changes in colony viability. The FS rearing system uses control charts for weekly assessment of critical performance traits such as percent hatch, percent pupation, number of larvae remaining, and percent females that lay eggs. In addition, a computerized indexing system has been recently developed that utilizes means and ranges to assess trends in performance traits. However, these quality control techniques were unable to detect the subtle increase in variation in percent pupation or number of larvae remaining, which would have indicated straggling, and investigation through established feedback loops failed to give any clue to cause or origin. It was necessary, then, to analyze the rearing system in a way that would provide information one could use to develop testable hypotheses for identifying cause and eliminating effect. That was the objective of an earlier draft of this paper and it was successful; in a 1989 workshop convened to develop research recommendations for investigating straggling, 50 federal, state, and university scientists identified and prioritized 12 problem areas and 42 specific studies using the information and format of the Problem Analysis.

The process of systematically examining and evaluating rearing procedures used in this problem analysis could serve as a model for problem solving in any rearing program regardless of scale. Optimally, this process should be used in the initial development of a rearing system, periodically updated to reflect new information about the insect's life history traits in the laboratory and in natural environments, and used to identify studies needed to evaluate proposed changes in the rearing process. The objective would be to avoid crisis management problems such as straggling as much as possible but to have in place the information (problem analysis) necessary to immediately initiate studies to solve them.

## Acknowledgments

The contributions of Raymond Pupedis and Jinnque Rho are greatly appreciated. Thanks also to Carol Butt ODell, FS Insectary Manager, and John Tanner, APHIS, for their continued support for NJSS improvement and their willingness to share information. Additionally, thanks to Kathleen Shields, Melody Keena, and Abdul Hamid for their constructive comments on an earlier draft of this manuscript.

## References

Adams, J. R., C. C. Beegle & C. J. Tompkins. 1982. A chlamydial-like organism isolated from insects in insect mass rearing facilities. 40th Ann. Proc. Election Microscopy Soc. Amer., Washington, D.C.

Bartlett, A. C. 1984. Genetic changes during insect domestication, pp. 2-8. *In* E. G. King & N. C. Leppla [eds.], Advances and challenges in insect rearing. Agric. Res. Service, USDA.

Bartlett, A. C. 1985. Guidelines for genetic diversity in laboratory colony establishment and maintenance, pp. 7-17. *In* P. Singh & R. F. Moore [eds.], Handbook of insect rearing, vol. 1. Elsevier Press, Amsterdam.

Behrenz, W. 1952. Experimentelle und histologische untersuchungen am weiblichen Genitalapparat von Lymantria dispar L. Zoologische Jahrbucher: Abteilung fur Antomie und ontogenie der Tiere 72: 147-215.

Bell, J. V., E. G. King & R. J. Hamalle. 1981a. Some microbial contaminants and control agents in a diet and larvae of *Heliothis zea*. J. Invertebr. Pathol. 37: 243-248.

Bell, R. A., D. C. Owens, M. Shapiro & J. R. Tardif. 1981b. Development of mass rearing technology, pp. 599-633. *In* C. C. Doane & M. L. McManus [eds.], The gypsy moth: Research toward integrated pest management. U.S. Dep. Agric. Tech. Bull. 1584.

Bell, R. A., M. Shapiro & O. T. Forrester. 1976. Gypsy moth mass rearing: evaluation and development of more economical diets and more efficient techniques for rearing and handling gypsy moths, pp. 54-61. APHIS Laboratory Report. Sept. 1976.

Bell, R. A., M. Shapiro & O. T. Forrester. 1977. Development of diet and efficient rearing techniques, pp. 77-83. APHIS Laboratory Report. Sept. 1977.

Bownes, M. 1986. Expression of genes coding for vitellogenin (yolk protein). Annu. Rev. Entomol. 31: 507-531.

Brewer, F. D. & O. Lindig, 1984. Ingredients for insect diets: Quality assurance, sources, and storage and handling, pp. 45-50. In E. G. King & N. C. Leppla [eds.], Advances and challenges in insect rearing. Agric. Res. Service, USDA.

Burton, R. L. & D. Perkins. 1984. Containerization of rearing insects, pp. 51-56. *In* E. G. King & N. C. Leppla [eds.], Advances and challenges in insect rearing. Agric. Res. Service, USDA.

Callahan, P. S. & J. B. Chapman. 1960. Morphology of the reproductive systems and mating in two representative members of the family *Noctuidae*,

*Pseudaletia unipuncta* and *Peridroma margaritosa*, with comparison to *Heliothis zea*. Ann. Entomol. Soc. Am. 53: 763-782.

Capinera, J. L., P. Barbosa & H. H. Hagedorn. 1977. Yolk and yolk depletion of gypsy moth eggs: Implications for population quality. Ann. Entomol. Soc. Amer. 70: 40-42.

Collins, A. M. 1985. Artificial selection of desired characteristics in insects, pp. 9-19. *In* E. G. King & N. C. Leppla [eds.], Advances and challenges in insect rearing. Agric. Res. Service, USDA.

Cothran, W. R. & G. G. Gyrisco. 1966. Influence of cold storage on the viability of alfalfa weevil eggs and feeding ability of hatching larvae. J. Econ. Entomol. 59: 1019-1020.

David, W. A. L., S. Ellaby & G. Taylor. 1972. The effect of reducing the content of certain ingredients in a semisynthetic diet on the indicence of granulosis virus disease in *Pieris brassicae*. J. Invertebr. Pathol. 19: 76-82.

Drummond III, B. A. 1984. Multiple mating and sperm competition in the Lepidoptera, pp. 291-370. *In* R. L. Smith [ed.], Sperm competition and the evolution of animal mating systems. Academic Press, New York.

Dutky, S. R. 1959. Insect microbiology. Adv. Appl. Micribiol. 1: 175-200.

Federici, B. A. 1980. Reproduction and morphogenisis of *Rickettsiella chironomi*, an unusual intracellular parasite of midge larvae. J. Bacteriol. 143: 995-1002.

Forrester, O. T. & R. A. Bell. 1977. Weight changes in gypsy moth egg masses during pre-diapause, diapause, and post-diapause development. APHIS Laboratory Report. March 1977: 87-88; March 1978: 86-91.

Gere, G. 1964. Change of weight, lipid and water content of *Lymantria dispar* L., with special regard to the chemical and energetic changes during insect metamorphosis and imaginal life. Acta Biol. Hungr. 15: 139-170.

Giebultowicz, J. M., R. A. Bell & R. B. Imberski. 1988. Circadian rhythm of sperm movement in the male reproductive tract of the gypsy moth, Lymantria dispar. J. Insect Physiol. 34: 527-532.

Giese, R. L. & M. L. Cittadino. 1977. Relationship of the gypsy moth to the physical environment. II. Diapause. Univ. Wisconsin, Madison. Staff Paper Series No. 6: 1-13.

Goldschmidt, R. 1934. Lymantria. Bibliogr. Genet. 11: 1-186.

Harrison, R. G., S. F. Wintermeyer & T. M. ODell. 1983. Patterns of genetic variation within and among gypsy moth, *Lymantria dispar* (Lep.: Lymantriidae) populations. Ann. Entomol. Soc. Am. 76: 652-656.

Henry, J. E., D. H. Street, E. A. Oma, & R. H. Goodwin. 1986. Ultrastructure of isolate of *Rickettsiella* from the African grasshopper, *Zonocerys varuegatys*. J. Invertebr. Pathol. 47: 203-213.

Hollander, H. L., C.-M. Yin & C. P. Schwalbe. 1982. Location, morphology and histology of sex pheromone glands of the female gypsy moth, *Lymantria dispar* (L.). J. Insect Physiol. 28: 513-518.

House, H. L. 1969. Effects of different proportions of nutrients on insects. Entomol. Exp. Appl. 12: 651-669.

Howell, J. F. 1970. Rearing the codling moth on an artificial diet. J. Econ. Entomol. 63: 1148-1150.

Hoy, M. A. 1977. Rapid response to selection for a non-diapausing gypsy moth. Science 196: 1462-1463.

Joslyn, D. J. 1985. Maintenance of genetic variability in reared insects, pp. 20-29. *In* E. G. King & N. C. Leppla [eds.], Advances and challenges in insect rearing. Agric. Res. Service, USDA.

Karmas, E. 1975. Nutritional aspects of food processing methods. Chp. 3, pp. 11-15. *In* R. F. Harris & E. Karmas [eds], Nutritional evaluation of food processing. AZI Publ. Co., Westport, CT.

Keller, W. R., D. F. Hoffman & R. A. Kwock. 1981. Wolbachia sp. (Rickettsiales: Rickettsiaceae) a symbiont of the almond moth, *Ephestia cautella*: ultrastructure and influence on host fertility. J. Invertebr. Pathol. 33: 273-283.

Keller, W. R., J. E. Lindegren & D. F. Hoffman. 1972. Developmental stages and structure of Rickettsiella in the naval orangeworm, *Paramyelois transitella* (Lepidoptera: Phycitidae). J. Invertebr. Pathol. 20: 143-199.

King, E. G. & N. C. Leppla 1984. Advances and challenges in insect rearing. Eds.: Agric. Res. Service, USDA.

Kishaba, A. N., T. J. Henneberry, R. Pandalgen & P. H. Tsao. 1968. Effects of mold inhibitors in larval diets on the biology of cabbage looper. J. Econ. Entomol. 61: 1189-1194.

Leonard, D. E. 1966. Differences in development of strains of the gypsy moth, *Porthetria dispar* (L.). Conn. Agric. Exp. Sta. Bull. 680: 1-31.

Leonard, D. E. 1970. Intrinsic factors causing qualitative changes in populations of Porthetria dispar (Lepidoptera: Lymantriidae). Can. Entomol. 102: 239-249.

Mackauer, M. 1976. Genetic problems in the production of biological control agents. Annu. Rev. Entomol. 21: 369-385.

Maksimovic, M. 1971. Effect of cobalt-60 irradiation of male pupae of the gypsy moth, *Lymantria dispar* L., on biological functions of male moths, pp. 15-22. *In* Symp. Sterility principle for insect control or eradication. Food Agric. Org./Int. At. Energy Assoc. (Athens, Greece, 1970).

Mastro, V. C., T. M. ODell & C. P. Schwalbe. 1989. Genetic control of Lymantriidae: Prospects for gypsy moth management, pp. 275-302. *In* W.

E. Wallner & K. A. McManus [eds.], Proc., Lymantriidae: A comparison of features of New and Old World tussock moths. U.S. Dep. Agric. Gen. Tech. Rep. NE-123.

Mithat, A. 1933. Experimentelle untersuchungen uber den Einfluss von Temperatur un Luftfenchtigkeit auf die Entwicklung des Schwammspinners, Porthetria dispar L. Z. Angew. Entomol. 20(3): 354-381.

Moore, R. F. 1985. Artificial diets: Development and improvement, pp. 67-83. *In* P. Singh & R. F. Moore [eds.], Handbook of insect rearing, vol. 1. Elsevier Press, Amsterdam.

Moore, R. F., T. M. ODell & C. O. Calkins. 1985. Quality assessment in laboratory-reared insects, pp. 107-135. *In* P. Singh & R. F. Moore [eds.], Handbook of insect rearing. vol. I. Elsevier Press, Amsterdam.

ODell, T. M., R. A. Bell, V. C. Mastro, J. A. Tanner & J. F. Kennedy. 1984. Production of the gypsy moth, *Lymantria dispar*, for research and biological control, pp. 156-166. *In* E. G. King & N. C. Leppla [eds.], Advances and challenges in insect rearing. Agric. Res. Service, USDA.

ODell, T. M., C. A. Butt & A. W. Bridgeforth 1985. *Lymantria dispar*, pp. 355-367. *In* P. Singh & R. F. Moore [eds.], Handbook of insect rearing. vol. II. Elsevier Press, Amsterdam.

ODell, T. M. & W. D. Rollinson 1966. A technique for rearing the gypsy moth, *Porthetria dispar* (L.), on an artificial diet. J. Econ. Entomol. 59: 741-742.

Ouye, M. T. 1962. Effects of antimicrobial agents on microorganisms and pink bollworm development. J. Econ. Entomol. 55: 854-857.

Raulston, J. R. & E. G. King. 1984. Rearing the tobacco budworm, *Heliothis virescens*, and corn earworm, *Heliothis zea*, pp. 167-175. *In* E. G. King & N. C. Leppla [eds.], Advances and challenges in insect rearing. Agric. Res. Service, USDA.

Riddle, R. A., R Sykes, L. J. Leffel & P. S. Dawson. 1985. Genetic basis of extremely slow development in a population of *Tribolium confusum*. Can. J. Genet. Cytol. 27: 650-654.

Rule, H. D., P. A. Godwin & W. E. Waters. 1965. Iradiation effects on spermatogenesis in the gypsy moth, *Porthetria dispar* (L.). J. Insect Physiol. 11: 369-378.

Roush, R. T. 1986. Inbreeding depression and laboratory adaptation in *Heliothis virescens* (Lepidoptera: Noctuidae). Ann. Entomol. Soc. Amer. 79: 583-387.

Shapiro, M. 1984. Microorganisms as contaminants and pathogens in insect rearing, pp. 130-142. *In* E. G. King & N. C. Leppla [eds.], Advances and challenges in insect rearing. Agric. Res. Service, USDA.

Shapiro, M., Bell, R. A. & C. D. Owen. 1981. In vivo mass production of gypsy moth nucleopolyhedrosis virus, pp. 633-655. *In* C. C. Doane & M.

L. McManus [eds.], The gypsy moth: Research toward integrated pest management. U.S. Dep. Agric. Tech. Bull. 1584.

Shermoen, A. & B. Kiefer. 1975. Regulation in rDNA-deficient *Drosophila melanogaster*. Cell 4: 275-280.

Singh, P. 1977. Artificial diets for insects, mites, and spiders. IFI/Plenum, New York.

Singh, P. & G. E. Bucher. 1971. Efficacy of safe levels of antimicrobial food additives to control microbial contaminants in a synthetic diet for Agria affinis larvae. Entomol. Exp. Appl. 14: 279-309.

Singh, P. & R. F. Moore (editors). 1985. Handbook of insect rearing. vol. I. Elsevier Press, Amsterdam.

Stewart, F. D. 1984. Mass rearing the pink bollworm, *Pectinophora gossypiella*, pp. 176-187. *In* E. G. King & N. C. Leppla [eds.], Advances and challenges in insect rearing. Agric. Res. Service, USDA.

Suitor, E. C., Jr. 1964. The relationship of *Wolbachia persica* Suitor and Weiss to its host. J. Insect Pathol. 6: 111-124.

Tanner, J. A. & J. J. Baker. 1985. Evaluating the development and reproduction of insects produced in the Otis Methods Development Rearing Facility. APHIS Laboratory Report, Sept. 1985: 117-121.

Tanner, J. A., J. J. Baker, M. Mathews, Jr. & others. 1988. Devlopment and evaluation to improve rearing techniques. APHIS Laboratory Report, Sept. 1988: 262-268.

Tanner, J. A. & T. Buck. 1979. Establishing standards. APHIS Laboratory Report, March 1979: 48-53.

Tanner, J .A. & L. F. Kennedy. 1975. Hatchability of gypsy moth eggs as influenced by the duration of embryonation prior to cold storage. APHIS Laboratory Report, June 1975: 114-116.

Tanner, J. A. & L. F. Kennedy. 1976a. Hatch rates of laboratory produced eggs as affected by duration of refrigeration. APHIS Laboratory Report, March 1976: 71-72.

Tanner, J. A. & L. F. Kennedy. 1976b. The hatch rate of *L. dispar* eggs as affected by the precooling period and the embryonation period. APHIS Laboratory Report, Sept. 1976: 45-48.

Tanner, J. A., J. G. R. Tardif, J. J. Baker & M. C. Flynn. 1985. Development and evaluation of improved rearing techniques. APHIS Laboratory Report, Sept. 1985: 122-135.

Tanner, J. A., J. G. R. Tardif, J. J. Baker, B. A. Fontes & M. Mathews, Jr. 1987. Development and evaluation of improved rearing techniques. APHIS Laboratory Report, Sept. 1987: 153-167.

Tanner, J. A., J. G. R. Tardif, J. J. Baker, H. Hamilton & B. A. Fontes. 1986. Development and evaluation of improved rearing techniques. APHIS Laboratory Report, Sept. 1986: 126-137.

Tanner, J. A., J. G. R. Tardif, J. J. Baker & B. P. Weeks. 1982. Development of mechanical egg infestation procedures. APHIS Laboratory Report, Sept. 1982: 153-168.

Tanner, J. A. & B. P. Weeks. 1980. Establishing standards. APHIS Laboratory Report, Sept. 1980: 167-175.

Tanner, J. A. & B. P. Weeks. 1981a. Evaluating the development and reproduction of insects produced in the Otis Methods Development Rearing Facility. APHIS Laboratory Report, Sept. 1981: 235-247.

Tanner, J.A. & B. P. Weeks. 1981. Development of mechanical egg infestation procedures. APHIS Laboratory Report, Sept. 1981: 248-256.

Tanner, J.A., B. P. Weeks & M. Palmieri. 1983. Development and evaluation of improved rearing techniques. APHIS Laboratory Report, Sept. 1983: 127-141.

Weiss, E. 1973. Growth and physiology of rickettsiae. Bact. Rev. 27: 259-283.

Weiss, E., G. A. Dasch & K. P. Chang. 198). Wolbachieae. *In* N. R. Krieg & J. G. Holt [eds.], Bergey's manual of systematic bacteriology, vol. 1. Williams and Wilkins, Baltimore.

Yen, J. H. & Barr, A. R. (1977). Incompatibility in *Culex pipiens*, pp. 97-118. *In* Pal & Whitten [eds], The use of genetics in insect control. Elsevier Press, Amsterdam.

# Appendix:
# Rearing Gypsy Moth

## Rearing Gypsy Moth Individually for Evaluation
## of Life-stage Parameters Without Diet or Container Changes

### General Considerations

Development and behavior can be monitored continuously without removing larva or changing diet. Developmental period from hatch to pupation, pupal weight, fecundity, and percent hatch are all within the standards set for normal colony production. Pupae should be removed from cup prior to eclosion to avoid adult wing deformities caused by inadequate moisture and space for wing expansion.

### Equipment and Materials
- Container: clear plastic, 59 ml, Comet (Chelmsford, MA)
- Lid: LP-2 - 2500, Comet (Chelmsford, MA)
- 15 ml diet (Bell)
- Environmental chamber set at: 25°C, 50-60% relative humidity, air speed 3-5 cfm, and a 16:8 (L:D) regimen

### Diet Ingredients (Bell et al. 1981)
| | |
|---|---|
| • Wheat germ | 1,200 g |
| • Casein | 250 g |
| • Salt mix | 80 g |
| • Sorbic acid | 20 g |
| | 1,550 g |

### Finished Diet
| | |
|---|---|
| • Dry mix | 1,550 g |
| • Water | 9,000 ml |
| • Methyl paraben | 10 g |
| • Vitamin premix | 100 g |
| • Agar (fine mix, 80-100 mesh) | 150 g |
| | 10,810 g |

*Brand Names and Sources of Ingredients for Bell Diet*

- Wheat Germ - Mennel Milling Co., Fostoria, OH
- Casein - New Zealand Milk Products, Petaluma, CA
- Salt Mix - Wesson's #902851
- Sorbic Acid - ICN Nutritional, Cleveland, OH
- Vitamin Premix #26862 - Hoffman LaRoche, Inc., Salisbury, MD
- Agar - Moorehead and Co., Van Nuys, CA
- Methyl paraben - Kalama Chemical, Inc., Garfield, NJ


## *Preparation for Rearing*

- Punch 2 holes the size of a normal dissection probe, one each on either side of the lid
- Pour 15 ml of Bell diet into each cup and allow to cool for 30-45 minutes
- Place larva in cup, cover, and place in environmental chamber.
  DO NOT invert cup or restrict air circulation over cup lid.

# 20

## New Technologies for Rearing *Epidinocarsis lopezi* (Hym., Encyrtidae), a Biological Control Agent Against the Cassava Mealybug, *Phenacoccus manihoti* (Hom., Pseudococcidae)

*P. Neuenschwander and T. Haug*

### Introduction

The cassava mealybug (CM) *Phenacoccus manihoti* Matile-Ferrero (Hom., Pseudococcidae) was accidentally introduced into Africa in the early 1970s, and it subsequently spread over most of the continent. Through its feeding damage and stunting of the tips, it dramatically reduced tuber yields, thereby becoming the most important pest of cassava (Neuenschwander & Herren 1988). To combat this new pest in collaboration with numerous national and international agencies, the Biological Control Program of the International Institute of Tropical Agriculture (IITA) was established (Herren 1987). Following extended exploration in South America (Yaseen 1986, Lohr et al. 1990) and quarantine at the International Institute of Biological Control (IIBC) in the United Kingdom, the solitary and host-specific wasp *Epidinocarsis lopezi* (De Santis) (Hym., Encyrtidae) was imported into Africa, reared, and first released in Nigeria in 1981 (Herren & Lema 1982). By 1988, it had been successfully established in 21 African countries and had spread over an area of over 1.5 million km$^2$ (Herren et al. 1987, Neuenschwander & Herren 1988). CM populations declined after the release and have remained low since (Hammond et al. 1987, Neuenschwander & Hammond 1988, Hammond & Neuenschwander 1990). Surveys of subsistence farms (Neuenschwander et al. 1989a), exclusion experiments (Neuenschwander et al. 1986), and a computer simulation model

(Gutierrez et al. 1988 a,b) have all documented the efficiency of E. *lopezi* in preventing CM outbreaks.    Overall, the project is considered an economic success (Norgaard 1988), though criticism from some countries like the Congo persists (Nenon & Fabres 1988, Le Ru et al. in press).    In Nigeria, ecological conditions where CM damage persisted (less than 5% of all fields) have now been characterized (Neuenschwander et al. 1990); similar conditions might prevail in the Congo.

Because of the size of the project, producing and delivering E. *lopezi* on time has been one of the main constraints.    In the 18 countries where E. *lopezi* has been released, ground releases were the rule, but for remote areas in some countries, aerial releases developed by IITA (Herren et al. 1987) have proven useful.    Timing of operations has often been influenced by administrative decisions in the various countries, leading to rather unpredictable requests for insects.    To satisfy this high and shifting demand for E. *lopezi,* several rearing systems have been developed that go beyond the classical insectary with potted plants.

This chapter reviews biological and ecological studies on cassava plants, CM, and E. *lopezi* that are relevant for rearing, while quality control measures have been discussed elsewhere (Neuenschwander et al. 1989b).    Classical insectary operations and new technologies developed for mass rearing of E. *lopezi* are described, together with their management in the day-to-day operation, leading to production and cost estimates for different rearing techniques.


## The Biological Basis

Rearing of E. *lopezi* started when the scientific description of the species was the only available information about this insect.    Observations on other encyrtids attacking mealybugs (Clausen 1942, De Bach & White 1960, Avidov et al. 1967, Viggiani 1975, Bartlett 1978) were therefore used as a starting point.    Similarly, the CM was not described as a new species until 1977 (Matile-Ferrero 1977).    Meanwhile, both organisms have been thoroughly studied and the results  progressively adapted and incorporated into the rearing procedures.

### The Host Plant

The CM has a very narrow host range, which in Africa comprises the introduced  Euphorbiaceae; cassava, *Manihot esculenta* Crantz; Ceara rubber, *Manihot glaziovii* (Mull.); and their hybrid.    Where CM populations on these hosts are extremely high, however, numerous other plant species from other families also become infested by CM and even allow it some reproduction. At IITA, all rearing of the CM was done on cassava, which was easily available.

Laboratories in temperate zones have also used alternate hosts like poinsettia, *Euphorbia pulcherrima* Wild. (A. Panis, personal communication), *Capsicum* sp. (T. Haug, personal communication), but continuous rearing on potatoes, a common replacement host for many mealybug species (Finney & Fisher 1964), is impossible. For rearing individuals, CM are sometimes kept in petri dishes on the fleshy sprouts of the Portulacaceae water leaf, *Talinum triangulare* (Jacq.) (Neuenschwander & Madojemu 1986), on detached cassava leaves on moist filter papers in a petri dish, or in leaf-clip cages on whole cassava plants (Schulthess et al. 1987, Lohr et al. 1989b).

Though cassava varieties with different degrees of tolerance to the CM are available (Hahn et al. 1979), all varieties can be used for CM production in the insectary because plants are infested with high numbers of CM crawlers. Under conditions of low light intensities, typical for an insectary, the farmers' varieties used for intercropping are preferred because they grow better (Neuenschwander et al. 1989b).

Cassava is multiplied vegetatively. Its production has been reviewed by Hahn et al. (1979). It can be grown from cut stems with one-half an eye only (Sykes & Harney 1972, Kamalam et al. 1977, Dahniya & Kallon 1984) or from excised buds in tissue culture (Ng 1984). From such a small amount of planting material, however, it takes a long time and correspondingly higher costs to produce a usable plant.

## The Phytophagous Host

The CM is a parthenogenetic species, with three instars, a mean developmental time from egg to adult of about four weeks, and an adult life span of another three weeks. From their laboratory experiments and those by several other authors (Fabres & Boussienguet 1981, Le Ru & Fabres 1984, Lema & Herren 1985, Nsiama She 1985), Schulthess et al. (1987) calculated the developmental rate (R) in the linear temperature (T) range from 14 to 27.5°C, as follows: $R = 0.0032 T - 0.047$, with an explained variance of 96.8%. The lower thermal threshold is 14.7°C, the upper thermal threshold about 35°C, and the thermal constant is 312.5 day-degrees. The mean generation time declines from 81 days at 20°C to 35 days at 32°C. The net reproductive rate and the intrinsic rate of increase are highest at 28°C, where each female lays about 500 eggs. These parameters are presented as day-degrees by Gutierrez et al. (1988b).

On drought-stressed potted plants, reproductive rate and intrinsic rate of increase are higher and the generation time is shorter than on nonstressed plants (Schulthess et al. 1987). This ecological preference for stressed plants has been observed in the field (Nwanze et al. 1979) and confirmed in a large-scale survey of areas with *E. lopezi,* where relatively high CM populations were restricted to cassava on unmulched sandy and leached-out soils (Neuenschwander et al. 1990).

## *The Parasitoid*

The basic life-table parameters of *E. lopezi* (Fig. 1) have been studied in detail and reviewed (Neuenschwander & Herren 1988, van Alphen et al. 1989), and the development data have been presented on the basis of day-degrees (Gutierrez et al. 1988b).

**Immatures.** *E. lopezi* is a solitary, internal parasitoid having four larval instars, as described by Lohr et al. (1989a). Before the parasitoid pupates, the dying host hides and its integument hardens to a so-called mummy, which lasts about the second half of the developmental time of immature *E. lopezi.* The period from oviposition to adult emergence is 14.2 days at 27°C (Lema & Herren 1982), and varies between 11 and 25 days at fluctuating temperatures (24-31°C) (Odebiyi & Bokonon-Ganta 1986). A small proportion, particularly from 2nd instar hosts, yields adult parasitoids that develop very slowly, and many of these late-developing parasitoid larvae die (Kraaijeveld & van Alphen 1986, Lohr et al. 1988). Similarly, all supernumerary larvae resulting from superparasitism die. Such larvae become encapsulated (Sullivan & Neuenschwander 1988). Under conditions of solitary parasitism, however, only 9% (Nenon et al. 1988) to 12.5% (Sullivan & Neuenschwander 1988) of all parasitoid larvae are melanized and cannot free themselves from the encapsulation sheet, although 26.7% was reported (Giordanengo & Nenon, in press).

**Adults.** Upon emergence, adult *E. lopezi* mate and are ready to lay eggs the same day. They are attracted to the host habitat by responding to synomones from cassava leaves of CM-infested plants (Nadel & van Alphen 1987). Once on the leaf, the females respond to kairomones emanating from the wax of the CM to find their host for oviposition (Langenbach & van Alphen 1986). In the field, males are attracted to virgin females (van Dijken et al. 1988).

The reproductive success of *E. lopezi* depends also on the host instar. The female encounters the large adult CM more frequently than the small younger stages (Lohr et al. 1988), but it prefers 3rd instars (Kraaijeveld & van Alphen 1986, Neuenschwander & Madojemu 1986, Biassangama et al. 1988, Lohr et al. 1988). Pre-ovipositing hosts give mainly female wasps. Third instar CM yield males and females in proportions that depend on the previous experience of the female (M. van Dijken, personal communication). Second instar hosts yield almost exclusively males. The CM grow larger before mummifying if

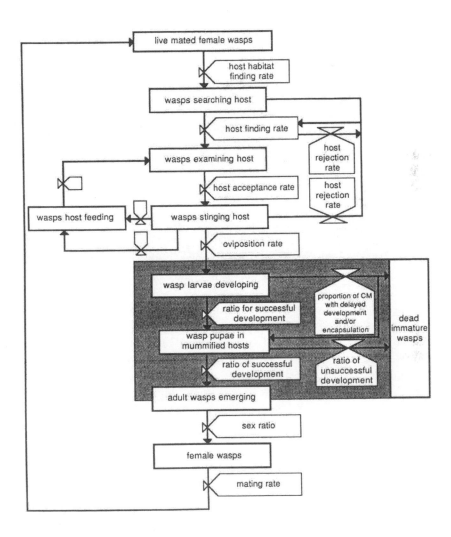

**Fig. 1. Basic life-table parameters of *E. lopezi*.**

parasitized by a female larva rather than having a male (Kraaijeveld & van Alphen 1986). As a result, adult females tend to be larger than males. Since size is correlated with fitness in females (Kraaijeveld & van Alphen 1986) but not in males, the observed host stage selection is advantageous. The choice is, however, plastic enough so that *E. lopezi* thrive on almost all combinations of host instars.

When observed under the binocular in the laboratory, not every encounter of an *E. lopezi* female with a host leads to oviposition. Fourth instar females, particularly those that have been stung before, often succeed in fending off a parasitoid attack by vigorously wiggling their abdomen, which reduces parasitization and, to some extent, superparasitism (Iziquel 1985). Second instar CM are often killed through host-feeding by the female parasitoid. These hosts do not usually receive a parasitoid egg. Many of the stung CM that have not received an egg nor been fed upon, die from mutilation by the ovipositor or, if they continue to live, realize only a small fraction of their usual oviposition capacity. On 2nd and 3rd instar CM, the nonreproductive mortality caused by host feeding and mutilation thus surpasses the death toll by parasitoid reproduction by a factor of 3.5 and 2.3, respectively (Neuenschwander & Madojemu 1986, Lohr et al. 1988).

*E. lopezi* adults live only a few days when kept in conditions simulating transport packages (Herren et al. 1987) and, on average, 13 days in the laboratory (Odebiyi & Bokonon-Ganta 1986). The short survival time in crowded conditions can be prolonged most by offering dry sugar grains, while sugar water, honey, honeydew, Wheast (a mixture of whey and yeasts [Hagen & Tassan 1970]), or the addition of CM hosts are less effective in prolonging life (Neuenschwander et al. 1989b). At 23°C, by contrast, females and males average 42.3 and 36.6 days, respectively (Biassangama et al. 1988). During this period, the females lay an average 208 eggs, though much lower values were obtained at similarly low temperatures in another laboratory (Lohr et al. 1989b). At 25-28°C, oviposition is considerably lower, namely 85 eggs (Iziquel 1985), and reproduction averages only 67 offspring per female or about 10 per day (Odebiyi & Bokonon-Ganta 1986, Lohr et al. 1989b).

Oviposition increases as the number of hosts offered increases, but percentage parasitism in these experiments tends to decrease (Odebiyi & Bokonon-Ganta 1986), except at very small host densities, when oviposition goes up with increasing host numbers (Lohr et al. 1988). This positive density-dependent reaction has been confirmed for low host densities in the field (Hammond & Neuenschwander 1990).

Superparasitism, even by the same female, is often observed under confined conditions, but also at low host densities in the field (Iziquel 1985, Biassangama et al. 1988, van Dijken et al. 1988), though females were shown to be able to distinguish between parasitized and unparasitized hosts (Kraaijeveld & van

Alphen 1986). Under conditions of superparasitism, male larvae seem to have higher mortalities, thus giving female-biased sex ratios (van Dijken et al. 1988).

The discrepancies in the data on developmental times and oviposition capacity are explained by differences in rearing, particularly the host densities offered, which in turn affect superparasitism. Sometimes considerable mortality of the larval stages occurs, leading authors to give different importance to encapsulation. Overall, *E. lopezi* reacts very strongly to the condition of its host population.

Reactions to the host plant, cassava, are much less well documented and are limited to two observations. Female *E. lopezi* are attracted to noninfested cassava leaves of a CM-infested plant (Nadel & van Alphen 1987) and in the field they prefer green, bushy plants over highly CM-infested ones with dropping leaves (Hammond & Neuenschwander 1990, Neuenschwander et al. 1990).

**Hyperparasitoids.** Before its importation from South America into Africa, *E. lopezi* passed through quarantine in the United Kingdom, where all hyperparasitoids from South America were removed from the culture. Hyperparasitoids of about 10 species, particularly *Prochiloneurus* spp. (Hym., Encyrtidae) and *Chartocerus* spp. (Hym.: Signiphoridae) are, however, commonly found on *E. lopezi* in Africa, sometimes already in the first samples following the release. They are polyphagous indigenous species, attacking *Anagyrus* spp. (Hym.: Encyrtidae) on several mealybugs on cassava and other hosts. *P. manihoti* is attacked only infrequently by these *Anagyrus* spp. (Fabres & Matile-Ferrero 1980, Boussienguet 1986, Neuenschwander et al. 1987, Boussienguet et al. in press). Since these hyperparasitoids react in a strongly density-dependent manner to their new host *E. lopezi* (Neuenschwander & Hammond 1988, Hammond & Neuenschwander 1990, Neuenschwander et al. 1990), they become disquietly abundant on high *E. lopezi* and CM populations in the field (Iziquel & Le Ru 1989). In an open insectary, they become an important threat to *E. lopezi* cultures.

### The Classical Insectary

In the common type of insectary, insects are kept in small cages. Substrates for the phytophagous hosts often are plant parts that can be stored easily, such as squashes and potatoes. Factitious plant, phytophagous hosts, or artificial diets are used for ease of rearing or for cutting production costs (Smith & Armitage 1931, Finney & Fisher 1964, King & Morrison 1984, Waage et al. 1985). If fresh plants are needed for rearing the phytophagous host, they are grown in pots in a greenhouse. Phytophagous hosts are reared separately and supplied to the entomophagous insects when needed. Beneficials are reared in continuous contact with their hosts in overlapping generations (Finney & Fisher 1964). This type of management is ideal for maintenance cultures, and it was used at IITA to rear *E. lopezi* on the CM on potted cassava plants.

## Production of Plants

Woody stems of 12- to 18-month-old plants (Hahn et al. 1979), including ratooned ones, are cut in pieces of about five internodes and used for plant production. This length assures sprouting of at least one bud to yield plants that can be infested by CM after 4 to 6 weeks, depending on the season.

On their dormant buds, these cuttings support infestations of CM and the cassava green mite, Mononychellus tanajoa (Bondar), sometimes for very long periods (Yaninek 1988). For plant production, cuttings are therefore disinfected with dimethoate 0.1% (Leuschner et al. 1980) or, better, by immersing them for five (5) minutes in 52°C hot water, which effectively kills all insects (Baker 1962, Neuenschwander et al. 1989b). After treatment, stems are dried. Cassava mosaic virus, contained as a systemic infestation in the cuttings (Hahn et al. 1979), is no problem in the insectary even on virus-susceptible varieties, provided temperatures remain relatively high, thus allowing plants to grow fast. Two cuttings are planted per 1 liter pots filled with well drained, preferably slightly acidic soil. Since cassava thrives under plastic cover (Kamalam et al. 1977) under high temperatures, with peaks above 40°C, provided the relative humidity stays high, little air-conditioning of the glass or screen houses used for plant production is needed. Usually, temperatures fluctuated between 27 and 35°C, with a luminosity of up to 60,000 lux.

To avoid attacks of the growing plants by the cassava green mite or whiteflies, *Bemisia* spp., the impact of phytosanitary treatments in the food chain was tested. With an acaricide, a small but deleterious effect on parasitoid production was noticed (Neuenschwander et al. 1989b), while treatment with natural pyrethrum proved to be useful for short periods (B. Megevand, personal communication). Usually, we found that no phytosanitary treatments were necessary and roguing of the few infested plants was sufficient.

## Production of Insects

**Insectary Cages.** Cultures of CM and *E. lopezi* are maintained in wooden cages (44 cm x 45 cm x 58 cm) with fine screen sides and a glass top on 30 cm high cassava plants from the greenhouse. Each cage accommodates four to five 1-liter pots. Polyester screen of 0.20 mm mesh size (Swiss Silk, Zurich, Switzerland), which does not allow passage of CM (though freshly hatched crawlers can pass) and also protects against entry of the tiny *Chartocerus* spp.

hyperparasitoids, proved to be superior to locally bought cotton tissue. For adequate protection of the cultures, plants and insects are handled through two sleeves in the door of the cage.

**Insect Collection Equipment.** For infestation of the plants with CM, crawlers are used because they are the only host instar certain to be free of *E. lopezi*. They are collected from ventilated jars covered with fine mesh and an inverted funnel (Haug et al. 1987), weighted (300 crawlers per mg), and sprinkled over the leaves with a vial resembling a salt shaker.

For collecting *E. lopezi,* either for inoculation or for later releases or experiments, normal entomological aspirators containing filter paper are used, with negligible mortality. To handle exceptionally large numbers of wasps, a battery-operated aspirator, as used for cleaning car seats, is occasionally employed, a procedure that kills only a small percentage.

**Mass Rearing of Parasitoids.** In each cage, four pots with two plants are infested with a total of several hundred second and third instar CM and inoculated with 50 to 70 wasps. One pair of infested plants per cage was added each week and the oldest plant that had collapsed was removed, thus leading to parasitoid cultures with overlapping generations. CM-infested stems and leaves from the removed plants were kept for last emergence of *E. lopezi.*

A relatively high host density proved necessary to reduce superparasitism and compensate for host feeding. *E. lopezi* females prefer, however, small clean CM colonies over large ones, which are sticky with honeydew. They also prefer green cassava tips over those with dry leaves. Care was therefore taken not to offer excessive host densities. If mixed host colonies containing all stages were offered, sex ratios were either even or slightly biased towards females.

When wasps were removed weekly for a large-scale aerial release program, average production from the wooden cages in the third week after inoculation was 211.5 E. lopezi ±16.7 S.E., N=39) per pot with two plants. Under conditions of rapid production, most plants stayed in the insectary for only 3 to 4 weeks, in addition to the 6 weeks in the greenhouse for plant growth and CM production. Usually, temperatures fluctuated between 25 and 32°C, with light intensities of 1,500 to 5,000 lux inside the cages and relative humidities of 50-85%.

## New Technologies

In view of the size of the CM-infested area in Africa, plans were made at the start of the project to improve the classical insectary. The main improvement in streamlining insect production was to come from a switch from soil in pots to some type of hydroponic culture (Hoagland & Arnon 1950, Resh 1978).

## Development of Hydroponic Cultures

To determine nutritional requirements and optimum environmental conditions for cassava grown in a nutrient solution, cuttings were placed in plastic bottles containing a diluted nutrient solution (Murashige & Skoog 1962, Forno et al. 1979), with or without aeration, or addition of $CO_2$ (Mahon et al. 1977). Plants in aerated hydroponic cultures produced 2.4 times as many leaves as nonaerated ones and 3.4 times as many as those grown in soil. Adding $CO_2$ did not improve growth (Neuenschwander et al. 1989b). These preliminary experiments showed that hydroponic culture of cassava was feasible.

To put this concept into practice, the following procedures were adopted: Root rot problems are reduced by treating cuttings with copper sulphate in a dip and by adding the organic fungicide (Roval, Rhone-Poulenc, Lyon, France) to the nutrient solution. To promote root formation, this treatment is followed by a quick dip into a solution of 2,000 ppm rooting hormone (indolebutyric acid) (Sykes & Harney 1974).

To save space and running time of expensive equipment, cuttings after drying are not placed directly in the hydroponic growth chambers. Instead, they are tightly packed in presprouting units consisting of a plastic tray under plastic cover and inundated two to three times per day up to 10 cm deep with a salt solution (Luwasa, Interhydro, Bern, Switzerland). After 1 to 1.5 weeks, the cuttings, which have short roots and newly formed sprouts, are brought to the hydroponic rearing unit, for which Luwasa solution (1g/liter) has been adopted as a standard nutrient for the entire growth period.

In large-scale hydroponic cultures, plants share the same nutrient solution. Despite careful treatment with fungicides, root rot problems cannot be avoided altogether under conditions of insect mass production. But by using a 50:50 mixture of hydrophilic and hydrophobic Rockwool (Rockwool/Grodan, Roermond, Holland) an inert and reusable substrate, the cleanliness of hydroponic cultures can be combined with the water-retention capacity of soil. In addition, such a system provides more stability against interruptions in irrigation.

For promoting plant growth and preventing root rot, Rockwool proved to be superior to other anorganic substrates. This was true for vermiculite even if good drainage was assured by adding styrofoam particles and water-retention capacity was increased by adding a water-resorbing gel (Aquastore, Cyanamid, Bradford, Great Britain). Both anorganic substrates, rockwool and vermiculite, were superior to organic ones such as wood spans, coconut shells, or palm kernel fibers. Rockwool is now used both in pots in the classical insectary and in the units using the new technology described below.

## Mechanized Cages

Systems of different technological levels have been developed for the soilless mass rearing of CM and its natural enemies. A prototype, developed from the Ruthner system (Resh 1978), consists of a stainless steel unit 2 m x 2 m x 4 m. Nine trays, holding a total of 594 cassava plants, are attached to a circular chain and rotate around a light source. After each rotation, they receive and empty a programmed amount of nutrient solution from a common tank. The nutrient solution, which also contains fungicides, is reused for about 1 week. The lights provide up to 34,000 lux at plant level and are contained in a tight cage of glass with forced-air ventilation. The whole system can be cooled by a normal air conditioner in the room with peaks of 32-35°C (Haug et al. 1987).

In these units, rearing of *E. lopezi* is managed as follows. The plants are infested with about 100 crawlers each when the sprouts are 5-10 cm long. After 1 week, a few leaves with third instar CM are placed onto each tray, and 2 days later 1,000 wasps are released in the cage. When their offspring, produced on the 3rd instars, emerge, the initial infestation with crawlers has reached the 3rd instar, ready to be stung. New crawlers are added when needed.

In 1988, a production cycle lasted an average of 65.6 days ($\pm$4.6 SE, N=9), in addition to the 10 days the cuttings spent in the presprouting unit. Harvested production averaged 40,800 ($\pm$3,754 SE) wasps per unit or 68.0 wasps per plant. But even during the most intensive use of *E. lopezi*, not all wasps that emerged could be recovered live with aspirators.

## Cassava Trees

Recently, a new rearing unit for plants, CM, and *E. lopezi* was developed, called the cassava "tree." Of simple design (Fig. 2), it is built from low-cost local materials. An oblong plastic bag filled with Rockwool is suspended in a metal structure consisting of galvanized conduit pipes of the type used by electricians. Cassava cuttings are inserted into the bag and irrigated through tubes like those used in drip irrigation with a nutrient solution from a tank. Mixing of nutrients by a fertilizer mixer (Gewa 20 liter by Volmatic, 2750 Ballerup, Denmark) and irrigation can be automated if electric power and water under pressure are available. The structure is enclosed by a polyester screen with a zippered entry. E. lopezi is collected in a container at the top, while the rest of the tree is darkened with a black cloth.

**Fig. 2. Cassava "tree" rearing unit for cassava plants, cassava mealybug, and *E. lopezi*.**

The technical advantages of the cassava tree over potted cassava in cages are: 1) the maintenance of all trophic levels in the same unit, thereby saving operational costs; 2) the more efficient use of space because of the vertical arrangement; and 3) improved plant growth and better sanitation because of hydroponic culture on Rockwool. Compared to the mechanized rearing cages, cassava trees are much cheaper and lighter, and can be constructed locally with a minimum of imported technology.

Cassava trees became fully functional in 1989 after the entire Biological Control Program had moved into its new facilities in Cotonou, Benin. Accurate statistics were kept on 130 trees planted between August 1988 and July 1989, among which the ones in 1988 suffered some initial problems when climatic control was difficult. Usually, temperatures were maintained at 27 to 35°C, relative humidities above 50%, and luminosity inside the screen varied between 5,000 and 15,000 lux. Eleven to 31 days after planting (mean 17.5 days), the plants were infested with 30,000 to 210,000 CM crawlers (mean 84,185), according to the condition of the plants. To improve adhesion of the crawlers to the plants, the leaves were sprayed with water before infestation. because of mortality and loss, only about 40,000 CM reached the third instar; 15.4 days after infestation with crawlers, the units were inoculated with 500 - 1,000 female E. lopezi. Collection of the newly emerged offspring started after a mean of 18.6 days and continued for 19.0 days. One production cycle lasted from 48 to 113 days and averaged 70.5 days. A unit yielded up to 28,750 E. *lopezi*, with an average of 7,870 ($\pm$535 SE) live wasps and a sex ratio of 49.0% females. As experience in rearing with trees increased, average production improved to close to 12,000 wasps in the last 20 cycles. During these cycles, dead wasps that had escaped collection were also counted, demonstrating that live recovery reached 74.9% of all E. *lopezi* that emerged.

All variables (X) that could have influenced E. *lopezi* production (Y) in these 130 trees were evaluated in a multiple regression analysis. Only variables that contributed significantly ($P < 0.05$) were maintained (Table 1), yielding an explained variance $R^2$ of 68%. The strongest influence on production came

**Table 1. Multiple regression analysis for predicting the number of *Epidinocarsis lopezi* produced per cassava tree rearing unit. Analysis of 130 trees planted between August 1988 and July 1989 in Cotonou, Benin. (b= partial regression coefficient, t = t-statistic, with \* if $P < 0.05$)**

| Variable | Regression statistics | |
|---|---|---|
| | b | t |
| Dependent variable (Y): | | |
|     Number of E. *lopezi* produced | | |
| Independent variables (X): | | |
|     N days wasps were collected | 277.7 | 12.2\* |
|     N CM crawlers seeded (x 1000) | 54.3 | 5.7\* |
|     N times Rockwool was used | 1,050.1 | 2.8\* |
|     N days between planting and CM infestation | 223.0 | 2.7\* |
|     Cassava variety | 772.2 | 2.2\* |

Intercept = - 6,324.2
Mean Square Error = 3,535.0 with 124 degrees of freedom
Explained variance $R^2$ = 0.68

from the duration of the collection (1 to 71 days, mean 19.0 days) and the number of CM crawlers added, which reflects the experimenters' perceptions of the quality of the plants. Production also depended significantly on the number of days the plants were grown before crawler infestation and, the cassava variety (five were tested). As Rockwool was used repeatedly, wasp production increased significantly from 6,677 on trees with new Rockwool to 11,229 on trees where the substrate had been used four times. Repeated use of Rockwool changed its consistency, perhaps increasing its content in organic matter and its acidity, thereby rendering it more suitable for cassava growth. All these factors point out that the most important factor in mass producing *E. lopezi* is the quality of plant growth. By contrast, the following variables played no important role: 1) the number of days the temperature surpassed 40°C during *E. lopezi* rearing (0 to 16 days, with a mean of 4 days), 2) the month of planting, 3) the *E. lopezi* noculum size, 4) the durations of CM development before inoculation, and 5) *E. lopezi* development before the first collection.

## Production Costs

In 1988, production of *E. lopezi* from the classical insectary and from the mechanized hydroponic units in Ibadan, Nigeria, was very high for several months with weekly shipments being sent for release in East Africa. Yearly production is extrapolated from these figures. Cost estimates are, however, difficult to obtain because rearing was done in old, refurbished buildings and building costs, as well as salaries, were paid in non-convertible local currency. The present cost estimate is therefore derived entirely from the known expenses for the newly constructed biological control center in Cotonou, Benin, which was inaugurated in December 1988. Construction of cassava trees and subsequent production of *E. lopezi* for release in 1989 was all done in Cotonou. Thus, all expenses, including salaries, are based on Franc CFA, a hard currency defined as 1/50 of a French Franc and converted at the rate of 300 per U.S. dollar.

### *The Yearly Production of* Epidinocarsis lopezi

All inputs and outputs of the different production systems were measured during 6 months to 1 year in order to estimate yearly production and costs. For the classical insectary, about 480 cuttings are planted in pots every week. Plants are used for infestation with CM 4 to 6 weeks after this planting, depending on the season. The CM culture for later production of *E. lopezi* requires about 30 cages, with four pots containing two plants each, maintained in the greenhouse for 2 weeks. Every week, plants from eight cages, which stayed in the greenhouse for 2 weeks, are brought to the rearing rooms for beneficials. The production is estimated at 250 wasps per pot during 4 weeks, that is 2 weeks of development and 2 weeks of collecting. With four pots per cage, the 54 cages in the large insectary room produce 540,000 *E. lopezi.* This assumes a productive period of 10 months, while a total of 2 months per year are needed for cleaning and repair.

For the mechanized cages, 600 cuttings are required weekly to plant one unit. Among the eight units, six are used to produce parasitoids, one is reserved for plants, and one for CM. With five cycles per year and 40,800 *E. lopezi* per cage, the annual production is estimated at 1.22 million wasps.

Six trees are planted weekly with a total of 500 cuttings. One tree is used for CM production and five for *E. lopezi,* yielding about 12,000 wasps each. This leads to an annual production of 3.12 million wasps.

## Yearly Expenses

For the three rearing systems, (cages, mechanized hydroponic units, and trees) costs for 1 year (Table 2) are estimated as follows using U.S. dollars:

**Personnel.** Plant production for the classical insectary is done by six technicians, CM rearing requires one technician, and the entomophagous insects are reared by three technicians and one research assistant. Operating the mechanized cages requires one assistant, a part-time electronic engineer, and three technicians. Cassava trees are maintained by six technicians and one research assistant.

In 1988, 40% of the time of all personnel in the green house and insectary using wooden cages was devoted to the production of *E. lopezi.* The rest of the time was used for the production of other exotic natural enemies of the CM not discussed here. For the mechanized hydroponic units and the trees, the corresponding percentages were 100% and 60%. The remaining 40% of the expenses for the trees were devoted to the production of *M. tanajoa* and its predators.

For supervision of rearing, fractions (between 10% and 80%) of the costs of principal staff (salary, benefits, administrative costs) are included. Total personnel costs are higher for trees than for the other systems because of the larger size of operation and the higher need for principal staff involvement.

**Capital Costs.** Several housing units are partially used for plant and insect rearing. Each consists of two offices and two laboratories, for a total cost of $120,000 in U.S. dollars; a specialized and climatized plastic house costs $200,000. A separate insectary building costs $200,000, and a workshop costs $150,000. The following fraction of these costs is used for the production of

**Table 2.** Cost estimates in U.S. dollars for three different systems for rearing *Epidinocarsis lopezi;* wooden cages, mechanized hydroponic units, and cassava trees, calculated for 1988/89 in Cotonou, Benin

| Item | Cages | Units | Trees |
|------|-------|-------|-------|
| Personnel | | | |
|   Technicians | $9,532 | $7,149 | $8,579 |
|   Research assistants | 3,928 | 9,821 | 5,893 |
|   Principal staff | 3,400 | 14,500 | 33,600 |
|     Subtotal | 16,860 | 31,470 | 48,072 |
| Capital | | | |
|   Buildings | 11,500 | 21,800 | 26,600 |
|   Cages | 1,040 | 24,000 | 4,500 |
|   Irrigation equipment | 0 | 0 | 1,000 |
|     Subtotal | 12,540 | 45,800 | 32,100 |
| Operating costs | | | |
|   Planting material | 420 | 520 | 440 |
|   Pots, Rockwool, | | | |
|     plastic bags, nutrients | 300 | 100 | 600 |
| Electricity, water | 1,500 | 16,000 | 4,000 |
|     Subtotal | 2,220 | 16,620 | 5,040 |
| Total costs per year | $31,620 | $93,890 | $85,212 |
| Total cost per | | | |
|   1,000 *E. lopezi* | $58.6 | $77.0 | $27.3 |

*E. lopezi*: for rearing in cages, 25% of a plastic house and 25% of the insectary rooms, as well as 10% of an office and 40% of a laboratory. The mechanized units are in a building of their own. Trees need one plastic house, 60% of one office and one laboratory, plus 20% of the main workshop. The initial cost of the buildings is depreciated linearly over 10 years.

**Equipment.** Costs for equipment include the prices for wooden cages (30 for rearing CM, 54 for *E. lopezi*, and 20 reserve, of $50 each), hydroponic units (8 for $15,000 each) and trees (100 units locally made, 50 of them with screens,

$225 each), which are depreciated over 5 years. As shown in Table 2, the cost of the mechanized cages is substantially higher than that of other systems.

## Operating Costs

Operating costs are prorated for the part used for *E. lopezi* and include electricity and water, expendable material such as pots, plastic bags, Rockwool, nutrient salts, fungicide, and cuttings (5 Franc CFA per piece). The value of the land (3 ha, half fallow) to produce these cuttings is not taken into account. Compared to personnel and capital costs, operating costs are minor.

**Cost per 1,000 *E. lopezi*.** From the total costs per year, the cost of 1,000 *E. lopezi* is calculated for the different rearing methods. Trees prove to be the most cost effective means of producing *E. lopezi*. They are, relative to production, less labor intensive than cages. Since they are made locally and contain relatively few expensive, high-technology components, they are considerably cheaper than the mechanized units on which insect rearing in hydroponic units was developed.

The estimate does not include development costs, delivery of the insects to the release sites, nor training of participants in the release program and the establishment of functioning national biological control programs (Herren 1987), all of which are considered essential to the success of this project.

A comparison with prices for parasitoids from commercial insectaries in a free-market economy (Dietrick 1981) is difficult for several reasons: 1) commercial insectaries sell beneficials exclusively for repeated inundative releases, 2) parasitoids are often reared on host plants and phytophagous insects which are alternative hosts chosen for ease of rearing, and 3) prices vary and are not usually published. Van Lenteren (1986) cites production costs of between U.S. $2 and $51 for four species, though widely produced *Trichogramma* spp. cost about 20 times less. Production costs for the ladybeetle *Cryptolaemus montrouzieri* Muls. were U.S. $2.50 as early as 1931 (Smith & Armitage 1931). Similarly, advertisements for weed control agents collected in the field offer U.S. $10 to $250 for 1,000 insects (IPM Practitioner 1989). It is concluded that production costs for *E. lopezi* are well within the range of those for other parasitoids from green plants and natural hosts but much higher than for those reared under artificial conditions.

## Conclusion

The impact of a biological control project through inoculative releases, such as the one with *E. lopezi,* is usually evaluated on the basis of the project costs and the savings incurred to the farmer (Dean et al. 1979, van den Bosch et al. 1982, Dover 1985, Headley 1985). Taking into account all expenses, the cassava mealybug project from its inception in 1979 up to the end of 1988 cost roughly 23 million U.S. dollars, of which about 60% was used for directly combatting the CM. During this period, *E. lopezi,* the most important and efficient natural enemy, was released and spread over about 1.5 million ha of cassava fields in Africa. This biological control program thus cost less than U.S. $10 per ha as a one-time expense to reduce the CM for all subsequent years. For the savannah zone in West Africa, it was shown that *E. lopezi's* presence reduced CM tuber yield loss on average by 2.5 ton per ha in 1985-86 alone (Neuenschwander et al. 1989a). Extrapolating these or similar figures leads to a very high rate of return (Norgaard 1988).

In the case of inundative releases, which have to be repeated every year from mass cultures of beneficials in insectaries, the cost per ha of the releases themselves can be calculated. Thus, in a developing country like China, costs for *Trichogramma* spp. releases amounted to U.S. $38 per ha on sugar cane (Cock 1985). For the inoculative releases with *E. lopezi,* and perhaps for all inoculative releases, such a question cannot be answered. Numbers of *E. lopezi* released varied from below 100 to several thousand per field (Herren et al. 1987), and all releases that could be followed were successful and led to the establishment and subsequent spread of *E. lopezi.* In view of the further propagation of *E. lopezi* in the field, production costs for the inoculum do not influence the economic impact of the project to a great extent. All that counts is the quality of the insects produced.

The following are considered key factors in maintaining the high quality of *E. lopezi* received from quarantine after its collection in the wild (Neuenschwander et al. 1989b): 1) rearing *E. lopezi* on its original host insect and host plant; 2) using careful rearing and good supervision, as often described, for the mass production of beneficials (Finney & Fisher 1964, King & Morrison 1984, Waage et al. 1985); and 3) incorporating the findings from ecological and bionomic studies of the phytophagous host and its beneficial. By developing new rearing technologies, such as the cassava tree, elements assuring good quality could be maintained and production costs lowered to a level matching those from commercial insectaries producing agents for inundative releases. This new technology has already been applied for the successful production on cassava of predatory mites and for the growth of maize, groundnuts, and cowpeas for experimental purposes. It has also been transferred to national biological control programs for producing their own control agents.

# References

Avidov, Z., Y. Roessler & D. Rosen. 1967. Studies on an Israel strain of *Anagyrus pseudococci* (Girault) (Hym.: Encyrtidae). II. Some biological aspects. Entomophaga 12: 111-118.

Baker, K. F. 1962. Thermotherapy of planting material. Phytopathology 52: 1244-1255.

Bartlett, B. R. 1978. Pseudococcidae. *In* C. P. Clausen [ed.], Introduced parasites and predators of arthropod pests and weeds: A world review, U.S. Department Agricultural Handbook 480: 137-170.

Biassangama, A., G. Fabres & J. P. Nenon. 1988. Parasitisme au laboratoire et au champ d'*Epidinocarsis* (Apoanagyrus) *lopezi* (Hym.: Encyrtidae) auxiliaire exotique introduit au Congo pour la regulation de l'abondance de *Phenacoccus manihoti* (Hom.: Pseudococcidae). Entomophaga 33: 453-465.

Boussienguet, J. 1986. Le complexe entomophage de la cochenille du manioc, *Phenacoccus manihoti* (Hom. Cocccoidea Pseudococcidae) au Gabon. I. Inventaire faunistique et relations trophiques. Ann. Soc. Entomol. Fr. (N.S.) 22: 35-44.

Boussienguet, J., P. Neuenschwander & H. R. Herren. Le complexe entomophage de la cochenilled du manioc, au Gabon. 4. Etablissement du parasitoide *Epidinocarsis lopezi*. Entomophaga (in press).

Clausen, C. P. 1942. Entomophagous insects. Haffner, New York.

Cock, M. J. W. 1985. The use of parasitoids for augmentative biological control of pests in the People's Republic of China. Biocontrol News Info. C.A.B. 6: 213-223.

Dahniya, N. T. & S. Kallon 1984. Rapid multiplication of cassava by direct planting. *In* E. R Terry, E. V. Doku, O. B. Arene & N. M. Mahungu [eds.], Tropical root crops: Production and uses in Africa. Proceedings Second Triennial Symp. Inter. Soc. Tropical Root Crops, Africa Branch, Douala, Cameroon.

De Bach, P. & E. B. White. 1960. Commercial mass culture of a California red scale parasite *Aphytis lignanensis*. Calif. Agric. Expt. Sta. Bull. 770: 58.

Dietrick, E. J. 1981. Commercial production of entomophagus insects and their successful use in agriculture, pp. 151-160. *In* G. C. Papavizas [ed.], Biological control in crop production. Beltsville Symp. Agric. Res., Allanheld, Osmum Granada.

Dean, H. A., M. F. Schuster, J. C. Boling & P. T. Riherd. 1979. Complete biological control of Antonina graminis in Texas with *Neodusmetia sangwani* (a classic example). ESA Bull. 25: 262-267.

Dover, M. J. 1985. A better mousetrap:  Improving pest management for agriculture. World Resources Institute, Holmes, PA.

Fabres, G. & J. Boussienguet. 1981. Bio-cologie de la cochenille du manioc (*Phenacoccus manihoti* Hom. Pseudococcidae) en Republique Populaire du Congo. Agron. Trop. 36: 82-89.

Fabres, G. & D. Matile-Ferrero. 1980. Les entomophages infeodes a la cochenille du manioc, *Phenacoccus manihoti* (Hom. Coccoidea Pseudococcidae) en Republique Populaire du Congo. I. Les composantes de l'entomocoenose et leurs inter-relations. Ann. Soc. Entomol. Fr. (N.S.) 16: 509-515.

Finney, G. L. & T. W. Fischer, 1964. Culture of entomophagous insects and their hosts. *In* P. De Bach & E. I. Schlinger [eds.], Biological control of insect pests and weeds. London: Chapman and Hall.

Forno, D. A., C. J. Asher & D. G. Edwards. 1979. Boron nutrition of cassava and the boron x temperature interaction. Field Crops Res. 2: 265-279.

Giordanengo, P. & J. P. Nenon. Melanisation and encapsulation of eggs and larvae of *Epidinocarsis lopezi* by host *Phenacoccus manihoti*: effects of superparasitism and egg laying pattern. Ent. Exp. Appl. (in press).

Gutierrez, A. P., B. Wermelinger, F. Schulthess, J. U. Baumgartner, H. R. Herren, C. K. Ellis & J. S. Yaninek. 1988a. Analysis of biological control of cassava pests in Africa. I. Simulation of carbon, nitrogen and water dynamics in cassava. J. Appl. Ecol. 25: 901-920.

Gutierrez, A. P., P. Neuenschwander, F. Schulthess, H. R. Herren, J. U.Baumgertner, B. Wermelinger, B. Lohr & C. K. Ellis. 1988b. Analysis of biological control of cassava pests in Africa. II. Cassava mealybug *Phenacoccus manihoti*. J. Appl. Ecol. 25: 921-940.

Hagen, K. S. & R. L.Tassan. 1970. The influence of food wheast and related *Saccharomyces fragilis* yeast products on the fecundity of *Chrysopa carnea* (Neuroptera: Chrysopidae). Can. Ent. 102: 806-811.

Hahn, S. K., E. R. Terry, K. Leuschner, I. O. Akobundu, C. Okali & R. Lal. 1979. Cassava improvement in Africa. Field Crops Res. 2: 193-226.

Hammond, W. N. O., P. Neuenschwander & H. R. Herren. 1987. Impact of the exotic parasitoid *Epidinocarsis lopezi* on cassava mealybug (*Phenacoccus manihoti*) populations. Insect Science Applic. 8: 887-891.

Hammond, W. O. & P. Neuenschwander. 1990. Sustained biological control of the cassava mealybug *Phenacoccus manihoti* (Hom.: Pseudococcidae) by *Epidinocarsis lopezi* (Hym.: Encyrtidae) in Nigeria. Entomophaga 35: (in press).

Haug, T., H. R. Herren, D. J. Nadel & J. B. Akinwumi. 1987. Technologies for mass-rearing of cassava mealybugs, cassava green mites and their natural enemies. Insect Science Applic. 8: 879-881.

Headley, J. C. 1985. Cost-benefit analysis: Defining research needs, pp. 53-63. *In* M. A. Hoy & D. C. Herzog [eds.], Biological control in agricultural IPM systems. Academic Press, Orlando.

Herren, H. R. 1987. A review of objectives and achievements. Insect Sci. Applic. 8: 837-840.

Herren, H. R. & K. M. Lema. 1982. CMB-first successfull releases. Biocontrol News Info., C.A.B. 3: 1.

Herren, H. R., P. Neuenschwander, R. D. Hennessey & W. O. Hammond. 1987. Introduction and dispersal of *Epidinocarsis lopezi* (Hym.: Encyrtidae), an exotic parasitoid of the cassava mealybug *Phenacoccus manihoti* (Hom.: Pseudococcidae), in Africa. Agric. Ecosystems Environ. 19: 131-144.

Hoagland, D. R. & D. I. Arnon. 1950. The water culture method for growing plants without soil. Calif. Agric. Exp. Stat. Circ. 347: 32.

IPM Practitioner. Monitoring the field of pest management, 1989. Commercial sources for biocontrol insects. IPM Practitioner 11: 9.

Iziquel, Y. 1985. Le parasitisme de la cochenille du manioc *Phenacoccus manihoti* par l'Encyrtidae *Apoanagyrus lopezi* (= *Epidinocarsis lopezi*): induction, modalites, et consequences agronomiques. Theses, Univ. Rennes.

Iziquel, Y. & B. Le Ru. 1989. Influence de l'hyperparasitisme sur les populations d'un hymenoptere Encyrtidae, *Epidinocarsis lopezi* (De Santis) parasitoide exotique de la cochenille du manioc *Phenacoccus manihoti* Mat.-Ferr. introduit au Congo. Entomol. Exp. Appl. 52: 239-247.

Kamalam, P., P. G. Rajendran & N. Hrishi. 1977. A new technique for the rapid propagation of cassava (*Manihot esculenta* Crantz). Tropic. Agric. (Trinidad) 54: 213-217.

Kraaijeveld, A. R. & J. J. M. van Alphen. 1986. Host-stage selection and sex allocation by *Epidinocarsis lopezi* (Hymenoptera: Encyrtidae), a parasitoid of the cassava mealybug, *Phenacoccus manihoti* (Homoptera: Pseudococcidae). Med. Fac. Landbouww. Rijksuniv. Gent. 51: 1067-1078.

King, E. G. & R. K. Morrison. 1984. Some systems for production of eight entomophagous arthropods, pp. 206-222. *In* E. G. King & N. C. Leppla [eds.], Advances and challenges in insect rearing. Agric. Res. Serv. U.S. Dept. of Agric., New Orleans.

Langenbach, G. E. J. & J. J. M. van Alphen. 1986. Searching behaviour of *Epidinocarsis lopezi* (Hymenoptera: Encyrtidae) on cassava: effect of leaf topography and a kairomone produced by its host, the cassava mealybug (*Phenacoccus manihoti*). Med. Fac. Landbouww. Rijksuniv. Gent. 51: 1057-1065.

Le Ru, B. & G. Fabres. 1984. Influence de la temperature et de l'hygrometrie relative sur la capacite d'accroissement et le profil des populations de la cochenille du manioc, *Phenacoccus manihoti* (Hom.: Pseudococcidae) au Congo. Acta Oecol. Oecol. Applic. 8: 165-174.

Le Ru, B., Y. Iziquel, A. Biassangama & A. Kiyindou. Comparaison des effectifs de la cochenille du manioc *Phenacoccus manihoti* avant et apres introduction d'Epidinocarsis lopezi Encyrtidae americain, au Congo en 1982. Ent. Exp. Appl. (in press).

Lema, K. M. & H. R. Herren. 1982. Temperature relationships of two imported natural enemies of CM. IITA Ann. Rep. 1981, 56-57.

Lema, K. M. & H. R. Herren. 1985. Release and establishment in Nigeria of *Epidinocarsis lopezi* a parasitoid of the cassava mealybug, *Phenacoccus manihoti*. Ent. Exp. Appl. 38: 171-175.

Leuschner, K., E. Terry & T. Akinlosotu. 1980. Field guide for identification and control of cassava pests and diseases in Nigeria. IITA Manual Series 3: 17.

Lohr, B., P. Neuenschwander, A. M. Varela & B. Santos. 1988. Interactions between the female parasitoid *Epidinocarsis lopezi* (Hym.: Encyrtidae) and its host the cassava mealybug, *Phenacoccus manihoti* (Hom.: Pseudococcidae). J. Ent. Appl. 105: 403-412.

Lohr, B., B. Santos & A. M. Varela. 1989a. Larval development and morphometry of *Epidinocarsis lopezi* (De Santis) (Hym.: Encyrtidae), parasitoid of the cassava mealybug, *Phenacoccus manihoti* Mat.-Ferr. (Hom.: Pseudococcidae). J. Appl. Ent. 107: 334-343.

Lohr, B., A. M. Varela & B. Santos. 1989b. Life-table studies of *Epidinocarsis lopezi* (De Santis) (Hym.: Encyrtidae), a parasitoid of the cassava mealybug, *Phenacoccus manihoti* Mat.-Ferr. (Homoptera: Pseudococcidae). J. Appl. Ent. 107: 425-434.

Lohr, B., A. M. Varela & B. Santos. 1990. Exploration for natural enemies of the cassava mealybug, *Phenacoccus manihoti* Matile-Ferrero (Hemiptera: Pseudococcidae), in South America for the biological control of this introduced pest in Africa. Bull Entomol. Res. (in press).

Mahon, J. D., S. B. Lowe, L. A. Hunt & M. Thiagarajah. 1977. Environmental effects on photosynthesis and transpiration in attached leaves of cassava (*Manihot esculenta* Crantz). Photosynthetica 11: 121-130.

Matile-Ferrero, D. 1977. Une cochenille nouvelle nuisible au manioc en Afrique equatoriale, *Phenacoccus manihoti* n. sp. (Hom.: Coccoidea Pseudococcidae). Ann. Soc. Entomol. Fr. (N.S.) 13: 145-152.

Murashige, T. & F. Skoog. 1962. A revised medium for rapid growth and bioassay with tobacco tissue culture. Physiologia Plantarum 15: 473-497.

Nadel, H. & J. J. M. van Alphen. 1987. The role of host- and host-plant odours in the attraction of a parasitoid, *Epidinocarsis lopezi* (Hymenoptera: Encyrtidae), to the habitat of its host, the cassava mealybug, *Phenacoccus manihoti*. Entomol. Exp. Appl. 45: 181-186.

Nenon, J. P., O. Guyomard & G. Hemon. 1988. Encapsulement des oeufs et des larves de l'Hymenoptere Encyrtidae *Epidinocarsis lopezi* (= Apoanagyrus) lopezi par son hote Pseudococcidae *Phenacoccus manihoti*; effet de la temperature et du superparasitisme. C. R. Acad. Sci. Paris 306, ser. 3: 325-331.

Nenon, J. P. & G. Fabres. 1988. Etude methodologique de l'efficacite parasitaire d'un Hymenoptere Encyrtidae *Epidinocarsis lopezi* introduit en Afrique pour lutter contre la cochenille du manioc *Phenacoccus manihoti*; bilan des travaux franco-congolais: 1982-1987. Proceedings VII Symposium Insect Sci. Appl. A.A.I.S. (in press).

Neuenschwander, P. & W. O. Hammond. 1988. Natural enemy activity following the introduction of *Epidinocarsis lopezi* (Hymenoptera: Encyrtidae) against the cassava mealybug *Phenacoccus manihoti* (Homoptera: Pseudococcidae), in southwestern Nigeria. Environ. Entomol. 17: 894-902.

Neuenschwander, P., W. O. Hammond, O. Ajuonu, A. Gado, N. Echendu, N., A. H. Bokonon-Ganta, R. Allomasso, I. Okon & T. A. Akinlosotu. 1990. Biological control of the cassava mealybug, *Phenacoccus manihoti* (Hom.: Pseudococcidae) by *Epidinocarsis lopezi* (Hym.: Encyrtidae) in West Africa, as influenced by climate and soil. Agric. Ecosystems Envir. (in press).

Neuenschwander, P., W. O. Hammond, A. P. Gutierrez, A. R. Cudjoe, J. U. Baumgertner, U. Regev & R. Adjakloe. 1989a. Impact assessment of the biological control of the cassava mealybug, *Phenacoccus manihoti* Matile-Ferrero (Hemiptera: Pseudococcidae) by the introduced parasitoid *Epidinocarsis lopezi* (De Santis) (Hymenoptera: Encyrtidae). Bull. Entomol. Res. 79: 579-594.

Neuenschwander, P., T. Haug, O. Ajuonu, H. Davis, B. Akinwumi & E. Madojemu. 1989b. Quality requirements in natural enemies used for inoculative release: Practical experience from a successful biological control programme. J. Appl. Ent. 108: 409-420.

Neuenschwander, P., R. D. Hennessey & H. R. Herren. 1987. Food web of insects associated with the cassava mealybug *Phenacoccus manihoti* Matile-Ferrero (Hemiptera: Pseudococcidae), and its introduced parasitoid *Epidinocarsis lopezi* (Hymenoptera: Encyrtidae), in Africa. Bull. ent. Res. 77: 177-189.

Neuenschwander, P. & H. R. Herren. 1988. Biological control of the cassava mealybug, *Phenacoccus manihoti*, by the exotic parasitoid *Epidinocarsis lopezi* in Africa. Phil. Trans. R. Soc. Lond. B. 318: 319-333.

Neuenschwander, P. & E. Madojemu. 1986. Mortality of the cassava mealybug *Phenacoccus manihoti* Mat.-Ferr. (Hom.: Pseudococcidae) associated with an attack by *Epidinocarsis lopezi* (Hym., Encyrtidae). Mitt. Schweiz. Ent. Ges. 59: 57-62.

Neuenschwander, P., F. Schulthess & E. Madojemu. 1986. Experimental evaluation of the efficiency of *Epidinocarsis lopezi*, a parasitoid introduced into Africa against the cassava mealybug *Phenacoccus manihoti*. Entomol. Exp. Appl. 42: 133-138.

Ng, S. Y. C. 1984. The application of tissue culture techniques in tuber crops at IITA. Central African Regional Root Crop Workshop, Brazzaville, Congo, 11-15 June 1984 (mimeograph).

Norgaard, R. B. 1988. The biological control of cassava mealybug in Africa. Amer. J. Agric. Econ. 70: 366-371.

Nsiama She, N. H. D. 1985. The biology of *Hyperaspis jucunda* (Col.: Coccinellidae) an exotic predator of the cassava mealybug *Phenacoccus manihoti* (Hom.: Pseudococcidae), Ph.D. thesis, Univ. Ibadan, Nigeria.

Nwanze, K. F., K. Leuschner & H. C. Ezumah. 1979. The cassava mealybug, *Phenacoccus* sp. in the Republic of Zaire. Pest Artic. News Summ. 25: 125-130.

Odebiyi, J. A. & A. H. Bokonon-Ganta. 1986. Biology of *Epidinocarsis* (=Apoanagyrus) *lopezi* (Hymenoptera: Encyrtidae) an exotic parasite of cassava mealybug, *Phenacoccus manihoti* (Homoptera: Pseudococcidae) in Nigeria. Entomophaga 31: 251-260.

Resh, H. M. 1978. Hydroponic food production. A definitive guidebook of soilless food growing methods. Woodbridge Press, Santa Barbara.

Schulthess, F., J. U. Baumgertner & H. R. Herren. 1987. Factors influencing the life table statistics of the cassava mealybug *Phenacoccus manihoti*. Insect Science Applic. 8: 851-856.

Smith, H. S. & H. M. Armitage. 1931. The biological control of mealybugs attacking citrus. Univ. Calif. Berkeley, Bull. 509: 74.

Sullivan, D. J. & P. Neuenschwander. 1988. Melanization of eggs and larvae of the parasitoid, *Epidinocarsis lopezi* (De Santis) (Hymenoptera: Encyrtidae), by the cassava mealybug, *Phenacoccus manihoti* Matile-Ferrero (Homoptera: Pseudococcidae). Can. Ent. 120: 63-71.

Sykes, J. T. & P. M. Harney. 1972. Rapid clonal multiplication of manioc from shoot and leaf-bud cuttings. J. R. Hort. Soc. 97: 530-534.

Sykes, J. T. & P. M. Harney. 1974. Cassava propagation: the effects of rooting medium and IBA on root initiation in hardwood cuttings. Trop. Agric. (Trinidad) 51: 13-21.

van Alphen, J. J. M., P. Neuenschwander, M. J. van Dijken, W. O. Hammond & H. R. Herren. 1989. Insect invasions: The case of the cassava mealybug and its natural enemies evaluated. The Entomologist 108: 38-55.

van den Bosch, R., P. S. Messenger & A. P. Gutierrez. 1982. An introduction to biological control. Plenum Press, New York.

van Dijken, M. J., P. van Stratum & J. J. M. van Alphen. 1988. Sex ratio studies in *Epidinocarsis lopezi*: local mate competition. Med. Fac. Landbouww. Rijksuniv. Gent 53: 1097-1108.

van Lenteren, J. C. 1986. Evaluation, mass production, quality control and release of entomophagous insects. Fortschritte Zool. 32: 31-56.

Viggiani, G. 1975. Possibilite di lotta biologica contro alcuni insetti degli agrumi (Planococcus citri Risso e Dialeurodes citri Ashm.). Boll. Lab. Ent. Agr. Filippo Silvestri, Portici 32: 3-10.

Waage, J. K., K. P. Carl, N. J. Mills & D. J. Greathead. 1985. Rearing entomophagous insects, pp. 45-66. *In* P. Singh & R. F. Moore [eds.], Handbook of insect rearing, vol. 1. Elsevier, Amsterdam.

Yaninek, J. S. 1988. Continental dispersal of the cassava green mite, an exotic pest in Africa, and implications for biological control. Exp. Appl. Acarol. 4: 211-224.

Yaseen, M. 1987. Exploration for *Phenacoccus manihoti* and *Mononychellus tanajoa* natural enemies: The challenge, the achievements, pp. 81-102. *In* R. D. Hennessey, H. R. Herren & R. Bitterli [eds.], Proceedings Workshop on Biological Control of the Cassava Mealybug and Green Spider Mites in Africa, 1982 December 6-10; Ibadan, Nigeria.

# 21

## Microhymenopterous Pupal Parasite Production for Controlling Muscoid Flies of Medical and Veterinary Importance

*Philip B. Morgan*

### Introduction

Although Legner and McCoy (1966) concluded that the common house fly, *Musca domestica* L., originated in Africa and was transported to the western hemisphere during pre-Columbian times, it has adapted to the western hemisphere and over the years has presented problems for farmers as urban development moved into the farming areas. The house fly developed resistance to chemical control and, with the failure of genetic engineering (Morgan et al. 1973, Wagoner et al. 1973) and the sterile insect technique (Meifert et al. 1967) to provide adequate control, investigators turned their attention to biological control.

Sustained releases of *Spalangia endius* Walker (Hymenoptera: Pteromalidae) have been conducted successfully against field populations of filth-breeding flies at beef, dairy, poultry, and swine installations (Morgan 1980a, 1981a; Morgan & Patterson 1977; Morgan et al. 1975a, 1975b, 1976a, 1981a, 1981b). A major prerequisite for successful augmentative or inoculative releases is selection of a host that can be produced both economically and efficiently for mass culturing the parasites and/or predators. The common house fly and stable fly, *Stomoxys calcitrans* (L.), are acceptable hosts. The life cycle of each species can be manipulated to provide immature stadia of the proper age (Morgan 1981c, 1985, 1986; Bailey et al. 1975). Stadia of both species have been used as a source of food for parasites and predators (Morgan 1980a, 1981b; Morgan & Patterson 1977; Morgan et al. 1978a, 1978b).


## Laboratory Studies

Both exotic and indigenous species of hymenopterous pupal parasites were maintained by the Mosquito and Fly Research Unit at the Insects Affecting Man and Animals Research Laboratory. Parasites of the genera *Spalangia, Muscidifurax,* and *Pachycrepoideus* (Hymenoptera: Pteromalidae) are ectoparasites of *M. domestica* and *S. calcitrans* (Diptera: Muscidae). *Coptera merceti* Kieffer is a solitary endoparasite of *S. calcitrans,* while *Trichopria* spp. (Hymenoptera: Diapriidae) are both gregarious and solitary endoparasites of *S. calcitrans,* and *Tachinaephagus* spp. (Hymenoptera: Encyrtidae) are gregarious endoparasites of *M. domestica* and *S. calcitrans* larvae.

The parasites were maintained in small Plexiglas cages that were held in isolation chambers (Fig. 1). The isolation chambers were held in controlled temperature rooms (two isolation chambers per controlled temperature room) (Fig. 1) located in laboratories maintained at 18.3-19.4°C. Temperatures ranging from 26-28°C were maintained in each controlled temperature room by actuating a supplementary portable, fan-forced heating unit. If the temperature exceeded 28°C, a thermostatically controlled fan on the roof of each controlled temperature room forced in cooler air from the laboratory. The temperature of each isolation chamber adjusted to that of each controlled temperature room. Relative humidity (70-80%) was maintained in each isolation chamber with a sponge floating in 500 ml of water in a glass container.

The inner surface of each isolation chamber was cleaned weekly with a 0.13% aqueous solution of zepharin chloride to control mold and bacteria. Mite infestations were controlled by dusting the inner surfaces of the isolation chambers and the outside of the Plexiglas cage with 1% sulfur dust. The inner surface of the Plexiglas cage was not dusted with the 1% sulfur dust since it was toxic to the parasites. Blacklight traps in each laboratory room attracted and killed escaped parasites (Morgan 1980b, 1981b, 1986; Morgan et al. 1975a, 1975b; Morgan & Patterson 1977, 1978).

The Pteromalidae were maintained normally on nonirradiated host pupae. However, they were reared successfully on irradiated host pupae. The endoparasitic Diapriidae required nonirradiated host pupae for survival. The host pupae were transported from the rearing facility to the isolation chambers in sealed nonwaxed paper containers (0.0625 liter) with 5,000-6,000 pupae per container. The containers of host pupae were placed in the parasite cages and opened, exposing the pupae to the parasites. If nonirradiated host pupae were exposed to the parasites, the carton was covered with aluminum window screen

**Fig. 1. Isolation chambers located in a walk-in free standing controlled temperature room.**

(25 squares/cm$^2$). This allowed the parasites free access to the pupae but prevented any nonparasitized flies from escaping from the carton. Twenty-four hours later the containers were resealed; labeled with the generation, colony number, and date; and held in the isolation chamber until the next parasite generation emerged. The following information was recorded for each generation for each parasite colony.

Since the Pteromalidae were maintained easily on irradiated pupae, all excess *M. domestica* and *S. calcitrans* 2-day-old pupae were exposed to 50 kR in groups of 24,000 in sealed 0.125-liter glass containers and held in a constant temperature cabinet (0.6-1.1°C). The irradiated pupae were suitable as hosts for 8 weeks (Morgan et al. 1986). The irradiation killed the host pre-imago, as well as any wild parasites that had parasitized the host pupae prior to separation from the larval medium. Parasite production was greater from freshly irradiated pupae than from nonirradiated 2-day-old *M. domestica* pupae. Irradiated *M. domestica* and *S. calcitrans* pupae were shipped overseas, exposed to *Spalangia cameroni* Perkins and *Muscidifurax raptor* G & S at the European Parasite Laboratory, USDA, Behoust, France, and transported to the United States with only a 10% reduction in parasite production. In the event of a reduction in host pupae production due to equipment failure or other problems, the irradiated pupae were used to supplement the nonirradiated host pupae. This assured an adequate supply of nonirradiated host pupae for the endoparasitic Diapriidae.

Each parasite generation was monitored for percentage parasitism and species purity. The males and females from 100 pupae were examined taxonomically (Boucek 1963, Peck et al. 1964) and then stored in 70% ethyl alcohol. Contamination of the *Spalangia* spp. by *M. raptor* or *Pachycrepoideus vindemiae* (Rondani) was controlled by holding the pupae from a *Spalangia* spp. colony in a constant temperature cabinet at 28-29°C and 70-80% relative humidity until the next generation of *Spalangia* spp. emerged. A pure culture of *Spalangia* spp. was produced in 21 days at this temperature, which was detrimental to the immature stages of both *M. raptor* and *P. vindemiae* (Morgan 1986). If a *Spalangia* spp. colony was contaminated by another species of Spalangia, the unwanted species was removed using a cold table (Fig. 2) and a vacuum pencil (Fig. 3). This was an effective but time-consuming procedure. Cross-contamination between *M. raptor* and *P. vindemiae* and contamination of either genera by *Spalangia* spp. required use of the cold table and vacuum pencil to obtain pure cultures.

Electrophoretic analysis was used to monitor the purity of the parasite colonies. *Spalangia endius* Walker, *S. cameroni* Perkins, and *M. raptor* were analyzed for isocitrate dehydrogenase (IDH), alpha glycerophosphate dehydrogenase (a-GPDH), phosphoglucomutase (PGM), malate dehydrogenase (MDH), glyceraldehyde-3-phosphate dehydrogenase (G-3-PDH), and creatine kinase (CK). Isocitric dehydrogenase patterns presented the clearest difference between *S. endius*, *S. cameroni*, and *M. raptor*. Malate dehydrogenase showed different bands between *S. endius* and *S. cameroni* but was a weak indicator for identifying *M. raptor*. The remaining enzymes were used to distinguish all three species (Morgan et al. 1988).

**Fig. 2. Equipment used to identify living parasites to species (cold table, stereoscope, fiber optic light, and vacuum pencil).**

Life history studies of the parasites were conducted, and a population model was developed for estimating the number of potential pupal hosts in a given area (Morgan & Patterson 1990, Weidhaas et al. 1977, Morgan et al. 1979a,b). Population host-parasite relationships were developed (Morgan et al. 1976b, 1978b, 1979b, 1989) so that a predetermined number of parasites could be released to attack a host fly population of a known size with a predicted degree of control (Morgan & Patterson 1990, Morgan et al. 1981a, Weidhaas et al. 1977).

**Fig. 3.** *Spalangia gemina* being held by the vacuum pencil.

## Mass Culturing *Spalangia endius* Walker

A special escape-proof Plexiglas parasite-host exposure cage 61 cm x 61 cm x 48 cm was used. Circular ports (19 cm in diameter) located in the top and side were covered with 0.0035 stainless steel cloth to aerate the cage. A front port 22 cm in diameter was covered with a 76-cm unbleached muslin tube and used for access. The exposure cage was held in a laboratory maintained at 23.3°C and 75 ± 5% relative humidity. Approximately 200,000 2-day-old *M. domestica* pupae were placed in aluminum trays (about 33,000/tray) in each cage, 4 days of each week (Monday through Thursday), and exposed to the female *S. endius* parasites at a parasite-to-host ratio of 1:5 to 1:10 for 18-24 h. Following the exposure period, the majority of the parasites were separated from the host pupae by attracting them to the ceiling of the cage with a fluorescent light and then to a lower corner of the cage with an incandescent light. The trays of pupae were emptied onto a screen (25 squares/cm$^2$) and the wasps not attracted to the fluorescent or incandescent lights were separated by hand from the host pupae. This procedure increased the efficiency of the mass culturing technique by retaining in the cage a majority of the surviving female parasites (Morgan 1986). The number of ovipositing females was kept at the correct parasite-to-host ratio by adding just enough females to compensate for the 33% daily loss rate (Morgan et al. 1976b). The parasitized pupae were packaged in groups of 20,000 in 1-liter paper cartons that were sealed and held in the parasite larvae development room. Two-thirds of the pupae were held at 27.8°C and one-third were held at 25.6°C, both with 75 ± 5% relative humidity. The parasites held at 27.8°C developed in 21 days, while the parasites held at 25.6° C developed in 25 days. This guaranteed parasite emergence Monday through Sunday with an overlap on Monday. Two-thirds of the emerging parasites were female.

This technique for mass culturing *S. endius* was used with slight modifications for other Pteromalidae and Diapriidae. Life history studies, such as daily loss rate, parasite-to-host ratio, sex ratio, and rate of increase, provided the data needed for mass culturing the different species.

## Mass Culturing *M. Domestica* and *S. Calcitrans*

The adults (about 12,000) were maintained in 45 x 35 x 35 cm cages constructed of aluminum framing covered with aluminum or plastic mesh (25 squares/cm$^2$) screening. A sleeve of orthopedic stockinet was attached at one end for easy access to the cage. The *M. domestica* adult diet consisted of 6 parts of granulated sugar, 6 parts of nonfat powdered milk, and 3 parts of powdered egg yolk. A crust that formed on the surface of the food due to high humidity and salivary secretions was broken up daily. Water in a 1-liter waxed container was also placed in each cage. The surface of the water was partially covered with styrofoam chips to provide greater access to the water by the flies and to prevent drowning. The water and food were replaced every other day (Morgan 1986). The *S. calcitrans* adult diet consisted of citrated bovine blood (3 g of sodium citrate ± 1 ml formalin/liter of blood) that was added to cotton balls to the point of saturation (Bailey et al. 1975). The adult fly rooms were maintained at 25°C and 75 ± 5% relative humidity with a 12:12 photoperiod.

Egg deposition by *M. domestica* was assured by increasing the temperature from 25°C and 75 ± 5% relative humidity to 27.7°C and 90 ± 5% relative humidity 1 hour prior to introducing the oviposition containers into the adult cages. The oviposition containers were prepared by wrapping aged larval medium in moistened unsized black cotton cloth (12 cm x 12 cm) and inserting it into a 0.35-ml waxed paper container. The black cloth provided ease of visibility and collection of eggs. Creases in the cloth induced more oviposition by the females. The eggs were carefully transferred to a 200-ml beaker of water and gently stirred to break up any egg masses. High embryonic mortality occurred if the eggs remained in the water longer than 15 min (Morgan 1986). The rearing room for *S. calcitrans* was maintained at 25°C and the females oviposited directly upon the surface of the cotton pad that had been saturated with bovine blood. The eggs were washed from the surface of the cotton pads and stored in distilled water for up to 48 h before being transferred to the larval medium (Bailey et al. 1975).

The larval medium for *M. domestica* was prepared by adding water to 2.6 kg of CSMA larval medium (Anonymous 1958). The amount of moisture in the dry larval medium, which varied with different shipments of medium, determined the amount of water that should be added. Normally, 1 liter of water was adequate if excess water could be obtained from a 100-g sample of medium. Three milliliters of eggs (10,000/ml) were added to the fresh larval medium. At 36 h, the fermenting larval medium peaked at 40°C, at 100 h pupation was initiated, and at 144 h 22,000 to 24,000 2-day-old pupae weighing 15 to 18 mg were produced. One hundred milligrams of medium were required for one larva to reach pupation. If fermentation was retarded, pupation was delayed, resulting in fewer and smaller pupae (Morgan 1986). The larval medium for *S. calcitrans* consisted of 3 parts of wheat bran, 1 part of bagasse, and 5 parts of water. Three milliliters of eggs (14,440/ml) added to the larval medium (2.26-2.72 kg) produced 22,324 pupae (13-14 mg) at 21°C and 70 ± 5% relative humidity 7 days following oviposition (Bailey et al. 1975). The pupae were separated easily from the dry surface of the medium by water flotation or by using a specially designed blower (Bailey 1970, Bailey et al. 1975).

## Activity of the Parasites in the Field

In evaluating the activity of parasites against *M. domestica, S. calcitrans, Fannia canicularis* (L.), and *F. femoralis* (Stein) at sites in the western hemisphere, Legner (1967) found that 92% of the house fly pupae collected in Uruguay were parasitized by six species of parasites *(M. raptor, S. cameroni, S. endius, S. nigroaenea* Perkins, *Tachinaephagus giraulti* Johnson and Tiegs, and *Trichopria n.* sp.); in New Brunswick, Canada, 90.7% of the house fly pupae were parasitized by *M. raptor* and *S. nigroaenea.* He also found that a host such as *S. calcitrans* with the habit of pupating nearer the surface in the breeding site was generally more heavily parasitized and that the parasites did not disperse rapidly from a release site. In southern California, Legner and Brydon (1966) evaluated parasitism at two poultry ranches over a period of 18 months. Six species of parasites were active; *M. raptor* and *S. endius* accounted for 95% of observed parasitism in *F. femoralis* and *Ophyra leucostoma* (Wiedemann).

Rutz (1986) monitored parasite populations with sentinel pupae at poultry installations near Ithaca, New York, and observed that *M. raptor* parasitized more sentinel pupae than did *S. cameroni, S. endius, S. nigroaenea, S. nigra,* or *P. vindemiae.* He concluded that the use of sentinel pupae favored *M. raptor* over the other parasites and thereby provided an accurate index of the predominant parasitoid species active in the manure, but periodic collections of naturally occurring pupae should be made to supplement data generated by sentinel pupae. Peterson (1986) evaluated the impact of parasites on filth fly populations associated with confined livestock installations in Nebraska and concluded that *S. nigra* was found more often in *S. calcitrans* pupae and *M. raptor* was more often recovered from *M. domestica* pupae; *M. raptor* were generally more often recovered from host pupae located near the surface of the breeding habitat.

Over a 2-year period, Hoyer (1986) recovered 22 species of parasites representing the families Pteromalidae, Braconidae, Diapriidae, Ichneuonidae, and Staphylinidae from field pupae collected in Europe and north Africa. The most frequent parasites collected were *S. cameroni* and *M. raptor,* with *S. endius* being recorded for the first time from France. A survey of Easter Island (Ripa 1986) demonstrated a complete absence of parasites, and the only predator present was *Teleogryllis oceanicus* (Le Guill). The author recovered both *S. endius* and *S. cameroni* at poultry installations in Hungary in 1986.

Thousands of *S. endius, M. raptor,* and *Tachinaephagus zealandicus* Ashmead were released from March through June 1970 at six poultry ranches to control *F. canicularais* and *F. femoralis.* Samples taken in June 1970 revealed a 6.5 times lower density (13.8 to 2.1) of these flies, and the parasitism percentage had almost doubled (12.9 to 22.5%) (Legner & Detrick 1972). Additional inoculative releases of *S. endius, M. zaraptor* Kogan and Legner, and *T. zealandicus* were made by Legner and Detrick (1974) over a 20-month period on poultry ranches located in southern California. They observed a significant reduction in average densities of Diptera: *M. domestica, Muscina stabulans* (Fallan), *F. canicularis, F. femoralis, O. leucostoma,* and *S. calcitrans.* Also, releases made during the spring had greater effect on fly populations than did similar releases in the summer. Olton and Legner (1975) made inoculative releases of *T. zealandicus, S. endius,* and *M. raptor* from December through April in an enclosed poultry house in southern California. The result was 46% parasitism of *M. domestica* but only 16% of *F. femoralis.* In addition, McCoy (1965) released *M. raptor* but found that parasitism of fly pupae never exceeded 25%.

Mourier (1972) also released *S. cameroni* and *M. raptor* on six farms (10,000 parasites per farm) in northern Denmark. Although the parasite population built up faster than normal, it was still insufficient to reduce host populations to an acceptable level. Monty (1972) released *S. endius, S. nigra, M. raptor, P. vindemiae,* and *Sphegigaster* spp. against populations of muscoid flies. *Spalangia* spp. was recovered in greater numbers than any other species released: 68% of the *M. domestica,* and 44.4% of the *S. calcitrans* were parasitized, and no parasites were recovered from *S. nigra* Marquart. Releases of *S. endius, S. cameroni,* and *M. raptor* against filth fly populations on Easter Island produced a low level of parasitism (Ripa 1986).

Augmentative releases of *S. endius* for 13 weeks against mixed populations of filth breeding flies at a 700-acre dairy in western Florida produced complete parasitism by the 12th week of all *M. domestica* and 99% to 100% parasitism of *S. calcitrans* that were recovered from the mixture of manure and silage. Parasitism of pupae collected from silage alone ranged from 33% to 88% (Morgan & Patterson 1977). Additional releases of *S. endius* at a small dairy in northern Florida reduced the fly population by 93% (Morgan et al. 1976a). Programmed releases of *S. endius* emerging four times each week against *M. domestica.* at a caged layer installation in northern Florida resulted in 100% parasitism by the eighth week, and an additional programmed release produced 100% parasitism when the parasites emerged seven times each week (Morgan et al. 1981a,b). Morgan and Patterson (1990) conducted augmentative releases of *S. endius* in conjunction with cyromazine against field populations of *M. domestica* in northern Florida and southern Georgia. The utilization of cyromazine in the food of the caged layers for 2 weeks prior to initiation of the parasite releases as well as good manure management resulted in a 99% reduction of the *M. domestica* population by the thirteenth week.

It can be concluded that the field-collected parasites of filth breeding flies can be successfully colonized in the laboratory. While earlier releases were less than successful (Legner & Detrick 1972, Mourier 1972), it has been demonstrated that programmed augmentative releases of *S. endius* against known populations of house flies can produce essentially complete parasitism of 1- and 2-day-old pupae and suppression of the fly population (Morgan et al. 1975a, 1975b, 1976a, 1979b, 1981a; Morgan & Patterson 1990; Weidhaas et al. 1977).

# References

Anonymous. 1958. Aerosol test method for flying insects, pp. 227-228. *In* Soap and chemical specialties blue book and catalog.

Bailey, D. L. 1970. Forced air for separating pupae of house flies from rearing medium. J. Econ. Entomol. 63: 331-33.

Bailey, D. L., T. L. Whitfield & G. C. LaBrecque. 1975. Laboratory biology and techniques for mass producing the stable fly, *Stomoxys calcitrans* (L.) (Diptera: Muscidae). J. Med. Entomol. 12: 189-193.

Boucek, Z. 1963. A taxonomic study in Spalangia Latr. (Hymenoptera, Chalcidoidea). Acta Entomol. Musei. Nationalis Pragae 35: 429-512.

Hoyer, H. 1986. Survey of Europe and North Africa for parasitoids that attack filth flies, pp. 35-38. *In* R. S. Patterson & D. A. Rutz [eds.], Biological control of muscoid flies. Entomol. Soc. Am. Misc. Publ. No. 61.

Legner, E. F. 1967. Behavior changes the reproduction of *Spalangia cameroni*, *S. endius*, *Muscidifurax raptor*, and *Nasonia vitripennis* (Hymenoptera: Pteromalidae) at increasing fly host densities. Ann. Entomol. Soc. Am. 60: 819-826.

Legner, E. F. & H. W. Brydon. 1966. Suppression of dung-inhabiting fly populations of pupal parasites. Ann. Entomol. Soc. Am. 59: 638-651.

Legner, E. F. & E. I. Dietrick. 1972. Inundation with parasitic insects to control filth breeding flies in California, pp. 129-130. *In* Proc. 40th Annu. Conf. Calif. Mosq. Control Assoc.

Legner, E. F. & E. I. Dietrick. 1974. Effectiveness of supervised control practices in lowering population densities of synanthropic flies on poultry ranches. Entomophaga 19: 467-478.

Legner, E. F. & C. W. McCoy 1966. The house fly, *Musca domestica* Linnaeus, an exotic species on the western hemisphere incited biological control studies. Can. Entomol. 98: 243-248.

McCoy, C. W. 1965. Biological control studies of *Musca domestica* and Fannia sp. on southern California poultry ranches, pp. 40-42. *In* Proc. 33rd Annu.

Meifert, D. W., G. D. LaBrecque, C. N. Smith & P. B. Morgan. 1967. Control of house flies on some West Indies Islands with metepa, apholate, and trichlorfon baits. J. Econ. Entomol. 60: 480-485. Conf. Calif. Mosq. Control Assoc.

Monty, J. 1972. A review of the stable fly problem in Mauritius. Rev. Agri. Sucriere Maurice 51: 13-29.

Morgan, P. B. 1980a. Sustained releases of *Spalangia endius* Walker (Hymenoptera: Pteromalidae) for the control of *Musca domestica* L. and *Stomoxys calcitrans* (L.) (Diptera: Muscidae). J. Kans. Entomol. Soc. 53: 367-372.

Morgan, P. B. 1980b. Mass culturing three species of microhymenopteran pupal parasites, *Spalangia endius* Walker, *Muscidifurax raptor* Girault and Sanders, and *Pachycrepoideus vindemiae* (Rondani) (Hymenoptera: Pteromalidae). *In* Proc. VIII Nat'l Meeting on Biological Control, Tecoman, Colima, Mexico.

Morgan, P. B. 1981a. The potential use of parasites to control *Musca domestica* L. and other filth breeding flies at agricultural installations in the southern

United States, pp. 11-25. *In* Status of biological control of filth flies. AR, SEA, USDA, New Orleans, Louisiana.

Morgan, P. B. 1981b. Mass production of *Spalangia endius* Walker for augmentative and/or inoculative field releases, pp. 185-188. *In* Status of biological control of filth flies. AR, SEA, USDA, New Orleans, Louisiana.

Morgan, P. B. 1981c. Mass production of *Musca domestica* L., pp. 189-191. *In* Status of biological control of filth flies. AR, SEA, USDA, New Orleans, Louisiana.

Morgan, P. B. 1985. *Musca domestica*, pp. 129-134. *In* P. Singh & R. F. Moore [eds.], Handbook of insect rearing, vol. 2. Elsevier, New York.

Morgan, P. B. 1986. Mass culturing microhymenopteran pupal parasites (Hymenoptera: Pteromalidae) of filth breeding flies, pp. 77-78. *In R. S.* Patterson & D. A. Rutz [eds.], Biological control of muscoid flies. Entomol. Soc. Am. Misc. Publ. No. 61.

Morgan, P.B., H. Hoyer, & R. S. Patterson. 1989. Life History of *Spalangia cameroni* (Hymenoptera: Pteromalidae), a microhymenopteran pupal parasite of muscoid flies (Diptera: Muscidae). J. Kans. Entomol. Soc. 62: 219-233.

Morgan, P. B., C. J. Jones, R. S. Patterson & D. Milne. 1988. Use of electrophoresis for monitoring purity of laboratory colonies of exotic parasitoids (Hymenoptera: Pteromalidae), pp. 525-31. *In* V. K. Gupta [ed.], Advances in parasitic hymenoptera research. E. J. Brill, Leiden/New York.

Morgan, P. B., G. C. LaBrecque & R. S. Patterson. 1978a. Mass culturing the microhymenopteran parasite, *Spalangia endius* (Hymenoptera: Pteromalidae). J. Med. Entomol. 14: 671-673.

Morgan, P. B., G. C. LaBrecque, D. E. Weidhaas & R. S. Patterson. 1979a. Interrelationship between two species of muscoid flies, and the pupal parasite, *Spalangia endius* (Hymenoptera: Pteromalidae). J. Med. Entomol. 16: 331-334.

Morgan, P. B. & R. S. Patterson. 1977. Sustained releases of *Spalangia endius* to parasitize field populations of three species of filth breeding flies. J. Econ. Entomol. 70: 450-452.

Morgan, P. B. & R. S. Patterson. 1978. Facilities for culturing microhymenopteran pupal parasitoids, pp. 32-33. *In* USDA Technical Bulletin 1576, Facilities for insect research and production.

Morgan, P. B. & R. S. Patterson. 1990. Efficiency of target formulations of pesticides plus augmentative releases of *Spalangia endius* Walker (Hymenoptera: Pteromalidae) to suppress populations of *Musca domestica* L. (Diptera: Muscidae) at poultry installations in the southeastern United States. *In* R. S. Patterson & D. A. Rutz [eds.], Biocontrol of arthropods affecting livestock and poultry. Entomol. Soc. Am. Misc. Publ. (in press).

Morgan, P. B., R. S. Patterson, G. C. LaBrecque, D. E. Weidhaas, A. Benton & T. Whitfield. 1975a. Rearing and release of the house fly pupal parasite *Spalangia endius* Walker. Environ. Entomol. 4: 609-611.

Morgan, P. B., R. S. Patterson, G. C. LaBrecque, D. E. Weidhaas & A. Benton. 1975b. Suppression of a field population of house flies with *Spalangia endius*. Science 189: 388-389.

Morgan, P. B., R. S. Patterson & G. C. LaBrecque. 1976a. Controlling house flies at a dairy installation by releasing a protelean parasitoid *Spalangia endius* (Hymenoptera: Pteromalidae). J. Georgia Entomol. Soc. 11: 39-43.

Morgan, P. B., R. S. Patterson & G. C. LaBrecque. 1976b. Host-parasitoid relationship of the house fly, *Musca domestica* L. and the protelean parasitoid, *Spalangia endius* Walker (Hymenoptera: Pteromalidae and Diptera: Muscidae). J. Kans. Entomol. Soc. 49: 483-488.

Morgan, P. B., B. J. Smittle & R. S. Patterson. 1986. Use of radiated pupae to mass culture the microhymenopteran pupal parasitoid *Spalangia endius* Walker (Hymenoptera: Pteromalidae). I. *Musca domestica* L. (Diptera: Muscidae). J. Entomol. Sci. 21: 222-227.

Morgan, P. B., D. E. Wagoner & R. L. Fye. 1973. Genetic manipulations used against a field population of house flies. 3. Males and females bearing a heterozygous translocation: Releases begun after initial seasonal peak population level reached. Environ. Entomol. 2: 779-782.

Morgan, P. B., D. E. Weidhaas & G. C. LaBrecque. 1978b. Host-parasite relationship of the house fly, *Musca domestica* L., and the microhymenopteran parasite, *Pachycrepoideus vindemiae* (Rondani). Southwest. Entomol. 3: 176-181.

Morgan, P. B., D. E. Weidhaas & G. C. LaBrecque. 1979b. Host-parasite relationship of the house fly *Musca domestica* L., and the microhymenopteran pupal parasite, *Muscidifurax raptor* Girault and Sanders (Diptera: Muscidae and Hymenoptera: Pteromalidae). J. Kans. Entomol. Soc. 52: 276-281.

Morgan, P. B., D. E. Weidhaas & R. S. Patterson. 1981a. Programmed releases of *Spalangia endius* and *Muscidifurax raptor* (Hymenoptera: Pteromalidae) against estimated populations of *Musca domestica* (Diptera: Muscidae). J. Med. Entomol. 18: 158-166.

Morgan, P. B., D. E. Weidhaas & R. S. Patterson. 1981b. Host-Parasite relationship: Augmentative releases of *Spalangia endius* Walker used in conjunction with population modeling to suppress field populations of *Musca domestica* L. (Hymenoptera: Pteromalidae and Diptera: Muscidae). J. Kans. Entomol. Soc. 54: 496-504.

Mouier, H. 1972. Release of native pupal parasitoids of house flies on Danish farms. Vidensk. Medd. Dan. Naturhist. Foren. 135: 129-137.

Olton, G. S. & E. F. Legner. 1975. Winter inoculative releases of parasitoids to reduce house flies in poultry manure. J. Econ. Entomol. 68: 35-38.

Peck, O., Z. Boucek & A. Hoffer. 1964. Keys to the Chalcidoidea of Czechoslovakia (Insects: Hymenoptera). Mem. Ent. Soc. Canada. 34: 1-170.

Peterson, J. J. 1986. Evaluating the Impact of Pteromalid Parasites on Filth Fly Populations Associated with Confined Livestock, pp. 52-56. *In* R. S. Patterson & D. A. Rutz [eds.], Biological control of muscoid flies. Entomol. Soc. Am. Misc. Publ. No. 61.

Ripa, R. 1986. Survey and use of Biological Control Agents on Easter Island and in Chile, pp. 39-44. *In* R. S. Patterson & D. A. Rutz (eds.), Biological control of muscoid flies. Entomol. Soc. Am. Misc. Publ. No. 61.

Rutz, D. A. 1986. Parasitoid monitoring and impact evaluation in the development of filth fly biological control programs for poultry farms, pp. 435-51. *In* R. S. Patterson & D. A. Rutz [eds.], Biological control of muscoid flies. Entomol. Soc. Am. Misc. Publ. No. 61.

Wagoner, D. E., P. B. Morgan, G. C. LaBrecque & O. A. Johnson. 1973. Genetic manipulations used against a field population of house flies. 1. Males bearing a heterozygous translocation. Environ. Entomol. 2: 128-134.

Weidhaas, D. E., D. G. Haile, P. B. Morgan & G. C. LaBrecque. 1977. A model to simulate control of house flies with a pupal parasite, *Spalangia endius*. Environ. Entomol. 6: 489-500.

# 22

# Rearing Systems for Screwworm Mass Production

## David B. Taylor

The screwworm, *Cochliomyia hominivorax* (Diptera: Calliphoridae), eradication effort employs the largest insect mass production program in the world. Approximately 400 billion insects have been reared, sterilized, and released since 1954, resulting in the eradication of screwworm from the United States and most of Mexico. The mass production factory, located at Tuxtla Gutierrez, Mexico, is designed to produce 500 million screwworm flies per week. When at peak production, the factory uses more than 450 metric tons of dry products to produce nearly 3.5 million liters of larval diet per month (Marroquin 1985). At that level of production, diet components cost $8.5 to $10 million per year. The Mexican eradication program achieved its goal of eliminating screwworm from Mexico west of the Isthmus of Tehuantepec in 1985 and switched from an eradication to barrier maintenance program. Production quotas for barrier maintenance have been between 200 and 300 million flies per week. Although extension of the eradication program into Central America (Belize and Guatemala) has already begun, increased production should not be needed (Snow et al. 1985, Krafsur et al. 1987).

The larval stage is the most critical part of the screwworm life cycle for mass production. The quality of the mass produced flies is determined primarily by the diet and care given during this stage, and the majority of consumable products, labor, energy, and physical facilities used for mass production is dedicated to the larvae. Therefore, improvements in larval rearing technology offer the greatest opportunity for improving the quality of the mass produced flies and reducing the cost of mass production.

The resources used for insect mass production are diet, labor, and facilities. The specific application of these resources to a mass rearing situation constitutes a rearing system. In most cases, facilities and labor are designed around the diet and the interaction of the biology of the insect with the diet. Major changes in diet, for example from meat to liquid or from liquid to gel, usually require modification of methods and facilities. When comparing diets for use in a mass rearing program, the interaction of the diet with the other resources must be considered. Optimal labor practices (techniques, schedules, etc.) and physical facilities (environment) should be used for each diet tested. Thus, rearing systems, not diets, must be compared.

The first artificial rearing system for screwworm used a diet consisting of lean beef, citrated beef blood, milk, water, and formalin (Melvin & Bushland 1936). Later, milk was eliminated from the diet (Melvin & Bushland 1940). The resulting rearing system was used for screwworm mass production with little modification until 1970 (Graham & Dudley 1959, Smith 1960, Baumhover et al. 1966). Several sources of meat were tested and used for varying periods of time, including cattle, horses, pigs, whales, and nutria (*Myocaster coypus*). Of these, horse meat was the most readily available, cheap, and conducive to good screwworm production. However, meat was expensive to store, and the supply was not reliable. Competition with fur ranchers and dog food manufacturers reduced the supply further and inflated the price of high quality meats (Gingrich et al. 1971). This prompted the search for alternative diets composed of stabilized proteins which could be shipped and stored without refrigeration. After testing hundreds of formulations, a liquid (hydroponic) diet, consisting of dried whole chicken egg, dried whole bovine blood, calf suckle (a milk substitute), sucrose, dried cottage cheese, and formalin suspended in water was developed (Gingrich et al. 1971). The hydroponic diet was initially suspended in cotton and later acetate fiber mats to provide support for the larvae. Waste hydroponic diet was vacuumed from the larval rearing vats every 4 h and replaced with fresh diet. The hydroponic rearing system, with minor modifications (Brown & Snow 1979), has been used for mass production up to the present time (Brown 1984, Goodenough et al. 1983, Marroquin 1985).

Though cheaper and easier to handle than meat, the hydroponic rearing system has several disadvantages, most importantly, it is labor dependent and intensive. Waste diet must be removed and replaced with fresh diet every 4 hours. Failure to properly remove waste diet results in accumulation of metabolic wastes, primarily ammonium bicarbonate (Brown & Snow 1978), forcing the larvae to leave the rearing vats before they are mature. Vacuuming and feeding are done by several crews dispersed throughout the larval rearing floor (about 4000 m$^2$), making supervision difficult. In addition, the environment on the larval rearing floor is less than conducive to quality workmanship. It is hot (about 35° C), humid (about 70% relative humidity) and ammonia concentrations in the air can be high (Merkle 1985).

There is little doubt that labor quality has a significant impact on the quantity and quality of screwworms produced with the hydroponic rearing system. During the past 3 years yield in the Tuxtla Gutierrez mass production facility has ranged from more than 9 liters to less than 3 liters of pupae per vat. During the same time period pupal weight has exceeded 50 mg and been below 40 mg. Pupal weight and yield are positively correlated ($r^2 = 0.78$), indicating that the variability is not due to competition. About 50% of the variability in pupal yield and weight during the last 3 years can be explained by one quantitative measure of labor quality (D. B. T. Taylor, unpublished data). Because of the limitations of the hydroponic system, a program was initiated to look for alternate diets and rearing systems.

Harris et al. (1984, 1985) tested nine gelling or solidifying agents for use in screwworm diets, six of which produced larvae of acceptable size (> 60 mg [Hightower et al. 1972]). A synthetic superabsorbent (poly [2-propenamide-co-2-propenoic acid, sodium salt]) marketed under the brand name Water-Lock G-400 yielded the most promising results. Initial tests indicated that the Water-Lock diet was roughly comparable with the meat diet being used for research rearing and the hydroponic diet used for mass production. Advantages of a gelled diet rearing system over the hydroponic rearing system were elimination of the acetate mats, reduced diet usage, and reduced labor.

The procedure used for rearing screwworm larvae on gelled diet in the research laboratory at Tuxtla Gutierrez, Mexico is as follows:

1. 100 mg of screwworm eggs (about 2,200 eggs) are incubated in a petri dish with a small piece of lean meat (5 g) 2 cm above a heated water bath (39° C) for 18 to 24 h (day 0).
2. First instar larvae with meat are transferred to a 19 by 16 by 9 cm deep plastic pan with 0.5 liters of gelled diet consisting of 70 g dried bovine blood, 30 g dried whole egg, 30 g nonfat dried milk, 1.2 ml formalin, 15 g Water-Lock G-400 gel, and 1000 ml of water. The pan is floated in a heated water bath (39° C) (day 1).
3. The second day after transferring the larvae to the diet, an additional 0.5 liters of diet is poured into the vats on top of the old diet and the larvae (day 3).
4. The next day one liter of gelled diet is added to the vat (day 4).
5. The following day the rearing vat is placed inside a larger plastic vat with 2 cm of sawdust in the bottom. Mature larvae crawl from the rearing vat and fall into the sawdust (day 5).
6. After 3 days of crawl-off, the rearing vat is discarded (day 8).
7. On day 10 the pupae are sifted from the sawdust. The volume of pupae recovered, weight of 10 ml of pupae, and number of pupae in 10 ml are recorded.
8. Pupae are then placed into cages. Adult eclosion usually begins on day 12.

Between 1,000 and 1,500 pupae (50 - 70% survival) weighing 50 - 55 mg are produced in each pan. A total of 2 liters of diet are used per plastic pan.

Factors critical to the successful use of the gelled diet are temperature of the water used to mix the diet, mixing the diet thoroughly, and the ambient environment, especially temperature and humidity. Mixing the diet with warm water (35° - 40° C) is necessary to get a gel of the proper consistency and avoid "chilling" the larvae with cold medium. Agitation of the diet throughout the gelling period (3 - 5 min) prevents the gel particles from sticking together and forming a hard, unsuitable gel. High temperatures (>38° C) and humidities (>90% relative humidity) cause excessive syneresis. The latter condition disrupts larval development and results in high mortality if the larvae are not transferred to fresh diet quickly. The gelled diet rearing system was compared with the meat rearing system used for research rearing by Taylor and Mangan (1987). They found that the gelled diet produced more consistent screwworms between rearing groups and over time than did the meat diet. Mean pupal weights and yields did not significantly differ between the two diets. Hydroponic diet was never adopted for research rearing of screwworms because it required 24 h per day attention and was not well suited to rearing many small discrete groups as was necessary in the laboratory. The Water-Lock diet offered advantages over the meat diet in that it was cheaper, the components were readily available and could be shipped and stored without refrigeration, while still being suited to rearing small discrete groups and fitting into the schedule of an 8 h per day operation. In addition, gelled diet was easier to handle, produced less odor, required less labor, and because there was no need to remove waste diet before refeeding, less opportunity for interstrain contamination. Gelled diet was adopted for all routine research rearing of screwworms at the Tuxtla Gutierrez laboratory in 1985.

Tests were initiated in 1986 to explore the use of gelled diet for screwworm mass production. Initial studies emphasized the use of gelled diet for initiation (first 2 days of development) because no modifications in equipment or procedures were necessary to implement gelled diet in this area of the rearing facility; whereas major modifications would be needed to introduce gelled diet onto the main rearing floor (last 3 days of larval development). Two obstacles were immediately encountered. At the high densities (8 grams of eggs per 5 liters of diet) being used in the starting rooms at that time, competition for nutrients and space had a greater effect on the development of the larvae than did the various diets tested (Taylor 1988). Once the densities in the starting rooms were reduced to 3 grams of eggs per 5 liters of diet, a second problem was encountered with the Water-Lock diet. The waste gelled diet clogged the vacuum heads when the larvae were transferred to the hydroponic vats on the main rearing floor. Inefficient removal of waste hydroponic diet resulted in reduced pupal yields and weights (R. Garcia, personal communication). A second gelling agent, carrageenan, which is soluble in warm hydroponic media was therefore adopted for use during initiation. Carrageenan starting diet produced smaller larvae at the end of the starting period, 4.7 mg verses 6.7 mg for Water-Lock, however, due to more efficient vacuuming on the hydroponic floor vats, carrageenan produced higher pupal weights and yields than did Water-Lock. Unfortunately, carrageenan was clearly inferior to Water-Lock for rearing screwworms through to pupation (Taylor 1988).

In mid 1987, tests were initiated to explore the use of Water-Lock gelled diet for rearing larvae in mass production from egg to pupation. Initial tests concentrated on using unmodified production floor vats. These vats are 91 by 152 by 4 cm deep and hold 55 liters of diet. In the first test, production of vats infested with 3 g of eggs and provided 55 liters of Walter-Lock diet were compared with production of vats infested with 6 g of eggs and provided 218 liters of hydroponic diet. The Water-Lock system produced an average of 32,400 pupae per vat (49% survival) with a mean weight of 50.3 mg while the hydroponic system produced an average of 78,900 pupae per vat (60% survival) with a mean weight of 53.6 mg. Though the hydroponic system produced higher pupal weights and yields, 27.8 liters of diet per 10,000 pupae (about 1 liter) produced were required for the hydroponic vats versus 17.3 liters of diet per 10,000 pupae with the Water-Lock system. In addition, the hydroponic system required 22 feedings occupying 21.25 min of labor versus 3 feedings for the Water-Lock system occupying 2.5 min of labor. After adjustment for yield per vat, 3.8 times more labor and 1.6 times more diet were required to produce a unit of pupae with the hydroponic system compared with the Water-Lock system. Experimental hydroponic vats were given better care than routine production vats. During the course of this experiment, production hydroponic rearing was averaging 49,500 pupae per vat (38% survival) with a mean weight of 45.9 mg. Calculations based on raw diet component consumption and direct observation indicate that production hydroponic vats received 107 liters of diet per vat (21.6 liters per 10,000 pupae) and 11.5 min of labor. Though the results of these tests were promising, especially with respect to diet usage and labor, survival and pupal weights were below what was expected based on the results of research rearing. Reducing the infestation rate from 3 to 2 g of eggs per vat increase survival to 59% and pupal weights to 52.6 mg. However, this also increased diet consumption (21 liters of diet per 10,000 pupae), labor, and required more space than was available in the mass production facility to maintain production quotas.

The most probable cause for the less than optimal production from the Water-Lock rearing system appeared to be the dimensions of the vats on the rearing floor. These vats were 4 cm deep whereas the vats used for research rearing were 9 cm deep. Therefore smaller, deeper vats, 66 by 46 by 9 cm deep, were tested. Each of these vats was infested with 1 g of screwworm eggs and provided 20 liters of diet. An average of 14,400 pupae (65% survival) weighing 53.5 mg were produced per vat. 13.9 liters of diet were required per

10,000 pupae. The smaller vats offered several advantages. They were not fixed to permanent supports and as such could be moved for feeding and cleaning, and they weighed less than 22 kg when full of diet so that one person could lift them. The ability to remove the vats from their support structure for feeding and cleaning allowed them to be packed closer together, thus reducing the space needed for larval rearing.

Temperature and humidity of the production larval rearing floor adversely affected the gelled diet rearing system as well. A test was conducted to compare rearing on the production floor (35°C) with rearing in the pupation room where the temperature was lower (30°C). Pupal yields did not significantly differ between the two rooms, however pupal weights increased from a mean of 50 mg on the production floor to 53.8 mg in the pupation room.

Based on these results, an area in the pupation room was modified to accommodate a pilot test for screwworm mass production with the Water-Lock rearing system. The pilot test was designed to produce 15 million pupae per week. This level of production was accomplished by preparing 160, 66 by 46 by 9 cm high plastic rearing pans. Each pan was infested with 1.25 g of screwworm eggs, provided 2 liters of diet, and placed in the first initiation room (39°C, 75%). An additional 2 and 6 liters of diet were added to the pans after 32 and 56 h. The pans were transferred to the main rearing floor at 56 h (35°C, 70%). An additional 10 liters of diet were added at 80 h and the pans were transferred to the pupation room (30°C, 60%). Mature larvae began to leave the rearing pans at approximately 96 h. The larvae that fell into metal pans under the rearing racks were collected every 4 h and placed in sawdust to pupate. For the 51 weeks during which the pilot test operated, an average of 19.2 million pupae averaging 47.1 mg (5.5 days postpupation) were produced each week. Yield averaged 1.38 liters of pupae per gram of eggs (63% survival) and diet usage was 11.7 liters per liter of pupae produced. During the same period, mass production pupae averaged 46.2 mg with a yield of 1.06 liters of pupae per gram of eggs (48% survival). Hydroponic diet usage was 25.3 liters per liter of pupae (Fig. 1). The gelled diet rearing system produced 2% heavier pupae, 32% higher yield, and consumed 54% less diet than mass production.

A cost comparison for hydroponic and gelled diet rearing systems is presented in Table 1. The cost for diet components to produce 300 million screwworm flies per week for one year with the hydroponic rearing system is $5,252,915 whereas the cost for the same number of flies using the Water-Lock rearing system is $3,805,312. Conversion to gelled diet will reduce the cost of diet components by $1,447,605 per year.

In addition to the savings in diet components, significant savings are foreseen in the costs of labor, facilities, and energy. Quantifying these savings is difficult. Conservative estimates indicate a 50% reduction in work force from 231 to 108 workers per day (H. Chris Hofmann, personal communication). Plant

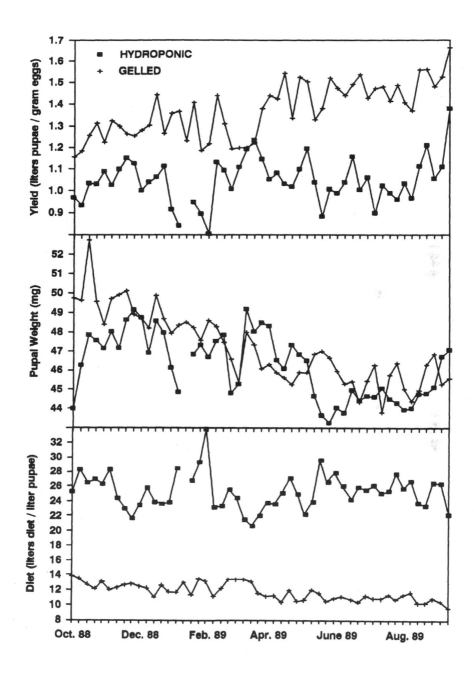

Fig. 1. Comparison of pupal yield (liters of pupae per gram of eggs), pupal weight, and diet consumption (liters of diet per liter of pupae produced) for screwworm mass production with hydroponic diet and a pilot test with gelled diet.

**Table 1. Cost of diet components for gelled and hydroponic diet rearing systems**

| | Diet Component | | | | | |
|---|---|---|---|---|---|---|
| Cost per: | Blood | Egg | Milk | Acetate | Water-Lock | Total |
| Kilogram | $0.95 | $1.54 | $1.03 | $4.73 | $7.01 | -- |
| **Gelled Diet System** | | | | | | |
| Liter diet | 0.057 | 0.046 | 0.015 | 0.000 | 0.091 | 0.209 |
| Vat[a] | 1.14 | 0.92 | 0.31 | 0.00 | 1.82 | 4.20 |
| Liter pupae[b] | 0.66 | 0.53 | 0.18 | 0.00 | 1.06 | 2.44 |
| **Hydroponic System** | | | | | | |
| Liter of diet | 0.057 | 0.046 | 0.015 | 0.00 | 0.00 | 0.118 |
| Vat[c] | 9.01 | 7.30 | 2.44 | 2.47 | 0.00 | 21.21 |
| Liter of pupae[d] | 1.43 | 1.16 | 0.39 | 0.39 | 0.00 | 3.37 |

Note: Diet usage and pupal yield data are for the 51 weeks beginning October 2, 1988 and ending September 17, 1989. Costs include purchase price and transportation to the rearing facility at Tuxtla Gutierrez, Mexico.
[a]20 liters of diet per vat.
[b]1.72 liters of pupae per vat.
[c]158 liters of diet per vat.
[d]6.3 liters of pupae per vat.

operation will be reduced from three full shifts per day to one full shift and two partial shifts. Energy and physical plant savings will be substantial, although insufficient information exists to make projections at this time.

Lower diet usage will reduce the amount of organic waste produced by the mass production facility. Waste gelled diet is semisolid, making handling, treatment, and disposal easier and reducing the organic load on the waste water treatment lagoons. This should improve the quality of the effluent being discharged. In addition, the waste diet still has a high protein content (Brown Snow 1978). Tests will be initiated soon to evaluate the use of waste screwworm diet as a protein supplement for livestock feed. Another potential use of the waste gelled diet is as a combination fertilizer-soil conditioner. Similar gelling agents are marketed for use as conditioners for sandy soil. Processing waste screwworm diet into a usable by-product will provide economic input to the local community as well as reduce the waste water treatment needs and environmental impact of the rearing facility.

Conversion of the mass production facility at Tuxtla Gutierrez from hydroponic to gelled diet will cost between $500,000 and $750,000. Modifications to the existing rearing floor, vat support structures, larval collection system, diet transport system, and feeding stations will cost approximately $35,000. Equipment for mixing the gelled diet, new pans and racks to hold them, and diet dispensers will cost about $675,000. Savings in expenditures for diet components during the first year should be sufficient to pay for the conversion.

A bilateral agreement was signed between the Mexican and American managers of the eradication program in June 1989 to convert the mass production facility to gelled diet. Modifications were initiated in October 1989 and conversion to the gelled diet rearing system was completed in April 1990.

Gelled diet produces screwworm pupae of equal or better quality than hydroponic diet with less labor and diet. Use of the gelled diet will improve the environment on the larval rearing floor and reduce the amount of time workers must spend on the floor, thus improving the overall working conditions in the mass production facility. Gelled diet will allow centralization of the feeding process and elimination of vacuuming, the two most critical steps in screwworm rearing. This will result in improved supervision and reduce the impact of labor on fly quality. Lower diet consumption will reduce waste production and improve the quality of the effluent discharged into the environment. Processing the waste diet into livestock feed supplements or soil conditioners will provide a badly needed boost to the local economy and improve the agricultural output of the region.

## Acknowledgment

Most of the experiments reported in this paper were done in cooperation with the Methods and Development Section of the Mexican-American Commission for the Eradication of Screwworms. Tom Ashley, Jim Bruce, and Rene Garcia provided administrative and technical support as well as data from Methods and Development tests. Don Bailey and Jim Mackley provided quality control data for mass production. H. Chris Hofmann provided data for the economic analyses. Marco Ocampo and his crew in the in-plant testing section provided technical assistance.

## References

Baumhover, A. H., C. N. Husman & A. J. Graham. 1966. Screwworms, pp. 53-554. *In* C. N. Smith [eds.], Insect colonization and mass production. Academic Press, New York.

Brown, H. E. 1984. Mass production of screwworm flies, *Cochliomyia hominivorax*, pp. 193-199. *In* E. G. King, & N. C. Leppla [eds.] Advances and Challenges in Insect Rearing. USDA, ARS, New Orleans.

Brown, H. E. & J. W. Snow. 1978. Protein utilization by screwworm larvae Diptera: Calliphoridae reared on liquid medium. J. Med. Entomol. 14: 531-533.

Brown, H. E. & J. W. Snow. 1979. Screwworms Diptera: Calliphoridae: a new liquid medium for rearing screwworm larvae. J. Med. Entomol. 16: 29-32.

Gingrich, R. E., A. J. Graham, & B. G. Hightower. 1971. Media containing liquified nutrients for mass-rearing larvae of the screwworm. J. Econ. Entomol. 64: 678-683.

Goodenough, J. L., H. E. Brown, L. E. Wendel, & F. H. Tannahill. 1983. Screwworm eradication program: a review of recent mass-rearing technology. Southwest. Entomol. 8: 16-31.

Graham, A. J. & F. H. Dudley. 1959. Culture methods for mass rearing of screwworm larvae. J. Econ. Entomol. 52: 1006-1008.

Harris, R. L., E. F. Gersabeck, C. Corso, & O. H. Graham. 1985. Screwworm larval production on gelled media. Southwest. Entomol. 10: 253-256.

Harris, R. L., R. D. Peterson, M. E. Vazquez-G., & O. H. Graham. 1984. Gelled media for the production of screwworm larvae. Southwest. Entomol. 9: 257-262.

Hightower, B. G., G. E. Spates, Jr. & J. J. Garcia. 1972. Relationship between weight of mature larvae, size of adults, and mating capability in medium-reared male screwworms. J. Econ. Entomol. 65: 1527-1528.

Krafsur, E. S., C. J. Whitten, & J. E. Novy. 1987. Screwworm eradication in North and Central America. Parasitol. Today 3: 131-137.

Marroquin, R. 1985. Mass production of screwworms in Mexico, pp. 31-36. *In* O. H. Graham [ed.], Symposium on eradication of the screwworm from the United States and Mexico. Misc. Publ. Entomol. Soc. Am. 62.

Melvin, R. & R. C. Bushland. 1936. A method of rearing *Cochliomyia americana* C. & P. on artificial media. U.S. Bur. Entomol. Plant Quar. [Rep.] ET-88, 2 pp.

Melvin, R. & R. C. Bushland. 1940. The nutritional requirements of screwworm larvae. J. Econ. Entomol. 33: 850-852.

Merkle, S. E. 1985. Ambient concentrations of formaldehyde and ammonia in a screwworm fly [*Cochliomyia hominivorax*] rearing facility. Am. Ind. Hyg. Assoc. J. 46: 336-340.

Smith, C. L. 1960. Mass production of screwworms *Callitroga hominivorax* for the eradication program in the southeastern States. J. Econ. Entomol. 53: 1110-1116.

Snow, J. W., C. J. Whitten, A. Salinas, J. Ferrer, & W. H. Sudlow. 1985. The screwworm, *Cochliomyia hominivorax* Diptera: Calliphoridae, in Central America and proposed plans for its eradication south to the Darien Gap in Panama. J. Med. Entomol. 22: 353-360.

Taylor, D. B. 1988. Comparison of two gelling agents for screwworm Diptera: Calliphoridae larval diets. J. Econ. Entomol. 81: 1414-1419.

Taylor, D. B. & R. L. Mangan. 1987. Comparison of gelled and meat diets for rearing screwworm, *Cochliomyia hominivorax* Diptera: Calliphoridae, larvae. J. Econ. Entomol. 80: 427-432.

# 23

## Mass Rearing Biology of Larval Parasitoids (Hymenoptera: Braconidae: Opiinae) of Tephritid Flies (Diptera: Tephritidae) in Hawaii

*Tim T. Y. Wong and Mohsen M. Ramadan*

### Introduction

The Mediterranean fruit fly, *Ceratitis capitata* (Wiedemann), the melon fly, *Dacus cucurbitae* Coquillett, and the oriental fruit fly, *Dacus dorsalis* Hendel, are major pests of fruits and vegetables in Hawaii. *D. cucurbitae* was introduced into Hawaii from the Orient in 1895 (Back & Pemberton 1917); *C. capitata* from Australia about 1910 (Back & Pemberton 1918); and *D. dorsalis* from the Pacific Islands in 1945 (Van Zwaluwenburg 1947). Immediately after the establishment of these fruit flies, importation of exotic parasitoids into Hawaii from Africa, Asia, and Australia resulted in outstanding biological control (Back & Pemberton 1918, Willard 1920, Nishida 1955, Bess & Haramoto 1961, Clausen et al. 1965). *Psyttalia fletcheri* (Silvestri), a larval parasitoid of *D. cucurbitae* was introduced into Hawaii from India in 1916 and by 1918 parasitization of more than 80% was obtained for *D. cucurbitae* larvae collected from cultivated cucurbits (Willard 1920). However, in recent years, parasitization of this species was about 40% (Newell et al. 1952, Nishida 1955). Willard and Mason (1937) reported that biological control of *C. capitata* in Hawaii before entry of *D. dorsalis* centered on the larval parasitoids, *Opius humilis* Silvestri and *Diachasmimorpha tryoni* (Cameron) (=*O. tryoni*). *O. humilis* attained its maximum parasitization in 1915, and *D. tryoni* was the dominant species from 1917 to 1933 (Willard & Mason 1937). Between 1947 and 1952, many parasitoid species were brought into Hawaii to control *D. dorsalis* (Clausen et al. 1965). However, only three became widespread and

abundant:    *Biosteres   arisanus*   (Sonan)   *(=Opius    oophilus* Fullaway),
*Diachasmimorpha longicaudata* (Ashmead) *(=O. longicaudatus)*, and *Biosteres*
*vandenboschi (Fullaway) (=O. vandenboschi)* (Van Den Bosch et al. 1951, Bess
& Haramoto 1961, Clausen et al. 1965). These three braconid species were also
effective in parasitizing *C. capitata* as there were no significant differences in
parasitization in cases where the tephritids present were all *D. dorsalis*, all *C.*
*capitata*, or a combination of the two (Bess & Haramoto 1961, Haramoto &
Bess 1970). Parasitization averaged 60% or more of tephritids obtained from
coffee and guava fruits on all of the Hawaiian islands since 1951 (Haramoto &
Bess 1970). Moreover, Haramoto and Bess (1970) reported that since 1951,
when *B. arisanus* became the dominant species, parasitization by *D.*
*longicaudata, D. tryoni,& B. vandenboschi* has accounted for less than 4% of
the total parasitization of the two tephritids on guava and coffee. However, in
more recent studies in the Kula area of Maui, Hawaii, Wong et al. (1984) and
Wong & Ramadan (1987) found that other opiine as well as *B. arisanus* are
important in regulating tephritid populations. In Kula during a 5-year study, *B.*
*arisanus* accounted for 74% of the total parasitization. However, *D.*
*longicaudata* exceeded 30% of the total parasitization in 1979 from infested
peaches and in 1985 from loquats; *D. tryoni* accounted for 33% in 1987 and
29% in 1985 from loquats (Wong & Ramadan 1987).

　　We have been rearing six species of these parasitoids of tephritids in the
laboratory since 1981. These are *B. arisanus, D. longicaudata, D. tryoni, B.*
*vandenboschi, P. fletcheri*, and Psyttalia incisi (Silvestri). Over a million wasps
per week were reared from *D. longicaudata, D. tryoni*, and *P. fletcheri* for use
in augmentative releases in the field to suppress *D. dorsalis, C. capitata*, and
*D. cucurbitae*, respectively. The other three species were reared in small
numbers primarily for basic biological studies and for exporting to other workers
throughout the world for use in suppression of various tephritids.

　　The present study describes the procedures we used for mass production and
reports on factors that affect the reproductive capacity of *D. longicaudata* reared
from *D. dorsalis, D. tryoni* from *C. capitata*, and *P. fletcheri* from *D.*
*cucurbitae* in order to achieve an optimum degree of efficiency in mass rearing.
Our study reports on  1) influence of host age on offspring production and
sex-ratio and 2) reproductive activity in relation to change in parasitoid age.


## Materials and Methods

　　Host larvae of the three tephritid species were mass cultured by methods
currently used by the Agricultural Research Service of the USDA (Tanaka 1965;
Tanaka & Steiner 1965; Tanaka et al. 1970, 1972; Mitchell et al. 1965; and
Vargas et al. 1986). A culture of *D. tryoni* had been established in our

laboratory in 1981 with the initial cohorts collected from Kula, Maui, Hawaii. The culture of *D. longicaudata* was initially reared from infested fruits collected from Oahu and Maui, Hawaii in 1981. Initial cohorts of *P. fletcheri* were collected from infested zucchini squash in Waimanalo, Oahu, Hawaii in 1985. *P. fletcheri, D. longicaudata,* and *D. tryoni* were continuously propagated in the laboratory on their primary hosts, *D. cucurbitae, D. dorsalis,* and *C. capitata,* respectively at 26°C and more than 60% relative humidity.

**General Procedure and Maintenance of Parasitoid Cultures.** Middle 3rd instar hosts (6-day-old) were used for insectary propagation of the three parasitoid species. Higher percentages of successful adult emergence were realized in the production of *D. tryoni* (Wong et al. 1990, Ramadan et al. 1989b) and *D. longicaudata* (Bess et al. 1950, Lawrence 1982) when young host larvae were used. To minimize the rate of superparasitism, unlimited numbers of host larvae (about 350-500 per oviposition unit) were exposed to parasitoids for 6 h daily. The oviposition unit was a modified Petri dish (9 cm in dia. and less than 0.5 cm deep (Fig. 1). Tight-fitted lids with organdy covers were used to hold the host larvae inside the dishes. Before packing the larvae in oviposition units, they were mixed while still in their larval medium (in plastic boxes 48 x 32 x 15 cm) with wheat shorts to make the larval medium drier. Moist oviposition units caused entanglement of the parasitoids at time of oviposition and appeared to be avoided by many other parasitoids. After 6 h exposure to parasitoids, host larvae were removed from units, dumped with their substrate media into plastic boxes (48 x 32 x 15 cm) and larval food-medium was mixed with enough water to maintain a soft, moist consistency. Host larvae tended to leave dry media prematurely or pupate inside the rearing trays. Host rearing trays (77 x 40 x 8 cm, adequate for larvae from about 100 oviposition units) were held in wooden holding cabinets (95 x 64 x 187 cm, Fig. 2). When mature, the host larvae vacated the larval food-medium and fell into water-filled fiberglass containers (48x 32 x 15 cm) located below. Collection of larvae in water for less than 24 h stopped their physiological development without harmful effect (Vargas et al. 1986), and they pupated normally when placed in vermiculite, the pupation substrate. After sifting from the water, the host larvae were distributed equally into pupation boxes (48 x 32 x 15 cm with 2 screened sides) provided with moist vermiculite and held at 26°C and more than 60% relative humidity. Host puparia during the first 2 days of pupal development were susceptible to desiccation (Langley et al. 1972); pupation without moist vermiculite resulted in small, short lived flies (Vargas et al. 1986). Environmental humidity was also important for parasitoid larval development, and the eclosion rate increased with increasing soil moisture (Delangue & Pralavorio 1977, Ashley et al. 1976). Host puparia were sifted from vermiculite on the 7th day after pupation. Pupal handling earlier than 6 days affected adult fly eclosion and caused a higher

**Fig. 1. Screened wooden cage containing parasitoids and oviposition units.**

percentage of uneclosed host puparia (Ozaki & Kobayashi 1981). Sifted puparia (for mechanical sifting, see Ozaki & Kobayashi 1982) were held on screened wooden trays (67 x 53 x 2 cm) in 4-sided screened emergence cabinets (76 x 63 x 155 cm). Flies emerged from unparasitized puparia about 10-13 days after pupation at 26°C and more than 60% relative humidity and died within 2 days after becoming adults; the remainder of the puparia contained the immature parasitoids. Empty puparia were blown away using a modified hair dryer (a regular hair dryer with the heater circuit removed). The puparia containing the immature parasitoids were divided into lots of 20 ml pupae, placed in paper containers, and distributed to the cohort cages. The cohort cage (Fig. 1) was of wooden construction (26 x 26 x 26 cm) with two sides and the top screened (fine-mesh saran filter cloth), a rear rubber side with circular opening closed with a tight fitting plastic cup, and the front side with sliding glass to enable washing. The bottom side of the cage was screened to interface with two oviposition units from outside the cage. Inserting oviposition units inside the cage increased the handling time and allowed the cohorts to escape. The cohort

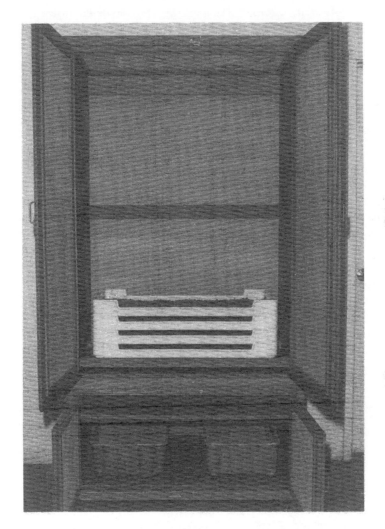

**Fig. 2. Host (fly larvae) rearing trays in a wooden holding cabinet.**

cages were provided with water bottles with cotton wicks and streaks of velvety-texture honey were applied on the top screen. The cohorts in cages that were not provided with water and honey suffered higher mortality than others. Water in cages raised humidity that diluted the honey and parasitoids were also seen chewing the wick fibers for sips of water. A constant difference of 2 days of developmental time was noticed between peaks of male and female emergence

(Ashley et al. 1976, Pemberton & Willard 1918a). More than 96% of all parasitoids eclosed within 7 days of the onset of adult parasitoid eclosion (16-20 days after host pupation). The portion of parasitoids that entered larval dormancy (up to several months if temperature is lower than 13°C) was not included in our results; however, at 24.7°C only 0.7% of *D. tryoni* entered larval dormancy (M. M. R., unpublished data). No premating periods were observed in these parasitoid species (Hagen 1953); more than 70% of the females were inseminated within 4 days after eclosion (M. M. R., unpublished data). As in other synovigenic species of Hymenoptera (Flanders 1950) cohort females of the three species were capable of ovigenesis throughout their lifetimes. Higher ovarian maturation and higher mating rates were achieved 4 days after female emergence (Ramadan et al. 1989a). Therefore, parasitoids were denied access to hosts until the 5th posteclosion day.

To determine the influence of cohort age on progeny production, newly emerged parasitoids (200 females and 200 males) were placed in cohort cages and exposed to host larvae on the 5th day following eclosion. One oviposition unit containing middle 3rd instar hosts was exposed to parasitoids for 6 h daily. Previous results showed that more than 90% of the daily ovipositing parasitoids died before 30 days. Therefore, tests were terminated at 30 days or when more than 95% of the cohorts died. After oviposition, exposed hosts were allowed to develop in plastic cups (900 ml) with screened tops at 26°C and more than 60% relative humidity until parasitoid eclosion. When parasitoids ceased to emerge (about 30 days after oviposition), parasitoid males and females, flies (host larvae that escaped parasitism), and uneclosed host puparia were counted. The number of dead parasitoids in the cages and the number of cohorts with females attracted to oviposition units were also counted once daily.

The effect of host age on parasitoid offspring sex ratio (percent females) and rates of uneclosed host puparia was determined. Host larvae of 1, 4, 5, 6, and 7 days at (26.7°C and more than 60% relative humidity) represented 1st instar, late 2nd instar, early 3rd instar, middle 3rd instar, and late 3rd instar, respectively. Tephritid larvae form a puparium after the 3rd instar and therefore are no longer available to the parasitoids. One-week-old parasitoids of the three species with previous oviposition experience (1-3 times exposed to hosts) were exposed to a series of oviposition units containing the different host instars. Host larvae were exposed to parasitoids for 2 h, prolonged to 6 h when 1st and 2nd instars were used, to allow parasitoids to encounter an adequate number of hosts. After exposure to parasitism, additional fresh food medium (Tanaka et al. 1972) was added to host-rearing containers to provide adequate nutrients for early larval instars to develop to maturity. The same procedure as in previous test was followed until adult parasitoids eclosed. The sex ratio of emerged progeny (percent females) and percentages of uneclosed host puparia were recorded for every category of host instar. The test was replicated at least 10 times for each variable. Analysis of variance was performed on the data to assess the potential significance ($p < 0.05$) of the reproductive attributes of the three parasitoid species at several successive cohort age intervals and the effect of host age on offspring sex ratio. Means were separated by Duncan's multiple range test at $p = 0.05$ level (SAS Institute 1985). Percentage data were transformed by arcsine proportion before analysis, but the untransformed means were presented for comparison.

## Results

Table 1 presents the results of progeny production of *P. fletcheri* as influenced by six maternal age intervals using middle 3rd instar of *D. cucurbitae*. An average of 11% of the initial cohorts remained alive at the last age interval (25-30 days). Cohort females reached their highest activity in the age interval (11-15 days) where 42.6% of the cohorts competed for oviposition on the host unit, however, this activity declined as the age of cohort females and mortality increased. A large number of hosts was presented to exposed parasitoids (about 400 per unit); consequently, only 34.7% of the hosts were encountered by the parasitoids during the 6 h exposure. Percentages of parasitism declined to 13% as the number of females attracted to the host unit declined in the last age interval (26-30 days) as well as parasitoid mortality. Total number of parasitoid progeny per oviposition unit was significantly high ($p < 0.05$) at first oviposition indicating that 5-day-old parasitoids were in their highest reproductive activity (230 parasitoid progeny per oviposition unit). Progeny production declined significantly but to an acceptable level until age group 21-25 days. Yield of last age interval was less than 50 parasitoids per unit. Yield of 91.4% of the total progeny was obtained during the reproductive period (5-25 days).

Fig. 3 presents the influence of cohort age intervals on the offspring sex ratio. The percentages of female offspring of *P. fletcheri* increased significantly reaching as high as 65.2% at cohort age interval (11-15 days) and then declined significantly as the cohort aged. However, progeny sex ratio consistently remained over 50% at all cohort age intervals.

Table 2 presents the results of progeny production of *D. longicaudata* as influenced by four maternal age intervals, using middle 3rd instar of *D. dorsalis* for 6 h exposure daily. By the end of the second week, after emergence, approximately three-fourths of the ovipositing parasitoids were dead, and only 4% of them survived until the last age interval (16 days). The highest percentages of females attracted to oviposition units (55.4%) were observed on the first oviposition day and this activity declined as cumulative mortality of parasitoids increased. As a result of this activity, significantly high percentages

Table 1. Influence of female parasitoid age on the production of *Psyttalia fletcheri* progeny using 3rd instar larvae (6-day-old) of *Dacus cucurbitae* exposed to parasitoid oviposition (initial cohort of 200 males and 200 females per cage) for 6 h ($\bar{x}$ ±SEM)[a]

| Age interval (days) | Replicates | Live females (%) | Females attracted to oviposition | Hosts/ oviposition unit | Parasitism[b] (%) | Parasitoid progeny/oviposition unit | | |
|---|---|---|---|---|---|---|---|---|
| | | | | | | ♂♂ | ♀♀ | Total |
| 5 | 5 | 95.1 ±1.6a | 11.0 ±1.8c | 668.2 ±30.2a | 34.7 ±2.6a | 112.0 ±9.6a | 118.2 ±8.5a | 230.2 ±14.6a |
| 6-10 | 25 | 85.8 ±1.3b | 22.8 ±2.3b | 445.7 ±27.3bc | 33.0 ±2.9a | 58.6 ±7.1b | 93.3 ±10.2b | 151.8 ±17.0b |
| 11-15 | 25 | 68.0 ±1.8c | 42.6 ±2.2a | 362.5 ±32.3c | 28.0 ±2.3a | 36.3 ±5.0bc | 61.6 ±5.9c | 97.9 ±10.4cd |
| 16-20 | 25 | 44.6 ±1.9d | 25.9 ±2.6b | 354.8 ±37.4c | 29.3 ±3.4a | 54.4 ±11.4b | 57.7 ±7.6c | 112.1 ±18.4bc |
| 21-25 | 25 | 24.8 ±1.6e | 15.7 ±1.4c | 536.4 ±30.5b | 16.3 ±1.9b | 47.7 ±7.8bc | 45.8 ±5.7cd | 93.4 ±13.0cd |
| 26-30 | 25 | 11.4 ±0.9f | 4.2 ±0.6d | 426.3 ±32.1bc | 13.0 ±1.9b | 21.2 ±2.7c | 25.6 ±3.1d | 46.8 ±5.4d |

[a]Means within columns followed by the same letter are not significantly different ($p < 0.05$; Duncan's multiple range test [SAS Institute 1985]).

[b]Average of 5 sample values of percent parasitism = total progeny - total host puparia.

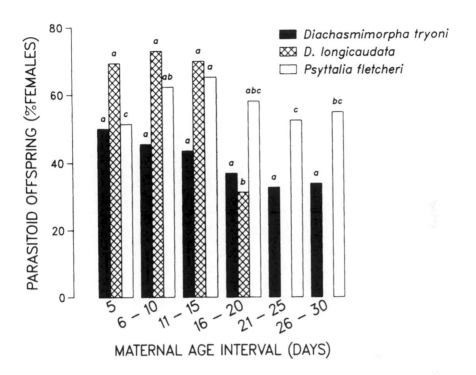

**Fig. 3. Influence of maternal age intervals (200 female parasitoids each) of *D. tryoni*, *D. longicaudata*, and *P. fletcheri* on the progeny sex (% emerged females). For each parasitoid species, mean percentages of females followed by the same letter are not significantly different (p < 0.05 Duncan's multiple range test [SAS Institute]).**

of hosts (69.6) were parasitized on the first oviposition day. The total number of parasitoid progeny per oviposition unit was grouped into three significantly different categories. The highest yield in age intervals was 5-10 days (270 parasitoids per unit), the medium yield in age interval was 11-15 days (165.8 parasitoids per unit) and the lowest yield in age interval was 16 days (less than 3 parasitoids per unit). Yields of 99.9% of the total progeny were obtained during parasitoid reproductive periods of 5-15 days. Progeny sex ratio (Fig. 3) of *D. longicaudata* was significantly high at age intervals 5-15 days. The maximum percentage of female offspring was 72.9% at age group 6-10 days, which declined thereafter to the lowest level (31.3% females) at age 16 days.

Table 2. Influence of female parasitoid age on the production of *Diachasmimorpha longicaudata* progeny using middle 3rd instar larvae (6-day-old) of *Dacus dorsalis* exposed to parasitoid oviposition (initial cohort of 200 males and 200 females per cage) for 6 h ($\bar{x}$ ±SEM)[a]

| Age interval (days) | Replicates | Live females (%) | Females attracted to oviposition | Hosts/ oviposition unit | Parasitism[b] (%) | Parasitoid progeny/oviposition unit | | |
|---|---|---|---|---|---|---|---|---|
| | | | | | | ♂♂ | ♀♀ | Total |
| 5 | 5 | 97.5 ± 0.7a | 55.4 ± 5.3a | 390.8 ± 29.1a | 69.6 ± 3.3a | 83.4 ± 9.0a | 187.4 ± 14.0a | 270.8 ± 21.1a |
| 6-10 | 25 | 85.3 ± 3.0b | 44.5 ± 2.5a | 424.8 ± 54.9a | 56.9 ± 3.3ab | 70.4 ±12.2a | 207.5 ± 36.9a | 277.8 ± 48.3a |
| 11-15 | 25 | 24.5 ± 3.9c | 15.3 ± 2.8b | 347.5 ± 27.7a | 41.9 ± 4.2b | 50.4 ± 8.4a | 115.4 ± 17.8ab | 165.8 ± 25.1ab |
| 16 | 5 | 4.0 ± 1.7d | 0.9 ± 0.2c | 212.2 ±10.6a | 1.3 ± 0.4c | 1.6 ± 0.4b | 1.0 ± 0.5b | 2.6 ± 0.7b |

[a]Means within columns followed by the same letter are not significantly different ($\underline{P} < 0.05$; Duncan's multiple range test [SAS Institute 1985]).

[b]Average of 5 sample values of percent parasitism = total progeny - total host puparia.

Table 3 presents the results of progeny production of *D. tryoni* as influenced by five maternal age intervals using middle 3rd instar of *C. capitata* for 6 h exposure daily. Survivorship patterns of the ovipositing females of *D. tryoni* were closer to *P. fletcheri* than to *D. longicaudata*. More than 80% of the females died before reaching age interval, 21-25 days, and only 4.7% were alive at the last age interval 26-30 days. Percentages of females attracted to oviposition units were significantly highest at age intervals 6-15 days, and then declined. Percentages of parasitism followed the pattern of female attraction to oviposition units. Maximum total progeny production was about 140 parasitoids per unit at the first age interval, which was optimum until age interval 16-20 days. Progeny production per unit at age intervals 21-30 days was significantly lowest (6-21 parasitoids per unit). Yields of 93.1% of the total progeny were obtained during parasitoid reproductive periods of 5-10 days. Fig. 3 shows that the progeny sex ratio of *D. tryoni* declined from 50% to 40% as the cohorts aged; however, differences in sex ratio in this test were not statistically significant.

Economy of progeny production per cohort cage of 200 females of the three parasitoid species are shown in Table 4. From the table, *D. longicaudata* had the significantly highest utilization of its host (47.1% parasitism), highest progeny production per unit (207.6 parasitoids per day), and the highest sex ratio (67.9% female progeny). Total parasitoid progeny production per cohort cage was 2740.6, 2491.4, and 1969.0 for *P. fletcheri, D. longicaudata, and D. tryoni* respectively. This represents a 10- to 13-fold increase in population growth of parasitoids during one generation (about one month per generation).

Fig. 4 summarizes the results of host age differences on progeny sex ratio of the three parasitoid species. In *D. tryoni* and *D. longicaudata*, progeny sex ratio increased significantly as their host larvae aged toward maturity. Mature late instars of *C. capitata* received the highest percentage of *D. tryoni* fertilized eggs (66.6 ± 6.1% females) whereas, the 1st instar received the highest percentage of unfertilized parasitoid eggs (93.8 ± 1.9% males). The same phenomenon occurred in *D. longicaudata*, where the late 3rd instar of *D. dorsalis* produced the highest percentage of emerged female offspring (66.7 ± 1.3% females) and significantly more male offspring emerged from 2nd instar (9.5 ± 1.5% females). Surprisingly, offspring sex ratio of *O. fletcheri* did not follow the same pattern in this test. The sex ratio obtained from 2nd instar *D. cucurbitae* (51.5 ± 1.9% females) was not significantly different from that obtained from late 3rd instars (63.3 ± 4.7% females).

Fig. 5 presents the results of the influence of host age on uneclosed host puparia. In *D. tryoni*, percentages of uneclosed host puparia increased significantly from 5.0 to 35.3% as host larvae exposed to cohorts matured. The mean percentage mortality in unparasitized host larvae reared under the same condition was 2.4%. In *D. longicaudata*, percentages of uneclosed host larvae

Table 3. Influence of female parasitoid age on the production of *Diachasmimorpha tryoni* progeny using middle 3rd instar larvae (6-day-old) of *Ceratitis capitata* exposed to parasitoid oviposition (initial cohort of 200 males and 200 females per cage) for 6 h ($\bar{x}$ ±SEM)[a]

| Age interval (days) | Replicates | Live females (%) | Females attracted to oviposition | Hosts/ oviposition unit | Parasitism[b] (%) | Parasitoid progeny/oviposition unit | | |
|---|---|---|---|---|---|---|---|---|
| | | | | | | ♂♂ | ♀♀ | Total |
| 6-10 | 5 | 80.6 ± 2.1b | 22.1 ± 3.6a | 587.4 ±132.3a | 27.2 ± 7.4a | 63.8 ± 13.9b | 57.8 ±15.8ab | 121.6 ±24.2ab |
| 11-15 | 5 | 64.5 ± 4.9c | 24.7 ± 2.9a | 550.2 ±123.9a | 22.9 ± 3.9a | 75.8 ± 21.5a | 63.8 ±17.9a | 139.6 ±37.1a |
| 16-20 | 5 | 36.8 ± 3.9d | 11.5 ± 1.5b | 509.2 ± 76.4a | 14.4 ± 3.0a | 41.8 ± 7.9ab | 25.2 ± 6.7bc | 67.0 ±13.3bc |
| 21-25 | 5 | 18.7 ± 2.4e | 5.0 ± 0.6c | 504.6 ± 46.5a | 4.0 ± 0.8b | 14.2 ± 4.0bc | 6.6 ± 1.6c | 20.8 ± 5.2c |
| 26-30 | 5 | 4.7 ± 0.9f | 2.1 ± 0.2c | 198.4 ± 11.1b | 3.3 ± 1.0b | 3.8 ± 0.7c | 2.6 ± 1.4c | 6.4 ± 2.0c |

[a]Means within columns followed by the same letter are not significantly different ($P < 0.05$; Duncan's multiple range test [SAS Institute 1985]).

[b]Average of 5 sample values of percent parasitism = total progeny - total host puparia.

Table 4. Summary of parasitoid progeny production per oviposition unit per day (6-h-exposure) and production per cohort cage of 200 females of *Psyttalia fletcheri*, *Diachasmimorpha longicaudata*, and *Diachasmimorpha tryoni* reared on middle 3rd instar larvae (6-day-old) of *Dacus cucurbitae*, *Dacus dorsalis*, and *Ceratitis capitata*, respectively (x̄ ±SEM)[a]

| Parasitoid | Hosts/ oviposition unit | Parasitism[b] (%) | Puparia uneclosed (%) | Parasitoid production cage of 200 | Parasitoid production per unit per day (6h) | | | |
|---|---|---|---|---|---|---|---|---|
| | | | | | ♂♂ | ♀♀ | Total | Sex ratio (% ♀♀) |
| P. fletcheri | 433.9 ±7.5b | 24.3 ±1.5b | 17.7 ±0.3b | 2740.6 ±191.2a | 46.3 ±3.3a | 59.2 ±4.7b | 105.4 ±4.7b | 58.3 ±1.8b |
| D. longicaudata | 372.0 ±3.4c | 47.1 ±2.3a | 13.7 ±0.7a | 2491.4 ±110.7a | 57.4 ±2.7a | 150.2 ±6.6a | 207.6 ±9.2a | 67.9 ±1.8a |
| D. tryoni | 480.1 ±46.0a | 14.8 ±2.5c | 7.1 ±1.1c | 1969.0 | 42. ±7.5a | 33.7 ±7.1b | 75.7 ±14.0b | 38.9 ±2.9c |

[a]Means within columns followed by the same letter are not significantly different (P < 0.05; Duncan's multiple range test [SAS Institute 1985]).

[b]Sample size equals 5 for *P. fletcheri*; 5 for *D. longicaudatus*; and 1 for *D. tryoni*.

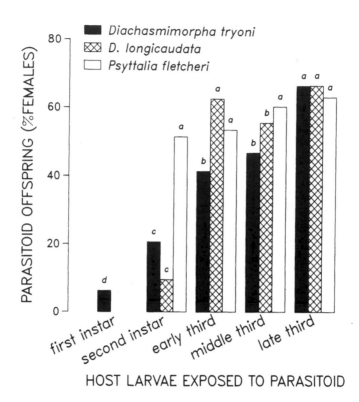

**Fig. 4. Influence of host larval instars of *C. capitata*, *D. dorsalis*, and *D. cucurbitae* on the progeny sex ratio (% emerged females) of *D. tryoni*, *D. longicaudata*, and *P. fletcheri*, respectively. For each parasitoid species, mean percentages of females followed by the same letter are not significantly different (p < 0.05 Duncan's multiple range test [SAS Institute 1985]).**

increased as the exposed hosts aged toward maturity but not to the same extent as in *C. capitata* exposed to *D. tryoni*. In *P. fletcheri* a significant increase (up to 27.2 ± 4.0%) in unclosed host puparia was observed in the late 3rd instar of *D. cucurbitae*.

## Discussion

Information on reproductive activity in relation to cohort maternal age is useful to determine optimum yield and knowledge of specific cohort age groups where parasitoid brood colonies should be discarded (Carey et al. 1988). Not

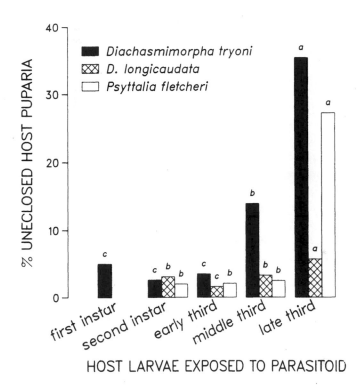

**Fig. 5. Influence of host larval instars of *C. capitata*, *D. dorsalis*, and *D. cucurbitae* exposed to *D. tryoni*, *D. longicaudata*, and *P. fletcheri* on the percentages of uneclosed host puparia. For each parasitoid species, mean percentages of uneclosed host puparia followed by the same letter are significantly different (p < 0.05 Duncan's multiple range test [SAS Institute 1985]).**

all the parasitoids were seen ovipositing, and they competed aggressively on the oviposition unit, especially at the early age intervals. This behavior maintained a proportion of the cohorts on the unit; as a result, not all of the host larvae were encountered and a proportion of the hosts always remained unparasitized. As with the other synovigenic species of Hymenoptera, the cohorts were able to achieve their highest oviposition rates a few days after eclosion. The highest egg maturation rates in *D. tryoni* (Ramadan et al. 1989a) and *D. longicaudata* (Lawrence et al. 1978a) were realized on the fifth day after eclosion. During this preoviposition period females were allowed to mature and mate before starting oviposition, thus achieving the highest possible production per unit.

The quality of hosts reflects significantly on the realized fecundity and longevity of parasitoids. Large, healthy larvae of *D. dorsalis* and *D. cucurbitae* would be expected to produce larger parasitoids than those emerged from *C. capitata*. Large parasitoids as in many other insect species tend to deposit more eggs than smaller ones (Hinton 1981). Therefore, the greater oviposition rates in *D. longicaudata* and *P. fletcheri* suggests that host species enhanced the parasitoid fecundity significantly. Proper oviposition exposure period is another important factor in mass-production of parasitoids in order to minimize the rates of superparasitization and achieve a high rate of host utilization. Four to eight hours exposure to parasitoids appeared optimum and was alternated according to the age of cohorts, number of oviposition parasitoids, and number and instar of hosts available in the oviposition unit (Ramadan et al. 1989b).

Yield per unit is an important criterion to determine when to discard cohorts. *D. longicaudata* tends to achieve a high rate of oviposition earlier than other species and also to die faster. Thus, cohorts older than 15 days should be discarded. Also *P. fletcheri* and *D. tryoni* females older than 25 days should be discarded.

Understanding the dynamics of progeny sex ratios of parasitoids is relevant to assist in improving their mass propagation in the insectary (Waage & Ming 1984). As gregarious parasitoids and parasitoids emerged from patchily-distributed hosts, progeny sex ratio can be expected to be influenced by theory of "Local Mate Competition" (Hamilton 1967; Werren 1980, 1983; Waage & Lane 1984). However, the concept of host quality (Charnov 1982, Van Den Assem et al. 1984) applies equally well to our parasitoids where the host age differentially affects emerged offspring sex ratio. Host qualities were widely reported in arrhenotokous parasitoids to influence the female's decision to fertilize (produce female) or not to fertilize (produce male) the deposited egg (Clausen 1939, Sumaroka 1967, Sandlan 1979, Waage 1982).

Quantitative changes in the hemolymph have been demonstrated as possible stimuli to the parasitoids for acceptance or rejection of hosts (Fisher 1971, Rajendram & Hagen 1974). King and Hopkins (1962) found that changing pH was the cue to sperm movement from the spermatheca. Late instars (more than younger tephritid hosts) apparently trigger the fertilization mechanism in *D. longicaudata* and *D. tryoni*.

Host cues can be detected by the parasitoid ovipositor, which is widely reported among endoparasitic Hymenoptera to contain sensory structures important in host selection (Gutierrez 1970, Vinson 1976). External dome-shape sensillae with sense organs, observed at the tip of the ovipositor in *D. longicaudata* (Greany et al. 1977), were proposed to function as chemosensory cells. We found similar sensillae in *D. tryoni* that may detect chemical stimuli (e.g., pH level, enzymes, metabolites, or hormones) that cause parasitoid females to permit or shut down fertilization of a particular egg.

Survival of immature parasitoids in the hosts was highly influenced by the host instars; hormonal interference was found to be involved (Lawrence et al. 1978b) that allowed 1st instar parasitoids to develop to the subsequent stages (Lawrence 1986). The phenomenon of unsuccessful development in older hosts is a function of endocrine interactions and a remarkable synchronization of development between endoparasitoids and their hosts (Bradley & Arbuthnot 1938, Nappi & Streams 1970, Omata 1984, Beckage 1985). However, physical competition among parasitoid 1st instars is another reason for the mortality among larvae of endoparasitic Hymenoptera (Salt 1961). Supernumerary 1st instars of *D. tryoni* competed with each other when the host was superparasitized (Pemberton & Willard 1918b). High titers of ecdysone and low juvenile hormone levels were necessary for 1st instar *D. longicaudata* to molt to subsequent instars (Lawrence 1982), and the same phenomenon was recorded in the related Opiinae, *Psyttalia* (=Opius) *concolor* (Szepligeti) (Cals-Usciati 1975). The fact that the first larval molt never occurred in these parasitoids until the host pupated (Willard 1920, Back & Pemberton 1918) supported the above hypothesis Even with high titers of host hormones, the 1st instar of *P. concolor* required 2 days to molt to the 2nd instar. This period of time was necessary for the 1st instar to feed and imbibe the hormone (Cals-Usciati 1975). The results of high parasitoid emergence from 2nd instar hosts agree with the previous explanation, as there is adequate time for the parasitoid egg to hatch and for the 1st instar to develop under an acceptable titer of host hormones.

We conclude from our results on the basic biology of these three parasitoids, that mass-production is feasible. The described procedures enabled us to produce several million per week of *D. tryoni* for shipments and pilot field tests against the Mediterranean fruit fly. *D. longicaudata* and *P. fletcheri* are even more amenable for mass production than *D. tryoni* in Hawaii because of the high quality of *D. dorsalis* and *D. cucurbitae* as hosts. Host larvae age and parasitoid maternal age were primary factors influencing the parasitoid yield per oviposition unit and offspring sex ratio. Thus, successful propagation of these valuable parasitoids depends on careful control of host quality.

## References

Ashley, T. R., P. D. Greaney & D. L. Chambers. 1976. Adult emergence in *Biosteres* (Opius) *longicaudatus* and *Anastrepha suspensa* in relation to the temperature and moisture concentration of the pupation medium. Florida Entomol. 59: 391-395.

Back, E. A. & C. E. Pemberton. 1917. The melon fly in Hawaii. U.S. Dept. Agric. Bull. 491.

Back, E. A. &C. E. Pemberton. 1918. The Mediterranean fruit fly in Hawaii. U.S. Dept. Agric. Bull. 536.

Beckage, N. E. 1985. Endocrine interactions between endoparasitic insects and their hosts. Ann. Rev. Entomol. 30: 371-413.

Bess, H. A. & F. H. Haramoto. 1961. Contributions to the biology and ecology of the oriental fruit fly, *Dacus dorsalis* Hendel (Diptera: Tephritidae) in Hawaii. Hawaii Agricultural Experiment Station, University of Hawaii, Hawaii Tech. Bull. 44.

Bess, H. A., R. Van Den Bosch & F. Haramoto. 1950. Progress and status of two recently introduced parasites of the oriental fruit fly, *Dacus dorsalis* Hendel, in Hawaii. Proc. Hawaiian Entomol. Soc. 14: 29-33.

Bradley, W. G. & K. D. Arbuthnot. 1938. The relation of host physiology to development of the braconid parasite, *Chelonus annulipes,* Wesmael. Ann. Entomol. Soc. Am. 31: 359-365.

Cals-Usciati, J. 1975. Repercussion de la modification due cycle normal de *Ceratitis capitata* (Wied.) (Diptera: Trypetidae) par irradiation (Y) et injection d'ecdysone sur le developpement de.son parasite *Opius concolor* (Hymenoptera: Braconidae). Comptes Rendus-Academie Science Paris, Serie D.281: 275-278.

Carey, J. R., T. T. Y. Wong & M. M. Ramadan. 1988. Demographic framework for parasitoid mass rearing: Case study of *Biosteres tryoni,* a larval parasitoid of tephritid fruit flies. Theoretical Population Biology. 34: 279-296.

Charnov, E. L. 1982. The theory of sex allocation. Princeton University Press, Princeton.

Clausen, C. P. 1939. The effect of host size upon sex ratio of hymenopterous parasites and its relation to methods of rearing and colonization. J.N.Y. Entomol. Soc. 47: 1-9.

Clausen, C. P., D. W. Clancy & Q. C. Chock. 1965. Biological control of the oriental fruit fly, *Dacus dorsalis* Hendel, and other fruit flies in Hawaii. U.S. Dept. Agric. Tech. Bull. 1322.

Delangue, P. & R. Pralavorio. 1977. Comparison of water requirement of *C. capitata* (Diptera: Tephritidae) and its internal parasite. *Opius concolor.* Szepl. (Hy: Braconidae) during the pupal stage of the host. Rev. Zool. Agric. Pathol. Veg. 76: 1-6.

Fisher, R. C. 1971. Aspects of the physiology of endoparasitic Hymenoptera. Biol. Rev. 46: 243-278.

Flanders, S. E. 1950. Regulation of ovulation and egg disposal in the parasitic Hymenoptera. Canadian Entomol. 82: 134-140.

Greany, P. D., S. D. Hawke, T. C. Carlysle & D. W. Anthony. 1977. Sense organs in the ovipositor of *Biosteres longicaudatus*, a parasite of the Caribbean fruit fly, *Anastrepha suspensa*. Ann. Entomol. Soc. Am. 70: 319-321.

Gutierrez, A. P. 1970. Studies on host selection and host specificity of aphid hyperparasite, *Charips victrix* (Hymenoptera: Cynipidae): 6. Description of sensory structures and synopsis of host selection and host specificity. Ann. Entomol. Soc. Am. 63: 1705-1709.

Hagen, K. S. 1953. A premating period in certain species of the genus *Opius* (Hymenoptera: Braconidae). Proc. Hawaiian Entomol. Soc. 15: 115-116.

Hamilton, W. D. 1967. Extraordinary sex ratios. Science 156: 477-488.

Haramoto, F. H. & H. A. Bess. 1970. Recent studies on the abundance of the oriental and Mediterranean fruit flies and the status of their parasites. Proc. Hawaii. Entomol. Soc. 20: 551-566.

Hinton, H. E. 1981. Biology of insect eggs (V.l.). Pergamon Press Ltd., New York.

King, P. E. & C. R. Hopkins. 1962. The structure and action of the ermatheca in *Nasonia vitripennis* (Walker) (Hymenoptera: Pteromalidae). Proc. R. Entomol. Soc. London. 37(A): 73-75.

Langley, P. A., H. Maly & F. Rhum. 1972. Application of the sterile principle for control of the Mediterranean fruit fly *(Ceratitis capitata*: pupal metabolism in relation to mass rearing techniques. Entomol. Exp. Appl. 15: 23-34.

Lawrence, P. O., P. D. Greany, J. L. Nation & R. M. Baranowski. 1978a. Oviposition behavior of *Biosteres longicaudatus*, a parasite of the Caribbean fruit fly *Anastrepha suspensa*. Ann. Entomol. Soc. Am. 17: 253-256.

Lawrence, P. O., P. D. Greany, J. L. Nation & H. Oberlander. 1978b. Influence of Hydroprene on Caribbean fruit fly suitability for parasite development. Florida Entomol. 61: 93-99.

Lawrence, P. O. 1982. Biosteres longicaudatus: Developmental dependence on host *(Anastrepha suspensa)* physiology. Exp. Parasitol. 53: 396-405.

Lawrence, P. O. 1986. The role of 20-hydroxyecdysone in the moulting of *Biosteres longicaudatus*, a parasite of the Caribbean fruit fly, *Anastrepha suspensa*. J. Insect Physiol. 32: 329-337.

Mitchell, S., N. Tanaka & L. F. Steiner. 1965. Methods of mass culturing melon flies, oriental and Mediterranean fruit flies, USDA/ARS, 33-104.

Nappi, A. J. & F. A. Streams. 1970. Abortive development of the Cynipid parasite, *Pseudeucoila bochei* (Hymenoptera) in species of the *Drosophila melanica* group. Ann. Entomol. Soc. Am. 63: 321-327.

Newell, I. M., W. C. Mitchell & F. L. Rathburn. 1952. Infestation norms for *Dacus cucurbitae* in *Momordica balsamina,* and seasonal differences in activity of the parasite, *Opius fletcheri.* Proc. Hawaii. Entomol. Soc. 14: 497-508.

Nishida, T. 1955. Natural enemies of the melon fly, *Dacus cucurbitae* Coq. in Hawaii. J. Econ. Entomol. 48: 171-178.

Omata, K. 1984. Influence of the physiological condition of the host *Papio xuthus* (Lepidoptera: Papilionidae) on the development of the parasite *Trogus mactator* Tosquinet (Hymenoptera: Ichnenmonidae). Applied Entomol. and Zool. 19: 430-435.

Ozaki, E. T. & R. M. Kobayashi. 1981. Effects of pupal handling during laboratory rearing on adult eclosion and flight capability in three tephritid species. J. Econ. Entomol. 74: 520-525.

Ozaki, E. T. & R. M. Kobayashi. 1982. Effects of duration and intensity of sifting pupae of various ages on adult eclosion and flight capability of the Mediterranean fruit fly (Diptera: Tephritidae). J. Econ. Entomol. 75: 773-776.

Pemberton, C. E. & H. F. Willard. 1918a. Contribution to the biology of fruit fly parasites in Hawaii. J. Agric. Res. 15: 419-466.

Pemberton, C. E. & H. F. Willard. 1918b. Work and parasitism of the Mediterranean fruit fly in Hawaii during 1917. J. Agric. Res. 14: 605-610.

Rajendram, G. F. & K. S. Hagen. 1974. Trichogramma oviposition into artificial substrates. Environ. Entomol. 3: 399-401.

Ramadan, M. M., T. T. Y. Wong & J. W. Beardsley. 1989a. Survivorship, potential, and realized fecundity of *Biosteres tryoni* (Hymenoptera: Braconidae), a larval parasitoid of *Ceratitis capitata* (Diptera: Tephritidae). Entomophaga 34: 291-297.

Ramadan, M. M., T. T. Y. Wong & J. W. Beardsley. 1989b. Insectary roduction of *Biosteres tryoni* (Hymenoptera:Braconidae), a larval parasitoid of *Ceratitis capitata* (Diptera:Tephritidae). Proc. Hawaiian Entomol. Soc. 29: 41-48.

Salt, G. 1961. Competition among insect parasitoids. Symp. Soc. Exp. Biol. 15: 96-119.

Sandlan, K. 1979. Sex ratio regulation in *Coccygomimus turionella* Linnaeus (Hymenoptera:Ichneumonoidea) and its ecological implications. Ecolog. Entomol. 4: 365-378.

SAS Institute. 1985. SAS User's Guide:Statistics. SAS Institute, Cary, N.C.

Sumaroka, A. F. 1967. Factors affecting the sex ratio of *Aphytis proclia* walk (Hymenoptera: Aphelinidae), external parasite on the San Jose Scale. Entomol. Rev. 46: 179-185.

Tanaka, N. 1965. Artificial egging receptacles for three species of tephritid fruit flies. J. Econ. Ent. 58: 177-178.

Tanaka, N. & L. F. Steiner. 1965. Methods of mass culturing melon flies, oriental and Mediterranean fruit flies. J. Econ. Entomol. 62: 967-968.

Tanaka, N., Okamoto & R. D. L. Chambers. 1970. Methods of mass rearing the Mediterranean fruit fly currently used by the U.S. Department of Agriculture, pp. 19-23. *In* IAEA Panel on "SIT to Control Fruit Flies." Vienna. 1969.

Tanaka, N., Hart, R. A., R. Y. Okamoto & L. F. Steiner. 1972. Control of excessive metabolic heat produced in diet by a high density of larvae of the Mediterranean fruit fly. J. Econ. Ent. 65: 866-867.

Van Den Assem, J. F., A. Putters & T. C. Prins. 1984. Host quality effects on sex ratio of the parasitic wasp, *Anisopteromalus calandrae* (Chalcidoidea, Pteromalidae). Netherlands Journal of Zoology 34: 33-62.

Van Den Bosch, R., H. A. Bess & F. H. Haramoto. 1951. Status of oriental fruit fly parasites in Hawaii. J. Econ. Entomol. 44: 753-759.

Van Zwaluwenburg, R. H. 1947. Notes and exhibitions. Proc. Hawaii. Entomol. Soc. 13: 8.

Vargas, R. I., H. B. C. Chang, M. Komura & D. S. Kawamoto. 1986. Evaluation of two pupation methods for mass production of Mediterranean fruit fly (Diptera: Tephritidae). J. Econ. Entomol. 79: 864-867.

Vinson, S. B. 1976. Host selection by insect parasitoids. Ann. Rev. Entomol. 21: 109-133.

Waage, J. K. 1982. Sex ratio and population dynamics of natural enemies: Some possible interactions. Ann. Appl. Biol. 101: 159-164.

Waage, J. K. & J. A. Lane. 1984. The reproductive strategy of a parasitic wasp II. Sex allocation and local mate competition in *Trichogramma evanescens*. J. Animal Ecology 53: 417-426.

Waage, J. K. & N. S. Ming. 1984. The reproductive strategy of a parasitic wasp I. Optimal progeny and sex allocation in *Trichogramma evanescens*. Journal of Animal Ecology 53: 401-416.

Werren, J. H. 1980. Sex ratio adaptations to local mate competition in a parasitic wasp. Science 208: 1157-1159.

Werren, J. H. 1983. Sex ratio evolution under local mate competition in a parasitic wasp. Evolution 37: 116-124.

Willard, H. J. 1920. *Opius fletcheri* as a parasite of the melon fly in Hawaii. J. Agric. Res. 20: 423-438.

Willard, H. F. & A. C. Mason. 1937. Parasitization of the Mediterranean fruit fly in Hawaii. 1914-1933. U.S. Dept. Agric. Circ. 439.

Wong, T. T. Y., N. Mochizuki & J. I. Nishimoto. 1984. Seasonal abundance of parasitoids of the Mediterranean and oriental fruit flies (Diptera: Tephritidae) in the Kula area of Maui, Hawaii. Environ. Entomol. 13: 140-145.

Wong, T. T. Y. & M. M. Ramadan. 1987. Parasitization of the Mediterranean and oriental fruit flies (Diptera: Tephritidae) in the Kula area of Maui, Hawaii. J. Econ. Entomol. 80: 77-80.

Wong, T. T. Y., M. M. Ramadan, D. O. McInnis & N. Mochizuki. 1990. Influence of cohort age and host age on oviposition activity and off-spring sex-ratio of *Biosteres tryoni* (Hymenoptera: Braconidae), a larval parasitoid of *Ceratitis capitata* (Diptera: Tephritidae). J. Econ. Entomol. 83: 779-783.

# 24

# Mass Rearing
of *Chrysoperla* Species

*Donald A. Nordlund and R. K. Morrison*

## Introduction

The genus *Chrysoperla* contains a number of important insect predators. *Chrysoperla carnea* and *Chrysoperla rufilabris* are two species that are widely distributed in North America (Bickley & MacLeod 1956, Agnew et al. 1981, Tauber & Tauber 1983). They are predaceous in the larval stage, attack a wide variety of pests (Hydorn 1971, and references therein) and are found in a wide variety of cropping systems (Agnew et al. 1981).

The effectiveness of *Chrysoperla* spp. as biological control agents has been demonstrated in field and orchard crops (Table 1) and in greenhouses (Harbaugh & Mattson 1973, Hagley & Miles 1987). Where both *C. carnea* and *C. rufilabris* are present, *C. carnea* is dominant early in the season, but *C. rufilabris* becomes dominant later in the season when pest problems tend to increase (Dickens 1970, Agnew et al. 1981). This is probably due to the photoperiodic ovipositional response shown by *C. rufilabris* (Nguyen et al. 1976, Elkarmi 1986, Elkarmi et al. 1987). Thus, in many cases *C. rufilabris* may be the more effective biological control agent.

Cost effective mass rearing systems for these two predators have been unavailable and thus have prevented large-scale practical application of them. However, a basic system for large-scale production of *Chrysoperla* spp. was first developed by Finney (1948, 1950) and then further refined by Morrison and co-workers (Ridgway et al. 1970, Morrison & Ridgway 1976, Nguyen et al. 1976, Morrison 1977a,b, King & Morrison 1984). The current rearing techniques for large quantities of *Chrysoperla* require considerable labor and are based on the use of lepidopteran eggs as the larval food source.

Therefore, we will review the evolution of techniques currently available for rearing *Chrysoperla* spp. and then discuss the advances necessary for large-scale practical application of these important biological control agents.

**Table 1. Examples of successful use of *Chrysoperla* spp. in release programs**

| Crop | Target Pests | References |
|------|--------------|------------|
| Cabbage<br>Pepper<br>Tomato | Aphids:<br>cowpea aphid and<br>green peach aphid | Adashkevich & Kuzina 1974<br>Beglyarov & Smetnik 1977<br>Radzivilovskaya &<br>Kaminova 1980 |
| Eggplant<br>Peas<br>Potato | pea aphid and<br>buckthorn aphid | Shands et al. 1972 |
| Eggplant<br>Potato | Colorado potato beetle | Adashkevich & Kuzina 1971<br>Shuvakhina 1974 |
| Apple | European red mite | Miszczak & Niemczyk 1978 |
| Pear | Grape mealybug | Doutt & Hagen 1949, 1950 |
| Mulberry<br>Catalpa | Comstock mealybug | Beglyarov & Smetnik 1977 |
| Cotton | Bollworm and tobacco<br>budworm | Lingrin et al. 1968<br>Ridgway & Jones 1968,<br>1969<br>Kinzer 1977<br>Ridgway et al. 1977<br>Anonymous 1982 |
| Cotton | Aphids | Anonymous 1982 |

## Rearing *Chrysoperla* Larvae

*Chrysoperla* are both highly predaceous and cannibalistic in the larval stage. This makes it difficult to hold and feed them in high-density cultures. Finney (1948, 1950) used wood trays (1.5 x 16 x 40 in.) covered with white muslin for larval rearing. Each tray was initially established with a sheet containing a large number of *Phthorimaea operculella* eggs plus a sheet of *C. carnea* eggs. After about 3 days an additional *P. operculella* egg sheet was added to the tray. Three days later 2,500-3,500 fully grown *P. operculla* larvae that first had been processed in hot water, sodium hypochlorite, and paraffin (Doutt & Finney 1947) were added to the tray. This process was repeated at 2-day intervals for a total of three feedings. Relative humidity was maintained at approximately 50% to prevent rapid deterioration of the processed larvae. Of the 850-900 *C. carnea* eggs initially placed in the tray, 450-500 (54%) pupated. Finney (1950) found that coddling the *P. operculla* eggs and larvae for 5 min at 135°F. destroyed a pathogenic protozoan, *Plistophora californica*, that was infecting the *C. carnea* colony. Finney's method relied on large larval rearing units and excess food, both of which are expensive in a mass rearing project, to minimize cannibalism. Hassan (1975) used similar techniques for larval rearing.

During the 1970s a number of advances occurred in *Chrysoperla* rearing. Ridgway et al. (1970) began using Hexcel for larval isolation. One side of the Hexcel was covered with organdy and the other was covered with a glass plate to which honey and Angoumois grain moth (AGM) *(Sitotroga cerealella)* eggs were attached. This technique required that the larvae be anesthetized with carbon dioxide at 2-day intervals while glass feeding plates were changed. Next, Morrison et al. (1975) used ornamental masonite for larval rearing, eliminating the need for anesthetizing the larvae. Three pieces of masonite separated by two pieces of organdy formed the cells. The initial feeding was provided by placing measured amounts of AGM and *C. carnea* eggs in the cells. Additional feedings of AGM eggs were placed on the organdy, exterior to the cells, and the *Chrysoperla* larvae fed on the eggs through the organdy. With this technique, 76% of the cells produced pupae, of which 82% emerged as adults. Then, when the manufacture of the masonite was discontinued, Morrison (1977a,b) began using Verticel, a product similar to Hexcel. Organdy was glued to both sides of the larval rearing unit with measured amounts of *Chrysoperla* and AGM eggs placed in the cells prior to covering the second side. Additional feedings were supplied at 3- or 4-day intervals by coating a glass plate with honey and AGM eggs and placing this on top of the unit, egg side down. When pupation was completed, the organdy was stripped from the unit, which was then placed in holding cages for adult emergence. This unit produced pupae in 93% of the cells, and adult emergence was 95%.

*Development of Larval Diets*

The predaceous larvae feed by sucking the body fluids from prey held in their sicklelike mouth parts. However, rearing systems that rely on insects as the food supply are expensive. Thus, efforts to develop artificial diets for *Chrysoperla* and techniques for presenting that diet to the larvae began quite early.

Hagen and Tassan (1965) were the first to report the use of artificial diets to rear reproductive adult *C. carnea* successfully from eggs. The diets (Table 2) were encapsulated in a thin layer of paraffin. Encapsulation permitted 1st instar *C. carnea* to feed without becoming stuck in the diet.

Vanderzant (1969) reported a diet that, when presented to larvae via saturated pieces of cellulose sponge, produced 50-65% yields of adults from larvae and 85% yields from AGM eggs. However, this diet was not satisfactory for machine encapsulation because some components were destroyed by heat sterilization, and the water- and fat-soluble materials did not completely emulsify.

**Table 2. Composition of two concentrated larval diets for *C. carnea* (Hagen & Tassan 1965)**

| Ingredients[a] | Quantity | |
|---|---|---|
| | Diet A | Diet B |
| Enzymatic Protein Hydrolysate of Yeast | 5.00 g | 5.00 g |
| Enzymatic Protein Hydrolysate of Casein | | 5.00 g |
| Chloine Chloride | 12.50 mg | 12.50 mg |
| Ascorbic Acid | 0.50 g | 0.50 g |
| Fructose | 8.75 g | 8.75 g |
| Water | 12.50 ml | 12.50 ml |

[a] To 10 ml of the diets add 30 ml of distilled water.

Pomonareva (1971) developed a larval diet for *C. carnea* that consisted of 5 g of dried ground adult *S. cerealella*, 2 ml of honey, 5 ml of autolyzed brewer's yeast and 3 ml of fresh milk. With this diet, larval development required 17 days and the yield of adults from larvae was 88.4%.

In 1973, Vanderzant reported an improved diet (Table 3) that could be heat sterilized and encapsulated. However, yields of adults from larvae fed via saturated cellulose sponges was only 39%. Bigler et al. (1976) found Vanderzant's 1973 diet to be superior to Pomonareva's.

Hassan and Hagen (1978) developed and tested a number of diets and found that a diet consisting of 5 g of honey, 5 g of sugar, 5 g of food yeast flakes, 6 g of yeast hydrolysate, 1 g of casein hydrolysate, 10 g of egg yolk, and 68 ml of water appeared to perform better than either Pomonareva's or Vanderzant's. They reported that larval development took 12 days and the yield of adults was 90% when the diet was presented in paraffin capsules. Yazlovetsky (in this volume) also described a diet for *C. carnea* and one for *Chrysopa septempunctata*.

Both Vanderzant's 1973 diet and Hassan and Hagen's diets appear to be adequate for producing *Chrysoperla*. The Hassen and Hagen diet given above is currently being used at Weslaco to rear *C. rufilabris*. The current procedure involves setting up the Verticel units as described by Morrison with *C. rufilabris* and AGM eggs that have first been frozen at least 24 hr and then thawed before use. After 3 days, a piece of cotton flannel is saturated with the liquid diet and placed under the Verticel unit to feed the larvae. Considerable work needs to be done to develop efficient and economical techniques for presenting the diet to the larvae, particularly to the 1st instars. However, the diet appears to be quite promising, although we have not conducted specific experiments on larval development. The flannel is currently changed on a daily basis. This method works, but it wastes a large percentage of the diet and requires considerable labor for the frequent changes.

Encapsulation of the diet in an "artificial egg" is one solution. Hagen and Tassan (1965) developed a hand-made wax capsule that was used to successfully rear *Chrysoperla* from egg to pupa, and Cohen (1983) further mechanized the technique. Total production, however, remained low. Ridgway (Anonymous 1971, Martin et al. 1978) was responsible for the development of a sophisticated device capable of encapsulating Vanderzant's 1973 diet in mass production quantities. Unfortunately, equipment and production area sterility was not achieved, which led to rapid spoilage of the diet within the capsules, even when they were held under refrigeration. In addition, the delicate 1st instar *Chrysoperla* had difficulty penetrating these capsules.

Thus, rearing continued to be based on an initial feeding of AGM eggs. With today's improved technology, this device used by Ridgway could be improved to provide quality capsules. In the People's Republic of China, a simple device has been used to produce paraffin-encapsulated diet at the rate of 2 kg/h. Large-scale rearing of *Chrysoperla sinica* and *Chrysopa septempunctata* utilizing these diet-filled capsules has been reported (Ma et al. 1986).

**Table 3.  Vanderzant's Improved Diet (Vanderzant 1973)**

| | |
|---|---|
| Casein hydrolysate, enzymatic | 5.00 g |
| Soy hydrolysate, enzymatic | 5.00 g |
| Yeast hydrolysate | 2.00 g |
| Sucrose | 15.00 g |
| Casein | 1.00 g |
| $K_2HPO_4$ | 0.16 g |
| $NaH_2PO_4H_2O$ | 0.08 g |
| $MgSO_4 7H_2O$ | 0.05 g |
| $FeSO_2 7H_2O$ | 0.005 g |
| Soybean lecithin and oil | 0.50 g |
| Cholesterol | 0.05 g |
| B-vitamins[a] | 2.0 ml |
| Choline | 0.50 g |
| Inositol | 0.20 g |
| Water to make | 120.00 ml |

[a]Amounts in mg per ml: nicotinamide 1.0, calcium pantothenate 1.0, thiamine-HC1 0.25, riboflavin 0.5, pyridoxine-HCL 0.25, folic acid 0.25, biotin 0.02, vitamin $B_{12}$ 0.002.

## Adult Handling

Finney (1948) used 1-gal cylindrical, waxed-cardboard containers as the adult feeding-oviposition units. This basic system remains the most common in use, though the size of the container has varied (Morrison & Ridgway 1976, Ridgway et al. 1970). One insectary uses cylinders made of styrene, which lasts longer than cardboard. A variety of materials, such as fiberglass screen or nylon net, have been used to cover the tops and bottoms of the containers. Kraft paper and poster board are used for liners. To change the feeding-oviposition unit and remove eggs, Finney anesthetized the adults with carbon dioxide. Morrison and Ridgway (1976) used a vacuum device to pull the adults down and hold them on the bottom of the container. We use this procedure at Weslaco. Adult *Chrysoperla* are not predaceous but feed on honeydew in nature (Sheldon & MacLeod 1971). Finney found that oviposition was reduced when *C. carnea* adults were fed only honey, so he fed them honey and *Planococcus citri* (Risso) honeydew. However, the time and effort required to maintain *P. citri* and collect the honeydew made economical production impossible. Hagen (1950) found that feeding *Chrysoperla* adults a protein hydrolysate of yeast resulted in oviposition rates as high as those obtained with *P. citri* honeydew. Hagen and Tassan (1966, 1970) further studied the effects of proteins from yeast and carbohydrates on oviposition by *C. carnea*. Based on their work, we are currently using Wheast and sucrose (1:1 by weight) and water to form an adult diet paste at Weslaco. The past diet can be applied directly to the liners (Morrison & Ridgway 1976). The usual procedure, however, is to smear the paste on a plastic strip and place it in the container. Water is provided by placing a piece of saturated sponge or cotton on top of the unit.

Adult *C. carnea* are nocturnal and solitary. Under the crowded conditions of the oviposition-feeding unit, they are extremely active during the scotophase. Finney (1950) found that a constant light regime moderated their activity to a slow steady movement so that very few adults became stuck and they appeared to be more healthy and vigorous. Thus, it became standard practice to maintain *C. carnea* in constant light. However, this procedure did not work with *C. rufilabris,* and Nguyen et al. (1976) reported that a 14:10 L:D regime was not successful in rearing *C. rufilabri* at Weslaco.

## Egg Collection

*Chrysoperla* eggs are deposited on slender stalks attached to the substrate. Egg collection thus presents some interesting opportunities for development of automated techniques. Finney (1950) used a diluted sodium hypochlorite solution for separating eggs from the substrate and removing the stalks. However, for a large-scale rearing program, such a system would require considerable time and labor to remove, rinse, dry, and recover the eggs. In addition, if not carefully performed, this procedure would result in the incomplete removal of eggs or in dissolving the egg corion to the point that the eggs are destroyed (Morrison 1977b). However, a sodium hypochlorite system for egg collection is being used in a *C. carnea* rearing program at the Centrale Ortofrutticola Alla Produzion in Cesena, Italy. Another method involves use of a loose ball of nylon netting rubbed gently across the substrate to remove the eggs mechanically (Ridgway et al. 1970). However, a significant portion of the eggs are destroyed, recovered eggs generally have attached stalks, and the method is laborious. Morrison (1977b) developed a hot wire-vacuum removal device that reduced destruction of the eggs and removed a significant portion of the stalk. With modifications and improvements in the adult feeding-oviposition units, this system might prove to be very effective.

### Needs for Development of *Chrysoperla* Mass Rearing

A mechanized mass rearing system is needed that utilizes artificial diet, rather than insect eggs, as a larval food source. It could produce all stages of *Chrysoperla* spp. at considerably lower cost than present systems. This technology, illustrated schematically in Fig. 1, would be applicable to a number of *Chrysoperla* species. The system would incorporate the following specific improvements currently under development at Weslaco:

1. **Improvements in larval food supply.** There are a number of adequate diets. However, techniques are needed for presenting the diet to 1st instars and for economical presentation to later instars. Thus, a low-cost technique for encapsulation of the larval diet is being developed. This encapsulated diet will be used for rearing and in the field release process.

2. **Adult feeding-oviposition unit.** The cardboard cylindrical-type containers used at present have a relatively low surface-to-volume ratio. A unit designed with an increased surface-to-volume ratio that also lends itself to mechanical egg collection is needed. An adult rearing unit made of plexiglass and paper and having a greatly increased surface-to-volume ratio is currently being tested. The design of this unit will permit mechanized egg collection.

3. **Mechanized egg collection and destalking.** Current methods result in the destruction of an appreciable number of eggs and in the retention of stalks. The hot-wire process developed by Morrison (1977b) is currently being modified and will be the basis for a mechanized egg collection system.

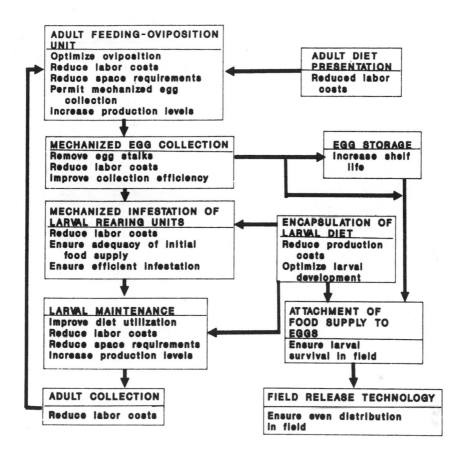

**Fig. 1.  Schematic of planned *Chrysoperla* production and utilization system with goals for each step.  Quality control criteria are applied at each step.**

4. **Mechanized system for larval rearing unit preparation.** The current method for preparing and infesting larval rearing units is a slow, laborious process.  The basic design of these units, however, is quite good.  A mechanized system that will take sheets of Verticel, attach the organdy, deposit *Chrysoperla* eggs and encapsulated diet in the cells, and seal the unit with organdy is currently being designed.

5. **Egg storage.** Techniques need to be developed for stockpiling *Chrysoperla* eggs.  At present, no work on egg storage is being conducted at Weslaco.  At some point in the future, this problem will be addressed.  The answer may lie in cryogenics.

**6. Field application.** Techniques for both aerial and ground application of *Chrysoperla* eggs, with an initial food supply attached, would significantly improve the feasibility of the widespread use of these predators in biological control programs. Attachment of a capsule of artificial diet to a *Chrysoperla* egg should not be a difficult technology to develop. At present, however, effort is being concentrated on development of the diet encapsulation system. Once that system is developed, work will begin on the attachment system.

The development of a cost effective system for mass production of *Chrysoperla* spp. would result in an effective system of biological control technology with broad application for a variety of crops and pests. These biological control agents may be particularly useful for high-value fruit and vegetable crops, where pesticide residues are a major concern, and against hard-to-control pests such as aphids, mites, and mealy bugs (Table 1) as well as other slow-moving soft- bodied pests, such as whiteflies, beetles, and lepidopteran eggs and larvae. *Chrysoperla* is a biological control agent that shows great promise for investment, because it can be used effectively in a wide variety of pest control situations. The development and adoption of a cost effective mass production system would significantly reduce the use of insecticides and their impact on nontarget organisms, improve food safety, and reduce groundwater contamination.

## References

Adashdevich, B. P. & N. P. Kuzina. 1971. *Chrysopa* against the Colorado potato beetle. Zashch. Rast. 12: 23 (in Russian).

Adashkevich, B. P. & N. P. Kuzina. 1974. Chrysopids on vegetable crops. Zashch. Rast. 9: 28-29.

Agnew, C. W., W. L. Sterling & D. A. Dean. 1981. Notes on the Chrysopidae and Hemerobiidae of eastern Texas with keys for their identification. Supplement to the Southwestern Entomologist No. 4.

Anonymous. 1971. Packaged meals for insects. Agric. Res. 19: 3-4.

Anonymous. 1982. Biological pests in China. China Program, Science and Technology Exchange Division, Office of International Cooperation and Development, USDA, Washington, D.C.

Beglyardov, G. A. & A. I. Smetnik. 1977. Seasonal colonization of entomophages in the USSR, pp. 283-328. *In* R. L. Ridgway & S. B. Vinson [eds.], Biological control by augmentation of natural enemies. Plenum, New York.

Bickley, W. E. & E. G. MacLeod. 1956. A synopsis of the nearctic Chrysopidae with a key to the genera. Proc. Entomol. Soc. Wash. 58: 177-202.

Bigler, F., A. Ferran & J. P. Lyon. 1976. L'elevage larvaire de deux predateurs aphidiphages *(Chrysopa carnea* Steph., *Chrysopa perla* L.) a l'aide de differents milieux artificiels. Ann. Zool. Ecol. Anim. 8: 5512-558.

Cohen, A. C. 1983. Improved method of encapsulating artificial diet for rearing predators of harmful insects. J. Econ. Entomol. 76: 957-959.

Dickins, R. L., J. R. Brazzel & C. A. Wilson. 1970. Species and relative abundance of *Chrysopa, Geocoris,* and *Nabis* in Mississippi cotton fields. J. Econ. Entomol. 63: 160-161.

Doutt, R. L. & G. L. Finney. 1947. Mass-culture techniques for *Dibrachys covus.* J. Econ. Entomol. 40: 57.

Doutt, R. L. & K. S. Hagen. 1949. Periodic colonization of *Chrysopa californica* as a possible control of mealybugs. J. Econ. Entomol. 42: 560.

Doutt, R. L. & K. S. Hagen. 1950. Biological control measures applied against *Pseudococcus maritimus* on pears. J. Econ. Entomol. 43: 94-96.

Elkarmi, L. A. 1986. Laboratory rearing of *Chrysoperla rufilabris* (Burmeister) (Neuroptera:Chrysopidae) a predator of pecan pest insects. M.S. thesis, Texas A&M Univ., College Station.

Elkarmi, L. A., M. K. Harris & R. K. Morrison. 1987. Laboratory rearing of *Chrysoperla rufilabris* (Burmeister), a predator of insect pests of pecans. Southwest. Entomol. 12: 73-78.

Finney, G. L. 1948. Culturing *Chrysopa californica* and obtaining eggs for field distribution. J. Econ. Entomol. 41: 719-721.

Finney, G. L. 1950. Mass culturing *Chrysopa californica* to obtain eggs for field distribution. J. Econ. Entomol. 43: 97-100.

Hagen, K. S. 1950. Fecundity of *Chrysopa californica* as affected by synthetic food. J. Econ. Entomol. 43: 101-104.

Hagen, K. S. & R. L. Tassan. 1965. A method of providing artificial diets to *Chrysopa* larvae. J. Econ. Entonol. 58: 99-1000.

Hagen, K. S. & R. L. Tasssan. 1966. The influence of protein hydrolysates of yeast and chemically defined diet upon the fecundity of *Chrysopa carnea* Steph. Vest. csl. Spol. Zool. 30: 216-227.

Hagen, K. S. & R. L. Tassan. 1970. The influence of food wheast and related *Saccharomyces fragilis* yeast products on the fecundity of *Chrysopa carnea* (Neuroptera:Chrysopidae). Can. Entomol. 107: 806-811.

Hagley, E. A. C. & N. Miles. 1987. Release of *Chrysoperla carnea* Stephens (Neuroptera:Chrysopidae) for control of *Tetranychus urticae* Koch (Acarina:Tetranychidae) on peach grown in a protected environment structure. Can. Entomol. 119: 205-206.

Harbaugh, B. K. & R. H. Mattson. 1973. Lacewing larvae control aphids on greenhouse snapdragons. J. Amer. Soc. Hort. Sci. 98: 306-309.

Hassan, S. A. 1975. ber de Massenzucht von *Chrysopa carnea* Steph. (Neuroptera, chrysopidae) Z. ang. Entomol. 79: 310-315.

Hassan, S. A. & K. S. Hagen. 1978. A new artificial diet for rearing *Chrysopa carnea* larvae (Neuroptera, Chrysopidae). Z. ang. Entomol. 86: 315-320.

Hydorn, S. B. 1971. Food preferences of *Chrysopa rufilabris* Burmeister in central Florida. Masters Thesis, University of Florida, Gainesville.

King, E. G. & R. K. Morrison. 1984. Some systems for production of eight entomophagous arthropods, pp. 206-222. *In* E. G. King & N. C. Leppla [eds.], Advances and challenges in insect rearing. USDA-ARS, Southern Region, New Orleans, LA.

Kinzer, R. E. 1977. Development and evaluation of techniques for using *Chrysopa carnea* Stephens to control *Heliothis* spp. on cotton. Ph.D. Dissertation, Texas A&M Univ., College Station.

Lingren, P. D., R. L. Ridgway & S. L. Jones. 1968. Consumption by several common predators of eggs and larvae of two *Heliothis* species that attack cotton. Ann. Entomol. Soc. Am. 61: 613-618.

Ma, A., X. Zhang & J. Zhao. 1986. A machine for making encapsulated diet for rearing *Chrysopa* spp. Chinese J. of Biol. Cont. 2: 145-147.

Martin, P. B., R. L. Ridgway & C. E. Schutze. 1978. Physical and biological evaluations of an encapsulated diet for rearing *Chrysopa carnea*. Fla. Entomol. 61: 145-152.

Miszczak, M. & E. Niemczyk. 1978. Green lacewing *Chrysopa carnea* Steph. (Neuroptera, Chrysopidae) as a predator of the European mite *Panonychus ulmi* Koch on apple trees. 2. The effectiveness of *Chrysopa carnea* larvae in biological control of *Panonychus ulmi* Koch. Fruit Sci. Red. 5(2): 21-31.

Morrison, R. K. 1977a. A simplified larval rearing unit for the common green lacewing. Southwest Entomol. 2: 188-190.

Morrison, R. K. 1977b. Developments in mass production of *Trichogramma* and *Chrysopa* spp., pp. 149-151. *In* Proceedings of the 1977 Beltwide Cotton Production Research Conference, National Cotton Council, Memphis, TN.

Morrison, R. K. & R. L. Ridgway. 1976. Improvements in techniques and equipment for production of a common green lacewing, *Chrysopa carnea*. ARS-S-143.

Morrison, R. K., V. S. House & R. L. Ridgway. 1975. Improved rearing unit for larvae of a common green lacewing. J. Econ. Entomol. 68: 821-822.

Nguyen, R. W., W. H. Witcomb & M. Murphy. 1976. Culturing of *Chrysopa rufilabris* (Neuroptera:Chrysopidae). Fla. Entomol. 59: 21-26.

Pomonareva, I. A. 1971. Artificial diet for the larvae of *Chrysopa carnea* Steph., pp. 77-90. *In* A. I. Sikura et al., Biological pest control on vegetable and fruit crops. Biol. Metod. Zasc. Plodov. i Ovosc. Kul'tur ot

Vredit., Bolezn. i Sornj,. Osnovy Integrirov. Sistem (Tezisy Dokl.). Okt. 1971. Minist. Sel'sk. Chozj. SSSR. Kisinev. (in Russian).

Radzivilovskaya, M. A. & D. B. Daminova. 1980. Prospects of using the aphis lion *(Chrysopa vulgaris)* against aphids in Dzhizak region, Uzbek, SSR. *In* A. D. Davletskina, T. S. Eremenko & S. A. Zhurauskaya [eds.]. Protecting Cotton in the Dzhizakskiy Oblast: Proceedings of a Scientific Research Conference. Tashkent: Fan (in Russian).

Ridgway, R. L. & S. L. Jones. 1968. Field-cage releases of *Chrysopa carnea* for suppression of populations of the bollworm and tobacco budworm on cotton. J. Econ. Entomol. 61: 895-898.

Ridgway, R. L. & S. L. Jones. 1969. Inundative releases of *Chrysopa carnea* for control of *Heliothis* on cotton. J. Econ. Entomol. 62: 177-180.

Ridgway, R. L., R. K. Morrison & M. Badgley. 1970. Mass rearing green lacewing. J. Econ. Entomol. 63: 834-836.

Ridgway, R. L., E. G. King & J. L. Carrillo. 1977. Augmentation of natural enemies for control of plant pests in the Western Hemisphere, pp. 379-416. *In* R. L. Ridgway & S. B. Vinson [eds.], Biological control by augmentation of natural enemies. Plenum, New York.

Shads, W. A., G. W. Simpson & M. H. Brunson. 1972. Insect Predators for controlling aphids on potatoes. 1. In small plots. J.Econ. Entomol. 65: 511-514.

Sheldon, J. K. & E. G. MacLeod. 1971. Studies on the biology of the Chrysopidae II. The feeding behavior of the adults of *Chrysopa carnea* (Neuroptera). Psyche. 78: 107-121.

Shuvakhina, E. Y. 1974. Lacewings and their utilization in controlling pests on agricultural crops, pp. 185-199. *In* Biological agents for plant protection. Moscow: Kolos (in Russian).

Tauber, M. J. & C. A. Tauber. 1983. Life histroy traits of *Chrysopa carnea* and *Chrysopa rufilabris* (Neuroptera: Chrysopidae): Influence of humidity. Ann. Entomol. Soc. Am. 76: 282-285.

Vanderzant, E. S. 1969. An artificial diet for larvae and adults of *Chrysopa carnea*, an insect predator of crop pests. J. Econ. Entomol. 62: 256-257.

Vanderzant, E. S. 1973. Improvements in the rearing diet for *Chrysopa carnea* and the amino acid requirements for growth. J.Econ. Entomol. 66: 336-338.

Yan, Y. 1981. Augmentation of *Chrysopa sinica* against spider mites *(Tetranychus vennenis* and *Panonychus ulmi)* in apple orchards. Bejing News Sci. Technol. Feb. 15, 1981. (in Chinese).

Yazlovetsky, I. G. 1992. Studies on the nutrition and digestion of entomophagous insects leading to the development of artificial diets. Advances in Insect Rearing for Research and Pest Management. Westview Press, Boulder (in this volume).

# 25

## Automated Mass Production System for Fruit Flies Based on the Melon Fly, *Dacus cucurbitae* Coquillett (Diptera: Tephritidae)

*H. Nakamori, H. Kakinohana, and M. Yamagishi*

Fruit flies are economically important not only because of their severe damage to fruits and vegetables but also because of quarantine and regulatory problems in many regions of the world.

Mass rearing of these insects is required for fundamental research in the fields of physiology, ecology, and genetics; and for insect release, hormone and pheromone manipulation, and integrated control programs (Singh 1977). Prior to 1955, most laboratory-reared fruit flies were used to breed parasites or to support research programs that required not more than a few thousand flies (Marlow 1934, Maruci & Clancy 1950, Maeda et al. 1953). Since the concept of eradication using sterile males was proposed by Knipling (1955), however, intensive efforts have been made to improve mass production methods for the Mediterranean fruit fly, *Ceratitis capitata*, the oriental fruit fly, *Dacus dorsalis*, and the melon fly, *Dacus cucurbitae*. The first large-scale rearing method that used artificial larval diet and adult media was developed by Finney (1956), Hagen (1953), and Mitchell et al. (1965). Establishment of mass-production methods to produce several million pupae per week made possible the first successful test of the sterile insect technique in the Mariana Islands (Steiner et al. 1962, 1956). Mass production capacity for the Mediterranean fruit fly was scaled up with the initiation of similar eradication programs in Costa Rica (Peleg et al. 1968) and Mexico (Patton 1980).

In Japan, mass production of the melon fly was begun to eradicate this species from Kume Island (Nakamori & Kakinohana 1980). Eradication from Kume was successful in 1977 (Iwahashi et al. 1977). In 1980, a large-scale project was started to eradicate the melon fly from all of Okinawa. Specifically for this project, a large-scale mass production facility was constructed that has produced 200 million flies per week since 1987. In this facility, automatic rearing systems were adopted for larval rearing, pupation, sifting of pupae, and irradiation. This paper provides an overview of the current status of fruit fly mass production and quality control methods.

## Mass Production and Sterilization Facility

The mass production facility is a three story building with a total area of 4265.6 m and divided into open and closed areas to prevent the escape of fruit flies. An open area on the first floor contains machinery, space for larval diet mixing, and a storage room. Ingredients of the larval medium are mixed in the diet-preparation room located in the open area and pumped into the larval rearing room in the closed area of the first floor (Fig. 1). All mass production and quality control operations are carried out in the closed area and workers are

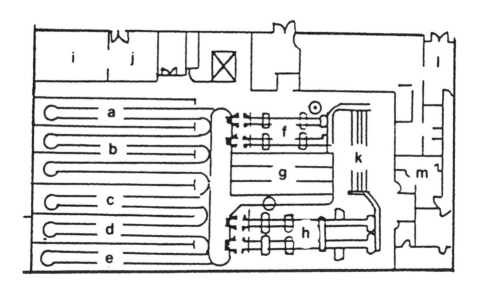

**Fig. 1. Larval rearing rooms (1st floor):** a-e, larval rearing area; **f**, spent diet disposal and tray washing area; **g**, tray carrier sand conveyer line; **h**, diet supply and egg seeding line; **i**, diet storage area; **j**, diet mixing area; **k**, tray stock; **l**, entrance; **m**, security area (to prevent the escape of flies); **n**, machinery area; **o**, maintenance.

required to change clothes when entering and leaving. Diet supply, egg seeding on the diet surface, larval rearing and collection, spent diet disposal, and tray washing are carried out in the closed area of the first floor. There are six rearing rooms, each capable of holding 960 rearing trays. It takes seven days for the eggs to hatch on the medium and grow to mature larvae. Mature larvae of Tephritid fruit flies emerge from infested fruit when they were subjected to light stimulation or exposed to lower temperatures.

Mature larvae are separated from the medium by taking advantage of this behavioral pattern. Larvae leaving the medium drop into a pan of water. The water and larvae are drained through a sieve, which collects the larvae. These larvae are sent to the 2nd floor, another closed area, for pupation.

Pupae are reared on the 2nd floor (Fig. 2). The pupation room (e) is capable of holding 5,400 pupation trays. Five or six days after pupation, the pupae are removed from the vermiculite with an automatic sifting device, automatically weighed, and placed on pupal trays. Development of the pupae is controlled by adjusting the room temperature. The pupae produced in the mass rearing facility are sterilized by 70 Gy of gamma irradiation from a cobalt-60 source, 3 days before emergence. The irradiation facility is located on the first floor of another building, and is connected by a conveyor system to the mass-rearing facility for automatic transfer of the pupae. The irradiation room is surrounded by a 2-m-thick concrete shield.

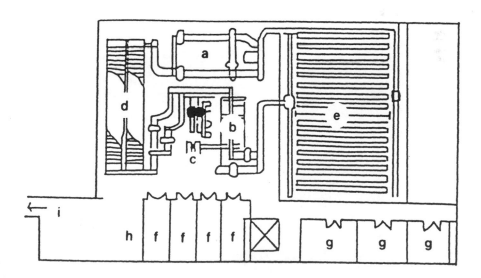

**Fig. 2. Pupal rearing room (2nd floor):** a, larval loading area; b, pupal sifting line; c, pupal supply line; d, tray stock; e, pupation area; f, pupal storage; g, quality control; h, pupal loading area; i, to sterilization area.

Adult flies are reared in two rooms on the third floor to provide eggs for mass rearing. They are reared in the closed area for 6 weeks after emergence. As flies reach the end of this period, they are replaced with newly emerged adult flies (Fig. 3). The eggs are collected by flushing the oviposition device with water each day. The eggs are suspended in tomato juice diluted with water to facilitate their even distribution on the diet.

## Mass Rearing Methods

Mass rearing and sterilization facility procedures are diagramed in Fig. 4. Fig. 5 summarizes the rearing periods, development stages of flies and the operation system for mass rearing, sterilization, and sterile release procedures. Specific rearing techniques are presented in this section according to the life stage of the fruit flies.

### Adult

Koidzumi (1933) and Keck (1951) reported that the most favorable temperature for rearing the melon fly was 27-28°C. Mitchell et al. (1965) maintained adult populations of *Dacus dorsalis*, *D. cucurbitae*, and *Ceratitis*

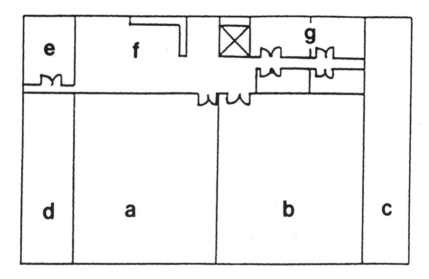

**Fig. 3. Adult rearing room (3rd floor):** a & b, adult rearing area; c & d, air conditioning equipment area; e, lounge area; f, egg collection area; g, new strain colonization area.

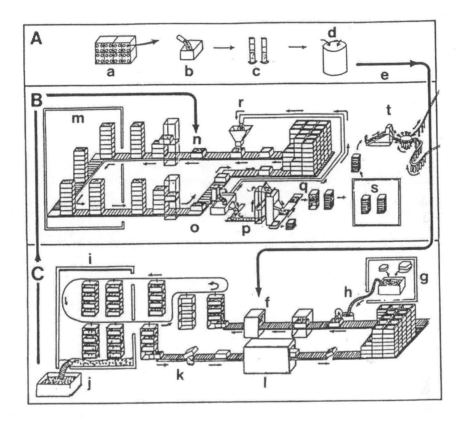

**Fig. 4. Outline of the mass production facility.**

*capitata* at the constant temperature of 27°C under a relative humidity of 80% or less in a cage initially containing about 22,000 to 26,000 adults. Therefore, our adult room is maintained to promote normal feeding and oviposition, with intervals of low illumination to stimulate mating (Steiner & Mitchell 1966). Photoperiod and light intensity are regulated to simulate natural conditions using a photosensor; except on the day of egg collection, when about 2,000 lux is maintained for 24 h. Adult flies are confined to a cage 90 x 60 x 120 cm divided into two sections. The front section has 10 holes covered with shutters to insert the artificial oviposition receptacles (Fig. 6). Each cage contained 50,000 flies, since adults under high density conditions show a higher death rate than those maintained under low density. Diet for adult flies developed by Hagen (1953), Finney (1956), and Mitchell et al., (1965) consisted of protein

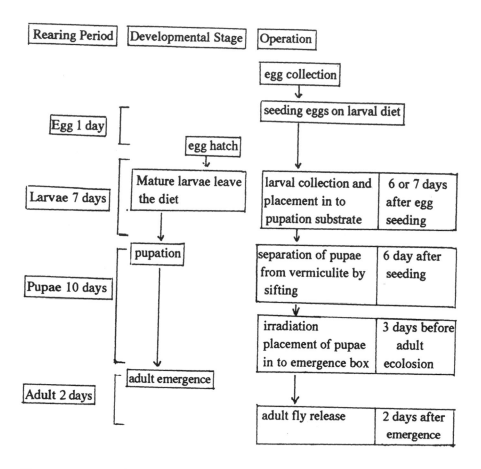

**Fig. 5. Procedures of mass rearing, sterilization, and release of sterile flies.**

hydrolysate and sugar. Sugimoto (1978) reported that a mixture of sugar and protein hydrolysate promoted fecundity. Tanaka et al. (1969) used the Amber BYF-Series 100 as protein hydrolysate, which was cheaper. Based on the results of previous studies, we used protein hydrolysate and sugar mixed in a ratio of 1:4. The artificial oviposition receptacles developed by Tanaka (1965) for the melon fly, oriental fruit fly, and Mediterranean fruit fly are cylindrical plastic containers perforated with 3,000 holes of about 0.3 mm in diameter (Fig. 6). A cellulose sponge saturated with a solution of 1:1 tomato juice and water was placed in the receptacle. Sugimoto (1978) used a sheet of tissue paper wetted with squash juice instead of the cellulose sponge and tomato juice. However, since the use of tissue paper required some skill, and a larger amount

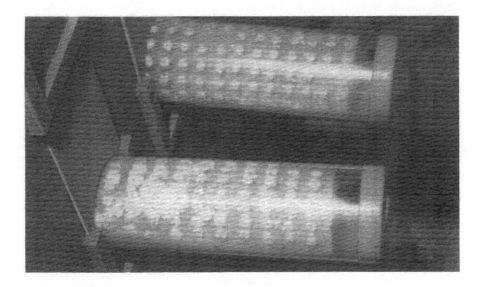

**Fig. 6. Oviposition device: Females insert their ovipositors through the small holes, 0.3 mm in diameter.**

of juice was needed, we used a 60-mesh nylon net to support the squash juice inside the oviposition receptacles. There was no significant difference in the volume of eggs oviposited when the nylon net was used compared to the sponge or tissue paper.

The periodicity and magnitude of oviposition in wild versus mass reared strains was tested by holding a cohort of 25,000 adults in cages 40 x 60 x 120 cm. The wild strain was obtained from pumpkin and bitter gourd in the field and reared in pumpkin for two additional generations, followed by one more generations on artificial diet. The mass reared strain orginiated from a culture maintained in the mass rearing facility for eight generations. The occurrence of oviposition in the wild strain was very different from that of the mass reared strain (Fig. 7). The length of the preoviposition period of the mass reared strain was shorter than that of the wild strain, and the total number of eggs oviposited by the former was greater than that of the latter. The difference in oviposition behavior between wild and mass reared strains seemed to be caused by selection to increase fertility in the mass reared culture. This is an important consideration, since selection for potentially detrimental behavioral traits often occurs inadvertently during the process of adaptation to mass rearing conditions.

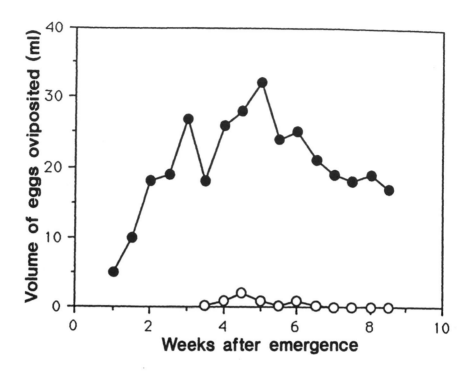

**Fig. 7. Difference in oviposition curves between mass reared (solid circles) and wild strains (open circles). (25,000 flies)**

Eggs were washed from the oviposition receptacles with water, gathered in a nylon gauze, measured volumetrically, and automatically dispensed in tomato juice diluted with water to facilitate their even distribution. An automatic dispenser with a 40 ml syringe distributed the egg-juice mixture onto the tray larval media (Fig. 8). Each tray (35 x 60 x 7 cm) loaded with 6 liters of medium, received 30 ml of the mixture containing about 90,000 eggs. The average yield was about 60,000 mature larvae.

### *Larvae*

Two brands of commercially available wheat bran were compared at different pH levels for their effect on pupal recovery. No. 1 Canadian Wheat, a dark bran, supported high pupal recovery over a wide range of pH. Western White, a white bran, yielded fewer pupae at a pH of less than 4.8. High pH of larval medium allowed mold to grow and had a harmful effect on larval development. Thus, No. 1 Canadian Wheat was incorporated into the medium

**Fig. 8. An automatic dispenser: Eggs were dispensed in tomato juice diluted with water to facilitate their even distribution.**

used for mass production (Table 1). This medium was prepared in 360 liter batches, using an electric mixer. Six liters of mixed medium were poured to a depth of 5 cm in each plastic tray (35 x 60 x 7 cm). A sheet of coarse tissue paper was placed on top of the medium to promote larval dispersal and to prevent larvae from drowning. The trays of inoculated medium were loaded into overhead conveyers held within an air conditioned room set at 27 ± 1.5 °C and 80 to 90% relative humidity with artificial lighting from 0700 to 1900 hours local time (12L:12D). Eggs hatched 24 h after oviposition under these conditions. Young larvae feed on the surface before entering the medium. Metabolic heat is produced in the medium by the crowded larvae, beginning 2 days after egg seeding. The maximum temperature occurs on the 5th day just before the emergence of mature larvae. The temperature in the rearing room was controlled to accommodate egg density of 20,000 per tray.

   In one test, the room temperature was maintained at 27°C for 7 days after egg seeding and lowered to 20°C for the last three hours to hasten larval emergence. The maximum temperature of the medium reached 40°C and had a harmful effect on larval development. In a second test the room temperature was maintained at 27°C for 2 days, 25°C for 1 day, 20°C for 3 days, 27°C for

**Table 1. Formula used at the mass rearing facility for preparing the media for the melon fly**

| Materials | Quantity to produce 10,000 pupae |
|---|---|
| Wheat bran | 153 g |
| de-fatted soybean meal | 33 g |
| Brewer's yeast | 33 g |
| Raw sugar | 66 g |
| Sodium benzoate | 1 g |
| Conc. HCl (3.5%) | 23 ml |
| Coarse tissue paper | 26 g |
| Water | 816 ml |

1 day, and then lowered to 20°C for the last 3 h. In this case, the temperature of the medium was maintained at less than 35°C during the rearing period. In order to synchronize larval emergence, room temperature and water spray were regulated. The effect of three treatments applied on the 7th day were compared: (A) 27°C; (B) 20°C; and (C) 20°C, plus water spraying to further decrease the temperature. The percentage of larvae remaining in the medium with treatments A or B were 5.9 and 4.3, respectively. In treatment C, only 2.3% larvae remained in the medium. Regardless of treatment, it took 5 to 6 h for all of the larvae to emerge. Thus, control of the rearing temperature is important to increase the efficiency of larval recovery.

## Pupae

Mature larvae are collected in water and held at 20°C for at least 24 h. Water containing the larvae is poured through a screen and the larvae are transferred to a plastic box, containing 10 liters of vermiculite with 40,000 larvae per box. Pupae are removed from the vermiculite with an automatic siever at 5 days after pupation. The sifting of younger 1 to 2 day old pupae reduced the ecolosion rate. Sifting of 3 to 4 day old pupae was harmful to the flight ability of the adult (Little et al. 1981, Soemori et al. 1982). The screen moves on parallel tracks with a 2.5 cm horizontal stroke at a speed of 273 strokes per min. After sifting, naked pupae are spread on a 10-mesh screen placed at the bottom of a 40 x 60 x 2 cm tray with a total of about 1.5 million pupae. Adult emergence averages more than 90%. Table 2 shows the mean pupal period, rate of development, total effective temperature and percentage of adult emergence. The regulation equation, R = 0.58t − 5.32, was calculated in considering the temperature (t) and rate of development (R). No adults emerged at 32°C and only a few at 14°C. The pupal development threshold, 9.16°C, can be obtained from the equation. The average total effective temperature required from the time larvae leave the diet to adult emergence is 172.9 degree days. Development was synchronized for the date of adult emergence so that pupae could be loaded into irradiation baskets at two days before ecolosion.

## Quality Control

Quality control determinations included the number of eggs per cage, percentage of egg hatch, number of matured larvae diet that pupated, pupal size, and adult emergence rate (Table 3). Egg volume averaged 62 ml per 50,000

**Table 2. Mean pupal period, rate of development, total effective temp-erature, and percent adult emergence under various temperature conditions**

| Temp (°C) | Mean pupal period (t) (days)[a] | Relative rate of development(R)[b] | Total effective temperature (degree day)[c] | Percent adult emergence |
|---|---|---|---|---|
| 12 | - | - | - | 0 |
| 14 | 34.6 | 2.9 | 167.5 | 8.7 |
| 16 | 27.5 | 3.6 | 188.1 | 52.9 |
| 18 | 18.3 | 5.5 | 161.8 | 44.9 |
| 20 | 18.3 | 5.5 | 197.3 | 96.6 |
| 22 | 12.9 | 7.8 | 165.6 | 95.0 |
| 24 | 11.0 | 9.1 | 164.7 | 83.6 |
| 25 | 10.4 | 9.6 | 167.9 | 95.6 |
| 26 | 10.2 | 9.8 | 171.8 | 77.8 |
| 27 | 9.1 | 11.0 | 162.9 | 94.3 |
| 28 | 9.1 | 11.0 | 171.4 | 65.1 |
| 30 | 8.2 | 12.2 | 170.9 | 34.8 |
| 32 | 8.0 | 12.5 | 182.7 | 13.1 |
| 34 | - | - | - | 0 |

[a] Pupal period extends from larval emergence from diet to adult ecolosion
[b] R = 100/t   [c] Development threshold is 9.16

**Table 3.** Quality investigation for each insect stage

| Insect Stage | Qualities Investigated |
|---|---|
| Adults for Egg Production | Number of generation, volume of eggs laid by age, total volume of eggs laid, survival rate. |
| Egg | Rate of egg hatch. |
| Larva | Temperature of rearing medium, percentage of larvae remaining in the medium at time of diet disposal. |
| Pupae | Individual pupal weight, total pupal weight, total number of pupae produced, diameter of the pupae, pupation rate. |
| Adult:   pre-irradiation | Emergence pattern, emergence rate, flight capability rate. |
|          post-irradiation | Emergence rate, flight capability rate. |
| Monitoring | Sterility check, survival rate, competitiveness. |

**Miyako Islands, Northern Okinawa**

| | |
|---|---|
| Arrival | Emergence rate, flight capability rate. |
| Release | Emergence rate, survival rate, flight capability rate, number released. |

**Central and Southern Okinawa**

| | |
|---|---|
| After Marking | Emergence pattern, emergence rate, flight capability rate, number released. |
| At Release | Emergence rate, survival rate, flight capability rate, number released. |

**Kume and Tarama Island**

| | |
|---|---|
| Release | Emergence rate, survival rate, flight rate, number released. |

flies, washability of eggs was about 90%, pupal recovery from egg to pupae was 65%, and adult emergence about 87%. Sexual aggregation tests were conducted in the laboratory on males from routine production (Koyama et al. 1986). Field tests for flight and dispersal ability (Nakamori & Shimizu 1983, Nakamori & Soemore 1981) and competitiveness of the released male (Iwahashi et al. 1983) were also conducted. Results showed that comparatively high quality flies had been produced. Many of these tests were also used to evaluate the effort of proposed changes in diet and handling.

# References

Finney, G. L. 1956. A fortified carrot medium for mass culture of the oriental fruit fly and certain other Tephritids. J. Econ. Entomol. 49: 134.

Hagen, K. S. 1953. Influence of adult nutrition upon the reproduction of three fruit fly species, pp. 72-76. Special report on control of oriental fruit fly *(Dacus dorsalis)* in Hawaii Islands. 3rd Senate of the State of California.

Iwahashi, O., Y. Ito, & M. Shiomi. 1983. A field evaluation of the sexual competitiveness of sterile melon fly, *Dacus (Zeugodacus) cucurbitae*. Ecological Entomol. 8: 43-48.

Keck, C. B. 1951. Effect of the temperature on the development and activity of the melon fly, *Dacus cucurbitae* Coquillett. J. Econ. Entomol. 44: 1001-1002.

Knipling, E. F. 1955. Possibilities of insect control or eradication through the use of sexually sterile males. J. Econ. Entomol. 48: 459-462.

Koidzumi, K. 1933. Experimental studies on the influence of the low temperature upon the development of fruit flies. J. Soc. Trop. Agric. Taiwan. 6: 687-696.

Koyama, J., H. Nakamori & H. Kuba. 1986. Mating behavior of wild and mass-reared strain of the melon fly, *Dacus cucurbitae* Coquillett (Diptera:Tephritidae) in a field cage. Appl. Ent. Zool. 21: 203-209.

Little, H. F., R. M. Kobayashi, E. T. Ozaki & R. T. Cuningham. 1981. Irreversible damage to flight muscles resulting from disturbance of pupae during rearing of the Mediterranean fruit fly, *Ceratitis capitata*. Ann. Ent. Soc. Am. 74: 520-525.

Maeda, S., K. S. Hagen & G. L. Finney. 1953. Artificial media and the control of microorganisms in the culture of Tephritid larvae (Diptera:Tephritidae). Proc. Hawaii Entomol. Soc. 15: 177-185.

Marlow, R. H. 1934. An artificial food medium for the Mediterranean fruit fly *(Ceratitis capitata)*. J. Econ. Entomol. 27: 1100.

Marucci, P. E. & D. W. Clancy. 1950. The artificial culture of fruit flies

*(Dacus dorsalis)* and their parasite *(Opius longicaudatus, O. persulatius).* Proc. Hawaii Entomol. Soc. 14: 163-166.

Mitchell, S., N. Tanaka & L. F. Steiner. 1965  Method for mass culturing oriental melon and Mediterranean fruit fly, pp. 1-22. USDA-ARS Ser.

Nakamori, H. & H. Kakinohana. 1980. Mass-production of the melon fly, *Dacus cucurbitae* Coquiullett, in Okinawa, Japan. Rev. Plant. Protec. Res. 13: 37-53.

Nakamori, H. & H. Soemori. 1981.  Comparison of dispersal ability and longevity for wild and mass-reared melon fly, *Dacus cucurbitae* Coquillett (Diptera:Tephritidae) under field conditions. Appl. Ent. Zool. 16: 321-327.

Nakamori, H. & K. Shimizu. 1983.  Comparison of flight ability between wild and mass-reared melon fly, *Dacus cucurbitae* Coquillett (Diptera: Tephritidae), using a flight mill. Appl. Ent. Zool. 18: 371-381.

Patton, N. C. 1980. Mediterranean fruit fly eradication in Mexico, pp. 81-84. Proc. Fruit Fly Problems, Kyoto and Naha, National Institute of Agricultural Science, Yatabe Ibaraki.

Peleg, B. A., R. H. Rhode & W. Calderson. 1968.  Mass-rearing of the Mediterranean fruit fly in Costa Rica, pp. 107- 110. *In* Radiation, radioisotope and rearing method in control of insect pest.  Proc. panel organized by the FAO/IAEA Division of Atomic Energy in Food and Agriculture. IAEA, Vienna.

Singh, P. 1977. Artificial diets for insects, mites and spiders. Plenum, New York.

Soemori, H., H. Kuba & I. Tsuji. 1982. Influence of sifting of pupae on the ecolosion rate and flight ability of the adult melon fly, *Dacus cucurbitae* Coquillett (Diptera:Tephritidae) during a mass-rearing procedure. J. Appl. Ent. Zool. 26: 196-198.

Steiner, L. F. & L. D. Christenson. 1956. Potential usefulness of the sterile fly release method in fruit fly eradication programs, pp. 17-18. Proc. Hawaii Acad. Soc.

Steiner, L. F., W. C. Mitchell & A. H. Boumhaver. 1962. Progress of fruit fly control by irradiation sterilization in Hawaii and Mariana Islands. International J. Appl. Radiation, Sterilization and Isotopes. 13: 427-434.

Steiner, L. F. & S. Mitchell. 1966. Tephritid fruit flies, pp. 555-583. *In* Insect colonization and mass-production.

Sugimoto, A. 1978. Egg collection method in mass-rearing the melon fly, *Dacus cucurbitae* Coquillett (Diptera:Tephritidae) J. Appl. Ent. Zool. 22: 219-227.

Tanaka, N. 1965. Artificial egging receptacles for the three species of Tephritid fruit flies. J. Econ. Entomol. 58: 178.

Tanaka, N., L. F. Steiner & K. Ohinata. 1969. Low-cost larval medium for mass-production of oriental and Mediterranean fruit flies, J. Econ. Entomol. 62: 967-968.

## PART FIVE

# Insect Rearing in the Marketplace

# 26

---

# The Establishment
# of Commercial Insectaries

*Peter L. Versoi and Lee K. French*

## Introduction

Many entomologists share the belief that their interest, experiences, and skills in rearing insects in laboratory colonies can be the foundation for a profitable, private business. However, very few commercial insectaries operate on the assumption that adequate financial income can be derived from the sale of insects. The names of some of them appear in published catalogs such as *Arthropod Species in Culture* (Edwards et al. 1987) and *Suppliers of Beneficial Organisms in North America* (Bezark 1989). Such enterprises deserve our admiration, their owners and operators our appreciation, as pioneers exploring the validity of insect rearing as a profit-generating prospect in a largely untapped market. The successful establishment of commercial insectaries should also be recognized as an important employment niche for entomologists and others with insect-rearing experience and interests.

## Getting Started in the Business

An insect-rearing specialist establishing a new commercial insectary must consider many of the essentials critical to any new business. Some of the more important basics are interdependent and include the following factors.

### Time

The task of getting started in this business and founding and maintaining all of the desired insect colonies requires a considerable amount of time. The vast amounts of time required during the start-up phase will test the specialist's faith in the new business.

## Space and Equipment

Fortunately there are excellent publications on the custom design and construction of insect rearing chambers, rooms, and larger facilities (Fisher & Leppla 1985, Leppla & Ashley 1978). During the commercial insectary start-up process, the insect-rearing specialist must select a suitable location, either a new or existing facility, and must design and equip rooms or other spaces in which to maintain a variety of insect colonies.

## Utilities

Once the appropriate location, facility, and other space requirements are acquired, the availability and suitability of utilities must be determined, including heating, cooling, air exchange and exhaust (ventilation), humidity and water (Goodenough & Parnell 1985). The safety aspects of the facility and its occupants must also receive critical examination when considering the utilities and other details of the operation (Wolf 1985).

## Capital

An unavoidable and sobering fact is that money will be needed to fund all aspects of the start-up process. Indeed, this process does not end until income from insect sales is received. Obtaining capital through financial institutions is almost impossible. Financial support must come from friends, associates, or others who appreciate the concepts and opportunities for rearing. Otherwise, financing will be a slow process, expanding as sales increase. Calculations completed by French Agricultural Research, Inc., indicate that the costs for starting the rearing of each insect are at least $20,000. This does not include building costs or other large fixed expenses but does include costs for specific equipment, diets, and time. It would take approximately 5 years to recover initial costs and obtain a profit. If the insectary is competing with other rearing organizations for sales, the time for recovery will increase.

## A Product to Sell

Insect rearing as a business is capable of producing many marketable products; namely, numbers of living or dead insects, their body parts, or insect by-products from laboratory culture. Given this broad range of possibilities, one can envision tremendous, varied, diverse markets to be explored or created. This is the bottom line in the success or failure of a commercial insectary.

*Potential Markets for Living Insects*

- Research in experimental insecticide laboratories, greenhouses, and the field.
- Mass release of parasites, predators, or disease vectors in greenhouses or field integrated pest management programs.
- Mass release as pollinators in orchards, floracultural, or agricultural crops.
- Fish food to hatcheries, food for small rodents and other mammals, and food for reptiles and amphibians.
- Hosts for the in vivo propagation of pest insect pathogens or parasites.
- Seed companies for host plant resistance research.
- Biotechnology companies.
- Biological demonstrations in elementary through college classrooms.
- Bait for sport fishermen.
- Starting stock to establish new colonies.

*Potential Markets for Dead Insects (Whole or Parts)*

- Pinned or otherwise mounted or preserved specimens for scientific presentation, individually or in kits.
- Specimens for dissection by students or for research in biology.
- Preparation of specialty human foods such as chocolate-covered ants, bees, and the agave (gusano) worm in tequila bottles.
- Pet fish food or sport fishing bait.
- Whole or in parts (such as butterfly wings) for artistic display.

*Potential Markets for Insect By-Products*

- Beeswax for candles and honey for food.
- Silk for fabrics.
- Frass as an organic fertilizer for house plants.
- Venom for medicinal sting and bite antidotes.

## A Buyer of the Product

There is no question that markets exist or can be created for commercial insectary products. But are these markets large or diverse enough to provide steady and predictable demand? Before getting started in the private commercial insectary business, the owner or operator must have some indication that one to several existing or new markets are available for the product(s). To enter such markets, the insect-rearing specialist must have potential customers and colonize insect species applicable to that clientele.

## *An Example:  Rearing for Industrial Insecticide Research*

Insecticide research and pest management programs generally require large numbers of pest or beneficial insects of a particular species, life stage, origin, health, or other specified standards, at regular intervals for use in tests in agriculture, forestry, health, hygiene, and other broad areas of study.  The purpose of the commercial insectary is to serve as the paid supplier of living insects from healthy, high-quality laboratory cultures so that tests can be performed according to the predetermined design and schedule.  From this standpoint, the private commercial insectary performs an essential support function vital to the larger purpose of generating test results.  Since insect rearing does not directly produce profits, the producer of insecticide products generally prefers not to invest the money, space, time, equipment, personnel, and supplies required to maintain an insectary.  It is often more expeditious to focus on the product research and development and therefore purchase the required insects.

## Competition

The private commercial insectary has a certain degree of competition in selling insects for insecticide research.

## *In-House Continuous Laboratory Cultures*

Most agricultural chemical companies, governmental agencies, and universities that conduct insecticide research have in-house insectaries, equipment, staffing, and know-how to maintain continuous or sporadic laboratory cultures of the most frequently required test insects.  These insectaries vary in size from one room to an entire facility and have from 1 to 50 or more cultures.  Apart from producing insects for their needs, some of these insectaries actually give or sell insects to outside researchers.

There are numerous benefits to "in-house" insectaries. In-house rearing provides a steady, predictable, year-round supply of insects in the appropriate life stages and in the required numbers.  Often testing schedules are based on the availability of these insects. Quality control is an inherent part of the insect production process.  The health and limitations of the insects are known because they are maintained in-house. Hands on experience is provided to employees. They are able to develop professionally through gaining an understanding of insect biology, behavior, life cycles, and more while learning to manage colonies.  Such on the job training is accompanied by encouraging initiative and innovation as needed in trouble-shooting problems, custom designing cages, refining handling techniques, and improving procedures. Satisfaction comes with independence from reliance on insects supplied by others. From the corporate standpoint, this independence goes beyond being able to declare that we can do it ourselves. It also maintains secrecy and security.

It is not clear whether it is relatively less or more expensive to rear insects in-house, than to obtain them from outside paid suppliers. This certainly depends on many different and changing factors such as the size and scope of the operation, employee costs, supply expenditures, capital purchases and depreciation. Calculating "task time" (Singh & Ashley 1985) and the cost per individual test insect helps to determine the cost effectiveness of the operation. However, the end worth of the operation in terms of the development of products for sale is not easy to evaluate.

## Field Collection

An alternative source of insects is field collection as the season and life cycle of the insects permits. However, insectary production allows for better prediction and planning, more defined quality and history, less disease and parasitism, and known degree of genetic resistance or susceptibility.

## Gratis Supply from Business Friends and Cooperators

This is a potential nightmare for the owner or operator of the private commercial insectary: Why would anyone buy your product if they can get it from another reliable source for free or from in-house? There are many valid reasons for purchasing from a commercial insectary.

Purchasing insects from an outside source leaves more time for other work. This work includes preparing, conducting and completing experiments which produce results closer to the goal of providing new products and marketable concepts. The consumption of other valuable resources, besides time and personnel, are also reduced when insects are purchases rather than reared in-house. These resources include equipment, expendable supplies, space, and utilities. Employees appreciate not having to respond to the demands of the insects, nonstop 7 days per week. Employers also benefit by not having to pay as much overtime and by not having to provide special safety measures and building security. Purchasing insects from an outside source must be less expensive than maintaining them in-house.

## Purchase from the USDA or
## Other Private Commercial Insectaries

Privately owned and operated commercial insectaries, however few, do exist as potential competition for the newcomer. Certain public insect rearing facilities are also able to sell various species and life stages of living insects without directly reaping the profits. However, these public institutions play a vital role by developing new rearing techniques that otherwise would be too costly for the commercial insectary.

## Incentives to Buy Insects

The owner or operator of the private commercial insectary must be able to convince clients with in-house rearing plans that there are clear advantages to the purchasing option:

1. The insects purchased are less expensive than they would be if alternatively reared in-house (i.e., the price is right).
2. A steady year-round supply is available, or alternatively, an occasional or sporadic, massive quantity is available on short notice.
3. The insect quality is consistent by whatever criteria specified by the customer, within and between shipments.
4. Delivery is timely in terms of experimental or other needs.
5. The recipient of the insects has the right to reveal or conceal the source. However, the supplier cannot identify the customer without consent.

## Pitfalls and Recommendations

In getting started, any commercial insectary owner must consider potential problems that could jeopardize steady, predictable business. Hire employees that are genuinely interested in insects. High turnover of personnel requires too much time training new individuals. Their training can take up to 1 year to master the rearing program. Each time a new insect colony is started, it will take 6 months to 2 years to have a good established colony. This will depend on your experience or the similarity of the insect to others already maintained in your insectary. Do not assume that a particular insect will be easy to rear just because you are rearing other insects. Customers may develop the ability to rear in-house and become potential competitors. It is, therefore, extremely important to maintain high standards of quality, timely delivery, and predictable supply. Make sure that wild-type genetics are introduced into laboratory colonies for the maintenance of genetic diversity. Make sure that your customers know how to handle the insects on receipt. It is also advisable to avoid abrupt or unexplained cost hikes that may encourage in-house rearing or use of other suppliers. Nonprofit competitors (university, private, industry, and other state, federal, or municipal agencies) may be able to supply the same insects at lower cost or even gratis. The private insectary operator must attempt to gain the respect of such competitors, so that prospective customers are actually referred to the commercial source.

To begin a commercial insectary, plan to specialize in one insect or group of insects before diversifying into other areas. When selecting, look at insects where research is concentrated and projected for several years in the future. This will allow the insectary to become established before a changing market occurs.

It is important to diversify to keep pace with changing markets. Insect-rearing services can be linked with other related services to create a package deal. For example, laboratory and field trials can be conducted rather than just providing insects. It is also possible to provide consulting services or classes in insect rearing; offer insect rearing biological supplies and equipment, and, incorporate pest management services. These forms of diversification may also attract investors while improving the likelihood of continued and increasing success.

## References

Bezark, L. G. 1989. Suppliers of beneficial organisms in North America. State of California, Biological Control Services Program, Publication No. BC 89-1, Sacramento, Calif.

Edwards, D. R., N. C. Leppla & W. A. Dickerson. 1987. Arthropod species in culture. Entomol. Soc. Amer., College Park, Maryland.

Fisher, W. R. & N. C. Leppla. 1985. Insectary design and operation, pp. 167-183. *In* P. Singh & R. F. Moore [eds.], Handbook of insect rearing, vol. 1. Elsevier, Amsterdam.

Goodenough, J. L. & C. B. Parnell. 1985. Basic engineering design requirements for ventilation, heating, cooling, and humidification of insect rearing facilities, pp. 137-155. *In* P. Singh & R. F. Moore [eds.], Handbook of insect rearing, vol. 1. Elsevier, Amsterdam.

Leppla, N. C. & T. R. Ashley (eds.). 1978. Facilities for insect research and production. USDA Tech. Bull. 1576.

Singh, P. & M. D. Ashley, 1985. Insect rearing management, pp. 185-215. *In* P. Singh & R. F. Moore [eds.], Handbook of insect rearing, vol. 1. Elsevier, Amsterdam.

Wolf, W. W. 1985. Recognition and prevention of health hazards associated with insect rearing, pp. 157-165. *In* P. Singh & R. F. Moore [eds.], Handbook of insect rearing, vol. 1. Elsevier, Amsterdam.

# 27

## Production and Utilization
## of Natural Enemies in
## Western European Glasshouse Crops

*W. J. Ravensberg*

### Historical Overview

The commercial application of biological control in European glasshouse crops has a history of nearly 25 years. Today, a high percentage of the growers use biological control in the three main glasshouse vegetable crops, i.e., tomato, cucumber, and sweet pepper. In this article the term biological control is used to identify integrated control programs utilizing beneficial organisms. The words parasite, predator, natural enemy, and beneficial organism are used more or less as synonyms. The words *prey* and *host* are used as synonyms, as are the words *insect* and *mite*, when mentioned in a general sense.

Earlier this century the parasite *Encarsia formosa* was used to control the greenhouse whitefly (*Trialeurodes vaporariorum*) in tomatoes. This was done in the United Kingdom (UK) in the 1930s on a small scale. However, with the develoment of synthetic pesticides after the second World War, interest in this method was lost.

Research in the Netherlands in the late fifties on the control of the twospotted spider mite (*Tetranychus urticae*) and the discovery of the predatory mite *Phytoseiulus persimilis* in 1958 led to new interest in biological control (Bravenboer & Dosse 1962).

*Encarsia formosa* was found again in the UK in a botanical garden, and a culture was started. Research was directed toward the development of a biological control program in glasshouse crops. Overviews of these early years of biological control are given by Hussey (1985) and van Lenteren (1990).

465

Undoubtedly, the initiative of a Dutch cucumber grower, Mr. Koppert, was an important milestone in the use of biological control in the whole of Europe. In 1967 he released a small number of predaceous mites, *Phytoseiulus persimilis*, in his cucumber crop, which was suffering from a heavy spider mite attack. Chemicals could no longer provide adequate control. Initial results with *Phytoseiulus persimilis* were encouraging, and in 1968 a few other growers successfully tried this method. In 1969, the systemic fungicide dimethyrimol became available for use against the most important fungal disease in cucumber, powdery mildew. The systemic application of this fungicide and the use of the predatory mite against the two key pests in cucumber dramatically reduced spraying in this crop. This program was used on 30 ha. In 1970 the program grew to 200 ha in the Netherlands.

Unfortunately, a sudden and tremendous outbreak of the greenhouse whitefly, *Trialeurodes vaporariorum*, resulted in the wide use of insecticides, causing a sharp decrease in biological control. Moreover, mildew was developing resistance to dimethyrimol, resulting in more spraying. This combination of events almost ruined biological control in greenhouses. A biological alternative was sought to control the whitefly. This started the cooperation between the small producer of the natural enemy, Mr. Koppert, and research institutes in the Netherlands and the UK. *Encarsia formosa* was again brought into production, and commercial sales were initiated in 1972 (Koppert 1978).

For a long time, *Phytoseiulus persimilis* and *Encarsia formosa* were the only two beneficials available (Table 1). Other pests were controlled by chemicals.

A secondary pest in tomatoes, the Dipteran leafminer *Liriomyza bryoniae*, caused many problems because chemical control interfered with the use of *Encarsia formosa*. In 1980 the parasite *Opius pallipes* was first sold for leafminer control (Ravensberg et al. 1983). In 1980 *Liriomyza trifolii*, the American serpentine leafminer, established itself in glasshouse vegetables in Europe, and *Opius pallipes* failed to control this imported species. Therefore, *Opius* production was stopped. Since then, *Dacnusa sibirica* has been produced and applied against both leafminers, and, since 1990, also against *Liriomyza huidobrensis*, another recently imported American species. Since control is not always satisfactory, especially at higher temperatures, another parasite, *Diglyphus isaea*, joined the forces in 1984.

Thrips are also difficult pests to control in integrated control programs, and biocontrol attempts started with the use of *Amblyseius* mites (Ramakers & Van Lieburg 1982). The use of *Amblyseius cucumeris* broke through in 1985 in sweet pepper against *Thrips tabaci* and years later in cucumber, when *Frankliniella occidentalis*, the Western flower thrips, was established as a severe pest (Ramakers et al. 1989). Predatory bugs (*Orius* spp.) have been available since 1991 for thrips control.

Control programs became more and more biological, although aphid control was still done by selective chemicals. However, the ineffectiveness of pirimicarb on *Aphis gossypii*, an increasing pest, made a biological solution more urgent. In 1989, the use of the predatory gall midge, *Aphidoletes aphidimyza* was initiated, together with the parasite *Aphidius matricariae* (Van Schelt et al. 1990). *Aphis gossypii* control with both natural enemies is not yet satisfactory, but *Myzus persicae* control is efficient. The use of aphid enemies is expected to increase in coming years.

Mealybug enemies *Cryptolaemus montrouzieri* and *Leptomasty dactylopii* are used on a relatively small scale, mainly in indoor plantscapes.

*Bacillus thuringiensis* is widely used against caterpillars and use of the insect-pathogenic fungus *Verticillium lecanii* (to control whitefly and aphids) is just starting (Ravensberg et al. 1990).

All these natural enemies are mainly used in cucumber, tomato, and sweet pepper. In other protected crops like eggplant, beans, melons, and strawberries, the use of biological control is limited to only a few tens of hectares. In glasshouse ornamentals, biocontrol is still very difficult, not because of the diseases and pests, but predominantly because the trade does not accept any damage at all, and a zero tolerance is very difficult to achieve with biocontrol. Problems with pesticide resistance and public concern for the environment will certainly open the door to increased biological control in ornamentals in the coming years.

## Production of Beneficial Arthropods

### *Development of a Mass Production System*

Once a natural enemy has been chosen, either after trial and error or an extensive selection program (van Lenteren 1980), it has to be reared in massive numbers for release. Before a biological control program can be called successful, it is essential to produce that natural enemy in large numbers economically and with availability upon demand.

Usually the beneficial organisms have to be reared on a small scale in order to perform small laboratory tests or semifield trials. These rearing systems are generally small and require a lot of attention. They are not meant to be enlarged into a mass production system. Investigations in this direction are seldom performed in scientific research programs; a producer of beneficial organisms has to develop the mass rearing systems personnally.

Enlarging a small laboratory rearing system into a commercial mass production system is not simply a matter of increasing the number of insect cages, plants, climatic rooms, and hours of labor. It requires a whole new approach, in which many problems are encountered that did not even exist in the small laboratory setting.

Until now, the most widely produced organisms are *Phytoseiulus persimilis*, to control the twospotted spider mite, and *Encarsia formosa*, to control the greenhouse whitefly. The examples presented here are mainly based on experience with the production of these two classical natural enemies, but they also hold for leafminer parasites, aphid parasites, aphid predators, and other insects. Since most of these enemies are specific, they have to be reared on the natural host (or on a very closely related species). This implies that a phytophagous host is needed in large numbers, requiring the production of food plants. There are at present no artificial rearing methods available to produce these phytophagous pests, at least not for large systems.

## Host Plant-Host System

The foundation of most predator or parasite mass rearing systems is the successful rearing of the host. This requires a detailed understanding of host plant-host systems, even before considering the predator or parasite.

Most host plants are easily obtainable as cultivated plants from a plant nursery. Proper cultivation of these plants is typically well known. These plants, propagated for growers or farmers, are also available to the insectary. Always check the supplier to ensure availability of the seeds, seedlings, or plants at any time and in any number. A standing contract with the nursery can guarantee availability. Check what kind of chemicals have been used on the plants. It is also useful to assess the suitability of different cultivars of the host plant.

The host plant should be rather easy to grow and tolerant of variation in daylength, temperature, and insect infestation. It should grow rather fast for easy propagation and to minimize the buildup of unwanted diseases and pests. If it takes three months to culture a plant to the size needed for insect production, and suddenly a series of plants dies because of some disease, it is very difficult to replace them, disrupting the production schedule. Use of a fast-growing host minimizes this problem.

In many cases there are big differences in host development on different plants or even on different cultivars of the same plant. It is known that eggplant is a favored host for the greenhouse whitefly, followed by cucumber, tomato, and sweet pepper (van Boxtel et al. 1978). This rating is derived from the rate of development, number of eggs laid, longevity, and mortality of the whitefly, characteristics that are also very important for the efficiency of a mass production system. Even with tomato there are big differences; beef tomato (cv. Dombito) is a far better host than the round-type tomato (cv. Calypso).

Structures on the host plant might hamper the parasite or predator. A leaf surface densely covered with hairs might slow down the activity of the natural enemy, trapping or even killing it. *Phytoseiulus persimilis* is killed by the glandular trichomes on the stem of the tomato plant (van Haren et al. 1987). Hooked trichomes on French beans capture and sometimes kill insects (Johnson 1953).

Plant and leaf shape may influence the efficient collection of beneficials. If it is necessary to check the number of beneficials on the leaves, especially on the underside of the leaves, it is easier if the plant has one big leaf instead of a compound leaf. Sometimes leaves are mechanically handled to collect natural enemies; a simple, flat leaf is better to handle than a segmented or curled leaf. In the case of mature larvae, which leave the plant to pupate in the soil, rosette-shaped plants are unwanted since many pupae are caught in the heart of the rosette. For instance, lettuce could be a good host plant for leafminer production, but its structure makes the collection of the pupae difficult; tomato is better in this case.

Once a host plant is widely grown as a year-round monoculture, it is bound to get diseases and pests other than the target species. Choose a host plant where these problems can be controlled by cultural methods (maintain a low relative humudity to prevent fungal diseases) or by other preventive measures that do not upset the whole system. Only pesticides with short residual effects can be used or are compatible, like many fungicides, with the predators or parasites. If beneficials are shipped on plant material, control of diseases and pests before the shipment leaves the production facility is essential.

In very specific parasites or predators, the range of hosts is limited, and often the target pest will also be used as the host in the mass production process. This is often the case with *Encarsia formosa*, which is produced on the greenhouse whitefly, *Trialeurodes vaporariorum*, and with the predatory mite *Phytoseiulus persimilis*, produced on the twospotted spider mite, *Tetranychus urticae*.

Sometimes a host that is totally different from the target pest is used. *Amblyseius* mites are used to control thrips species, but they are reared on bran mites (Ramakers & van Lieburg 1982). Lacewings are reared on moth eggs (Morrison 1977) and used mainly against aphids.

Another advantage of using a nontarget host for the beneficial production is that there is no chance of shipping a pest species with the biological control product. Growers do not accept pest organisms with their beneficials, even if there is no risk involved. In case of export, custom rules are even more strict. If the substitute host is no threat to the plants on which biocontrol is applied, this makes the beneficial collection much less critical.

Alternate hosts can be used only as long as the natural enemy does not lose its capability of finding, accepting, and controlling the target pest. This must be checked regularly and routinely.

The host is the source on which the multiplication of the natural enemy is based. Simply said, a large number of hosts will give a large number of enemies. But large numbers are not the only consideration. The size of the host may influence the size of the beneficial, especially in the case of a parasite. High host densities on a plant may lead to smaller hosts and thus to smaller parasites. Generally, smaller insects are of a lower quality.

In parasite production it is often essential to have the right stage of the host available. *Diglyphus isaea*, a larval ectoparasite of leafminers, feeds on young leafminer larvae. If the larvae are too large, they will soon be mature and leave the plant to pupate. Exact timing of the host production is absolutely critical for maximum parasitisation.

## Manipulation of a Tri-trophic System

Rearing beneficial arthropods requires manipulation of a tri-trophic system. Tri-trophic systems can be pretty complex; variables include determining the right moments of infestation of the host plant, introduction and yield of the natural enemy, and regulation of the environmental conditions. This process requires complete knowledge of the tri-trophic system. Successful manipulation of the system becomes an art, a feeling, an intuition, to be learned only by experience. Successful manipulation of the plants, hosts, and beneficials will achieve a balance between the input of labor, energy, space, and time and the output in yield of beneficials.

When to infest the host plant is critical. The plant should survive the attack by the phytophage long enough to maximize natural enemy production. This can be achieved by regulating the numbers of the host and their distribution over the plant. Generally, an inoculative release of the pest will increase to serve as food for the beneficial.

Care should be taken that the host plant and large numbers of hosts survive. Early plant death will decrease yield. Careful measures must be taken to ensure that hosts and predators are not introduced into the rearing system too early. Proper containment is essential.

Obviously, environmental conditions can influence the dynamics of a mass rearing system. Both plants and animals are influenced by temperature, relative humidity, and light conditions. Each of the organisms may have different optimal environmental conditions. It is necessary to regulate these abiotic conditions for optimal development, dispersal, oviposition, foraging behavior, etc.

When three clearly separated phases (plant, host, and beneficial) can be distinguished in time or location, it is possible to set a climate for each phase. This is the case in the production of *Phytoseiulus persimilis*. First the host plants, French beans, are sown. After growth to the proper stage, the climate is changed to be optimal for the twospotted spider mite. Even during the cultivation of the host plant, conditions are changed; germination demands a higher temperature and relative humidity than leaf development. The spider mites thrive at a low relative humidity, but the predatory mites are produced at a high relative humidity to set optimal conditions for rapid multiplication. When all three organisms are reared at the same time, a compromise between their demands must be found.

Environmental conditions can also be influenced by the irrigation system. Plants irrigated by an overhead raining system are subjected to a change in temperature and relative humidity in the rearing unit, also changing the plant temperature and its activity. Irrigation by means of a drip system might be a better solution. Both plants and insects will show biorythm changes. When this occurs, a viable solution may be simply increasing the scale of production.

The beneficial organisms can be harvested several times per production cycle, more or less continuously, or only once when high numbers are present at the end of the production. Selection of the most efficient method depends on whether the predator or parasite can be collected separately from the host or if all hosts are eaten or parasitized before collection. No strict rules can be given; each situation requires its own solution.

Small-scale breeding of insects for research is often done by professionel entomologists or well-trained assistants. In large-scale operations, however, typically entomologically untrained employees do most of the work, supervised by trained people. But the experience, enthusiasm, insight, responsibility, and discipline of the workers determines the success of the operation. They should get a feeling for the job from experience over a long time. Production is not just a sequence of facts; many things interact with each other, so a feeling for the whole system is important. Faults, problems, and other unexpected matters must be anticipated or discovered early to prevent disasters.

## Contamination

Contamination is an ever-present danger in large-scale insect production. At all levels of the system, contaminants are able to cause severe problems or even destroy production. The range of contaminants include chemical vapors from building materials; insect, mite, and other pests; and microbiological pathogens in both hosts and natural enemies. Many of these contaminants are very difficult to trace. In many rearings, contaminants will sooner or later show up. They are often part of the growing pains of the production system. The challenge of a large production is to cope with these problems, overcome them, and prevent them as much as possible in the future.

All kinds of materials may produce vapors that are unfavorable or even harmful for the natural enemies. Some examples are plastic containers or trays, paints, glues, screens, rope, even (treated) soil, or an excess of $CO_x$ or $NO_x$ of the insects or the heating system. In general, new synthetic materials are suspect! Efficient air exchange may prevent a lot of problems. Contaminants can be brought in with plants purchased for production. Chemical residues can be detrimental for insects, and only after the growth of new leaves can the plants be used. Plant pathogens can also enter via purchased plants. If plants are grown in containers, the soil should be disinfected.

Field-collected insects are a prime source of contaminants. Collecting only the needed insect species is sometimes difficult, especially if large numbers are needed. If possible, a selective capture system should be used, or at least species should be carefully separated after they have been collected. Insect or mite pathogens are difficult to detect, and only through rearing under quarantine conditions for several generations can these contaminants be identified. Larval or pupal parasites are more easily reared out in this way.

Pathogens can be very difficult to eliminate. Starting over is often the best solution. When more knowledge about the pathogen becomes available, it can be kept at a low level with adequate measures. In some cases, it is possible to maintain a colony at an acceptable level even with contamination. Contaminants can be prevented or suppressed by regularly cleaning and disinfecting the rearing rooms and materials before they are used again. Always investigate the long-term effects of disinfectants or chemical control compounds, and be aware of sublethal effects. A contaminant monitoring system should be maintained. Whenever possible, work with disposable materials. Even the natural enemy is sometimes a contaminant in the rearing process and establishment must be prevented until the right moment. Many problems are transmitted by people. Measures such as providing strict work directions; insisting upon clean clothes, shoes, and equipment; and regular cleaning can prevent a lot of problems. Discipline is the key factor.

## *Quality Control*

In insect mass production, quality control can be divided into production control (performance of rearing procedures), process control (rearing process quality), and product control (end product quality). A detailed analysis is given by Leppla & Fisher (1989).

Generally, producers of beneficial arthropods carefully follow rearing procedures and processes. Success in these areas can be measured as production efficiency. Quality of the end product, however, is more difficult to determine. Emergence rate, mortality rate, size, and longevity can be measured in the laboratory. The effects of mechanical handling, packaging, and shipping can be measured in terms of these parameters. But the performance of the product in the field is very difficult to check. Searching behavior, dispersal capacity, longevity, oviposition rate, and tolerance to variable environmental conditions determine the effectiveness of the natural enemy. However, these parameters are often difficult to measure.

More detailed information on components affecting colony quality are given by Bigler (1989) and van Lenteren (1986).

Because of the difficulty of assessing colony quality, most producers of natural enemies are inclined to check their insects' quality only when problems or complaints occur.

*Phytoseiulus persimilis* and *Encarsia formosa* have been used for about 20 years without severe quality problems. These natural enemies are produced on their natural hosts and under conditions (glasshouse systems) that are relatively close to those of their use. Presumably, this keeps them fit enough for successful biological control. In the case of artificial rearing conditions, such as use of artificial diets or factitious hosts, quality is more likely to be affected. A well-documented case involving *Trichogramma* parasites is described by Van Bergeijk et al. (1989).

Quality standards are essential for commercial biocontrol agents. Growers and authorities will demand uniform products of a certain standard. Producers and scientists must come up with reliable testing methods leading to biocontrol agents of a standard, acceptable quality. The development of quantifiable standards will make it easier for producers to document the quality of their products.

## Production of the Greenhouse Whitefly and its Parasite Encarsia formosa—*An Example*

A first attempt to rear the whitefly was made on poinsettia. This plant caused a high mortality in the larval stages of the whitefly, it is relatively slow growing, and it is a small plant with little leaf surface. Moreover, the idea of using this plant in the production of *Encarsia formosa* was to use it as a banker plant: an "open rearing unit" from which both pest and beneficial insect migrate into the crop and establish a balance at a low, predictable level. This method, however, never reached commercial application. For some years whiteflies were produced on cucumber. Cucumber can be grown year round, grows quickly, has a big leaf surface, and is not harmed if leaves are picked from it. It proved to be a very good host plant for the greenhouse whitefly and is an acceptable plant for the parasite, although the hairs do hamper *Encarsia formosa* (Hulspas-Jordaan & van Lenteren 1978).

Many diseases and pests occur on cucumber, however, and they will complicate production. Powdery mildew, such as *Botrytis* and *Mycosphaerella*, easily infect cucumber. Twospotted spider mites may also cause problems.

*Encarsia formosa* was supplied for biological control in the pupal stage on leaf pieces. Cucumber leaves mold easily, especially in closed boxes. Because of this, leaf pieces were often unusable upon receipt by the customer. For these reasons, tobacco is now used as a host plant in the production of *Encarsia formosa*. This plant has few diseases and pests (at least in greenhouse cultures), and it has big, flat leaves that are easily checked for numbers of hosts and parasites. Also the parasite's behavior is not hampered by hairs on its leaf. Many cultivars exist, and cultivation knowledge is readily available. Moreover, *Encarsia formosa* pupae are now generally supplied on paper cards, and tobacco leaves are more suitable for mechanical removal of the pupae than cucumber leaves. Although tobacco is not as good a host plant for the whitefly as cucumber, it has proved to be a good alternative.

The choice of a cultivar can be very crucial. For one season the cultivar "White Burley" was used with good results. Suddenly in autumn (when daylength becomes shorter) whitefly mortality increased dramatically. It is possible that slight changes in the plants' physiology led to detrimental effects on the whitefly. A change to cv. "Xanthi" solved these problems.

The choice of a host plant will always be a compromise between many of the factors described above. Only experience can provide the answer.

## Utilization

### Marketing System

A key factor for the success of biological control in European glasshouse crops is the system of combined sale and advice (Ravensberg 1987).

Beneficials have a very short shelflife, quite different from conventional chemical compounds, and they also have a different way of controlling pests. Because of this, direct contact between supplier and grower is a necessity. Only in this way can information and material reach the grower in a timely way.

In most cases, this contact is between the grower and the local (horticultural) supplier, and between the latter and the producer of natural enemies. Suppliers regularly equip growers with all kinds of horticultural material and equipment, including pesticides. In their product portfolio, suppliers also have natural enemies for biological control, and, typically, one or more employees specialized in biological control. These specialists are trained by the producer of natural enemies, are constantly informed about new develoments, and are supported in case of difficulties. In this way, information transfer between the user and the producer is very rapid. After observation of a pest, the beneficials are delivered within 2 or 3 days, an essential condition for biological control.

Natural enemies are not just sold by the number per square meter; they are sold as an integrated control package. The grower pays both for the beneficials and for advice. This means that a biological control specialist visits the grower regularly. At first sight of the pest, the specialist decides on the use of natural enemies and the moment of the (first) introduction. After establishment of the parasites or predators, the specialist visits once or twice in the following weeks and later visits the grower upon request. In this way, an integrated control program is maintained during the whole season. This close support is especially important for growers who have little experience with this control method.

## Method of Release

The method of release for mass-reared natural enemies depends on the developmental stage at which they are introduced into the crops. They can be released as:

- eggs, (e.g., *Chrysoperla*);
- larvae or nymphs, (e.g., *Chrysoperla*, *Orius*, and entomopathogenic nematodes);
- pupae, (e.g., *Encarsia*, and *Trichogramma*);
- adults, (e.g., *Dacnusa*, *Diglyphus*, and *Cryptolaemus*);
- all stages, (e.g., *Amblyseius*, and *Phytoseiulus*).

The choice of the developmental stage that will be released depends on:

- which stage is best to collect from the production process;
- which stage is least vulnerable in handling;
- which stage is least vulnerable in transport;
- which stage is easiest to release;
- which stage is best for establishment.

The "release stage" can be prepared for shipping and introduction in several ways:

- on leaves of the plant on which they are produced (all stages, pupae);
- glued onto paper cards (eggs, pupae);
- in containers (adults);
- in a dispersal medium like bran or vermiculite (all stages, pupae);
- loose (pupae, aerial release);
- as "open rearing units" (banker plants with host and beneficial).

When insects are shipped to foreign countries or even to other continents, efforts be made to deliver the products cleanly, without leaves, soil, or pest organisms. Generally the preferable stage for shipment is the egg or pupal stage, since they are the least vulnerable to mechanical handling.

## Storage

Storage of natural enemies is often necessary since production and demand are often difficult to synchronize. Short-term storage is common. Many beneficials are stored up to 2 weeks at 4-14°C, depending on the insect and growth stage. For pupal stages, lower temperatures can be maintained than for mobile stages. Cold storage merely retards development. After long cold storage, a loss of fitness results. This can often be measured in mortality or emergence rates, but a decline in field performance may be difficult to trace.

Long-term storage usually focuses on induction of diapause. This is sometimes possible, but emergence from diapause is often unpredictable. Emergence usually takes place over a long period. For example, it is quite easy to get *Aphidoletes aphidimyza* into diapause by rearing them under short day conditions and low temperatures (Gilkeson & Hill 1986), but emergence from diapause is very gradual and takes several weeks before all emerge. The same problem occurs with *Dacnusa sibirica*. Only in the case of *Trichogramma* are there some promising results for long-term storage (Van Schelt & Ravensberg, in press). Solutions for long-term storage have to be found. However, much more research is needed before adequate methods for maintaining quality of the insects are available.

## Shipping of Natural Enemies

Shipment may lead to negative effects on the natural enemies and even to high mortality rates. The carefully reared and packed insects are out of sight of the producer and user during shipping, and as this time gets longer, the risks increase. Generally the transporter is not aware of the vulnerability of the package. Producers or suppliers should know all details of handling and the time involved in transport so that they can react to possible risks. Packaging is very important. Avoid overcrowding (stress, heat, or carbon dioxide production by the insects), ship the least vulnerable developmental stage whenever possible, take care that emergence cannot take place during the time of transport, add extra food in case mobile stages are shipped, and allow some air exchange.

Temperature is a critical factor during transport and is often the cause of problems. Use packing materials with good insulation properties and/or ice elements to make packages shock-proof and easy to handle. Pack so that special handling procedures and the possibility of errors are minimized. The local supplier and user must expect the shipment and be knowledgeable in its handling so that further transport and the actual release can be executed as soon as possible.

## *The Introduction*

The activity and the efficacy of the natural enemy depends to a great extent on the methods of introduction. Many aspects discussed below must be taken into account for successful establishment and control.

**Mode of Activity:** Factors to be considered are the way in which the pest is controlled, which life stages are attacked, how the effect can be seen, when the first results can be expected, how long the control lasts, whether multiple introductions are needed, and whether the natural enemy is compatible with chemicals (integration).

**Moment of Release:** A monitoring system (pheromones, light traps, sticky color traps, optical observation, or leaf samples) should be used to observe the pest and time the release. It is important to determine whether the stages on which the beneficial is active are present, whether its density is right, whether any toxic residues are present, and whether environmental conditions are right.

**Handling of Material:** The material should be easy to handle, and quick to apply. It should help prevent the risks of crushing pupae or other stages, and the places where it has been released should be visible. The material should be applied as soon as possible upon arrival.

**Frequency of Releases:** Factors to consider include whether there will be one or multiple introductions, what the intervals will be between introductions, whether failed introductions will be repeated, and whether the introduction will be on a spot or blanket basis.

**Dosage:** Variables include whether the release is to be inoculative or inundative, whether one or more developmental stages of the natural enemy will be released, the variability in emergence dates, and the number to be released per plant or per square meter.

**Distribution:** Decisions must be made about the number of release points per surface or per plant, whether the natural enemies will be released in spots or evenly placed over the whole surface, and where in or on the plants or on the soil they will be placed.

**Releasing Person:** Releases can be made by the supplier, the grower, or the grower's employees. Sufficient information should be available for correct handling by any of them.

Many of these aspects are empirically determined and learned by evaluating the results. Also, the user should know and understand what the critical issues are and where the decision points lie, underscoring the need for close contact between the grower and the supplier or producer.

## Biological Control in Sweet Pepper—An Example

Sweet peppers were grown in the Netherlands on about 180-200 ha in the seventies and early eighties. Pests and diseases were controlled mainly with chemical compounds. An IPM scheme (Ravensberg et al. 1983) based on the OP (organo-phosphorous) resistant *Phytoseiulus persimilis* (against spider mites), low-dose applications of tetrachlorvinphos (against *Thrips tabaci*), pirimicarb (against *Myzus persicae*), and *Bacillus thuringiensis* (against caterpillars) was applied by a slowly increasing number of growers (Table 1).

**Table 1. Area (ha) on which natural enemies have been used against the main pests of sweet pepper in the Netherlands**

| Year | Total area | P.p.[1] | A.c.[2] | A.a./A.m.[3] |
|------|-----------|---------|---------|--------------|
| 1972 | 75  | -   | -   | -  |
| 1974 | 160 | 12  | -   | -  |
| 1976 | 160 | 20  | -   | -  |
| 1978 | 180 | 35  | -   | -  |
| 1980 | 200 | 35  | -   | -  |
| 1982 | 220 | 40  | -   | -  |
| 1984 | 240 | 60  | 7   | -  |
| 1985 | 290 | 100 | 60  | -  |
| 1986 | 250 | 145 | 140 | -  |
| 1987 | 310 | 270 | 250 | -  |
| 1988 | 360 | 340 | 320 | -  |
| 1989 | 450 | 390 | 400 | 50 |
| 1990 | 60  | 580 | 580 | 80 |

[1]P.p. = *Phytoseiulus persimilis*
[2]A.c. = *Amblyseius cucumeris*
[3]A.a. = *Aphidoletes aphidimyza;* A.m. = *Aphidius matricariae*
(Adapted from van Lenteren et al. 1980; Ravensberg et al. 1983; Ramakers et al. 1989.)

A key chemical for thrips control, tetrachlorvinphos, was expected to become unavailable in 1986 and restrictions on residues (from a.o. tetrachlorvinphos and some fungicides) on fruits for export, especially to the U.S., necessitated revision of pest control methods in this crop. Research on biocontrol of thrips had started for cucumber and sweet pepper (Ramakers 1980, 1983), and the first experiments in sweet pepper in 1982 to 1984 with the predatory mite *Amblyseius cucumeris* showed promising results. The export market for sweet pepper fruit to the U.S. grew considerably, together with the total acreage of sweet pepper. With this increasing high-value market, the growers organization insisted on scaling up the biocontrol of thrips because of export restrictions. In 1985, 60 ha (20%) were treated with *Amblyseius cucumeris* and *Phytoseiulus persimilis*. The whole program was supported by the Auctions, the Extension Service, the Glasshouse Crops and Research Centre in Naaldwijk, and the producers of natural enemies. A monitoring program was set up for 2 years in order to follow and check the applications and results. The results were very good (De Klerk & Ramakers 1986), and a year later 60% of the growers applied this system (Ravensberg & Altena 1987).

In this case a combined effort of growers, researchers, and producers resulted in effective biological control of thrips, stimulated by market demands and a lack of effective chemicals. The acreage under this IPM program has reached nearly 100% (580 ha out of 600 ha in 1990, Table 1) although a new pest, *Frankliniella occidentalis,* the Western flower thrips, makes the system less secure. This insect was imported into the Netherlands in 1983 (Mantel & Van de Vrie 1988) and is causing severe problems in existing IPM programs in sweet pepper and cucumber, where it is now the key pest. In sweet pepper, adjustment of the number of introductions and number of predatory mites managed to keep thrips under control. Still, research has started on the use of *Orius* species, predatory flower bugs (Anthocoridae). These bugs enter the greenhouse in the summer and are a valuable aid in thrips control.

Aphid control, mainly against *Myzus persicae,* used pirimicarb, a rather selective insecticide. However, repeated applications disturb the *Amblyseius* population, and therefore biological control of aphids is needed. On a small scale, the predatory gall midge *Aphidoletes aphidimyza* has been used since 1989 in combination with the parasite *Aphidius matricariae* against *Myzus persicae.* Presumably, this practice will increase in coming years, especially because the cotton aphid, *Aphis gossypii,* is an increasing problem and this species cannot be controlled by pirimicarb.

The greenhouse whitefly is a relatively new pest in sweet pepper. With the expansion of the acreage, new cultivars were introduced that are more susceptable to whitefly. Biocontrol with *Encarsia formosa* (on white, yellow, and orange cv's) is satisfactory. Leafminers are a minor pest and can be

**Table 2. Integrated control program in sweet pepper**

| Pest or disease | Control method | Remarks/additional control |
|---|---|---|
| Thrips | *Amblyseius cucumeris* | Dichlorvos before start of biocontrol program. |
| Spider mites | *Phytoseiulus persimilis* | Fenbutatin oxide, if necessary. |
| Aphids | *Aphidoletes aphidimyza, Aphidius matricariae* | Pirimicarb, heptenophos against *Aphis gossypii.* |
| Whitefly | *Encarsia formosa* | Pest of minor importance. |
| Leafminers | *Dacnusa sibirica, Diglyphus isaea* | Pest of minor importance. |
| Caterpillars | *Bacillus thuringiensis* | |
| Tarsonemid mites | Dicofol | Pest of minor importance. |
| Powdery mildew | Bitertanol | Increasing problem. |
| Botrytis | Tolylfluanid | Pest of minor importance. |

controlled by releasing the parasites *Dacnusa sibirica* and *Diglyphus isaea*. Caterpillars are well controlled by *Bacillus thuringiensis* and some fungal diseases with relatively harmless fungicides. The IPM program in sweet pepper is a relatively complicated program involving the introduction of many different natural enemies (Table 2). Nevertheless, growers adopted this system completely and are happy with it. With so many natural enemies, costs may be higher than with broad-spectrum pesticides, but residue-free produce will compensate for this and offers growers a leading position in the present market.

## Current Markets and Future Opportunities

### Commercially Available Natural Enemies

The natural enemies mentioned are applied on a large part of the western European market and are considered "established" natural enemies (Table 3). Beneficials that are applied on only a limited scale (a few hectares) for a short period in a specific market or country are not mentioned. The commercial application of biological control started with the predatory mite *Phytoseiulus persimilis* against the twospotted spider mite and three years later with *Encarsia formosa* against the greenhouse whitefly (van Lenteren et al. 1980). For a long time these two beneficials were the only ones available.

In the early 1980s leafminer parasites became available, and their use increased to control the new pest *Liriomyza trifolii*. The use of *Amblyseius* mites against thrips began in 1981 but increased only in the mid-1980s, especially after the establishment of the Western flower thrips in Europe. Later in that decade the emphasis changed to aphid control, mainly because of increasing difficulties with *Aphis gossypii*, and aphid enemies came onto the market. The most recent development is the production of predatory bugs, and since 1991 *Orius* spp. are available for thrips control

The number of commercially available natural enemies is steadily increasing. "New" enemies are investigated and several are waiting for practical use, such as *Chrysoperla, Orius,* and parasites and predators of mealybugs, soft and hard scales, thrips, and whitefly. Insect parasitic nematodes are increasingly available and used to control black vine weevil and Sciarid larvae. The use of

**Table 3. The large-scale application of natural enemies in western European glasshouse crops**

| Used since | Natural enemy | Target pest |
|---|---|---|
| 1969 | *Phytoseiulus persimilis* | *Tetranychus urticae* |
| 1972 | *Encarsia formosa* | *Trialeurodes vaporariorum*[1] |
| 1980 | *Opius pallipes*[2] | *Liriomyza bryoniae* |
| 1981 | *Dacnusa sibirica* | *Liriomyza bryoniae/ L.trifolii*[3] |
| 1981 | *Amblyseius barkeri* | *Thrips tabaci*[4,5] |
| 1984 | *Diglyphus isaea* | *Liriomyza bryoniae/ L.trifolii*[3] |
| 1985 | *Amblyseius cucumeris* | *Thrips tabaci*[5] |
| 1989 | *Aphidoletes aphidimyza* | *Myzus persicae* e.o. |
| 1990 | *Aphidius matricariae* | *Myzus persicae* e.o. |
| 1991 | *Orius* | *Frankliniella occidentalis* e.o. |

[1]Since 1988, also used against *Bemisia tabaci*.
[2]Available only until 1982.
[3]Since 1990, also used against *Liriomyza huidobrensis*.
[4]Available only until 1990.
[5]Since 1986, also used against *Frankliniella occidentalis*.

microorganisms as control agents is increasing. Of course, *Bacillus thuringiensis* is extensively used against caterpillars. Promising agents are *Verticillium lecanii* against whitefly and aphids (Ravensberg et al. 1990), *Metarhizium anosipliae* against the black vine weevil (Reinecke et al. 1990), and *Bacillus thuringiensis* var. *israelensis* against Sciarid larvae (Osborne et al. 1985). Chemical companies show great interest in developing microbial pesticides on the basis of these organisms.

## Growth of European Markets for Natural Enemies

Natural enemies are used in almost all countries in western Europe (van Lenteren 1985). Over 80% of biological control in protected crops is applied to cucumber, tomato, and sweet pepper. Other crops in which natural enemies are used are strawberries, eggplant, melon, watermelon, beans, and several cutflowers and ornamental plants. Biocontrol is predominantly applied in heated glasshouses where crops are cultivated for periods of several (4 to 10) months. Some applications are done in plastic tunnels and in field crops.

In Table 4 the markets for specific natural enemies are given for the last decade. If *Encarsia formosa* is used in a crop that also uses *Phytoseiulus persimilis*, both are mentioned, so the areas are overlapping to some extent. The total area with protected crops in western Europe is about 65,000 ha, out of which 17,000 ha are in glasshouses (van Lenteren & Woets 1988). Biocontrol is now applied on an estimated area of about 4,000 ha. Large areas are still under chemical control, and it is expected that this will slowly but steadily change into biological control.

**Table 4. Area (ha) of protected crops in western Europe on which natural enemies are used**

|                                                    | 1982 | 1985 | 1988 | 1990 |
|----------------------------------------------------|------|------|------|------|
| *Encarsia formosa*                                 | 1150 | 1500 | 2800 | 3200 |
| *Phytoseiulus persimilis*                          | 1000 | 1200 | 2500 | 2900 |
| *Amblyseius cucumeris*                             | 30   | 140  | 800  | 1100 |
| *Aphidoletes aphidimyza / Aphidius matricariae*    | 5    | 30   | 75   | 150  |
| *Dacnusa sibirica / Diglyphus isaea*               | 40   | 460  | 600  | 900  |

Adapted from van Lenteren 1983; van Lenteren 1985; van Lenteren & Woets 1988; Ravensberg et al. 1983.

## Challenges and Opportunities for Future Biological Control

The glasshouse industry is an area of rapid change: There are new cultural methods (such as growing on Rockwool, new types of glasshouses, and computer control of environmental conditions), new pesticides, new cultivars, and new imported pests. These changes constantly require adjustments in biocontrol systems. For more information on specific challenges for biocontrol in the glasshouse industry, see van Lenteren & Woets (1988).

A number of new pests and diseases have been accidentally imported into Europe during the last two decades (van Lenteren et al. 1987). With the ever-increasing worldwide trade of plants these risks are increasing, despite intensive phytosanitary inspections. If these pests establish in crops with a well-developed IPM scheme, it often disturbs the whole program. Growers will first try to battle the unknown problem with chemicals, but since little information on the new pest is available, this approach often fails. Moreover, most new pests show a high degree of resistance to many chemicals, including those that are compatible with the beneficial organisms. This leaves no option other than biocontrol.

The occurrence of newly introduced pests stimulates new funds for research and accelerates the application of natural enemies. Examples are *Liriomyza trifolii, Liriomyza huidobrensis, Frankliniella occidentalis, Bemisia tabaci, Aphis gossypii*, and *Spodoptera exigua*, all pests introduced into Europe during the last decade. Biocontrol possibilities are being investigated, and for most of these pests natural enemies are available for release.

Many countries set a zero tolerance for the presence and damage of arthropods on imported plant material. In some cases plants even have to be free of beneficial organisms. This leaves little or no room for biological control in potplants or cutflowers since the whole plant or a large part of the plant is being sold. In greenhouse vegetables growers face the same requirements, but in that case only the fruits (e.g., tomatoes) are being sold, and some leaf damage can be tolerated. For ornamental crops these restrictions present a big impediment to biological control.

However, natural enemies have been used in ornamental crops for many years in situations where products are grown for domestic markets. Here zero tolerance demands have not been applicable. Recently in Denmark, Germany, and Sweden, however, biocontrol has been applied to crops meant for export. Results are satisfactory, export demands have been met, and the use is expanding (Albert 1990). In other countries the attitudes are slowly changing also, and the first steps are carefully set. The cotton whitefly, *Bemisia tabaci*, was introduced into Europe on poinsettia cuttings and chemical control is very difficult. *Encarsia formosa* is now used with good results (preventive releases), and this application is growing (Albert & Schneller 1989; Wardlow 1989).

The development of new pesticides is getting more and more expensive, partly because of environmental concerns (registration questions). This means that there will be very few new pesticides developed for the small glasshouse market and thus more opportunities for biocontrol.

Growers' attitudes toward chemical control are also changing. If there is a possibility for a biological solution that is as effective and not more expensive than a chemical solution, they will certainly use it. Moreover, the consumer is demanding residue-free produce and, in some cases, will even pay more for it. The environmental concern is a political issue. Governments want to reduce the use of chemicals. Funds for research are becoming available, and in the near future nonconventional control methods will find increasing favor.

Besides the professional glasshouse market, biocontrol is also applied in several other areas including types of interior landscapes, big offices, botanical gardens, and zoos. Even in small consumer glasshouses some natural enemies can be used. In the hardy plants nursery, interest in natural enemies is awakening, especially for plants in unheated glasshouses. In warmer areas biocontrol is also being applied in field crops, and the use is increasing in crops such as strawberries, melons, watermelons, and others in France, Italy, Spain, and Greece. In these countries vast areas of horticultural and fruit crops are awaiting biocontrol.

## References

Albert, R. 1990. Experiences with biological control measures in glasshouses in Southwest Germany. IOBC/WPRS Bull. XIII/5: 1-5.

Albert, R. & H. Schneller. 1989. Biologische Schadlingsbekampfung in Zierpflanzenbau. I. Poinsettien (*Euphorbia pulcherrima* willd. ex. klotzsch.) Med. Fac. Landbouww. Rijksuniv. Gent 54/3a: 873-882.

Bergeijk, van, K. E., F. Bigler, N. K. Kaashoek & G. A. Pak. 1989. Changes in host acceptance and host suitability as an effect of rearing *Trichogramma maidis* on a factitious host. Entomol. Exp. Appl. 52: 229-238.

Bigler, F. 1989. Quality assessment and control in entomophagous insects used for biological control. J. Appl. Entomol. 108: 390-400.

Bravenboer, L. & G. Dosse. 1962. *Phytoseiulus riegeli* Dosse als Predator einiger Schildmilben aus der *Tetranychus urticae* gruppe. Entomol. Exp. Appl. 5: 291-304.

Boxtel, van, W., J. Woets & J. C. van Lenteren. 1978. Determination of host plant quality of eggplant (*Solanum melongena* L.), cucumber (*Cucumeris sativus* L.), tomato (*Lycopersicon exculentum* L.) and paprika (*Capsicum annuum* L.) for the greenhouse white fly (*Trialeurodes vaporariorum*

(Westwood) (Homoptera: Aleyrodidae). Med. Fac. Landbouww. Rijksuniv. Gent. 43: 397-408.

Gilkeson, L. A. & S. B. Hill. 1986. Diapause prevention in *Aphidoletes aphidimyza* (Diptera: Cecidomyiidae) by low-intensity light. Environ. Entomol. 15: 1067-1069.

Haren, van, R. J. F., M. M. Steenhuis, M. W. Sabelis & O. M. B. De Ponti. 1987. Tomato stem trichomes and dispersal success of *Phytoseiulus persimilis* relative to its prey *Tetranychus urticae*. Exp. and Appl. Acarol. 3: 115-121.

Hulspas-Jordaan, P. M. & J. C. van Lenteren. 1978. The relationship between host-plant leaf structure and parasitization efficiency of the parasitic wasp *Encarsia formosa* Gahan (Hymenoptera: Aphelinidae). Med. Fac. Landbouww. Rijksuniv. Gent. 43: 431-440.

Hussey, N. W. 1985. History of biological control in protected culture in western Europe, pp. 11-22. *In* N. W. Hussey & N. E. A. Scopes [eds.], Biological pest control, the glasshouse experience. Pool, Dorset, Blandford Press.

Johnson, B. 1953. The injurious effects of the hooked epidermal hairs of French beans (*Phaseolus vulgaris* L.) on *Aphis craccivora*. Koch. Bull. Entomol. Res. 44: 779-788.

Klerk, de M. L. & P. M. J. Ramakers. 1986. Monitoring population densities of the phytoseiid predator *Amblyseius cucumeris* and its prey after large scale introductions to control *Thrips tabaci* on sweet pepper. Med. Fac. Landbouww. Rijksuniv. Gent. 51/3a: 1045-1048.

Koppert, J. P. 1978. Ten years of biological control in glasshouses in the Netherlands. Med. Fac. Landbouww. Rijksuniv. Gent. 43/2: 373-378.

Lenteren, van, J. C. 1980. Evaluation of control capabilities of natural enemies: does art have to become science? Neth. J. Zool. 30: 369-381.

Lenteren, van, J. C. 1985. Sting, Newsl. Biol. Control Greenhouses, Vol. 8: 32.

Lenteren, van, J. C. 1986. Evaluation, mass production, quality control and release of entomophagous insects, pp. 31-56. *In* J. M. Franz [ed.], Biological plant and health protection. Stuttgart, Fischer.

Lenteren, van, J. C. 1990. A century of biological control in West Europe. Proc. Exper. & Appl. Entomol. N.E.V. Amsterdam, Vol. 1: 3-12.

Lenteren, van, J. C., P. M. J. Ramakers & J. Woets. 1980. Integrated control of vegetable pests in greenhouses, pp. 109-118. *In* A. K. Minks & P. Gruys [eds.], Integrated control of insect pests in the Netherlands. Wageningen, Pudoc.

Lenteren, van, J. C., J. Woets, P. Grijpsma, S. A. Ulenberg & O. P. J. M. Minkenberg. 1987. Invasions of pest and beneficial insects in the Netherlands. Entomology Proceedings C 90 (1), March 30: 51-58.

Lenteren, van, J. C. & J. Woets. 1988. Biological and integrated pest control in greenhouses. Ann. Rev. Entomol. 33: 239-269.

Leppla, N. C. & W. R. Fisher. 1989. Total quality control in insect mass production for insect pest management. J. Appl. Entomol. 108: 452-461.

Mantel, W. P. & M. van de Vrie. 1988. The Western flower thrips, *Frankliniella occidentalis*, a new thrips species causing damage in protected cultures in the Netherlands. Ent. Ber. Amst. 48(9): 140-144.

Morrison, R. K. 1977. A simplified larval rearing unit for the common green lacewing. Southwest. Entomol. 2: 188-190.

Osborne, L. S., D. G. Boucias & R. K. Lindquist. 1985. Activity of *Bacillus thuringiensis* var. *israelensis* on *Bradysia coprophila* (Diptera: Sciaridae). J. Econ. Ent. 78: 922-925.

Ramakers, P. J. M. 1980. Biological control of *Thrips tabaci* (Thysanoptera: Thripidae) with *Amblyseius* spp. (Acari: Phytoseiidae). IOBC/WPRS/Bull. III/3: 203-207.

Ramakers, P. J. M. 1983. Mass production and introduction of *Amblyseius mckenziei* and *A. cucumeris*. IOBC/WPRS/Bull. VI/3: 203-206.

Ramakers, P. J. M. & M. J. van Lieburg. 1982. Start of commercial production and introduction of *Amblyseius mckenziei* Sch. & Pr. (Acarina: Phytoseiidae) for the control of *Thrips tabaci* Lind. (Thysanoptera: Thripidae) in glasshouses. Med. Fac. Landbouww. Rijksuniv. Gent 47/2: 541-545.

Ramakers, P. J. M., M. Dissevelt & K. Peeters. 1989. Large scale introductions of phytoseiid predators to control thrips on cucumber. Med. Fac. Landbouww. Rijksuniv. Gent 54/3a: 923-929.

Ravensberg, W. J. 1987. Mass production of beneficial arthropods: commercialization and other problems. Biologischer Pflanzenschutz. Schriftenreihe des Bundesministers fr Ernhrung, Landwirtschaft und Forsten, Reihe A: Angewandte Wissenschaft, Heft 344: 259-268.

Ravensberg, W. J. & K. Altena. 1987. Recent developments in the control of thrips in sweet pepper and cucumber. Bull. IOBC/WPRS X/2: 160-164.

Ravensberg, W. J., J. C. van Lenteren & J. Woets. 1983. Developments in application of biological control in greenhouse vegetables in the Netherlands since 1979. Bull. IOBC/WPRS VI/3: 36-48.

Ravensberg, W. J., M. Malais & D. A. van der Schaaf. 1990. Applications of *Verticillium lecanii* in tomatoes and cucumbers to control whitefly and thrips. IOBC/WPRS Bull. XIII/5: 173-178.

Reinecke, P., V. Andersch, K. Stenzel & J. Hartwig. 1990. Bio1020, a new microbial insecticide for horticultural crops. Brighton Crop Protection Conf. Pest and Diseases, 49-54.

Schelt, van, J., J. B. Douma & W. J. Ravensberg. 1990. Recent developments in the control of aphids in sweet peppers and cucumbers. IOBC/WPRS Bull. XIII/5: 190-193.b

Schelt, van, J. & W. J. Ravensberg, in press. Some aspects on the storage and application of *Trichogramma maidis* in corn. (Trichogramma and other egg parasites, 3rd Symposium San Antonio, September 1990.)

Wardlow, L. R. 1989. Integrated pest management in poinsettias grown under glass. Med. Fac. Landbouww. Rijksuniv. Gent 54/3a: 867-872.

# 28

## Mass Rearing of
## Phytoseiid Mites for Testing
## and Commercial Application

*L. A. Gilkeson*

### Introduction

Mass production of phytoseiid mites has changed radically in both scale and efficiency since *Phytoseiulus persimilis* Athias-Henriot was first reared experimentally in the 1960s. *P. persimilis* is the most widely produced and applied phytoseiid, used on an estimated 517 ha of crops in 1985 (van Lenteren & Woets 1988) and currently reared by the millions weekly in insectaries in the Netherlands, England, France, Canada, the United States, and other countries. For this species and others applied commercially, the rise in production capacity was made possible by continuous improvements in mass-rearing methods over the last 20 years.

For the purpose of discussing mass production, phytoseiids can be divided into two main groups: those species such as *P. persimilis* or *Metaseiulus (=Typhlodromus) occidentalis* (Nesbitt) that must be reared on their natural hosts, the Tetranychidae, and those polyphagous species of *Typhlodromus*, *Amblyseius* and *Eusieus*, that can be reared on pollens, alternate prey, or even artificial diets. In this paper, a review of phytoseiid rearing methods will be followed by a description of current commercial production methods for the obligate spider mite predator, *P. persimilis*, and for a phytoseiid reared commercially on alternate prey, *Amblyseius cucumeris* Oudemans. Problems that arise in commercial mass production, such as maintaining continuity of production and quality of predators, will be discussed, followed by a brief review of commercial prospects for application of phytoseiids.

## Review of Rearing Techniques

### *Rearing on Arenas*

Laboratory rearing methods for phytoseiids have been reviewed in detail by Overmeer (1985a). Most early rearing was conducted on small arenas made from excised leaves or artificial materials; they are reviewed here because of their usefulness for maintaining research colonies. The greatest advantage of arena rearing is that it is the most reliable way to avoid contamination of colonies with other species because a mite-repellant barrier isolates each group of mites. Also, because most arenas use water as the barrier, it is simple to maintain a high relative humidity, which is ideal for phytoseiid rearing (McMurtry & Scriven 1965). Many species of phytoseiids have been successfully reared on arenas (*P. persimilis, P. macropilus* [Banks]; *M. occidentalis, Typhlodromus pyri* Scheuten; *Amblyseius fallacis* [Garman]; *A. potentillae* [Garman]; and *A. cucumeris*) (Overmeer 1985b). However, some species, such as *Eusius hibisci* (Chant) that normally feed on plant juices (McMurtry & Scriven 1964) or *A. finlandicus* Oudemans that readily run off the arenas have been difficult to rear (Overmeer 1985b).

One of the first arena designs, developed for rearing *A. fallacis,* was cut bean leaves laid on wet filter paper in petri dishes (Ristich 1956). A thin line of sticky material around the leaf perimeter prevented predators from being lost. Gilstrap (1977) used a variation of this arena with bean leaves on moist cotton. The cut edge of the bean petiole was pressed into the wet cotton, and the leaf blade was supported on a small square of bent wire mesh to allow mites access to both sides. A compact, enclosed rearing unit for a leaf arena on moist cotton, described by Abou-Seta and Childers (1987), provides a deep water reservoir with a wick to the cotton pad and ventilated lid to maintain optimum humidity.

An artificial arena for large-scale rearing of several phytoseiid species was designed by McMurtry and Scriven (1965). The arena was made of heavy construction paper sprayed with black paint and placed on a layer of wet polyurethane foam in a steel pan. A metal tile could also be used (McMurtry & Scriven 1975). A wet strip of cotton around the edge of the arena kept predators on the arena. This design was later modified by Overmeer (1985b) to incorporate a thicker foam sponge topped with a tile the same size as the sponge. Four wide strips of tissue paper laid along the edges of the arena and allowed to hang down the side of the sponge to the tray of water acted as a water barrier, which was reinforced by an extra barrier of sticky material applied along the edge of the wet paper. Kamburov (1966) reared phytoseiids on plastic plates made with a narrow channel around the edge to hold an oil barrier. An ingenious variation on using a water barrier was developed by Theaker and Tonks (1977), who reared *P. persimilis* on blotting papers in plastic lids floating in trays of water. Each lid was kept centered by a magnet on the bottom of the lid, which was attracted to another magnet glued to the center of the water tray. An entirely disposable arena made of circles of black plastic garbage bags resting on wet cotton in a petri dish was devised by Ball (1980). A similar idea for an arena has been successfully used for rearing *A. cucumeris* on 10 cm arenas cut with a broadly bevelled edge from thin rigid plastic foam sheets (Morewood & Gilkeson, in press). The arena is anchored in a petri dish of water by a thin, conical pad of wet cotton that allows the float to rise and fall with the changing level of water in the dish. Indentations scored in the top surface of the foam arenas provide refuges for the mites.

On artificial arenas, mites require some sort of shelter and oviposition site. If detached leaf pieces are not used, other shelters should be provided—for example, a cover slip resting on strands of cotton (McMurtry & Scriven 1975), small roof-shaped pieces of clear acetate sheets (Overmeer 1985b), or 25 mm squares of black felt. The latter is particularly useful for collecting phytoseiid eggs because they show against the dark background.

## Rearing in Cages

Cages have been used successfully to mass produce phytoseiids. A small, domed plastic rearing cage containing a dish of 2% agar to supply water and raise humidity was developed by McMurtry and Scriven (1962). Spider mite prey was added through a port in the side. A successful cage for rearing *M. occidentalis* developed by Tanigoshi et al. (1975) was made from two cardboard cartons with the bottoms cut out and replaced by window screen. The cartons were fastened together, bottom to bottom, and lids with holes in the top covered by parachute cloth were placed on the cartons. The top chamber was filled with beans infested with spider mites, then inoculated with predators. After 3 days, the cage was inverted and fresh leaves and prey were placed in the upper chamber; as leaves deteriorated, the mites migrated upward to the new food supply. A similar cage designed by Fournier et al. (1985) for *P. persimilis* production was made of 30 cm-diameter x 30 cm-high plastic cylinders with large mesh screens in the bottom. Cut bean leaves infested with spider mites were placed in the cage with the predators; as the leaves dried, the predators moved upward and a new cage with fresh prey was added to the top of the stack.

## Rearing on Plants

Today's mass production systems for phytoseiids reared on tetranychid prey on potted host plants are basically expanded versions of a method first outlined by Ristich (1956). He reared spider mites on flats of kidney bean plants, then reared *A. fallacis* on the plants once they were well infested with spider mites. Bean plants have been used universally for rearing spider mites, although the preferred variety varies among kidney beans (Ristich 1956), lima beans (Naegele & McEnroe 1963, Jacklin & Smith 1966), broad beans (Anonymous 1975) and snap beans (Hendrickson 1980). Early rearing systems, such as the first mass-production guide published by the Glasshouse Crops Research Institute in England (Anonymous 1975) were based on moving the clean broad bean plants to spider mite rearing rooms, then using these plants to inoculate the phytoseiid rearing plants (dwarf French beans) with spider mites. When well infested these plants were moved to phytoseiid rearing sections and inoculated with *P. persimilis* for the last stage of rearing. Using this system, each pot of beans produced 1,500-4,000 predators, or about 22,500-60,000 weekly from a 7.5-m$^2$ greenhouse.

## Rearing Outdoors

Phytoseiids have also been reared on a large scale outdoors. Field et al. (1979) reported successful production of a pesticide resistant strain of *M. occidentalis* in Australia in large plots of soybeans and on apple trees for release to apple and peach growers. In California, Hoy et al. (1982) reared 62 million *M. occidentalis* in a 0.2 ha field of soybeans from June to September. A pure product was obtained from outdoor rearing because the strain of *M. occidentalis* was resistant to various pesticides that were used to eliminate pests and wild lines of phytoseiids while maintaining resistance in the predators. Outdoor rearing was found to be more cost effective than greenhouse rearing, but greenhouse production had the advantage of providing an earlier and longer harvest period (Hoy et al. 1982).

A disadvantage of field rearing revealed in subsequent years was that it was not predictable from year to year. In Australia, a successful, long-term program based on seasonal outdoor rearing of phytoseiids has been implemented over the last 10 years by CSIRO and local departments of agriculture to establish integrated mite programs for apple growers (Bower & Thwaite 1982). *Typhlodromus pyri* are reared on apple trees that have been left unpruned to provide numerous shoots, which are harvested in summer when predators are numerous. Government personnel or growers cut the shoots on appointed days and use them to inoculate orchard blocks to control the European red mite, *Panonychus ulmi* (Bower, personal communication).

## Rearing on Artificial and Alternate Diets

The advantages of rearing phytoseiids on alternate diets are that it can be cheaper and more predictable and that less labor and space is required than rearing on tetranychids. Instead of balancing three phases of production (clean host plants, spider mites, and predators), the producer need only measure out the diet for the predator. In addition to Tetranychidae, the food spectrum of phytoseiids includes stored products mites (Tyroglyphidae), eriophyid mites, tarsonemid mites, larvae and eggs of insects, pollen, honeydew, and plant juices (Overmeer 1985a, Karg et al. 1987). The polyphagous species will feed on most of these; of 15 species studied by Karg et al. (1987), all except *P. persimilis* fed on at least two of these groups. While there is little use for an alternate diet of tarsonemid mites, the use of tyroglyphid mites, which are simple to rear in large quantities, has been a breakthrough for mass production of two species. *A. cucumeris* and *Amblyseius barkeri* (Hughes) (*=mckenziei*). These are now very inexpensively mass produced on *Acarus* and *Tyrophagus* species for application against thrips in greenhouse crops (Ramakers 1984). Another predator, *Amblyseius gossipi* Elbadry, can also be reared on the tyroglyphid *Tyrophagus casei* Oudemans mixed with pollen (Ramsy et al. 1987).

Although there has been a great deal of interest in developing artificial diets for the mass production of phytoseiids, so far few species and of these only the most polyphagous - have been maintained for long periods on diets lacking prey. McMurtry and Scriven (1966) found that a 20% yeast and 20% sucrose mixture provided enough nutrients to sustain oviposition in *A. limonicus* Garman & McGregor and *E. hibisci* and that the diet also allowed some individuals to develop to maturity. Kennett and Hamai (1980) were able to rear seven species of phytoseiids on an artificial diet of honey, sugar, yeast, yeast hydrolysate, casein hydrolysate, egg yolk, and water. An artificial liquid diet containing milk powder, honey, egg yolk, Wesson's salt, and water was developed for *Amblyseius teke* Pritchard and Baker (Ochieng et al. 1987). Mites were reared for 25 generations on this diet, with somewhat slower development but comparable fecundity and longevity as on their natural red spider mite prey.

Supplemental diets, which sustain the predators but do not support reproduction, are useful for species such as *P. persimilis* and *M. occidentalis* during shipment to reduce mortality and maintain the mites in good condition. A 20% molasses or sucrose diet increased the longevity of *M. occidentalis* and *Typhlodromus rickeri* Chant (McMurtry & Scriven 1966). The provision of a water source alone has been shown to improve the survival of *P. persimilis* (Ashihara et al. 1978), and when a 10% sucrose solution or droplet of honey was provided, the mites lived much longer (over 44 days). McMurtry and Scriven (1962) found that phytoseiids could use 2% agar in the shipping container as a water source and that the addition of 10% yeast hydrolysate and 6% honey to the agar increased the oviposition rate of *E. hibisci.*

## Rearing on Natural Hosts: *Phytoseiulus persimilis*

In 1959, Dosse found *P. persimilis* on flowers imported into Germany from Chile (Hussey 1985). He propagated *P. persimilis* and distributed them to workers in several countries, who studied the species for control of twospotted mites in greenhouse crops (Hussey 1985). He also sent them to the United States for control of spider mites in strawberries (Oatman & McMurtry 1966). Commercial mass production began in 1970 in the Netherlands and has continued to expand ever since.

Although alternate prey for *P. persimilis* has been studied extensively (Ashihara et al. 1978, Kennett & Hamai 1980, Karg et al. 1987), so far the species can be successfully reared only on *Tetranyshus* spp. Other prey may be nutritionally adequate, but the mites refuse to feed on them, possibly because important phagostimulants are absent (Kennett & Hamai 1980). Although most commercial producers rear *P. persimilis* on *Tetranychus urticae* Koch, one U.S. company uses the Pacific spider mite, *T. pacificus* (McGregor) (Scriven & McMurtry 1971), and in Japan they are reared on *T. kanzawai* Kishida (Hamamura et al. 1976).

Mass-rearing systems all begin with the production of bean plants for spider mite production. They are grown in large trays, soil-filled open plastic bags, or directly in beds or benches in the greenhouse. As discovered by Jacklin and Smith (1966), a dual system of spider mite rearing is necessary for continuous production. A pure spider mite culture, maintained with maximum isolation in separate controlled environment rooms or greenhouses to prevent contamination by phytoseiids, is necessary for inoculating the predator host plants. Personnel involved in rearing the pure spider mite culture should not be involved in the phytoseiid production because of the risk of transporting predators on clothing to the pure cultures. The host plant culture of spider mites is used to produce the large quantities necessary to feed the predators. At this stage, contamination by a few phytoseiids is not as serious a problem as it would be in the pure culture.

Most commercial mass-production systems conserve labor by leaving plants in place throughout the rearing cycle, either after one move from a clean host plant greenhouse to the mite production section or permanently in place from seeding until predators are harvested. When bean plants are well grown, leaves with spider mites from the pure culture are distributed through the plant canopy and conditions are kept warm (22-30°C) and relatively dry to speed reproduction of spider mites. A series of greenhouse sections or benches are inoculated at weekly intervals to provide a continuous supply.

When host plants are well infested with spider mites (7-14 days, depending on temperatures and quantities used for inoculation), the plants are inoculated with predators and grown for 2-3 weeks. At this stage, the temperature should be 22-26°C and relative humidity 75-85% to ensure the highest rate of reproduction. At higher temperatures, *P. persimilis* lays proportionally fewer eggs; higher mortality results from rearing at temperatures over 30-32°C (McClanahan 1968, Hamamura et al. 1976).

A different mass-rearing system that is also used for large-scale commercial production is described by Scriven and McMurtry (1971). Pacific spider mites (*T. pacificus*) are reared on bean plants as above, after which the plants are cut and washed in a separating tub that floats mites and eggs from the leaves. The mites are collected and washed through a graduated series of mesh screens, dried, weighed, and fed to phytoseiids on large arenas (McMurtry & Scriven 1965, 1975). The spider mite eggs are mixed with ground corn cob on the arenas, giving the predators shelter as well as providing them with a convenient substrate for shipping and distribution. As many as 4,000 female predators could be reared per arena in 6-10 weeks, an estimated 15,000 predators per gram of spider mite eggs (McMurtry & Scriven 1975). This system, although arguably more labor intensive than greenhouse rearing systems, is an excellent method for insectaries rearing several different species of phytoseiids because it ensures a high degree of isolation between cultures.

Once the maximum number of *P. persimilis* are produced in a given section, the mites are harvested, counted, packaged, and shipped to growers. Some companies still send *P. persimilis* on cut bean leaves, which usually provides a good quality product because the high humidity and presence of prey on leaves enables predators to arrive in good condition, ready to resume oviposition. Most growers do not like this method, however, because distribution of predators through the crop is laborious and because there is the risk of importing pests on the leaves. Western flower thrips is currently a serious concern. More popular with growers is the system originally employed on a large scale in the Netherlands and now also used by insectaries in Canada and England, which is to package phytoseiids in a granular material (e.g., vermiculite or coarse wheat bran). Growers distribute the predators by shaking the granular material carrying the mites onto the plants or by using leaf blowers or blowers mounted on backpacks (Ables et al. 1979, Grossman 1989). For large-scale field applications, *P. persimilis* has also been distributed over cornfields from airplanes (Pickett et al. 1987).

Several methods have been developed for getting mites into granular carriers. The only published method consists of pouring moist bran past leaves with predators on them to carry the mites into the bran (Fournier et al. 1985).

This system does not work well enough to be useful on a large scale: it requires too much labor and leaves too many predators on the plants. Although designs for equipment used to extract predators from leaves are not divulged by commercial producers, most take advantage of the observation that phytoseiids leave the foliage voluntarily when the leaves dry and prey becomes scarce, employing the concept embodied in rearing cages developed by Fournier et al. (1985) and by Tanigoshi et al. (1975), who harvested mites from uppermost cage lids every 3 days. The upward movement can be enhanced by lighting to attract *P. persimilis* (Ewert, unpublished data). By allowing foliage to dry, the mites are collected as they move upward into a collecting container where they can be counted volumetrically (van Lenteren & Woets 1988) and packaged in containers of bran. Some suppliers ensure that predators have water in transit by incorporating agar into the container lid (McMurtry & Scriven 1962). The method of rearing phytoseiids on arenas with spider mite eggs mixed in ground corn cobs (Scriven & McMurtry 1971) also has the advantage of being easy to harvest while providing mites already in a granular carrier along with a food supply.

## Rearing on Alternate Hosts: *Amblyseius cucumeris*

Both *A. cucumeris* and *A. barkeri(=mckenziei)* are currently used in European and North American greenhouse crops to control thrips *(Thrips tabaci* Lindeman and *Frankliniella occidentalis* (Pergande). *A. barkeri* (Ramakers 1988) is the most widely produced of the two species. A breakthrough in mass production was made when P. J. M. Ramakers and colleagues at the Glasshouse Crops Research and Experimental Station in the Netherlands developed a way to rear this species on stored product mites (Ramakers & van Leigburg 1982). The first rearing systems were vials containing 200 ml of wheat bran inside larger vials of saturated salt solutions to maintain high humidity. The pasteurized bran was inoculated with bran mites *(Acarus farris* (Oudemans), which fed on fungus mycelia growing on the bran. When grain mites were well established the predators were added and reared on the alternate diet. The commercial prototype of this system incorporated 20-liter carboys with humidified air forced through the bran to aerate and humidify it (Ramakers & van Leiburg 1982). Using this method, up to 120,000 predators could be produced per liter. A refinement of this system used by one commercial insectary employs 10-liter plastic buckets fitted with tight lids, air inlets covered with mite screen, and outlets closed with a water seal, similar to a fermentation air lock, to trap dust and prevent contamination of cultures. With this type of forced air system it is important to manage conditions to prevent condensation by eliminating temperature differentials (23-25°C is optimum) and pressure drops in the lines (Hansen & Geyti 1985). Condensation in the buckets results in moldy places in the bran; condensation in air lines can eventually block the air flow. Commercial refinements to this system include development of handling systems for pasteurizing, mixing bran and mites in bulk, and improvement of the diet for the prey mites. Hansen & Geyti (1985) added 250 g of yeast to each 25-liter container of bran, and Jakobsen (1989) added yeast and wheat germ to the bran in the proportions of 1:10:89, respectively. Different species of tyroglyphids have been used for mass rearing: *A. farris* was used in Ramaker's early work, *A. siro* Linnaeus by Jakobsen (1989), and *Tyrophagus putrescentiae* (Schr.) by commercial producers in Canada and England. Often, cultures are a mixture of several species.

Although the forced air system described has been used successfully since 1985 in commercial production, some large companies now use passively aerated or open systems to rear mites. These systems do not have the condensation problem, nor are they as vulnerable to electrical interruption as the forced air systems. The mites can be reared inside 3-liter bags made of nonwoven terylene on one side and plastic on the other, placed above water in a larger box to maintain correct relative humidity (Jakobsen 1989). Karg et al. (1987) describe an open rearing system for mass production of *A. barkeri* in cabinets that hold racks of 30 x 25 x 6-cm rearing trays, each set in a slightly larger tray of water. The entire cabinet is covered with plastic and humidified air is forced between the racks to maintain 90-95% relative humidity. The tyroglyphids are reared in a similar system of racks; however, they are kept in a separate rearing container placed inside the rearing tray above water. The rearing tray is covered to maintain high humidity.

In all systems, the commercial product is stored in the bran, including both grain mites and the predators. The number of predators in the bran is determined by taking measured samples and either counting the number of phytoseiids in the dry bran or counting them after the sample is washed through a graduated series of screens (400, 200, and 63 mu) to catch the phytoseiids. Because predators are shipped along with their prey, they usually survive cold storage (10°C) and shipment very well.

## Problems in Commercial Mass Production of Phytoseiids

### Continuous Supply

The central problem in commercial mass production is ensuring a continuous, predictable supply of predators, starting as early in the season as growers need them. For example, in the North American market, California strawberry growers release *P. persimilis* from December to March. Most greenhouse crops require predators from January through August, with peak demand in early spring. Interior plantscapes and conservatories release several phytoseiid species year-round (Steiner & Elliott 1987). To sustain the continuity of phytoseiid supply, regular production must be maintained as well as a certain percentage of overproduction to act as a buffer in weeks of unexpectedly high demand.

Continuous production is achieved by initiating new predator cultures weekly so that harvest can continue regularly. Most large commercial insectaries now rear *P. persimilis* year-round, with some seasonal variation in numbers to match the market demand. Production schedules for spider mite host plants must be adjusted seasonally to allow for slower growth in winter and accelerated growth in summer.

For phytoseiids used in field crops, such as *M. occidentalis* or *T. pyri,* which are released inoculatively one year to become established for spider mite control the following year, seasonal production in greenhouses or outdoors during the summer is feasible and more economical than maintaining winter production in greenhouses (Hoy et al. 1982). Production in greenhouses during the winter entails high costs for heating and supplemental lighting to compensate for short days. Depending on the climate, however, rearing in greenhouses during the summer may be expensive because of the cooling and humidifying required. Contamination of production cultures with phytophagous pests, such as thrips and whiteflies or even native phytoseiids, is a more serious problem in the summer when ventilation is increased.

### Quality of Predators

The second problem in commercial mass production is providing a high-quality product. This problem is divisible into two main components: the quality of the mites (their health, vigor, fecundity, and genetic composition) and the quality of the packaged product (accuracy of counts and survival in transit, packaging, and handling).

The quality of the mites depends primarily on rearing them in optimum conditions: moderate temperatures (22-26°C), high relatively humidity (70-85%) and plentiful food supply. When temperatures during summer production exceed 30°C, mortality in most lines of *P. persimilis* is very high (McClanahan 1968, Hamamura et al. 1976) and survivors live only a few days, often barely long enough to arrive at their destination. Quantity and quality of the phytoseiid food supply starts with healthy host plants for those predator species reared on spider mites and with good environmental conditions in culture containers for those reared on tyroglyphid mites. Mites under the stress of poor rearing conditions do not produce well and are susceptible to diseases. For example, warm, wet rearing conditions are conducive to the spread of fungus diseases in spider mites (van der Geest 1985) and poor rearing conditions for tyroglyphids has frequently caused disease problems for producers of *A. cucumeris* and *A. barkeri*. When mites are overcrowded or when poor ventilation permits gases from decomposition to accumulate in containers, tyroglyphids are susceptible to a protozoan disease that appears to be latent in the cultures. When the phytoseiids feed on the sick prey, the protozoa apparently collect in their guts (Giffiths, personal communication), which become white and opaque. The predators become thin and pale and stop reproducing, although when fed healthy prey the protozoa are excreted and predators quickly recover and resume oviposition. Although not well known, diseases do occur in phytoseiid cultures stressed by poor rearing conditions. Infected predators have short lifespans and poor dispersal ability once released and may be sterile. Hess and Hoy (1982) isolated two unknown microorganisms from diseased *M. occidentalis* cultures. Recently, an unidentified disease *of A. cucumeris*, in which the entire idiosoma contents turn cloudy white caused losses in production in at least one commercial insectary.

As the application of genetically improved strains of phytoseiids becomes more widespread (Hoy 1985), rigorous quality control programs will be required to ensure the genetic consistency of the strains. Appropriate challenges to maintain selection pressure, whether from target pesticides for pesticide resistant strains, rearing under high temperatures for heat tolerant strains, or exposure to diapause-inducing conditions for nondiapausing strains, are necessary to maintain a high frequency of desirable traits in the population. Follow-up studies in the field are also necessary to determine the persistence of improved strains (Hoy & Knop 1981, Roush & Hoy 1981). Currently, there are few guidelines for this aspect of quality control; producers may claim attributes for predators that could no longer exist because of extensive mass breeding. For example, the "OP" line of *P. persimilis*, originally recovered from greenhouses in the Netherlands and marketed after it showed tolerance to several common organophosphorus pesticides, has been distributed to several insectaries and mass produced for over 10 years. Not only is there confusion among producers about the range of chemicals and levels of resistance in this line (Ooman 1983), but most producers do not claim to have testing programs to check or maintain resistance. Unfortunately, commercial insectaries, most of which are small companies, are rarely able to afford the scientific research and development effort required to support their rearing programs.

In addition to rearing high-quality mites, commercial producers must ensure a high-quality product. The benefits of optimum rearing conditions can be negated by poor handling and shipping after harvest. For example, *P. persimilis* removed from plants for packing in granular carriers are especially vulnerable to delays during handling, resulting in mortality from cannibalism or desiccation. Aside from periodic quality control checks to ensure that the carriers actually contain the number of mites stated on the label, the producer must make sure that containers provide good environmental conditions for the mites. For mites packaged in granular materials, containers of waxed or foil-lined cardboard or plastic bottles with screened vents in the lid are favored because they maintain high relative humidity, yet allow for some air exchange. Packages must also be easy to use and carry sufficient information for the grower.

Once packaged, phytoseiids can be cold stored at 5-10°C and 90% relative humidity (Hamamura et al. 1978), but the length of time depends on whether a water source, live prey, or a supplementary diet are available to them. In one test, about half of *P. persimilis* females survived cold storage for 25 days at 100% relative humidity; this time extended up to 70 days when they were packaged on leaves with spider mite prey (Hamamura et al. 1978). *P. persimilis* shipped in bran without prey become cannibalistic within 12 hours and start feeding on eggs and nymphs (Markkula et al. 1987). The unusually high number of failures in biological control of spider mites in Finnish greenhouses in 1986 was blamed on the lack of food for *P. persimilis* during shipping. This resulted in starving predators that did not resume a normal oviposition rate until 3 days after release (Markkula et al. 1987). Mori and Chant (1966) found that *P. persimilis* kept without food for several days subsequently ate fewer spider mites than those fed continuously, although Amano and Chant (1978) showed that starvation for 3 days did not have a significant long-term effect on later fecundity.

The quality of information, extension service, and training supplied by the commercial producer is an important part of the product because biological control is a new method to most growers. They usually need initial assistance to determine the timing and rates of release as well as to monitor the progress of their biological control program.

## Prospects for Commercial Application

The application of phytoseiids in greenhouses and field crops, especially deciduous fruits and berries, has expanded enormously in the last decade. It is not difficult to predict that their use will continue to increase, especially as the public becomes more interested in alternatives to pesticides. As McMurtry (1983) pointed out, new applications for known predators have been highly successful, but we are using very few of the many potential species (McMurtry 1982), and this is still true today. The commercial market is currently relatively unsophisticated, with only *P. persimilis* produced in massive numbers. A few different biotypes or strains of this species are sold for application in climates and crops as diverse as greenhouse cucumbers in the Netherlands and field strawberries in California. Further research on other biotypes, genetically improved strains, other species, or even mixtures of species is needed to broaden the applicability of phytoseiids to a wider variety of crops.

As seen recently in California, a market for phytoseiids can materialize very quickly in commercial crops. When registration for the miticide Plictran was removed for strawberries in 1987, there was a sudden demand for predator mites, of which *P. persimilis* was the most well known and widely available. An estimated 500 million mites were needed for release in late 1988 and early 1989, much more than local insectaries could produce. A global search for predators ensued, and mites were shipped into California from the Netherlands, England, and Canada. Although the potential application of phytoseiids to control spider mites in strawberries has been investigated since the 1960s (Oatman & McMurtry 1966), and applied on a large scale by a few progressive growers and pest managers (Grossman 1989), it was the restriction of miticides that turned growers en masse to biological control. Increasing interest in the application of phytoseiids to floriculture and ornamental crops reflects the decreasing range of registered pesticides available as well as their loss in effectiveness due to the development of resistance in pest populations.

During the last 10 years there have been many changes in commercial biological control. The largest worldwide producer of biological controls, which became well-established in the Netherlands during the 1970s, has been joined by growing companies in England, Canada, the United States, Germany, France, and other countries. Suppliers of horticultural products and services have begun adding biological control product lines to their catalogs and investing in biocontrol companies. As production capacity and exposure to larger markets expands, and as growers gain experience and confidence in using biological controls, the demand for phytoseiids can only expand.

Expansion in application of phytoseiids depends on the education and training of growers, extension agents, and pest managers, who must learn how to use biological controls successfully. Expansion also depends on the supply of mites, which depends on the continued development of mass-rearing methods, not just for the few species now produced on a large scale *(P. persimilis, A. cucumeris, A. barkeri, and M. occidentalis)* but for others, such as *T. pyri, A. fallacis,* and *E. hibisci,* which have been studied extensively, as well as for other less well-known species. Particularly promising are production systems based on alternate or artificial diets because of the potential for economical rearing. Through the continuous improvements in production methods and the development of competition between commercial suppliers, the cost of *P. persimilis* to growers declined considerably over the last decade. This species now sold wholesale in California for as little as U.S. $6.50 per thousand. *A. cucumeris* is sold for as little as $1.00 per thousand. As mass-rearing methods become more efficient, a greater number of species will become available at prices competitive with pesticides.

# References

Ables, J. R., B. G. Reeves, R. K. Morrison, R. E. Kinzer, S. L. Jones, R. L. Ridgway & D. L. Bull. 1979. Methods for the field release of insect parasites and predators. Trans. Amer. Soc. Agric. Engineers. 22(1): 59-62.

Abou-Setta, M. M. & C. C. Childers. 1987. A modified leaf arena technique for rearing phytoseiid or tetranychid mites for biological studies. Fla. Entomol. 70: 245-248.

Amano, H. & D. A. Chant. 1978. Some factors affecting reproduction and sex ratios in two species of predacious mites *Phytoseiulus persimilis* Athias-Henriot and *Amblyseius andersoni* (Chant) Acarina: Phytoseiidae. Can. J. Zool. 56: 1593-1607.

Anonymous. 1975. Biological pest control: Rearing parasites and predators. Growers' Bull. 2. Glasshouse Crops Research Institute, Littlehampton, Sussex.

Ashihara, W., T. Hamamura & N. Shinkaji. 1978. Feeding, reproduction, and development of *Phytoseiulus persimilis* Athias-Henriot (Acarina: Phytoseiidae) on various food substances. Bull. Fruit Tree Res. Stn., Japan, E2: 91-98.

Ball, J. C. 1980. Development, fecundity, and prey consumption of four species of predaceous mites (Phytoseiidae) at two constant temperatures. Environ. Entomol. 9: 298-303.

Bower, C. C. & W. G. Thwaite. 1982. Development and implementation of integrated control of orchard mites in New South Wales, pp. 177-190. *In* P. J. Cameron, C. H. Wearing & W. M. Kain [eds.], Proceedings Australasian Workshop on Development and Implementation of IPM. Gov. Printer, Auckland, New Zealand.

Field, R. P., W. J. Webster & D. S. Morris. 1979. Mass rearing *Typhlodromus occidentalis* Nesbitt (Acarina: Phytoseiidae) for release in orchards. J. Aust. Entomol. Soc. 18: 213-215.

Fournier, D., P. Millot & M. Pralavorio. 1985. Rearing and mass production of the predatory mite *Phytoseiulus persimilis*, Entomol. Exp. Appl. 38: 97-100.

Gilstrap, F. E. 1977. Table-top production of tetranychid mites (Acarina) and their phytoseiid natural enemies. J. Kansas Entomol. Soc. 50(2): 229-233.

Grossman, J. 1989. Update: Strawberry IPM features biological and mechanical control. The IPM Practitioner 11(5): 1-4.

Hamamura, T. N. Shinkaji & W. Ashihara. 1976. The relationship between temperature and developmental period, and oviposition of *Phytoseiulus persimilis* Athais-Henriot (Acarina: Phytoseiidae). Bull. Fruit Tree Res. St. Japan E. l: 117-125.

Hamamura, T. N. Shinkaji & W. Ashihara. 1978. Studies on the low temperature storage of *Phytoseiulus persimilis* Athias-Henriot (Acarina: Phytoseiidae) Bull. Fruit Tree Res. Stn. Japan E. 2: 83-90.

Hansen, L. S. & J. Geyti. 1985. Possibilities and limitations of the use of *Amblyseius mckenziei* Sch. & Pr. for biological control of thrips *(Thrips tabaci* Lind) on glasshouse crops of cucumber. *In* Proceedings, Biological Control in Protected Crops, 1985 April 24-29, Heraklion, Greece.

Hendrickson, R. M., Jr. 1980. Continuous production of predacious mites in the greenhouse. N.Y. Entomol. Soc. 88(4): 252-256.

Hess, R. T. & M. A. Hoy. 1982. Microorganisms associated with the spider mite predator *Metaseiulus(= Typhlodromus) occidentalis:* Electron microscope observations. J. Invert. Pathol. 40: 989-106.

Hoy, M. A. 1985. Recent advances in genetics and genetic improvement of the Phytoseiidae. Annu. Rev. Entomol. 30: 345-370.

Hoy, M. A., D. Castro & D. Cahn. 1982. Two methods for large scale production of pesticide-resistant strains of the spider mite predator *Metaseiulus occidentalis* (Nesbitt) (Acarina, Phytoseiidae). Z. Angew Entomol. 94: 1-9.

Hoy, M. A. & N. F. Knop. 1981. Selection for and genetic analysis of permethrin resistance in *Metaseiulus occidentailis*. Genetic improvement of a biological control agent. Entomol. Exp. Appl. 30: 10-18.

Hussey, N. W. 1985. History of biological control in protected culture, pp. 11-22. *In* N. W. Hussey & N. Scopes [eds.], Biological pest control: The glasshouse experience. Cornell Univ. Press, Ithaca, New York.

Jacklin, S. W. & F. F. Smith. 1986. Phytophagous mites, pp. 445-449. *In* C. M. Smith [ed.], Insect colonization and mass production. Academic Press, New York.

Jakobsen, J. 1989. Masse produktion of rovmiden *Amblyseius barkeri* till tripsbekaempelse (Mass rearing of the predatory mite *Amblyseius barkeri* for control of thrips), pp. 77-82. 6th Danish Plant Protection Conference Pests and Diseases.

Kamburov, S. S. 1966. Methods of rearing and transporting predacious mites. J. Econ. Entomol. 59(4): 875-877.

Karg, W., S. Mack & B. Baier. 1987. Advantages of oligophagous predatory mites for biological control. Bull. SROP/WPRS 10(2): 66-73.

Kennett, C. E. & J. Hamai. 1980. Oviposition and development in predaceous mites fed with artiticial and natural diets (Acari: Phytoseiidae). Entomol. Exp. Appl. 28: 116-122.

McClanahan, R. J. 1968. Influence of temperature on the reproductive potential of two mite predators of the two-spotted spider mite. Can. Entomol. 100(5): 549-556.

McMurtry, J. A. 1982. The use of phytoseiids for biological control: Progress and future prospects, pp. 23-48. *In* M. A. Hoy [ed.], Recent advances in knowledge of the Phytoseiidae. Div. Agric. Sci., U. of Calif., Berkeley, California.

McMurtry, J. A. 1983. Phytoseiid predators in orchard systems: A classical biological control success story, pp. 21-26. *In* M. A. Hoy, G. L. Cunningham & L. Knutson [eds.], Biological control of pests by mites. Div. Agric. & Nat. Resources, U. of Calif., Berkeley, CA.

McMurtry, J. A. & G. T. Scriven. 1962. The use of agar media in transporting and rearing phytoseiid mites. J. Econ. Entomol. 55: 412-414.

McMurtry, J. A. & G. T. Scriven. 1964. Studies on the feeding, reproduction, and development of *Amblyseius hibisci* (Acarina: Phytoseiidae) on various food substances. Ann. Entomol. Soc. Am. 57: 649-655.

McMurtry, J. A. & G. T. Scriven. 1965. Insectary production of phytoseiid mites. J. Econ. Entomol. 58(2): 282-4.

McMurtry, J. A. & G. T. Scriven. 1966. Effects of artificial foods on reproduction and development of four species of phytoseiid mites. Ann. Entomol. Soc. Am. 59: 267-269.

McMurtry, J. A. & G. T. Scriven. 1975. Population increase of *Phytoseiulus persimilis* on different insectary feeding programs. J. Econ. Entomol. 68: 319-321.

Markkula, M., K. Tittanen & H. M. T. Hokkanen. 1987. Failures in biological control of spider mites due to predatory mites or their users? Bull. SROP/WPRS. 10(2): 108-110.

Morewood, W. D. & L. A. Gilkeson. Diapause induction in the thrips predator *Amblyseius cucumeris*. (Acarina: phytoseiidue) under greenhouse conditons. Entomophaga: (in press).

Mori, H. & D. A. Chant. 1966. The influence of prey density, relative humidity, and starvation on the predacious behavior of *Phytoseiulus persimilis* Athias-Henriot (Acarina: Phytoseiidae). Can. J. Zool. 44: 483-491.

Naegele, J. A. & W. D. McEnroe. 1963. Mass rearing of the two-spotted spider mite, pp. 191-192. *In* J. A. Naegele [ed.], Recent advances in acarology, vol. 1, Cornell Univ., Ithaca, New York.

Oatman, E. R. & J. A. McMurtry. 1966. Biological control of the two-spotted spider mite on strawberry in southern California. J. Econ. Entomol. 59: 433-439.

Ochieng, R. S., G. W. Oloo & E. O. Amboga. 1987. An artificial diet for rearing the phytoseiid mite. *Amblyseius teke*. Pritchard and Baker. Exp. App. Acarol. 3: 169-173.

Oomen, P. A. 1983. Resistance and sensitivity of *Phytoseiulus persimilis* to different pesticides—a survey among producers. *In* J. Woets & J. van

Lenteren [eds.], Sting: Newsletter on Biological Control in Greenhouses 6: 1-15.

Overmeer, W. P. J. 1985a. Alternative prey and other food resources, pp. 131-137. *In* W. Helle & M. W. Sabelis [eds.], Spider mites: Their biology, natural enemies and control, vol. 1B. Elsevier, Amsterdam.

Overmeer, W. P. J. 1985b. Rearing and handling, pp. 161-170. *In* W. Helle & M. W. Sabelis [eds.], Spider mites: Their biology, natural enemies and control, vol. 1B. Elsevier. Amsterdam.

Pickett, C. H., F. E. Gilstrap, R. K. Morrison & L. F. Bouse. 1987. Release of predatory mites (Acari: Phytoseiidae) by aircraft for the biological control of spider mites (Acari: Tetranychidae) infesting corn. J. Econ. Entomol. 80: 906-910.

Ramakers, P. M. J. 1984. Mass production and introduction of *Amblyseius mckenziei* and *A. cucumeris*. OILB SROP/WPRS Bull. 6(3): 203-210.

Ramakers, P. M. J. 1988. Population dynamics of the thrips predators *Amblyseius mckenziei* and *Amblyseius cucumeris* (Acarina: Phytoseiidae) on sweet pepper. Neth. J. Agric. Sci. 36: 247-252.

Ramakers, P. M. J. & M. J. van Lieburg. 1982. Start of commercial production and introduction of *Amblyseius mckenziei* Sch. & Pr. (Acarina: Phytoseiidae) for the control of *Thrips tabaci* Lind. (Thysanoptera: Thripidae) in glasshouses. Med. Fac. Landbouww. Rijksuniv. Gent. 47(2): 540-545.

Ramsy, A. H., M. E. Elbagoury & A.S. Reda. 1987. A new diet for reproduction of two predacious mites *Amblyseius gossips* and *Agistemus exsertus* (Acari: Phytoseiidae, Stigmaeide). Entomophaga 32: 277-280.

Ristich, S. S. 1956. Mass rearing and testing techniques of *Typhlodromus fallacis* (Gar.) J. Econ. Entomol. 49(4): 476-479.

Roush, R. T. & M. A. Hoy. 1981. Laboratory, glasshouse, and field studies of artificially selected carbaryl resistance in *Metaseiulus occidentalis*. J. Econ. Entomol. 74: 142-147.

Scriven, G. T. & J. A. McMurtry. 1971. Quantitiative production and processing tetranychid mites for large-scale testing or predator production. J. Econ. Entomol. 64: 1255-1257.

Steiner, M. Y. & D. P. Elliott. 1987. Biological pest management for interior plantscapes, 2nd ed. Alberta Environmental Centre, Vegreville, Alberta.

Tanigoshi, L. K., S. C. Hoyt, R. W. Browne & J. A. Logan. 1975. Influence of temperature on population increase of *Metaseiulus occidentailis* (Acarina: Phytoseiidae) Ann. Entomol. Soc. Am. 68: 979-986.

Theaker, T. L. & N. V. Tonks. 1977. A method for rearing the predaceous mite, *Phytoseiulus persimilis* (Acarina: Phytoseiidae). J. Entomol. Soc. B.C. 74: 8-9.

van der Geest, L. P. S. 1985. Pathogens of spider mites, pp. 247-256. *In* W. Helle and M. W. Sabelis [eds.], Spider mites: Their biology, natural enemies and control, vol. 1 B. Elsevier, Amsterdam.

van Lenteren, J. C. & J. Woets. 1988. Biological and integrated pest control in greenhouses. Annu. Rev. Entomol. 33: 239-269.

# 29

## Gypsy Moth Parasites:
## Commercial Production and Profitability

*Mark Ticehurst*

### Introduction

The United States Department of Agriculture (USDA) and other governmental labs may employ the largest number of insect production workers, but private enterprise may provide a prize worth pursuing. I will describe a commercial operation, based on the National Gypsy Moth Management Group, Inc., that blends science and business, and I will attempt to stimulate thoughts of—and perhaps action toward—developing a business in insect production. The chapter is designed to answer the following questions: Why start a business in insect production? Is there a market for natural enemies? Can the production of natural enemies be profitable?

Many of us who produce insects have scientific training that is useful in most aspects of production, from the formulation of diets to the manipulation of diapause. Although we tend to be strong in our scientific capabilities, many of us are deficient in the business side of insect production. Our training consisted of botany, biology, microbiology, entomology, chemistry, physiology, and most other scientific "ologies" in the college catalog. We had little time and even less incentive to pursue economics, business management, accounting, and marketing. Today, most of our major decisions require an analysis of business, not scientific, matters.

The gypsy moth is a unique pest problem. Most arthropod pests are considered to be problems because they induce economic losses. Several agricultural pests such as bollworms, root worms, weevils, and aphids destroy cash crops or render them less valuable. Economic thresholds are established to pinpoint the pest level at which control is necessary. Economics are the primary, if not the only, factors that trigger control tactics. The gypsy moth also destroys its host, trees (Houston 1981, Quimby 1986), but tree defoliation and mortality are only the tip of the iceberg. The gypsy moth impacts people (Ticehurst & Finley 1988). The gypsy moth infests trees around homes. Infestations typically produce several hundred thousands to millions of highly mobile caterpillars per acre. Frass expelled from caterpillars in tree canopies can sound like rain. During daylight a majority of the caterpillar population seek shelter from the sun and heat. It is not uncommon to discover thousands of caterpillars under the eves and on the sides of homes. People have used snow shovels to remove caterpillars from their sidewalks. This human impact extends the gypsy moth problem beyond the realm of economics alone.

## Why Start a Business in Insect Production?

The first question to ask may be "Why start any business?" Most people have, at one time or another, considered working for themselves. Many entrepreneurs begin by hustling to shovel snow and mow lawns for neighbors. However, most people tend to start businesses when one or more of the following occur:

1. They have, or think they have, a good idea for a business.
2. They desire greater control over their future.
3. They lack employment.
4. They desire a change in current employment.

We began the business moth parasite production in 1982 for several reasons:

1. The gypsy moth, an introduced pest, was expanding its range at about 500,000 acres per year. This represented a *significant and increasing market*.
2. *Lack of natural enemies*, as well as an abundant food supply, were key factors that fueled the gypsy moths' rapid dispersal and explosive population growth.
3. Federal funds that supported state parasite production facilities were sharply cut in the early 1980s. The *reduced supply of parasites* combined with the increasing need pinpointed the niche.
4. The establishment of natural enemies could *reduce the use of chemical insecticides*.
5. I was *knowledgeable* of gypsy moth management and the production of parasites.
6. I wanted greater *control* over my future.

When the decision is made to start a business, the questions of what, where, and when will follow. There are several sources of assistance to help answer these questions. The federal Small Business Administration (SBA) and states' Small Business Development Centers (SBDC) can provide free or low-cost help. Private business counseling centers can also provide help for a fee. In addition, attorneys, accountants, and associates or friends in the business community can provide information. Despite helpful hints and suggestions, you will need to complete a business plan. The format for the plan can be obtained from the SBA or an SBDC. Completion of this rigorous document will force you to consider factors such as competition, cost, capital requirements, and business organization.

Our business began in 1982 as a part-time sole proprietorship with two employees, my wife and I, and has evolved to a full-time Chapter S corporation with 7 permanent and 20 seasonal employees. Initially, we thought that insect rearing would be good part-time job to supplement my regular job. We reared the insects in a 12ft x 12ft room that I constructed in our home. Work began at 4:00 am and continued until 7:00 am, when I left for my full-time position. After supper it was insect rearing again until I couldn't see. Fortunately, it was necessary to maintain this schedule for only 2 to 3 months each year.

Despite the high failure rate and often long hours of start-up businesses, establishing an insect production business has been a positive experience. In 1985 when the sale of parasites and research grants exceeded wages from my full-time job, I jumped into the uncertainty of the business world. The process of starting and maintaining a business can extend personal and economic stability for the scientist. A basic understanding of business processes can improve most scientists' ability to manage any organization where budgets, not necessarily biological factors, determine the limitations and success of programs.

## Is There a Market for Gypsy Moth Parasites?

The market for gypsy moth parasites, as for radial tires, is dependent on the need for and competition to supply the product. Although their specific role is not well understood, natural enemies, including parasites, are thought to be key factors in regulating gypsy moth throughout much of the world (Reardon 1981). This was the premise for overseas exploration and importation of gypsy moth parasites by government agencies. These efforts began in the early 1900s and continue today (Reardon 1981). The primary objective of this work is the establishment of exotic parasites in gypsy-moth-infested areas.

Government has been involved with gypsy moth control for more than 100 years, close to the time of its first accidental introduction in 1869. State agencies attempted to eradicate the gypsy moth with chemical pesticides. Early efforts by the USDA were focused on exploration and importation of natural enemies. Today the role of both state and USDA agencies have expanded. Most states emphasize suppression of moderate to dense gypsy moth infestations with aerial spraying. A portion of these funds are provided by USDA. The role of the USDA is multifaceted and directly and indirectly associated with research, long- and short-term management, quarantine, and other areas. USDA has also taken the leadership position in parasite exploration and importation.

Currently, 10 or possibly 11 species of parasites have been successfully imported and established throughout much of the gypsy moth infestation area in the eastern United States. These parasites were imported and released by the USDA. Some state agencies participated in the release program. The efficacy of gypsy moth parasites has been difficult to determine. Much of the problem is associated with the physical requirements of sampling the host, which occupies the forest canopy to heights of 100 feet. Another factor is the behavior of the gypsy moth, which is influenced by its density. Competition for food and space affect mobility and therefore the sampling universe. Commonly, gypsy moth life stages are collected in the field and reared in the laboratory at weekly intervals to assess the species composition and relative importance of parasites.

Ticehurst et al. (1978) reported weekly mean rates of 49.7% parasitism by *Parasetigena silvestris* (Robineau-Desvoidy) (Diptera: Tachinidae) throughout evaluation sites in Pennsylvania. Parasitism rates in New York range from 30 to 76% and were based on peak parasitism totals by all parasites of larvae and pupae (Tigner 1974, Tigner et al. 1974). Reardon (1981) noted that parasitism varied between 4 and 17% for all larvae and pupae collected in Massachusetts, New York, and New Jersey in 1972 and between 3 and 18% in 1973.

In addition, Ticehurst et al. (1982) showed that parasitism by the gypsy moth larval parasite *Cotesia (=Apanteles) melanoscela* (Hymenoptera: Braconidae) increased from 4.4% in check areas to 32.3% in areas treated with parasites in combination with *Bacillus thuringiensis*. This has been the biological basis for the utilization of parasites in combination with *B. thuringiensis* for the gypsy moth IPM program (Ticehurst & Finley 1988).

Despite file cabinets of data on rates of parasitism, there is a paucity of cause-and-effect information. Only recently, Liebhold and Elkinton (1989) demonstrated a density-dependent relationship (parasitism increases as host populations increase) relationship between the tachinid, *Compsilura consinnata* and the gypsy moth. ODell (personal communication) also detected high rates of parasitism by *C. concinnata* when gypsy moth populations were artificially increased. However, most parasites do not appear to be density-dependent. Despite the inconsistent data, a parasite that induces variable rates of mortality of 50 to 60% would appear to be a key factor in the population dynamics of its host.

One reason for the lack of pertinent information is related to the sampling universe. The gypsy moth occupies the forest canopy, a difficult sampling arena. Incomplete information on the population dynamics of the gypsy moth has contributed to the scarcity of information on the influence of natural enemies on their arboreal host.

Is there a need for parasites even if their effectiveness has not been quantified? Traditionally, scientists would suggest that more research is needed before parasites should be used to control the gypsy moth. Alternatively, Tom Peters, in his bestseller *In Search of Excellence* (1982), suggests another approach, "Ready, Fire, Aim" rather than "Ready, Aim, Fire." He urges us to do something, NOW! Modifications and improvements can continue as we manufacture, market, and use the product. Opportunities and benefits to both the consumer and business can be lost while waiting for all the answers. Perfection is rarely achieved in science or business. Since parasites kill gypsy moths and killing pests is a valid objective in pest management programs, we have determined that there is a need for gypsy moth parasites.

More than 12 organizations produce gypsy moth parasites. All but one are nonprofit institutions and nearly all are government agencies. The NGMMG is the largest producer of these parasites. It is likely that our market would increase substantially if the other nonprofit producers would purchase instead of produce. This multiplication of effort greatly increases production costs. Each lab must purchase and maintain environmental chambers, diet preparation equipment, and other equipment and facilities that constitute capital expenditures. An economic as well as biological evaluation of the need to develop and operate a gypsy moth parasite lab in each state could reduce the current number.

The White House Conference on Small Business surveyed thousands of businesses throughout the country in 1986 to identify problem areas and to solicit solutions. Sixty major problems were described and prioritized. Competition with nonprofit organizations, including government and university, was third on this list. Final recommendations from this conference were that "Government at all levels has failed to protect small business from damaging levels of unfair competition. At the federal, state, and local levels, therefore, laws, regulations, and policies should:

a. Prohibit unfair competition in which nonprofit tax-exempt organizations use their tax-exempt status and other advantages in selling products and services offered by small businesses.

b. Prohibit direct, government competition in which government organizations perform commercial services.

The travel agency business offers another example of competition between profit-making and nonprofit organizations. Many governmental and university entities maintain their own travel agency, despite the availability of local private companies. A presumed argument for maintenance of the nonprofit travel agency is lower cost and greater control over the quality of service. It is tempting to generalize about the relative productivity and quality of performance of workers in a nonprofit bureaucracy, but there are too many examples of excellence in the public sector. Nevertheless there is greater incentive to do the job faster, more efficiently, and at lower cost in a private business.

Our largest market is in the public (government) sector. Most sales are made to county and local governments as part of their IPM (Integrated Pest Management) programs. We also sell to a state agency that supplements its production with ours. About 20% of our parasites are used as components in IPM programs for large landowners. We do not sell parasites to individuals. Generally, individuals have unrealistic expectations concerning the influence of natural enemies. Some of the expectations can be credited to unsubstantiated claims by the producer. In other cases the disappointment is related to the user's enthusiastic but uninformed use of a natural enemy purchased from a catalog.

## Can Gypsy Moth Parasites Be Sold for a Profit?

Simply stated, profit is the difference between the cost of production plus overhead and income generated from sales. Production costs for gypsy moth parasites include supplies (e.g. cups and diet), labor, and equipment (e.g. microscopes and environmental chambers). Overhead expenses include rental or purchase of facilities; employee benefits such as health insurance, vacations, and pensions; travel; and taxes on employees, including social security, workmans' compensation, and unemployment compensation. In our business, overhead is approximately 90% of salaries and wages.

Most of our production costs are directly related to rearing the host gypsy moth. Much of the required technology was developed at the USDA Methods Development Center at Otis AFB, Mass., by Animal Plant Health Inspection Service (APHIS) and Agricultural Research Service (ARS) scientists. This USDA center has supported our production of gypsy moth by sharing technology and colonies of insects. Despite this assistance, the cost of producing the gypsy moth has increased during 1988 and 1989 because of a problem that is characterized by irregular growth and development. Since specific causes of this condition are not known, additional hosts must be produced to better ensure the availability of specific numbers of suitable larvae each day. Current gypsy moth rearing problems are being addressed, and production will be optimized at the center. This research and development program will directly benefit our company.

The generational period, egg to egg, for gypsy moth is about 200 days. Consequently, planning for production must begin at least 1 year in advance. We have utilized another source for hosts when the pressure is applied. Collection of gypsy moth egg masses from the field can supplement a temporary need, but this method introduces considerable risk. Field eggs may contain a variety of pathogens, including the dreaded nucleopolyhedrosis virus (NPV). This virus is commonly responsible for epizootics that result in up to 99.9% mortality and collapse of infestation in the field. The development of an epizootic in the lab would be catastrophic.

Gypsy moth and its parasites must be reared 52 weeks per year to target releases during a 3- to 6-week period in the spring, synchronous with the occurrence of gypsy moth larvae in the field. Releases made prior to or after that period would be completely useless. Consequently, we accelerate production in the fall, attain peak levels in winter and early spring, and change to colony maintenance in late spring and summer. This roller coaster schedule requires a flexible and available work force. Most of our lab workers are either women with children in school or retired individuals. Much of the production work is routine and requires considerable attention and speed. A college education is neither required nor desired for this work. College graduates are generally less productive, less satisfied, and more expensive than nongraduates.

The sale of gypsy moth parasites has increased over the past few years but remains unstable. Most sales are made to government agencies. Programs and goals are at the mercy of managers and budgets. Changes in managers and modifications in budgets have directly influenced the sale of parasites. For example, we provided 300,000 to 500,000 parasites each year for 3 years to a state agency, our largest client. When the manager of this project left, parasites were eliminated as a component of their gypsy moth management program.

An analysis of the gypsy moth parasite business based on risk and profit would not have the venture capitalists waiting in line; the risk is high and the profits are marginal. The narrow market niche is not attractive to major commercial enterprises. Therefore, we have adopted a strategy that attempts to spread the risk and increase the return—diversification. We are now directly involved with contract research through the federal Small Business Innovative Research (SBIR) program, the Ben Franklin Partnership, an initiative sponsored by the Pennsylvania Department of Commerce and several medium to large corporations. We also conduct a complete "turn-key" gypsy moth management operation that is specifically designed for municipalities. This IPM program includes public information and education, survey and evaluation, recommendations, and implementation of the control tactics, which may include aerial spraying of biorational insecticides and the release of parasites. Consequently, gypsy moth parasites can be sold for a profit by increasing their value through utilization in research and management programs.

## Summary

The real and potential market for natural enemies appears to be strong. This assumption is based on the growing concern associated with some pesticides and the public's willingness to accept biological controls that may be more expensive and less effective than the chemical alternative. In addition, current and future legislation may further restrict the use of chemicals and thus expand the market for alternative control techniques. Research is needed to better define the impact of natural enemies, including gypsy moth parasites, on their hosts so that this choice can be better evaluated and utilized.

The production of gypsy parasites and many other insects is an art and a science. Selling insects requires skills beyond the training of most biologists. The ability to sell insects for a profit is dependent on the cost of production, the market for the insects, and competition. Competition with government agencies can be the most serious threat to a successful business. Ironically, the government also represents the greatest market for gypsy moth parasites. In addition, our business might have failed without the assistance of and cooperation of government agencies. Consultation with responsible state and federal personnel, who may represent competition, should be an early step in developing the business. A frank discussion of program goals, objectives, limitations, cooperation, and competition should precede the first draw on a business loan. A partnership between government and industry makes much more economic, biological sense than a competitive relationship.

I have answered the following questions: Why start a business in insect production? Is there a market for gypsy moth parasites? Can gypsy moth parasites be sold for a profit? The answers were both generic and specific to the business of producing and selling gypsy moth parasites. The following critical steps should be considered when establishing a business in insect production:

- List reasons, both economic and personal, for starting a business.
- Compare the benefits and costs of the new business with those of your current position.
- Evaluate the need for the new business now and in the future.
- Seek a specific market niche, but plan for diversification.
- Consider the current and potential influence of competition in the private and public sector.
- Consider federal and state research and business development grants or low-interest loans.
- Seek free advice from the federal Small Business Administration and state Small Business Development Centers. Supplement this with assistance from trusted business professionals.

The economic and environmental climate for the successful establishment of an insect production business is positive. There has never been a better time to start!

## References

Liebhold, A. M. & J. S. Elkinton. 1989. Elevated parasitism in artificially augmented populations of *Lymantira dispar* (Lepidoptera:Lymantriidae). Environ. Entomol. 18(6): 986-995.

Houston, D. R. 1981. Effects of defoliation on trees and stand, pp. 217-297. *In* C. C. Doane & M. L. McManus [eds.], The gypsy moth: Research toward integrated pest management. U.S. Dept. Agric. Tech. Bull. 1584.

Peters, T. J. & R. H. Waterman, Jr. 1982. In Search of Excellence. Warner Books, New York.

Quimby, J. W. 1986. Impact of gypsy moth defoliation on forest stands, pp. 186-193. *In* Proceedings, Gypsy Moth Annual Review-1986. Norfolk, Virginia.

Reardon, R. C. 1981. Population dynamics, pp. 86-96. *In* C. C. Doane & M. L. McManus [eds.], The gypsy moth: Research toward integrated pest management. U.S. Dept. Agric. Tech. Bull 1584.

Ticehurst, M. & S. Finley. 1988. An urban forest integrated pest management program for gypsy moth: An example. J. Arborol. 17(4): 172-175.

Ticehurst, M., R. A. Fusco & E. M. Blumenthal. 1982. Effects of reduced rates of Dipel 4L, Dylox 1.5 Oil, and Dimilin 25W on *Lymantria dispar* (L.) (Lepidoptera: Lymantriidae), parasitism, and defoliation. Environ. Entomol. 11(5): 1058-1062.

Ticehurst, M., R. A. Fusco, R. P. Kling & J. Unger. 1978. Observations on parasites of the gypsy moth in first cycle infestations in Pennsylvania from 1974-1977. Environ. Entomol. 7(3):355-358.

Tigner, T. C. 1974. Local distributions of gypsy moth larvae and larval parasitoid. N. Y. State Univ., Coll. Environ. Sci. For., Appl. For. Res. Inst. Syracuse, Rep. 16.

Tigner, T. C., C. Palm & J. Jackson. 1974. Gypsy moth parasitism under and outside burlap skirts at two heights. N. Y. State University, Coll. Environ. Sci. For., Appl. For. Res. Inst. Syracuse Rep. 20.

# Contributors

**T. E. Anderson,** BASF Corporation, P. O. Box 13528, Research Triangle Park, NC 27709 USA

**M. J. Berlinger,** Entomology Laboratory, Gilat Regional Experiment Station Mobile Post, Negev 2, 85-280 Israel

**F. Bourgeois,** Ciba Geigy Ag, 2,721 CH-4002, Basel, Switzerland

**G. K. Clare,** Entomology Division, Department of Scientific and Industrial Research, Private Bag, Auckland, New Zealand

**A. C. Cohen,** USDA, ARS, 2000 E. Allen Road Tucson, AZ 85719 USA

**F. M. Davis,** USDA, ARS, P. O. Box 5367, Mississippi State, MS 39762 USA

**W. A. Dickerson,** North Carolina Department of Agriculture, P. O. Box 27647, Raleigh, NC 27611 USA

**V. Flueck,** Ciba-Geigy AG, Basel AG, 2,721 CH-4002, Basel,Switzerland

**L. K. French,** French Agricultural Res. Service, RR 2, P. O. Box 161A, Lamberton, MN 56152 USA

**L. A. Gilkeson,** Pesticide Management Branch, B. C. Environment, 4th Floor, 737 Courtney St., Victoria, BC V8V 1X5 Canada

**W. D. Guthrie,** R D No. 2, Nevada, Iowa 50201 USA

**A. M. Handler,** USDA, ARS, Insect Attractants, Behavior, and Basic Biology Research Laboratory, 1700 SW 23rd Drive, P. O. Box 14565, Gainesville, FL 32604 USA

**G. G. Hartley,** Southern Field Crop Insect Management Laboratory, USDA-ARS, Stoneville, MS 28776.

**T. Haug,** International Institute of Tropical Agriculture, Oyo Road, PNB 5320, Ibadan, Nigeria

**N. C. Hinkle,** USDA, ARS, Insects Affecting Man and Animals Laboratory, P. O. Box 14565, Gainesville, FL 32604 USA

**H. Kakinohana,** Fruit Flies Laboratory, Okinawa Prefectual Agricultural Experiment Station, 4-222 Sakiyama-Cho, Naha, Okinawa, 903 Japan

**E. G. King, Jr.,** USDA, ARS, Subtropical Agricultural Research Laboratory, P. O. Box 267, Farm Road 1015 South, Weslaco, TX 78596 USA

**P. G. Koehler,** USDA, ARS, Insects Affecting Man and Animals Laboratory, P. O. Box 14565, Gainesville, FL 32604 USA

**N. C. Leppla,** USDA, ARS, Science and Technology Methods Development, Federal Building, Room 540, Hyattsville, MD 20782 USA

**R. L. Mangan,** USDA, ARS, Subtropical Agricultural Research Laboratory, 2301 S. International Blvd., Farm Road 1015, South, Weslaco, TX 68596 USA

**P. G. Marrone,** Entotech Inc., 1497 Drew Avenue, Davis, CA 95616-4880 USA

**L. H. Mihsfeldt,** Departmento de Fitotechnia, Fundacão Faculadade de Agronomia "Luiz Meneghel," Banierantes, Paraná, Brazil

**P. B. Morgan,** USDA, ARS, Insects Affecting Man and Animals Laboratory, P. O. 14565, Gainesville, FL 32604 USA

**R. K. Morrison,** USDA, ARS (Retired) College Station, TX USA

**H. Nakamori,** Department of Environmental Biology, National Institute of Agro-Environmental Science, Kannonndai, Tsukuba, Ibaraki, 305 Japan

**P. Neuenschwander,** International Institute of Tropical Agriculture, Oyo Road, PNB 5320 Ibadan, Nigeria

**D. A. Nordlund,** USDA, ARS, Subtropical Agricultural Research Laboratory, 2413 E. Highway 83, Weslaco, TX 78596 USA

**T. M. ODell,** USDA, Forest Service, Center for Biological Control of Northeastern Forest Insects and Diseases, 51 Mill Pond Road, Hamden, CT 06514 USA

**J. R. P. Parra,** ESALQ, Dept. de Entomologia, Universidade de São Paulo, Piracicaba 13400 SP, Brazil

**R. S. Patterson,** USDA, ARS, Insects Affecting Man and Animals Laboratory, Gainesville, FL 32604 USA

**M. M. Ramadan,** USDA, ARS,Tropical Fruit and Vegetable Research Laboratory, P. O. Box 917, Hilo, HI 97620 USA

**W. J. Ravensberg,** Koppert B.Y. Veilingweg 64, NL-2651 BE Berkel en Rodenrjs, The Netherlands

**J. P. Shapiro,** USDA, ARS, US Horticultural Research Laboratory, 2120 Camden Road, Orlando, FL 32803 USA

**P. Singh,** Entomology Division, Department of Scientific and Industrial Research, Private Bag, Auckland, New Zealand

**S. R. Sims,** Monsanto Agricultural Products Co., 700 Chesterfield Village Parkway, Chesterfield, MO 63198 USA

**G. G. Soares, Jr.,** Mycogen Corporation, 5451 Oberlin Road, San Diego, CA 92121 USA

**P. Stoeklin,** Ciba Geigy Ag, 2,721 CH-4002, Basel, Switzerland

**T. B. Stone,** Monsanto Agricultural Products Co., 700 Chesterfield Village Parkway, Chesterfield, MO 63198 USA

**D. B. Taylor,** USDA, ARS, Bioscience Research Laboratory, P. O. Box 5674, Fargo, ND 58105 USA

**M. Ticehurst,** National Gypsy Moth Management Group, Inc., RD 1, Box 715, Landisburg, PA 17040 USA

**G. J. Tsiropoulos,** Ministry of Industry, Energy, and Technology, General Secretariat of Research and Technology, National Research Center for Physical Sciences, "Demokritos" GR-153 10 AG Paraskevi, PO Box 60228, Athens, Greece

**P. L. Versoi,** BASF Corporation, P. O. Box 13528, Research Triangle Park, NC 27709 USA

**Wang Zhenying,** Institute of Plant Protection, Chinese Academy of Agricultural Sciences West Yaun Ming Yaun Road, Beijing, People's Republic of China

**T. Y. Wong,** USDA, ARS, Tropical Fruit and Vegetable Research Laboratory, P. O. Box 917, Hilo, HI 97620 USA

**M. Yamagishi,** Fruit Flies Laboratory, Okinawa Prefectural Agricultural Experiment Station, 4-222 Sakiyama-Cho, Naha, Okinawa, 903 Japan

**I. G. Yazlovetsky,** All Union Institute of Biological Control, Prospeckt Mira 58, Kishinev 277072, Moldavia

**Ye Zhihua,** Institute of Plant Protection, Chinese Academy of Agricultural Sciences, West Yaun Ming Yaun Road, Beijing, People's Republic of China

**Zhou Darong,** Institute of Plant Protection, Chinese Academy of Agricultural Sciences, West Yaun Ming Yaun Road, Beijing, People's Republic of China

# About the Book and Editors

The efficient production of large numbers of high-quality insects is a concern both for basic research and for the success of control programs for pests of agricultural and medical significance. This volume provides a comprehensive overview of this important issue, identifying the major applications for insect-rearing technology. The chapters, international in scope, cover genetics and molecular biology; insect rearing and the development of bioengineered crops; nutrition, digestion, and artificial diets; and the practical concerns of commercial insect rearing.

**Thomas E. Anderson** is the Entomology Research Group leader at the BASF Corporation North American Agricultural Research Center. **Norman C. Leppla** is director of methods development, Plant Protection and Quarantine, USDA-APHIS.

Printed and bound by CPI Group (UK) Ltd, Croydon, CR0 4YY

30/10/2024

01781019-0001